A FIRST COURSE IN THE
QUALITATIVE THEORY
OF
DIFFERENTIAL
EQUATIONS

James Hetao Liu
James Madison University

Prentice
Hall

Pearson Education, Inc., Upper Saddle River, New Jersey 07458

Library of Congress Cataloging-in-Publication Data

Liu, James Hetao.
 A first course in the qualitative theory of differential equations / James Hetao Liu.
    p. cm.
 Includes bibliographical references and index.
 ISBN 0-13-008380-1
 1. Differential equations, Nonlinear. I. Title.

 QA372.L765   2003
 515'.355–dc21   2002032370

Editor-in-Chief: *Sally Yagan*
Acquisitions Editor: *George Lobell*
Vice President/Director of Production and Manufacturing: *David W. Riccardi*
Executive Managing Editor: *Kathleen Schiaparelli*
Senior Managing Editor: *Linda Mihatov Behrens*
Production Editor: *Bob Walters*
Manufacturing Buyer: *Michael Bell*
Manufacturing Manager: *Trudy Pisciotti*
Marketing Assistant: *Rachel Beckman*
Editorial Assistant: *Jennifer Brady*
Art Director: *Jayne Conte*
Creative Director: *Carole Anson*
Director of Creative Services: *Paul Belfanti*
Cover Design: *Bruce Kenselaar*
Cover Photo: Artist: *Gordon Huether*, Photo: *Michael Bruk*
Art Studio: *MacroTeX*

© 2003 Pearson Education, Inc.
Pearson Education, Inc.
Upper Saddle River, New Jersey 07458

Printed in the United States of America

10 9 8 7 6 5 4 3 2 1

ISBN 0-13-008380-1

Pearson Education LTD., *London*
Pearson Education Australia PTY, Limited, *Sydney*
Pearson Education, Singapore, Pte. Ltd.
Pearson Education North Asia Ltd., *Hong Kong*
Pearson Education Canada, Ltd., *Toronto*
Pearson Educaciûn de Mexico S.A. de C.V.
Pearson Education – Japan, *Tokyo*
Pearson Education Malaysia, Pte. Ltd.

TO YOU

the readers

who would like to start

a journey

with me

Contents

Preface

Why should we learn some qualitative theory of differential equations?

Differential equations are mainly used to describe the change of quantities or behavior of certain systems in applications, such as those governed by Newton's laws in physics.

When the differential equations under study are linear, the conventional methods, such as the Laplace transform method and the power series solutions, can be used to solve the differential equations analytically, that is, the solutions can be **written out** using formulas.

When the differential equations under study are nonlinear, analytical solutions cannot, in general, be found; that is, solutions cannot be written out using formulas. In those cases, one approach is to use **numerical approximations**. In fact, the recent advances in computer technology make the numerical approximation classes very popular because powerful software allows students to quickly approximate solutions of nonlinear differential equations and visualize, even in 3-D, their properties.

However, in most applications in biology, chemistry, and physics modeled by nonlinear differential equations where analytical solutions may be unavailable, people are interested in the questions related to the so-called **qualitative properties**, such as: will the system have at least one solution? will the system have at most one solution? can certain behavior of the system be controlled or stabilized? or will the system exhibit some periodicity? If these questions can be answered without solving the differential equations, especially when analytical solutions are unavailable, we can still get a very good understanding of the system. Therefore, besides learning some numerical methods, it is also important and beneficial to learn how to analyze some qualitative properties, such as the existence and uniqueness of solutions, phase portraits analysis, dynamics of systems, stability, bifurcations, chaos, boundedness, and periodicity of differential equations without solving them analytically or numerically.

This makes learning the qualitative theory of differential equations very valuable, as it helps students get well equipped with tools they can use when they apply the knowledge of differential equations in their future studies and careers. For example, when taking a numerical methods course, before a numerical approximation is carried out, the existence and uniqueness of solutions should be checked to make sure that there does exist one and only one solution to be approximated. Otherwise, how does one know what one is approximating? A related remark is that even though numerical solutions can be carried out to **suggest** certain properties, they are obtained through discretization on finite intervals. They reveal certain properties that are only valid for the limited numerical solutions on finite intervals and, therefore, cannot be used to **determine** the qualitative properties on the whole interval of all solutions of the original differential equations.

Based on the above remarks, we conclude that in order to have a more complete knowledge of differential equations, and be able to analyze differential equations without solving them analytically or numerically, we should learn some qualitative theory of differential equations.

To whom is this book written?

This book is written for upper level undergraduates (second undergraduate course in ODEs) and beginning graduate students. To be more specific, Chapters 1–7 are for upper level undergraduate students, where the basic qualitative properties concerning existence and uniqueness, structures of solutions, phase portraits, stability, bifurcation, and chaos are discussed. Chapters 8–12, together with Chapters 1–7, are for beginning graduate students, where some additional subjects on stability, dynamical systems, bounded and periodic solutions are covered.

Another reason for writing this book is that nowadays it is a popular trend for upper level undergraduate students and beginning graduate students to get involved in some research activity. Compared to other more abstract subjects in mathematics, qualitative analysis of differential equations is readily accessible to upper level undergraduates and beginning graduate students. It is a vast hunting field in which students will get an opportunity to combine and apply their knowledge in linear algebra, elementary differential equations, advanced calculus, and others to "hunt some prey."

Furthermore, the qualitative analysis of differential equations is on the border line of applied mathematics and pure mathematics, so it can attract students interested in either discipline.

How does this book differ from other ODE books?

It is often the case that in a book written at the graduate level or even at the upper undergraduate level, there are "jumps" in the reasonings or in the proofs, as evidenced by words like "obviously" or "clearly." However, inexperienced undergraduate and beginning graduate students need more careful and detailed guidance to help them learn the material and gain maturity on the subject.

In this book, I selected only the subjects that are of fundamental importance, that are accessible to upper level undergraduate and beginning graduate students, and that are related to current research in the field. Then, for each selected topic, I provided a complete analysis that is suitable for the targeted audience, and filled in the details and gaps which are missing from some other books. Sometimes, I produced elementary proofs using calculus and linear algebra for certain results that are treated in a more abstract frame in other books. Also, examples and reasons are given before introducing many concepts and results.

Therefore, this book is different from other ODE books in that it is more detailed, and, as the title of this book indicates, the level of this book is lower than most books for graduate students, and higher than the books for elementary differential equations. Moreover, this book contains many interesting pure and applied topics that can be used for one or two semesters.

What topics are covered in this book?

Chapter 1. A Brief Description. We first give a brief treatment of some subjects covered in an elementary differential equations course. Then we introduce some terminology and describe some qualitative properties of differential equations that we are going to study in this book. We use the geometric and physical arguments to show why certain qualitative properties are plausible and why sometimes we pursue a qualitative analysis rather than solving differential equations analytically or numerically. This will give readers an opportunity to become familiar with the objective and terminology of qualitative analysis in a somewhat familiar setting.

Chapter 2. Existence and Uniqueness. In Section 1, we use examples from applications to define general first order differential equations in $\Re^n$. In Section 2, we study existence and uniqueness of solutions, that is, we examine under what conditions a differential equation has solutions and how the solutions are uniquely determined, without solving the differential equation analytically. A condition called "Lipschitz condition" is utilized. In Section 3, we show under certain conditions that solutions are continuous and differentiable with respect to initial data and parameters. In Section 4, we determine structures of the maximal intervals of existence for solutions, and study properties of solutions with respect to the maximal intervals of existence. In Section 5 (which may be optional), we introduce the Fixed Point Method. We use the contraction mapping principle to derive existence and uniqueness of solutions if a local Lipschitz condition is satisfied. Then, when a local Lipschitz condition is not assumed, we use Schauder's second fixed point theorem to obtain existence of solutions, in which case, uniqueness is not guaranteed.

Chapter 3. Linear Differential Equations. In Section 1, we make some definitions concerning linear differential equations. In Section 2, we study general nonhomogeneous linear differential equations and obtain the fundamental matrix solutions and verify that they satisfy the "evolution system property." Then we derive the variation of parameters formula using the fundamental matrix solutions and observe what these solutions should look like. In Section 3, we look at equations with constant coefficients and examine detailed structure of solutions in terms of eigenvalues of the leading constant matrix, using the Jordan canonical form theorem. In addition we derive the Putzer algorithm that can be used to actually solve or compute solutions for equations with constant coefficients. This result will appeal to readers interested in computation. In Section 4, we look at equations with periodic coefficients and study Floquet theory, which allows us to transform equations with periodic coefficients into equations with constant coefficients. The results of Section 3 can then be applied to the transformed equations. The concept of Liapunov exponents is also briefly introduced in Section 4.

Chapter 4. Autonomous Differential Equations in $\Re^2$. In Section 1, we introduce the concept of dynamical systems, discuss possible trajectories in phase planes for two-dimensional autonomous equations, and outline the relationship between nonlinear differential equations and their linearizations. In Section 2, we provide a complete analysis for linear autonomous differential equations in $\Re^2$ and draw all phase portraits for the different cases according to eigenvalues of the coefficient matrix. We also introduce some terminology, including stability of solutions, according to the properties revealed, which leads us to detailed study of the same subject later for general differential equations in $\Re^n, n \geq 1$. In Section 3, we examine the conditions which ensure that solutions of autonomous differential equations and their linearizations have essentially the same local geometric and qualitative properties near the origin. In Section 4, we apply the results to analyze an equation of a simple pendulum. In Section 5, we generalize the ideas of a simple pendulum and study the Hamiltonian systems and gradient systems.

Chapter 5. Stability. Part I. In Section 1, we introduce the notion of stabilities in the sense of Liapunov for general differential equations in $\Re^n$, which are based on some consideration in physics and the planar differential equations studied in Chapter 4. In Section 2, we study stabilities for linear differential equations with constant coefficients and show that eigenvalues of the coefficient matrices determine stability properties. In Section 3, stabilities of linear equations with linear or nonlinear perturbations are studied using the variation of parameters formula and Gronwall's inequality. The results include some planar autonomous nonlinear differential equations studied in Chapter 4 as special cases. Therefore, some unproven results in Chapter 4 can now get a partial proof. In Section 4, linear periodic differential equations are treated. The Floquet theory from Chapter 3 is used to transform linear periodic equations into linear equations with constant coefficients and the results from Section 2 can then be applied. In Section 5, we introduce Liapunov's method for autonomous nonlinear differential equations and prove their stability properties under the assumption that there exist appropriate Liapunov functions. Thus, we can obtain stabilities without explicitly solving differential equations. In Section 6, we provide examples to demonstrate how the Liapunov theory is applied by constructing Liapunov functions in specific applications. Liapunov's method for general (nonautonomous) differential equations will be given in Chapter 9.

Chapter 6. Bifurcation. In Section 1, we use examples, including Euler's buckling beam, to introduce the concept of bifurcation of critical points of differential equations when some parameters are varied. In Section 2, we study saddle-node bifurcations and use examples to explain why saddle and node appear for this type of bifurcations. We analyze the geometric aspects of some scalar differential equations that undergo saddle-node bifurcations and use them to formulate and prove a result concerning saddle-node bifurcations for scalar differential equations. In Section 3, we study transcritical bifurcations and apply them to a solid-state laser in physics. Again, the geometric aspects of some examples are analyzed and used to formulate and prove a result concerning transcritical bifurcations for scalar differential equations. In Section 4, we study pitchfork bifurcations and apply them to Euler's buckling beam and calculate Euler's first buckling load, which is the value the buckling takes place. The hysteresis effect with applications in physics is also discussed. A result concerning pitchfork bifurcations for scalar differential equations is formulated using the geometric interpretation. In Section 5, we analyze the situations where a pair of two conjugate complex eigenvalues cross the pure imaginary axis when some parameters are varied. We introduce the Poincaré-Andronov-Hopf bifurcation theorem and apply it to van der Pol's oscillator in physics.

Chapter 7. Chaos. In Section 1, we use examples, such as some discrete maps and the Lorenz system, to introduce the concept of chaos. In Section 2, we study recursion relations, also called maps, and their bifurcation properties by finding the similarities to the bifurcations of critical points of differential equations, hence the results in Chapter 6 can be carried over. In Section 3, we look at a phenomenon called period-doubling bifurcations cascade, which provides a route to chaos. In Section 4, we introduce some universality results concerning one-dimensional maps. In Section 5, we study some properties of the Lorenz system and introduce the notion of strange attractors. In Section 6, we study the Smale horseshoe which provides an example of a strange invariant set possessing chaotic dynamics.

Chapter 8. Dynamical Systems. In Section 1, we discuss the need to study the global properties concerning the geometrical relationship between critical points, periodic orbits, and nonintersecting curves. In Section 2, we study the dynamics in $\Re^2$ and prove the Poincaré-Bendixson theorem. In Section 3, we use the Poincaré-Bendixson theorem, together with other results, to obtain existence and nonexistence of limit cycles, which in turn help us determine the global properties of planar systems. In Section 4, we apply the results to a Lotka-Volterra competition equation. In Section 5, we study invariant manifolds and the Hartman-Grobman theorem, which generalize certain results for planar equations in Chapter 4 to differential equations in $\Re^n$.

Chapter 9. Stability. Part II. In Section 1, we prove a result concerning the equivalence of "stability" (or "asymptotic stability") and "uniform stability" (or "uniform asymptotic stability") for autonomous differential equations. In Section 2, we use the results from Chapter 3 to derive stability properties for general linear differential equations, and prove that they are determined by the fundamental matrix solutions. The results here include those derived in Chapter 5 for linear differential equations with constant or periodic coefficients as special cases. Stability properties of general linear differential equations with linear or nonlinear perturbations are also studied using the variation of parameters formula and Gronwall's inequality. In Section 3, we introduce Liapunov's method for general (nonautonomous) differential equations and derive their stability properties, which extends the study of stabilities in Chapter 5 for autonomous differential equations.

Chapter 10. Bounded Solutions. In Section 1, we make some definitions and discuss the relationship between boundedness and ultimate boundedness. In Section 2, we derive boundedness results for general linear differential equations by using the results from Chapter 9. It will be seen that stability and boundedness are almost equivalent for linear homogeneous differential equations, and they are determined by the fundamental matrix solutions. For nonlinear differential equations, examples will be given to show that the concepts of stability and boundedness are not equivalent. In Section 3, we look at the case when the coefficient matrix is a constant matrix, and verify that the eigenvalues of the coefficient matrix determine boundedness properties. In Section 4, the case of a periodic coefficient matrix is treated. The Floquet theory from Chapter 3 is used to transform the equation with a periodic coefficient matrix into an equation with a constant coefficient matrix. Therefore, the results from Section 3 can be applied. In Section 5, we use Liapunov's method to study boundedness properties for general nonlinear differential equations.

Chapter 11. Periodic Solutions. In Section 1, we give some basic results concerning the search of periodic solutions and indicate that it is appropriate to use a fixed point approach. In Section 2, we derive the existence of periodic solutions for general linear differential equations. First, we derive periodic solutions using the eigenvalues of $U(T, 0)$, where $U(t, s)$ is the fundamental matrix solution of linear homogeneous differential equations. Then we derive periodic solutions from the bounded solutions. Periodic solutions of linear differential equations with linear and nonlinear perturbations are also given. In Section 3, we look at general nonlinear differential equations. Since using eigenvalues is not applicable now, we extend the idea of deriving periodic solutions using the boundedness. First, we present some Massera-type results for one-dimensional and two-dimensional differential equations, whose proofs are generally not extendible to higher dimensional cases. Then, for general n-dimensional differential equations, we apply Horn's fixed point theorem to obtain fixed points, and hence periodic solutions, under the assumption that the solutions are equi-ultimate bounded.

Chapter 12. Some New Types of Equations. In this chapter, we use applications, such as those in biology and physics, to introduce some new types of differential equations, which are extensions and improvements of the differential equations discussed in the previous chapters. They include finite delay differential equations, infinite delay differential equations, integrodifferential equations, impulsive differential equations, differential equations with nonlocal conditions, impulsive differential equations with nonlocal conditions, and abstract differential equations. For each new type of differential equations mentioned above, we use one section to describe some of their important features. For example, for integrodifferential equations, we outline a method which can reformulate an integrodifferential equation as a differential equation in a product space; and for abstract differential and integrodifferential equations, we introduce the semigroup and resolvent operator approaches. The purpose of this chapter is to provide some remarks and references for the recent advancement in differential equations, which will help readers to access the frontline research, so they may be able to contribute their own findings in the research of differential equations and other related areas.

How to use this book?

For an upper level undergraduate course. The material in Chapters 1–7 is enough. Moreover, if there are time constraints, then some results, such as the following, can be mentioned without detailed proofs: in Chapter 2, the proofs concerning existence and existence without uniqueness of solutions, the dependence on initial data and parameters, and the maximal interval of existence; in Chapter 3, differential equations with periodic coefficients and Floquet theory; in Chapter 5, the proofs concerning Liapunov's method; in Chapters 6–7, certain proofs concerning bifurcations and chaos. (Note that Section 2.5 concerning the Fixed Point Method is optional.)

For a beginning graduate course. Chapters 1–11 provide a sufficient resource for different selections of subjects to be covered. If time permits, Chapter 12 can provide some direction for further reading and/or research in the qualitative theory of differential equations.

One more thing we would like to point out is that Chapters 6 through 12 are rather independent of each other and the instructors may choose among them to fit the last part of the course to their particular needs.

Exercises and notations. Most questions in the Exercises are quite important and should be assigned to give the students a good understanding of the subjects.

In Theorem x.y.z, x indicates the chapter number, y the section number, and z the number of the result in section y. The same numbering system holds true for Lemma x.y.z, Example x.y.z, etc.

Acknowledgments

First, I thank Professors ZuXiu Zheng and Ronald Grimmer for their inspirations and for directing my Master thesis and Ph.D. dissertation, respectively, in the area of Differential Equations. Then I thank my department and college for supporting me during the planning and writing of the book.

This book grew from my class notes, so for their valuable comments I thank my students: Kathleen Bellino, Paul Dostert, Roxana Karimianpour, Robert Knapik, Justin Lacy, Florin Nedelciuc, Rebecca Wasyk, and Bruce Whalen. My colleagues Carter Lyons and Esther Stenson read certain parts of the first version and made some modifications which greatly improved the exposition, so I thank them for their help. I also thank my colleague Bo Zhang for informing me of some key reference books, and colleagues Carl Droms, Jim Sochacki, and Paul Warne for helping with LaTex and Maple.

I sincerely thank the following reviewers of the manuscript:

Nguyen Cac, University of Iowa,
Michael Kirby, Colorado State University,
Przemo Kranz, University of Mississippi,
Jens Lorenz, University of New Mexico,
Martin Sambarino, University of Maryland,
Anonymous, Brigham Young University,
Anonymous, Lafayette College.

During the writing of the book, I was sometimes at "critical points" (meaning "directionless" in ODEs). The comments and suggestions of the reviewers gave me the direction I needed. Some reviewers pointed out errors and confusing statements, and made specific recommendations to correct them, which greatly improved the presentation of the book. However, I am solely responsible for the remaining errors, if any, and invite the readers to contact me with comments, recommendations and corrections, using the address or email given below.

I also thank George Lobell, Acquisitions Editor of Prentice Hall, for accepting my humble first version and encouraging me to expand the first version to include additional subjects.

The production phase has been more involved than I anticipated. However, with the help of my production editor Bob Walters, copyeditor Elaine Swillinger, LaTex expert Adam Lewenberg, and Adobe software expert Bayani DeLeon of Prentice Hall, it went smoothly. I thank them for giving detailed instructions and for doing an excellent job in helping to make this book a reality.

Finally, I thank my wife Tina and daughter Linda for their understanding, support, and help.

James Hetao Liu

Department of Mathematics and Statistics
James Madison University
Harrisonburg, VA 22807

liujh@jmu.edu

Chapter 1

A Brief Description

We first begin with a brief treatment of some subjects covered in an Elementary Differential Equations course that we assume you have taken. Then we describe some qualitative properties of differential equations that we are going to study in this book. Many of the descriptions here will be done with the geometric and physical arguments to help you see why certain qualitative properties are plausible and why sometimes we pursue a qualitative analysis rather than solving differential equations analytically or numerically. This will give you an opportunity to become familiar with the objective and terminology of qualitative analysis in a somewhat familiar setting.

1.1 Linear Differential Equations

To provide a background for our discussions, let's begin with some examples.

Example 1.1.1 To mathematically model the population growth of, say, a university, the simplest assumption we can make is to assume that the population grows at a rate proportional (with a proportional constant k) to its current population of that year. For example, k may be 0.05, which means the population grows 5% per year. If we use t for the time and $x(t)$

for the population at time t, and if we know the population at a time, say t_0, to be x_0, (for example, the population is $x_0 = 18,000$ in year $t_0 = 2001$), then we can set up the following equation

$$x'(t) = kx(t), \quad x(t_0) = x_0. \qquad \spadesuit \qquad\qquad (1.1)$$

Eq. (1.1) is an equation involving the derivative of an unknown function $x(t)$ that we want to solve. Therefore, we define an **ordinary differential equation** as an equation involving derivatives of an unknown function with one variable.

The **order** of an ordinary differential equation is the highest derivative of the unknown function that appears in the equation. For example, Eq. (1.1) is a first-order ordinary differential equation.

A **solution** of an ordinary differential equation is a function that satisfies the ordinary differential equation. $x(t_0) = x_0$ in Eq. (1.1) is referred to as an **initial condition**, an **initial value**, or an **initial data**, and Eq. (1.1) is also called an **initial value problem**.

Sometimes, we only consider $t \geq t_0$ in Eq. (1.1) because we are only concerned with the development in the future time of t_0. Next, $x(t)$ in Eq. (1.1) is a number, thus we say that Eq. (1.1) is a **differential equation in** $\Re$ (or $\Re^1$), where $\Re = (-\infty, \infty)$. We also say that Eq. (1.1) is a **scalar equation**.

Since the study in this book doesn't involve partial differential equations, sometimes we will use "differential equations" or just "equations" to mean "ordinary differential equations."

The **direction field** consisting of direction vectors (or slope vectors) for Eq. (1.1) with $k > 0$ is given in **Figure 1.1**.

For Eq. (1.1), $x(t) \equiv 0$ is a solution (with its initial value being zero), and is called a **constant solution**. Otherwise, we assume $x(t) \neq 0$ such that Eq. (1.1) can be written as

$$\frac{x'(t)}{x(t)} = k. \qquad\qquad (1.2)$$

Now, we can use the method of **separation of variables** to solve Eq. (1.2). That is, we separate the variables x and t and write Eq. (1.2) as

$$\frac{1}{x}dx = kdt,$$

then solve

$$\int \frac{1}{x}dx = \int kdt,$$

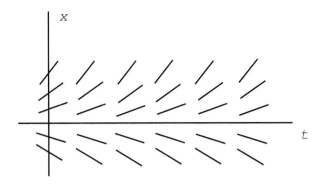

Figure 1.1: Direction field of Eq. (1.1) with $k > 0$

and obtain

$$\ln|x| = kt + C.$$

Finally, we derive the solution of Eq. (1.1), given by

$$x(t) = x_0 e^{k(t-t_0)}. \tag{1.3}$$

For the solution given in (1.3), we have the pictures in **Figure 1.2** and **Figure 1.3**.

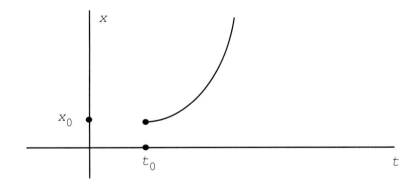

Figure 1.2: Solutions of Eq. (1.1) with $k > 0$

Accordingly, we say that in Eq. (1.1), the population **grows exponentially** when $k > 0$, and the population **decays exponentially** when $k < 0$.

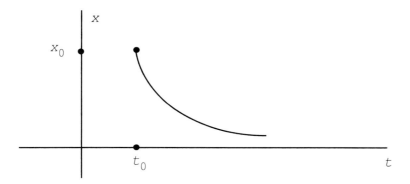

Figure 1.3: Solutions of Eq. (1.1) with $k < 0$

Now, if we do not solve Eq. (1.1), but, instead, start at (t_0, x_0) and **flow** along the directions of the direction field in Figure 1.1 as t increases, then the curve obtained, see **Figure 1.4**, matches well with the picture of solutions in Figure 1.2. The point of view of regarding solutions as curves flowing in a direction field will be very useful for understanding some results in differential equations, especially when we study the geometric aspects of differential equations.

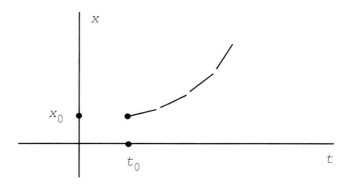

Figure 1.4: A curve obtained using the direction field in Figure 1.1

In Eq. (1.1), if we take t_0 to be 0, (for example, treat year 2001 as year 0), then Eq. (1.1) becomes

$$x'(t) = kx(t), \ x(0) = x_0, \tag{1.4}$$

and the solution is now given by

$$x(t) = x_0 e^{kt}.$$

Consider Eq. (1.1). If we assume further that other factors, such as those from the environment, are also involved in the population growth, then we may replace Eq. (1.1) by

$$x'(t) = kx(t) + f(t), \ x(t_0) = x_0, \tag{1.5}$$

for some continuous "factor" function $f(t)$.

In some applications, the proportional constant k may change with the time t, thus in those cases we need to replace k by a continuous function in t, say $k(t)$. Then Eq. (1.5) becomes

$$x'(t) = k(t)x(t) + f(t), \ x(t_0) = x_0. \tag{1.6}$$

Sometimes, other forms of differential equations are also encounted in applications, as in the following examples.

Example 1.1.2 (Restricted population growth) In many applications, it is assumed that the population $(x(t))$ does not exceed some number C, called the **carrying capacity** of the environment; it is also assumed that the population grows at a rate proportional (with a constant k) to the difference between C and the population at that time. Then $x(t)$ satisfies

$$x'(t) = k[C - x(t)], \ x(t_0) = x_0. \quad \spadesuit \tag{1.7}$$

Example 1.1.3 (Newton's law of cooling) Newton's law of cooling states that the temperature of a subject $(T(t))$ changes at a rate proportional (with a constant k) to the difference between the temperature of the subject and the temperature of the surrounding medium (T_m). Then we have

$$T'(t) = k[T_m - T(t)], \ T(t_0) = T_0. \quad \spadesuit \tag{1.8}$$

Eq. (1.7) and Eq. (1.8) are of the same form, so we only need to look at Eq. (1.7). Now, $x(t) = C$ is a constant solution. If $x(t) \neq C$, then, using separation of variables, we solve

$$\frac{1}{C - x} dx = kdt, \tag{1.9}$$

then the solution of Eq. (1.7) is given as

$$x(t) = C - [C - x_0]e^{-k(t-t_0)}, \quad t \geq t_0. \tag{1.10}$$

The direction field and the picture of the solutions for Eq. (1.7) are given in **Figure 1.5** and **Figure 1.6**. Again, they match well.

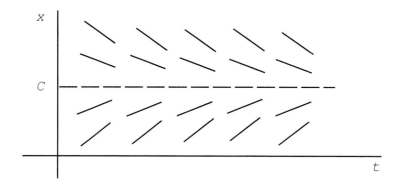

Figure 1.5: Direction field of Eq. (1.7) with $k > 0$

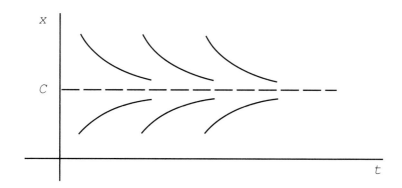

Figure 1.6: Solutions of Eq. (1.7) with $k > 0$

Eq. (1.7) and Eq. (1.8) are solved as above. However, we point out that they can also be formulated as Eq. (1.5) or Eq. (1.6), which has more applications. For example, Eq. (1.8) can be formulated as $T'(t) = -kT(t) + kT_m$.

Now, let's solve Eq. (1.5). We note that when $f = 0$, Eq. (1.5) becomes Eq. (1.1), whose solution is given by $x_0 e^{k(t-t_0)}$. Then we use the method of **variation of parameters**, that is, we determine the conditions on an unknown function $C(t)$ such that $C(t)e^{k(t-t_0)}$ is a solution of Eq. (1.5). This leads to

$$C'(t)e^{k(t-t_0)} + C(t)ke^{k(t-t_0)} = kC(t)e^{k(t-t_0)} + f(t),$$

hence,

$$C(t) = x_0 + \int_{t_0}^t e^{-k(s-t_0)} f(s)ds.$$

Therefore, we obtain the solution of Eq. (1.5), given by the **variation of parameters formula**

$$
\begin{aligned}
x(t) &= e^{k(t-t_0)}\left[x_0 + \int_{t_0}^t e^{-k(s-t_0)} f(s)ds\right] \\
&= e^{k(t-t_0)}x_0 + \int_{t_0}^t e^{k(t-s)} f(s)ds.
\end{aligned}
\tag{1.11}
$$

If we define $T(t) = e^{kt}$, then (1.11) can be written as

$$x(t) = T(t-t_0)x_0 + \int_{t_0}^t T(t-s)f(s)ds.\tag{1.12}$$

Note that $T(t) = e^{kt}$ satisfies the following property:

(S1). $T(0) = 1$,

(S2). $T(t)T(s) = T(t+s)$, $t, s \geq 0$.

In some literature, this property is called the "**semigroup property**."

To solve Eq. (1.6), we can use the idea of variation of parameters again (see an exercise) and let $f = 0$ first and derive the solution $x_0 e^{\int_{t_0}^t k(s)ds}$. We then determine the conditions on $C(t)e^{\int_{t_0}^t k(s)ds}$ from

$$C'(t)e^{\int_{t_0}^t k(s)ds} + C(t)k(t)e^{\int_{t_0}^t k(s)ds} = k(t)C(t)e^{\int_{t_0}^t k(s)ds} + f(t),$$

and obtain the solution of Eq. (1.6), given by another **variation of parameters formula**

$$
\begin{aligned}
x(t) &= e^{\int_{t_0}^{t} k(s)ds}\left[x_0 + \int_{t_0}^{t} e^{-\int_{t_0}^{s} k(h)dh} f(s)ds\right] \\
&= e^{\int_{t_0}^{t} k(s)ds} x_0 + \int_{t_0}^{t} e^{\int_{s}^{t} k(h)dh} f(s)ds.
\end{aligned}
\tag{1.13}
$$

In this case, if we define $U(t,s) = e^{\int_{s}^{t} k(h)dh}$, then (1.13) can be written as

$$
x(t) = U(t,t_0)x_0 + \int_{t_0}^{t} U(t,s)f(s)ds.
\tag{1.14}
$$

Now, $U(t,s) = e^{\int_{s}^{t} k(h)dh}$ satisfies the following property:

(E1). $U(t,t) = 1$, $t \geq t_0$,

(E2). $U(t,r)U(r,s) = U(t,s)$, $t_0 \leq s \leq r \leq t$.

This property is called the "**evolution system property**" in some literature.

Some higher order differential equations can also be treated in a similar way. One example is given below.

Example 1.1.4 Consider the second-order differential equation

$$
x''(t) + a_1 x'(t) + a_2 x(t) = f(t).
$$

Besides using the characteristic equations, we can define

$$
x_1(t) = x(t), \ x_2(t) = x'(t),
$$

then,

$$
\begin{cases}
x_1'(t) = x'(t) = x_2(t), \\
x_2'(t) = x''(t) = -a_2 x(t) - a_1 x'(t) + f(t) = -a_2 x_1(t) - a_1 x_2(t) + f(t).
\end{cases}
\tag{1.15}
$$

Thus, writing in matrix and vector notations, we obtain

$$
\begin{bmatrix} x_1(t) \\ x_2(t) \end{bmatrix}' = \begin{bmatrix} 0 & 1 \\ -a_2 & -a_1 \end{bmatrix} \begin{bmatrix} x_1(t) \\ x_2(t) \end{bmatrix} + \begin{bmatrix} 0 \\ f(t) \end{bmatrix}.
\tag{1.16}
$$

Eq. (1.16) is called a **differential equation in $\Re^2$**, which looks like Eq. (1.5) when k in Eq. (1.5) is regarded as a 2×2 matrix and x as a vector in $\Re^2$. ♠

For equations in $\Re^2$, solutions should be viewed in the (t, x_1, x_2) space, and the direction field should be drawn in the (x_1, x_2) space, as in the following example.

Example 1.1.5 Consider

$$\begin{cases} x_1' & = x_2, \\ x_2' & = -x_1. \end{cases} \tag{1.17}$$

We find that $x_1(t) = \sin t$ and $x_2(t) = \cos t$ is a solution, and the picture in the (t, x_1, x_2) space is shown in **Figure 1.7**. The direction field of Eq. (1.17) is shown in **Figure 1.8**.

(To get Figure 1.8, you can check a few points. For example, at the point $[x_1, x_2]^T = [0, 1]^T$ (here T means the transpose, so $[0, 1]^T$ is a 2×1 vector $\begin{bmatrix} 0 \\ 1 \end{bmatrix}$ in $\Re^2$), the direction in the field is $[x_1', x_2']^T = [x_2, -x_1]^T = [1, 0]^T$; at the point $[x_1, x_2]^T = [1, 1]^T$, the direction in the field is $[x_2, -x_1]^T = [1, -1]^T$; at the point $[x_1, x_2]^T = [1, 0]^T$, the direction in the field is $[x_2, -x_1]^T = [0, -1]^T$, and so on. Thus it goes like a circle in the clockwise direction. Again, the picture of the solution $x_1(t) = \sin t$, $x_2(t) = \cos t$ and the direction field in Figure 1.8 match well.) ♠

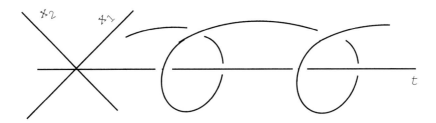

Figure 1.7: Solutions of Eq. (1.17) in the (t, x_1, x_2) space

---

Observe that when the right-hand side of Eq. (1.5) or Eq. (1.6) is regarded as a function in (t, x), the term kx or $k(t)x$ involving x is **linear in** x. Thus, in this sense, Eq. (1.5) and Eq. (1.6) are called **linear differential equations**. When k and $k(t)$ in Eq. (1.5) and Eq. (1.6) are regarded as $n \times n$ matrices and x as an $n \times 1$ vector, Eq. (1.5) and Eq. (1.6) are called

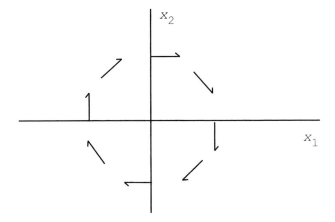

Figure 1.8: Direction field of Eq. (1.17) in the (x_1, x_2) space

linear differential equations in $\Re^n$. For equations in $\Re^n$, $n \geq 3$, if the t direction is also added, then we end up with something that is at least of four dimensions; hence, we lose geometric view because humans can only see objects of at most three dimensions. In those cases, especially in determining how close a solution is to the zero solution (when the zero is a solution), or in determining the distance from a point on the solution curve to the t-axis, we simply draw pictures in a plane, and treat $\Re^n$ as one dimensional, or use the vertical direction to denote the distance of a solution to the zero solution (t-axis, when the zero is a solution), as shown in **Figure 1.9**.

We will see in Chapter 3 that the solution formulas (1.12) and (1.14), as well as the semigroup and evolution system properties derived for one-dimensional equations are also valid for all linear differential equations in $\Re^n, n \geq 1$, and thereby constitute a complete and elegant theory for linear differential equations.

1.2   The Need for Qualitative Analysis

Our problems are almost solved, at least for finding solutions, if we only need to deal with linear differential equations. However, we are living in a complex world, and in most applications, such as those in biology, chemistry, and physics, we have to deal with **nonlinear differential equations**, that is, the differential equations where the terms involving the unknown function

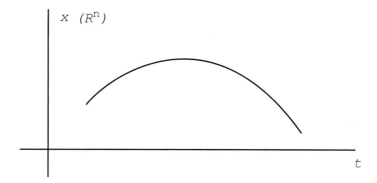

Figure 1.9: A solution in the $\Re \times \Re^n$ space viewed in the (t, x) plane where $\Re^n$ is treated as the x direction

---

x are not linear in x, as in the following examples.

Example 1.2.1 (Logistic equation) The exponential growth of $x' = kx$, which was studied around 1800, was used by some economists to argue that human misery is inevitable because population grows exponentially fast and supplies cannot keep up. In 1845, the Belgian mathematician P. Verhulst argued that to get better models, the proportional constant k of $x' = kx$ should be replaced by $C - x$, where C is the carrying capacity; and proposed the following equation,

$$x'(t) = ax(t)[C - x(t)], \quad x(t_0) = x_0, \ t \geq t_0, \tag{2.1}$$

where a and C are positive constants. The model is used to accommodate the situations that when the population $x(t)$ is small, the rate $x'(t) \approx aCx(t)$, thus the population grows exponentially; when $x(t)$ approaches C but is still less than C, the rate $x'(t)$ decreases and is still positive, thus the population is still growing but at a slow rate; finally, when $x(t)$ is large enough $(x > C)$, the rate $x'(t) < 0$, therefore the population decreases. These can be seen from the graph of the function $f(x) = ax[C - x] = -ax^2 + aCx$ in **Figure 1.10**. Verhulst called the solution curves of Eq. (2.1) "logistic curves," from a Greek word meaning "skilled in computation." Nowadays, equations of the form of Eq. (2.1) are called "logistic equations." After a change in function $\frac{x}{C} \to x$, Eq. (2.1) can be replaced by

$$x'(t) = rx(t)[1 - x(t)], \quad x(t_0) = x_0, \ t \geq t_0, \tag{2.2}$$

where $r = aC$. (Some analysis of Eq. (2.1) and Eq. (2.2) will be given later.) ♠

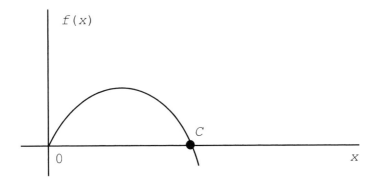

Figure 1.10: The graph of the function $f(x) = ax[C - x] = -ax^2 + aCx$

Example 1.2.2 Let's look at Eq. (1.6) again, but now assume that the function $f(t)$ is also determined by the unknown function x. Then we need to replace $f(t)$ by a function $f(t, x)$ that may be nonlinear in x, such as $f(t, x) = \sin(tx)$. Now we have

$$x'(t) = k(t)x(t) + \sin(tx(t)). \qquad ♠ \qquad (2.3)$$

Example 1.2.3 (Lotka-Volterra competition equation) Lotka-Volterra competition equation states that

$$\begin{cases} x_1' = \beta_1 x_1(K_1 - x_1 - \mu_1 x_2), \\ x_2' = \beta_2 x_2(K_2 - x_2 - \mu_2 x_1), \\ \quad x_1(0) \geq 0, \ x_2(0) \geq 0, \end{cases} \qquad (2.4)$$

where β_i, K_i, μ_i, $i = 1, 2$, are positive constants and $x_1(t)$, $x_2(t)$ are two populations. If the populations x_1 and x_2 grow and decay independently of each other, then the constants μ_1 and μ_2 will not appear in Eq. (2.4), resulting in two independent differential equations where each is of the form of a logistic equation. However, if the two populations compete for a shared limited resource (space or a nutrient, for example), and each interferes with the other's utilization of it, then the growth or decay of one population will affect the well-being or fate of the other one. Now μ_1 and μ_2 will appear in

Eq. (2.4), and this explains why Eq. (2.4) is proposed. For detailed studies in this area, see for example, May [1973] and Smith [1974]. ♠

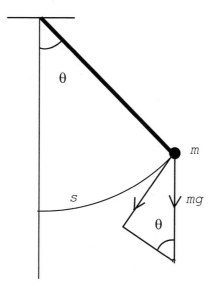

Figure 1.11: Motion of a simple pendulum

Example 1.2.4 (A simple pendulum) Figure 1.11 shows a rigid simple pendulum of length l oscillating around the vertical downward position.

It is assumed that the mass of the rod of the pendulum is negligible with respect to the mass m at the end of the pendulum. Let $\theta = \theta(t)$ be a function in the time variable t measuring the angle formed by the pendulum and the vertical downward direction, and let s be the arc length in the figure formed by the path of the end of the pendulum starting from the vertical downward position. Then $s = s(t)$ is a function in t and $s'(t)$ is the velocity of the end of the pendulum along the arc s (or in a direction tangential to the arc s). Now, the gravity of the pendulum is mg, hence, from the small triangle in the figure, the component of gravity in the direction tangential to the arc s is

$$-mg\sin\theta,$$

(the minus sign is needed because, for example, when the pendulum moves away, the force of gravity will try to drag the pendulum back). Next, assume that the damping or resistance is linear and is in the opposite direction of the velocity $s'(t)$, given in the form of

$$-\tau s'(t)$$

for some constant $\tau \geq 0$. Then, from Newton's second law of motion, which says, in this case, that

> (m)(the tangential acceleration)
>
> = the tangential component of the gravitational force
>
>   + the damping,

we derive

$$ms''(t) = -mg \sin \theta(t) - \tau s'(t). \tag{2.5}$$

Since $\frac{s}{\theta} = \frac{2\pi l}{2\pi}$, we get $s = l\theta$. Hence Eq. (2.5) becomes

$$ml\theta''(t) = -mg \sin \theta(t) - \tau l\theta'(t). \tag{2.6}$$

Simplifying, we get the motion of a **simple pendulum**, given by the following differential equation,

$$\theta''(t) + k\theta'(t) + q \sin \theta(t) = 0, \tag{2.7}$$

where $k \geq 0$, $q > 0$ are constants, with k related to a damping term, and $\theta(t)$ measures the angle formed by the pendulum and the vertical downward position. ♠

In general, for the equations in physics governed by Newton's second law of motion, such as the motions concerning oscillations, the following type of second-order differential equations

$$x'' + f(t, x, x')x' + g(x) = p(t) \tag{2.8}$$

are subjects of intensive studies. Here, f usually represents a damping or friction term, such as k in Eq. (2.7); g represents a restoring force, such as $q \sin \theta(t)$ in Eq. (2.7); and p is an externally applied force.

Eq. (2.8) includes the famous **Lienard-type equations**

$$x'' + f(x)x' + g(x) = 0, \tag{2.9}$$

where $xg(x) > 0$ if $x \neq 0$, which includes the well-known **van der Pol equation**

$$x'' + (x^2 - 1)x' + x = 0, \qquad (2.10)$$

named after Lienard [1928] and van der Pol [1927] for their important contributions in the analysis of the equations and their applications concerning sustained oscillations, the modeling of the voltage in a triode circuit and also the human heartbeat.

In Examples 1.2.1–1.2.4, the unknown function x **appears nonlinearly**, such as x^2 and $\sin(tx)$, thus those differential equations are called **nonlinear differential equations**. Certain nonlinear differential equations can be **solved analytically**, meaning that the formulas for solutions can be derived analytically. For example, the logistic equation in Example 1.2.1 can be solved analytically by rewriting the equation as

$$\frac{x'}{ax[C - x]} = 1, \quad ([C - x]x \neq 0),$$

then, using separation of variables and partial fractions, one obtains (see an exercise)

$$x(t) = \frac{Cx_0}{x_0 + [C - x_0]e^{-aC(t - t_0)}}, \quad t \geq t_0. \qquad (2.11)$$

However, **most nonlinear differential equations, such as those in Examples 1.2.2–1.2.4, cannot be solved analytically, that is, no formulas for solutions are available.** (Try it to see why.)

Now, the question is: **What do we do for general nonlinear differential equations?** It is true that in most applications, differential equations are handled by numerical approximations with the help of powerful computers, and evidently courses in numerical methods are very popular nowadays. Students who have taken such courses may want to use numerical approximation methods to approximate solutions of nonlinear differential equations they cannot solve analytically.

But wait a minute and think about this: If we do not even know that a solution exists in the first place, then what are we approximating? Another question to ask is: Suppose an approximation gives one solution, and we then use a different way to make an approximation, are we sure that we will get the same solution? If we don't get the same solution, then which solution do we want to use in order to explain the physical situation that we are modeling using the differential equation?

A further question to consider is that to determine some asymptotic properties (properties of solutions for large time variable t), even though numerical solutions can be carried out to **suggest** certain properties, they are obtained through discretization on finite intervals. They reveal certain properties that are only valid for the limited numerical solutions on finite intervals and, therefore, cannot be used to **prove** the properties on the whole interval of all solutions of the original differential equations.

These questions and remarks give the reason why, besides learning some numerical methods, we also need a theory on **qualitatively analyzing differential equations**, that is, deriving certain properties qualitatively without solving differential equations analytically or numerically.

1.3   Description and Terminology

To get some basic idea of what do we mean by **qualitative analysis** or **qualitative theory**, let's look at the following analysis of the logistic equation (2.1).

Look at Figure 1.10. We see that $f(x) = ax[C - x] = 0$ has two roots: $x = 0$ and $x = C$. Now, define $x_1(t) = 0$, $x_2(t) = C$, $t \geq t_0$, then $x_1(t)$ and $x_2(t)$ are both constant solutions of Eq. (2.1) (with their corresponding initial values). Since x_1 and x_2 are constants, or will "stay put" for all $t \geq t_0$, they are also called **steady solutions**, **critical points**, or **equilibrium points**.

Furthermore, assume $x(t) > 0$ is another solution (we need $x \geq 0$ to represent the population). Then we have $x' = ax[C - x] > 0$ if $0 < x < C$; and $x' = ax[C - x] < 0$ if $x > C$. Therefore, on the x-axis, the solutions with initial values in $(0, C)$ will "**flow**" monotonously to C from the left-hand side of C as t increases; and solutions with initial values bigger than C will flow monotonously to C from the right-hand side of C as t increases, see **Figure 1.12**. (Think of a basketball that is rolling on the ground.)

Accordingly, we have the picture in **Figure 1.13**, which tells us very roughly what the solutions should look like.

The concavity in Figure 1.13 is determined based on the increasing or decreasing of $x'(t)$. For example, when $0 < x(t) < \frac{C}{2}$, $x(t)$ increases as t increases (because in Figure 1.12, x moves to the right on the x-axis when $0 < x < C$); hence, $x'(t) = ax(t)[C - x(t)]$ increases in t (because now $f(x) = ax[C - x]$ increases in Figure 1.12), thus the function $x(t)$ is concave up. When $\frac{C}{2} < x(t) < C$, $x(t)$ increases in t; hence, $x'(t) = ax(t)[C - $

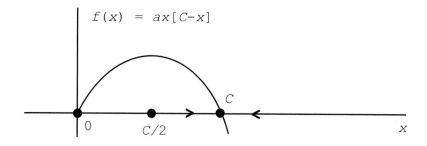

Figure 1.12: Flows of the solutions of a logistic equation on the x-axis

$x(t)$] decreases in t (because now $f(x) = ax[C - x]$ decreases), therefore the function $x(t)$ is concave down. The case for $x(t) > C$ can also be determined similarly, (see an exercise).

Of course, in this special case, Eq. (2.1) can be solved and the formula (2.11) for the solutions can be used to check these properties. But we do want to point out that it is much easier to determine the concavity of the solutions from Figure 1.12 than by looking at the formula (2.11). To see why, you should take the second derivative of $x(t)$ given in (2.11) and then see how difficult it is to determine its signs, (see an exercise).

In Figure 1.12, $x_1(t) = 0$ "sends other solutions away," hence it is called a **repeller** or **source**, or we say that **the critical point (or the constant solution)** $x_1 = 0$ **is unstable**. However, $x_2(t) = C$ **attracts** other solutions, thus it is called a **sink** or an **attractor**, or we say that **the critical point (or the constant solution)** $x_2 = C$ **is stable**. Articles can be found, for example, in Krebs [1972] and Murray [1989], indicating that the logistic equation (2.1) provides a good match for the experiments done with colonies of bacteria, yeast, or other simple organisms in conditions of constant food supply, climate, and with no predators. However, results of experiments done with fruit flies, flour beetles, and other organisms that have complex life cycles are more complex and do not match well with the logistic equation, because other facts are involved, including age structure and the time-delay effect.

In the above, we derived certain properties, including stabilities, of the solutions of the logistic equation (2.1) without solving it. These properties are called **qualitative properties** because they only tell us the certain **ten-**

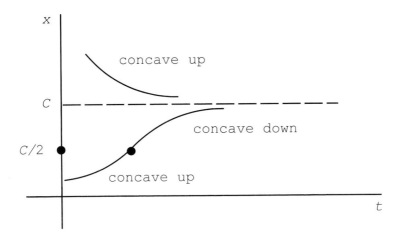

Figure 1.13: A rough sketch of the solutions of a logistic equation according to the flows

---

dency of how the solutions will behave, and no quantitative information is given. However, in many applications in sciences, the qualitative properties of the equations are the only things we care about, especially when analytical solutions are not available. Therefore, it is very valuable to learn some qualitative analysis of differential equations, as it adds to your knowledge of the subject and helps you get well equipped with tools useful in your applications of differential equations in your future studies and careers.

The first qualitative property we will study is existence and uniqueness theory. This can be used to verify that some differential equations have solutions and these solutions are uniquely determined without solving the differential equations analytically. This theory will also provide a foundation for numerical approximation methods. Based on existence and uniqueness of solutions, we will study other qualitative properties in this book, such as bifurcation, chaos, stability, boundedness, and periodicity, as we explain in the following.

Example 1.3.1 (Euler's buckling beam) A famous example in physics used to introduce the notion of bifurcation is Euler's buckling beam studied by Euler [1744]. If a small weight is placed on the top of a beam shown in **Figure 1.14**, then the beam can support the weight and stay vertical. When the weight increases a little, the position of the beam will change a

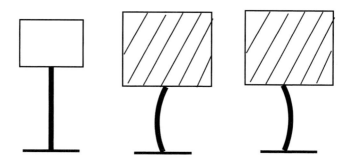

Figure 1.14: Euler's buckling beam

little and remain nearly vertical. Accordingly, this change in position of the beam is called a **quantitative change**. However, if you keep increasing the weight, then there will be a moment that the beam cannot take it any more and will **buckle**, or there is a critical value such that when the weight increases beyond that value the beam will buckle, see Figure 1.14.

Now, the difference is that the beam had undergone a **qualitative change**: from nearly vertical to a buckling position. And with some symmetry assumption, the beam can buckle in all directions.

Therefore, we find that for some systems, when some **parameters**, such as the weight here, are varied and pass some critical values, the systems may experience some abrupt changes, or undergo some qualitative changes. These qualitative changes are generally called **bifurcations**, and the parameter values at which bifurcations occur are called **bifurcation points** or **bifurcation values**. Euler's buckling beam will be analyzed in some detail in Chapter 6 (Bifurcation), where a differential equation describing the motion of the beam will be given, and the bifurcation value, called **Euler's first buckling load**, will be calculated. ♠

Let's look at one more example, which can also explain why the word "bifurcation" is used.

Example 1.3.2 Consider the scalar differential equation

$$x' = \mu - x^2, \tag{3.1}$$

where $\mu \in \Re$ is a parameter. If $\mu < 0$, then Eq. (3.1) has no critical point (that is, $x' = \mu - x^2 = 0$ has no solution), or the curve $y = \mu - x^2$ will not intersect the x-axis, see **Figure 1.15**.

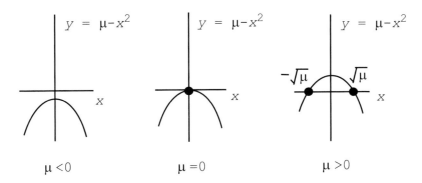

Figure 1.15: Graph of $y = \mu - x^2$ and the critical points of $x' = \mu - x^2$

When μ increases to 0 from below, the graph of $y = \mu - x^2$ moves up and intersects the x-axis when $\mu = 0$, in which case, one critical point appears at $x = 0$. If μ continues to increase, then $\mu > 0$ and hence the graph of $y = \mu - x^2$ will cross the x-axis and then two critical points appear at

$$x = \sqrt{\mu} \quad \text{and} \quad x = -\sqrt{\mu}. \tag{3.2}$$

Now, if we treat μ as the independent variable and treat the corresponding critical point x (if any) as a function of μ, and graph those functions in one (μ, x) plane, then we get **Figure 1.16**, from which we find that for $\mu < 0$, there is no critical point (or function $x(\mu)$ is not defined for $\mu < 0$); however, when μ increases and passes 0, then suddenly, **two branches** of critical points appear according to $x = \sqrt{\mu}$ and $x = -\sqrt{\mu}$, or a "**bi**"-furcation takes place. This explains why the word "bifurcation" is used. In Eq. (3.1), the total number of critical points is also a qualitative property of the system, therefore, when the parameter μ is varied and passes 0, the system undergoes a qualitative change: the number of critical points changes from 0 to 2. Thus, we say that for Eq. (3.1), when the parameter μ is varied, a bifurcation occurs at the bifurcation value $\mu = 0$.   ♠

Example 1.3.3 Let x_0 be any fixed number in $[0, 1]$ and consider a recursion relation

$$x_1 = r \sin \pi x_0, \ x_2 = r \sin \pi x_1, \ \cdots, \ x_{m+1} = r \sin \pi x_m, \quad m = 0, 1, 2, \cdots,$$

where $r \in [0, 1]$ is regarded as a parameter. Let's do it with $x_0 = 0.5$, $r = 0.6$,

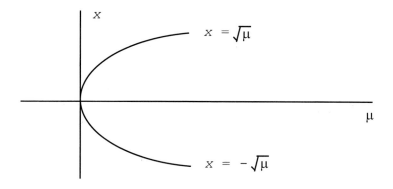

Figure 1.16: The branching of two sets of critical points determined by $x = \pm\sqrt{\mu}$, or a bifurcation takes place

and use a software called Maple, then the code

```
x[0]:=0.5;
for i from 1 by 1 to 100 do
x[i]:=evalf(0.6*sin(Pi*x[i-1]));
od;
```

gives the following result (note that we use $x[m]$ for x_m because it is what you will see in Maple).

x[1] = .6, x[2] = .570633, x[3] = .585288, x[4] = .578590, x[5] = .581804, x[6] = .580294, x[7] = .581011, x[8] = .580672, x[9] = .580833, x[10] = .580757, x[11] = .580793, x[12] = .580776, x[13] = .580784, x[14] = .580780, x[15] = .580782, x[16] = .580781, x[17] = .580781, x[18] = .580781, x[19] = .580781, x[20] = .580781, $\cdots\cdots$
x[81] = .580781, x[82] = .580781, x[83] = .580781, x[84] = .580781, x[85] = .580781, x[86] = .580781, x[87] = .580781, x[88] = .580781, x[89] = .580781, x[90] = .580781, x[91] = .580781, x[92] = .580781, x[93] = .580781, x[94] = .580781, x[95] = .580781, x[96] = .580781, x[97] = .580781, x[98] = .580781, x[99] = .580781, x[100] = .580781.

Accordingly, we find that

$$x_m \to .580781 \quad \text{as} \quad m \to \infty.$$

Or you can imagine that when you say "Order!" the numbers x_m will listen to you and order themselves to approach .580781.

Next, let's still use $x_0 = 0.5$ but replace $r = 0.6$ with $r = 0.77$ in the above code, then we get

x[1] = .77, x[2] = .509210, x[3] = .769677, x[4] = .509794, x[5] = .769635, x[6] = .509871, x[7] = .769629, x[8] = .509881, x[9] = .769628, x[10] = .509882, x[11] = .769628, x[12] = .509883, x[13] = .769628, x[14] = .509883, x[15] = .769628, x[16] = .509883, x[17] = .769628, x[18] = .509883, x[19] = .769628, x[20] = .509883, ······
x[81] = .769628, x[82] = .509883, x[83] = .769628, x[84] = .509883, x[85] = .769628, x[86] = .509883, x[87] = .769628, x[88] = .509883, x[89] = .769628, x[90] = .509883, x[91] = .769628, x[92] = .509883, x[93] = .769628, x[94] = .509883, x[95] = .769628, x[96] = .509883, x[97] = .769628, x[98] = .509883, x[99] = .769628, x[100] = .509883.

In this case, the numbers x_m will "pile up" at the two values

$$\{.509883, \quad .769628\},$$

or the sequence $\{x_m\}$ repeats each of the two values after every two iterations, in which case the set of the two values $\{.509883, .769628\}$ looks like a "**cycle**" with period 2.

Finally, let $x_0 = 0.5$ and $r = 0.9$, then we get

x[1] = .9, x[2] = .278115, x[3] = .690053, x[4] = .744288, x[5] = .647712, x[6] = .804821, x[7] = .517917, x[8] = .898574, x[9] = .281945, x[10] = .696955, x[11] = .733141, x[12] = .669193, x[13] = .775825, x[14] = .582725, x[15] = .869776, x[16] = .358013, x[17] = .811936, x[18] = .501337, x[19] = .899992, x[20] = .278136, ······
x[81] = .899132, x[82] = .280447, x[83] = .694267, x[84] = .737523, x[85] = .660844, x[86] = .787522, x[87] = .557134, x[88] = .885540, x[89] = .316696, x[90] = .754849, x[91] = .626626, x[92] = .829721, x[93] = .458815, x[94] = .892477, x[95] = .298264, x[96] = .725220, x[97] = .683961, x[98] = .753835, x[99] = .628681, x[100] = .827452.

Now, the placement of those numbers are so complex and unpredictable that no matter how loud you shout "Order!!" nobody will listen! So you may want to say "It is chaotic!" If you do, then you are right, because that is exactly the word we are going to use to describe the situation. Of course, you probably want to ask "what does this have to do with differential equations?" We will explain in Chapter 7 that the recursion relation $x_{m+1} = r \sin \pi x_m$, or in general $x_{m+1} = f(x_m)$, defines a "difference equation," or a "map," which is a discrete-time version of a differential equation. ♠

Example 1.3.3 indicates that for some differential equations, the behavior of solutions are very complicated and showing "no orders," therefore, they are generally described as **chaos**. Think about how strange things are in Example 1.3.3 because the maps $x_{m+1} = 0.6 \sin \pi x_m$, $x_{m+1} = 0.77 \sin \pi x_m$, and $x_{m+1} = 0.9 \sin \pi x_m$ look "almost the same," so how could the small difference in the coefficients of 0.6, 0.77, and 0.9 make the sequences $\{x_m\}$ of the iterations behave so differently? This is, in fact, the key to understand bifurcation and chaos: When parameters are different, the corresponding systems could behave completely differently.

Besides the above discrete maps, solutions of continuous systems (that is, differential equations) can also be chaotic. Especially for differential equations in $\Re^n$, $n \geq 3$, solutions are moving in space and could get twisted and twisted and become complex and strange. A famous equation is given by the **Lorenz system**,

$$\begin{cases} \frac{dx}{dt} & = & 10(y - x), \\ \frac{dy}{dt} & = & 28x - y - xz, \\ \frac{dz}{dt} & = & xy - (8/3)z, \end{cases} \tag{3.3}$$

in a milestone paper of Lorenz [1963] (in fact, the paper was reprinted in SPIE Milestone Series, 1994). The system was used to model the weather forecast (see Chapter 7 for some details). Despite of its innocuous looks, the numerical experiments of Lorenz [1963] showed that the solutions of Eq. (3.3) behave in a very complex and strange fashion. For example, the (x, z) plane projection of a three-dimensional solution of the Lorenz system is given in **Figure 1.17**.

The solution in Figure 1.17 does not intersect itself in $\Re^3$, so the crossings in Figure 1.17 are the result of projection in $\Re^2$. Here, the solution will cruise a few circuits on one side, then suddenly moves to the other side and cruises a few circuits, and then suddenly moves back $\cdots$. This process will continue forever, such that the solution will wind around the two sides infinitely many times without ever settling down. The solution also moves around the two sides in an unpredictable fashion. Lorenz showed with numerical experiments that the system (3.3) has an attractor whose properties are so strange and

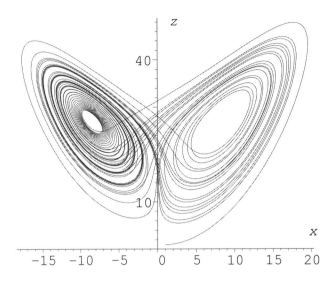

Figure 1.17: The (x, z) plane projection of a three-dimensional solution of the Lorenz system (3.3)

---

complex that it is called a **strange attractor**, a very important subject in the study of chaos.

However, we also need to point out that solutions of differential equations in $\Re^2$ do behave in an "orderly" or "predictable" fashion, due to another milestone result: the Poincaré-Bendixson theorem in $\Re^2$, to be studied in Chapter 8.

For the subject on stability, let's look at the following examples.

Example 1.3.4 Consider the scalar differential equation $x'(t) = 0$, $x(t_0) = x_0$. The solution is given by $x(t) = x_0$, $t \geq t_0$, see **Figure 1.18**.

In particular, $\phi(t) = 0$, $t \geq 0$, is a solution (with the initial value being zero). Now, for any $t_0 \geq 0$ and any other initial value x_0 that is close to ϕ, the corresponding solution $x(t) = x_0$ will stay close to ϕ for $t \geq t_0$.   ♠

According to the situation in Example 1.3.4, we say that the **solution** ϕ **is stable**, which we define as follows for a general solution ϕ that may be nonzero:

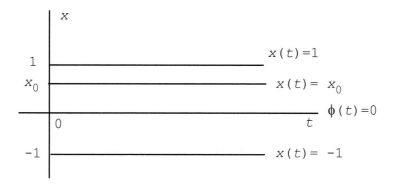

Figure 1.18: Solutions of $x'(t) = 0$ concerning stability

*A solution ϕ defined on $[t_\phi, \infty)$ is said to be **stable** if for any $\varepsilon > 0$ and any $t_0 \geq t_\phi$, there exists a $\delta = \delta(\varepsilon, t_0) > 0$ ($\delta(\varepsilon, t_0)$ means δ is determined by ε and t_0; typically $\delta \leq \varepsilon$), such that if the initial value x_0 satisfies $|x_0 - \phi(t_0)| \leq \delta$, then the corresponding solution $x(t)$ starting from t_0 satisfies $|x(t) - \phi(t)| \leq \varepsilon$ for $t \geq t_0$. See **Figure 1.19**.*

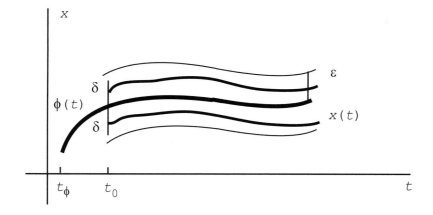

Figure 1.19: Definition of a stable solution ϕ

In physics applications, a solution ϕ may represent certain behavior or property of some physical experiment, and x_0 may be the initial measurement of certain quantity. For example, in some experiment, we need to put one gallon of acid initially to create a certain reaction, which inevitably involves some errors in measurements or approximations. That is, in applications, real data always have some inherent uncertainty, and initial values taken from real data are never known precisely. Now, ϕ being stable means that the corresponding behavior or property is stable in the sense that a small change in initial measurement will result in a small change in the behavior or property for future time. This idea can be seen further in the following example.

Example 1.3.5 Consider again the motion of a simple pendulum given by

$$\theta''(t) + k\theta'(t) + q\sin\theta(t) = 0, \quad t \geq t_0, \qquad (3.4)$$

where $k \geq 0$, $q > 0$ are constants, with k related to a damping term. If we place this pendulum in honey or any viscous fluid, and if the inertia term (related to θ'') is relatively small compared to the strong damping (related to $k\theta'$) of the viscous fluid, and if the angle $\theta(t)$ is also small, then we can neglect the $\theta''(t)$ term and approximate $\sin\theta$ with θ and then consider the differential equation

$$\theta'(t) = -(\frac{q}{k})\theta(t), \quad t \geq t_0, \qquad (3.5)$$

where we have assumed $k > 0$ since a damping exists. The solution of Eq. (3.5) is given by

$$\theta(t) = \theta_0 e^{-\frac{q}{k}(t-t_0)}, \quad t \geq t_0,$$

and we have

$$\lim_{t\to\infty} \theta(t) = \lim_{t\to\infty} \theta_0 e^{-\frac{q}{k}(t-t_0)} = 0.$$

Now, the interpretation in physics is that $\phi(t) = 0$ is a solution corresponding to the steady state (downward vertical position), if the pendulum is moved slightly from the downward vertical position, then, due to the strong damping of the medium, the pendulum tends to the downward vertical position but will not cross the downward vertical position, that is, no oscillations will occur. See **Figure 1.20**.

That is, in this case, for any initial value θ_0 that is close to $\phi = 0$, the corresponding solution $\theta(t) = \theta_0 e^{-\frac{q}{k}(t-t_0)}$ will not only stay close to ϕ for $t \geq t_0$, but we also have $\lim_{t\to\infty} \theta(t) = \lim_{t\to\infty} \theta_0 e^{-\frac{q}{k}(t-t_0)} = 0 = \phi$, or

$$\lim_{t\to\infty} |\theta(t) - \phi(t)| = 0. \qquad \spadesuit \qquad (3.6)$$

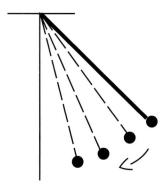

Figure 1.20: The motion of a pendulum going back to the downward vertical position without oscillations

In this sense, we say that for Eq. (3.5) of Example 1.3.5, the solution ϕ **attracts** other solutions, or the **solution ϕ is asymptotically stable**, see **Figure 1.21**.

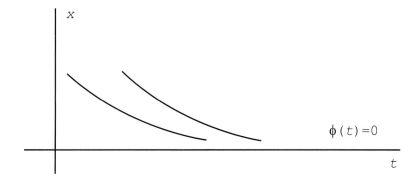

Figure 1.21: The solution $\phi = 0$ is asymptotically stable

A solution ϕ being asymptotically stable means, roughly, that ϕ is stable, and in addition, one has

$$\lim_{t \to \infty} |x(t) - \phi(t)| = 0,$$

where x is any solution whose initial value is close to ϕ. Evidently, $\phi = 0$ in

Example 1.3.4 is stable but not asymptotically stable, because solutions are given by constants there.

Example 1.3.6 Consider the scalar differential equation $x'(t) = 3x(t)$, $x(0) = x_0$. The solution is given by $x(t) = x_0e^{3t}$, $t \geq 0$, see **Figure 1.22**.

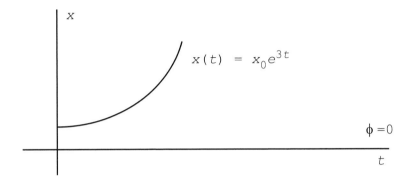

Figure 1.22: Solutions of $x'(t) = 3x(t)$ concerning instability

Now, we also have $\phi(t) = 0$ as a solution. For any other initial value $x_0 \neq 0$, no matter how close it is to ϕ, the corresponding solution $x(t) = x_0e^{3t}$ will not stay close to ϕ for $t \geq 0$. In this sense, we say that the **solution ϕ is unstable**. For example, $\phi = 0$ in the logistic equation (2.1) is unstable.

♠

Next, we examine boundedness properties. In Example 1.3.4, where the solutions are given by $x(t) = x_0$, $t \geq t_0$, if we specify a range B_1 first (that is, let $B_1 > 0$), then we are able to find a **bound $B_2 > 0$** (typically $B_2 \geq B_1$) such that when an initial value x_0 is in the range of B_1 (that is, $|x_0| \leq B_1$), then we can use B_2 to bound or control the corresponding solution for $t \geq t_0$. In fact, in this case, we can take $B_2 = B_1$, such that for the solution $x(t) = x_0$, $t \geq t_0$,

$$|x_0| \leq B_1 \quad \text{implies} \quad |x(t)| = |x_0| \leq B_1 = B_2, \quad t \geq t_0.$$

Accordingly, we say that in Example 1.3.4, the **solutions are uniformly bounded**, which is defined as (see **Figure 1.23**):

The solutions of a differential equation are said to be **uniformly bounded** *if for any $B_1 > 0$, there exists a $B_2 > 0$ such that if an initial value $|x_0| \leq B_1$, then the corresponding solution starting from t_0 satisfies $|x(t)| \leq B_2$ for $t \geq t_0$.*

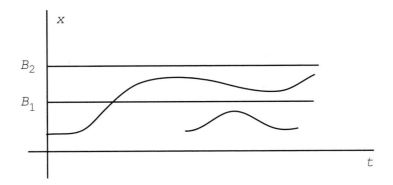

Figure 1.23: The solutions are uniformly bounded

The solutions of Eq. (3.5) in Example 1.3.5 are also uniformly bounded, but the solutions in Example 1.3.6 are not uniformly bounded.

Next, let's consider a related concept. We prescribe a bound B to begin with, and allow the initial value x_0 (at time t_0) to be in a range B_1 that is arbitrary, say, maybe $B_1 > B$. Now, in general, B cannot be used to bound the corresponding solution for $t \geq t_0$, because when $B_1 > B$, B cannot even be used to bound the initial value x_0 for which $|x_0| = B_1 > B$, see **Figure 1.24**.

Thus, it only makes sense to require that B can be used to bound the corresponding solution when t is large, say for example, when $t \geq t_0 + T$, where $T > 0$ is a constant, see **Figure 1.25**.

The requirement that "solutions be bounded by B when t is large" cannot be met by the solutions in Example 1.3.4, because the solution is given by $x(t) = x_0$ there, thus when $|x_0| = B_1 > B$, one has $|x(t)| = |x_0| = B_1 > B$, $t \geq t_0$. But this requirement can be met by the solutions of Eq. (3.5) in Example 1.3.5, because the solution is given by $x(t) = x_0 e^{-\frac{q}{k}(t-t_0)}$ there, thus when $|x_0| \leq B_1$,

$$|x(t)| = |x_0 e^{-\frac{q}{k}(t-t_0)}| \leq B_1 e^{-\frac{q}{k}(t-t_0)} \leq B, \quad t \geq t_0 + T,$$

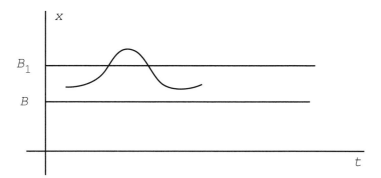

Figure 1.24: B cannot be used to bound the solutions on $[t_0, \infty)$

is true if we solve $T > 0$ in such a way that if $B_1 \leq B$, then let $T = 1$ (or any positive number); if $B_1 > B$, then solve T from $B_1 e^{-\frac{q}{k}T} = B$ and obtain $T = -\frac{k}{q} \ln \frac{B}{B_1} > 0$.

Accordingly, we say that for Eq. (3.5) in Example 1.3.5, the **solutions are uniformly ultimately bounded**, which is defined as:

The solutions of a differential equation are said to be **uniformly ultimately bounded** *if there is an (independent or generic) constant $B > 0$ such that for any $B_1 > 0$, there exists a $T > 0$ such that if an initial value $|x_0| \leq B_1$, then the corresponding solution starting from t_0 satisfies $|x(t)| \leq B$ for $t \geq t_0 + T$. (See Figure 1.25.)*

Therefore, the solutions in Example 1.3.4 and Example 1.3.6 are not uniformly ultimately bounded.

Notice the difference between uniform boundedness and uniform ultimate boundedness. In uniform boundedness, the bound B_2 can be chosen later after the initial range B_1 is fixed. However, in uniform ultimate boundedness, the bound B is fixed first, and the initial range B_1 can be chosen arbitrarily later and, of course, can be bigger than B, therefore, B may be used to bound the solutions only when t is large.

For the study of periodic solutions, consider the following examples.

Example 1.3.7 Consider the scalar differential equation

$$x''(t) + x(t) = 0. \tag{3.7}$$

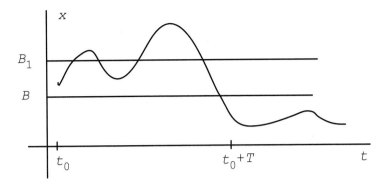

Figure 1.25: B may be used to bound the solutions on $[t_0 + T, \infty)$ for a large T

We find that $x(t) = \sin t$ is a solution. Now,

$$x(t + 2\pi) = x(t).$$

Thus we say that Eq. (3.7) has a **periodic solution** of period 2π. ♠

Example 1.3.8 For the scalar differential equation $x'(t) = 4$, the solutions are straight lines with slope 4, hence the equation has no periodic solution. In general, let's consider the scalar differential equation

$$x' = f(x), \tag{3.8}$$

where f is a continuous function in x. If $f(x) > 0$ (or $f(x) < 0$) for all x, then any solution $x(t)$ (if exists) is strictly increasing (or decreasing) in t, thus Eq. (3.8) has no periodic solutions. If $f(x) = 0$ has some real roots, for example, when the curve of $f(x)$ is given in **Figure 1.26**, then $x_1(t) = \alpha$ and $x_2(t) = \beta$, $t \geq t_0$, are two constant solutions, hence they are periodic solutions (with periods being any positive numbers). Now, x_1 and x_2 are the **only** periodic solutions of Eq. (3.8). Because, for example, if the initial value of a solution is from (α, β), then, similar to the analysis of the logistic equation (2.1), the solution will flow toward the critical point β and will never come back to where it started, thus it cannot be periodic. ♠

This geometric interpretation matches well with some experiments in physics. For example, consider Example 1.3.5 where a pendulum is placed

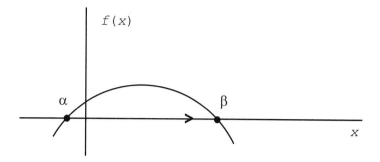

Figure 1.26: The graph of $f(x)$ showing no periodic solutions other than the constant solutions

in honey or any viscous fluid and the motion of the pendulum is approximated by the first-order differential equation (3.5), which is governed by Eq. (3.8). We find in Example 1.3.5 that no oscillations can occur there, that is, nonconstant periodic solutions do not exist there.

The analysis of Eq. (3.8) indicates that when studying periodic solutions, other forms of differential equations should be considered. For example, differential equations in $\Re^n$, $n \geq 2$, may have periodic solutions because now a solution may follow a "circle" and comes back to where it started.

Now, we have introduced some qualitative properties concerning existence and uniqueness of solutions, bifurcation, chaos, stability, boundedness, and periodicity that we will study in this book. Next, we briefly describe how to derive these properties for some simple differential equations. This will help you get ready to the ideas and methods that we will use in the rest of the book for general differential equations.

Example 1.3.9 For existence and uniqueness of solutions, let's consider the scalar differential equation

$$x'(t) = x(t), \quad x(0) = 1, \quad t \geq 0. \tag{3.9}$$

To get some idea of what to expect for general (nonlinear) differential equations, we define the right-hand side of Eq. (3.9) as $f(t,x) = x$, and consider the **Picard approximations** given by

$$\begin{cases} x_0(t) = x_0 = 1, \\ x_1(t) = x_0 + \int_0^t f(s, x_0(s))ds = 1 + \int_0^t x_0(s)ds = 1 + \int_0^t ds = 1 + t, \\ x_2(t) = x_0 + \int_0^t f(s, x_1(s))ds = 1 + \int_0^t x_1(s)ds = 1 + \int_0^t (1+s)ds, \\ \cdots \\ x_n(t) = x_0 + \int_0^t f(s, x_{n-1}(s))ds. \end{cases}$$

Now, an induction shows that

$$x_n(t) = 1 + t + \frac{t^2}{2} + \cdots + \frac{t^n}{n!},$$

consequently, we have

$$\lim_{n \to \infty} x_n(t) = e^t,$$

and we can take a derivative to check that e^t is really a solution of Eq. (3.9). ♠

Thus, for a general differential equation, we will use the Picard approximations to define a sequence of functions on a certain interval. Then, we verify, under some conditions this sequence converges to a function that gives rise to a solution of the equation.

For bifurcations, we will demonstrate, using examples and geometrical analysis, that under certain circumstances, the implicit function theorem fails to apply, thus singularities may exist and some qualitative properties of solutions may change abruptly when some parameters are varied, such as the creation and disappearance of critical points, or the exchange of stabilities of critical points.

For chaos, we will look at the discrete maps and the Lorenz system, and discuss their qualitative property changes, such as the period-doubling bifurcation cascades and their routes to chaos.

For stability and boundedness, if the differential equations are linear, then the structure of solutions using the semigroup and evolution system properties, as given by the variation of parameters formulas (1.12) and (1.14), can be used to derive the properties. When **eigenvalues** are available, they can be used directly to derive the properties. For example, in Example 1.3.4, the eigenvalue is 0, thus $\phi = 0$ is stable but not asymptotically stable; in Eq. (3.5) of Example 1.3.5, the eigenvalue is $-\frac{q}{k} < 0$, thus $\phi = 0$ is asymptotically stable; and the eigenvalue in Example 1.3.6 is $3 > 0$, thus $\phi = 0$ is unstable.

Otherwise, for general nonlinear differential equations, to determine the stability properties of a solution ϕ (maybe nonzero in general), we define a function V (called a **Liapunov function**) that is related to the distance between another solution and the solution ϕ. (Typically, we make a transformation, after which ϕ is regarded as the zero solution.) Then we take a derivative (in time t) of V by plugging in the differential equation and argue, with certain conditions, that the derivative is $V' \leq 0$. This indicates that the distance of another solution and ϕ is decreasing, which may reveal the desired properties.

In fact, this idea is already used in the analysis of the logistic equation (2.1). For example, for the critical point $\phi = C$ there, when $x(t) > C$, the distance of $x(t)$ and ϕ is $V = x(t) - C$. Now $\frac{d}{dt} V = x'(t) = ax[C - x] < 0$, thus $x(t)$ flows to $\phi = C$ from the right-hand side. When $0 < x(t) < C$, the distance of $x(t)$ and ϕ is $V = C - x(t)$. Now $\frac{d}{dt} V = -x'(t) = -ax[C - x] < 0$, thus $x(t)$ flows to $\phi = C$ from the left-hand side. Therefore, the distance of $x(t)$ and $\phi = C$ is always decreasing and hence $\phi = C$ is stable, which is already obtained. As for the critical point $\phi = 0$ there, the distance of $x(t) > 0$ and ϕ is $V = x(t)$. Now, $\frac{d}{dt} V = x'(t) = ax[C - x] > 0$ for $x(t) \in (0, C)$, thus the distance of $x(t)$ and $\phi = 0$ is increasing and hence $\phi = 0$ is unstable, which is also already obtained.

To further demonstrate this idea, let's look at the following example.

Example 1.3.10 Consider the scalar differential equation

$$x'(t) = -t^2 x^3(t), \quad t \geq 0. \tag{3.10}$$

Now, $\phi = 0$ is a solution. To determine the stability of $\phi = 0$, we define

$$V(t, x) = [x - 0]^2 = x^2. \tag{3.11}$$

Let $x(t)$ be a solution of Eq. (3.10), then $V(t, x(t)) = x^2(t)$, which is related to the distance of the solution $x(t)$ and $\phi = 0$. Now, taking a derivative of $V(t, x(t))$ in t and plugging in Eq. (3.10), we obtain

$$
\begin{aligned}
\frac{d}{dt} V(t, x(t)) &= \frac{d}{dt} x^2(t) = 2x(t)x'(t) = 2x(t)[-t^2 x^3(t)] \\
&= -2t^2 x^4(t) < 0, \quad \text{if } t > 0, \ x(t) \neq 0. \tag{3.12}
\end{aligned}
$$

Thus we expect that $|x(t) - \phi(t)| = |x(t)| \to 0$ as $t \to \infty$ (which in fact is true using the so-called **Liapunov's method** that we will introduce later).

♠

In the analysis of the logistic equation (2.1) and Example 1.3.10, the key idea is that we can **obtain certain results without solving the differential equations**. Of course, in the special case of Example 1.3.10, we can actually solve Eq. (3.10) using separation of variables (see an exercise) and obtain that

$$x(t) = \frac{1}{\sqrt{\frac{2}{3}t^3 + c}} \longrightarrow 0, \ \ t \to \infty, \tag{3.13}$$

where c is a positive constant.

This idea of deriving certain qualitative properties without solving the differential equations can also be found in applications in physics.

Example 1.3.11 Consider the scalar differential equation

$$u'' + g(u) = 0, \quad u = u(t) \in \Re, \tag{3.14}$$

where g is nonlinear, $g(0) = 0$, and satisfies some other conditions. (Note that Eq. (2.7) of a simple pendulum is a special case of Eq. (3.14) when $k = 0$, that is, damping is ignored.) Now, $u = 0$ is a constant solution or an equilibrium. In physics, we can think of $g(u)$ as the restoring force acting on a particle at a displacement u from the equilibrium $u = 0$, and of u' as the velocity of the particle. Then the potential energy at a displacement u from equilibrium is $\int_0^u g(s)ds$, and the kinetic energy is $\frac{1}{2}(u')^2$. Thus, the **total energy** is

$$V(t) = \frac{1}{2}[u'(t)]^2 + \int_0^{u(t)} g(s)ds. \tag{3.15}$$

Now, the **law of conservation of energy** in physics indicates that $V(t)$ is a constant, or $\frac{d}{dt}V(t) = 0$. Indeed, we have

$$\frac{d}{dt}V(t) = u'u'' + g(u)u' = u'[u'' + g(u)] = 0. \tag{3.16}$$

That is, without solving Eq. (3.14), we can define a function V in (3.15) and obtain that the total energy of Eq. (3.14) is a constant, or $\frac{d}{dt}V(t) = 0$. This shows the compelling connection of the method of using a Liapunov function in **mathematics** and the conservation of energy in **physics**. Later, we will verify that this V function for Eq. (3.14) is related to the distance of the solution u and the equilibrium $u = 0$, thus some properties can be derived in this direction, and applications in physics can be carried out. ♠

For periodicity, we see from Example 1.3.7 that Eq. (3.7) has a periodic solution of period 2π. Now, for a general differential equation, to find a periodic solution on interval $[0, \infty)$ of period, say for example $T > 0$, we need a solution $x(t)$ such that

$$x(t + T) = x(t), \quad t \geq 0.$$

In particular, when $t = 0$, we need

$$x(T) = x(0).$$

Accordingly, we define a **mapping** P such that if $x(t)$ is the unique solution corresponding to the initial value $x(0) = x_0$, then we let

$$P(x_0) = x(T),$$

see **Figure 1.27**.

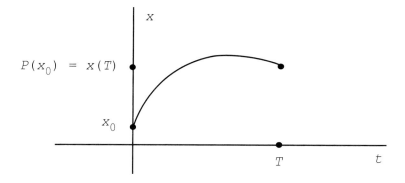

Figure 1.27: The mapping $P : P(x_0) = x(T)$

Notice that if P has a **fixed point**, that is, if there exists an $\overline{x}_0$ such that

$$P(\overline{x}_0) = \overline{x}_0,$$

then we obtain, for the unique solution $x(t)$ with the initial value $x(0) = \overline{x}_0$,

$$x(T) = P(\overline{x}_0) = \overline{x}_0 = x(0).$$

Based on this, other results can be used to verify that

$$x(t + T) = x(t), \quad t \geq 0,$$

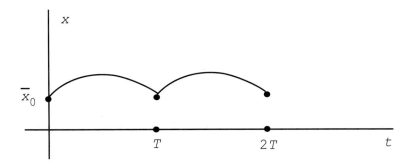

Figure 1.28: A T-periodic solution

therefore, $x(t)$ is a periodic solution of period T, see **Figure 1.28**.

Now, we have briefly described the qualitative properties that we will study in this book. We hope this will inspire your interest and curiosity enough to ask some questions, such as "are those concepts about stability and boundedness the same or what are their differences?" or "how could things like these be done for general nonlinear differential equations?" or even questions like "how bad could solutions of differential equations get?" In doing so, you will be motivated to study the following chapters.

Chapters 1–7 are for upper level undergraduate students, thus we only present the proofs for those results and theorems where some elementary arguments using calculus and linear algebra can be produced. (Notice that the subject on Fixed Point Method in Chapter 2 is optional.) Thus the proofs are accessible and will give these students a chance to use their knowledge in calculus and linear algebra to solve some problems before completing their undergraduate study. For other results whose proofs are too complex and/or involve other subjects not covered here, we do not prove them; instead, we argue their plausibility using geometric and physical interpretation.

Of course, if there are time constraints, the following results can be mentioned without detailed proofs: in Chapter 2, the proofs concerning existence and existence without uniqueness of solutions, the dependence on parameters and the maximal interval of existence; in Chapter 3, differential equations with periodic coefficients and Floquet theory; in Chapter 5, the proofs concerning Liapunov's method; in chapters 6–7, certain proofs concerning bifurcation and chaos.

Chapters 8–12, together with chapters 1–7, are for beginning graduate

students. Therefore the treatment and proofs of some results are quite involved. Certain results and their analysis are closely related to current research in differential equations, (in fact, a few are taken from some recent
research papers), thereby preparing you to access or even do some research
in qualitative theory of differential equations.

One more thing we would like to point out is that chapters 6–12 are
rather independent of each other and instructors may choose among them
to best fit the last part of the course.

Chapter 1 Exercises

1. Derive (1.10) in Example 1.1.3.

2. Solve the following differential equations.

(a) $x'(t) = 5x(t)$, $x(1) = 4$.

(b) $x'(t) = -4x(t) + t$, $x(2) = 5$.

(c) $x'(t) = tx(t) + 7t$, $x(3) = 6$.

(The integration techniques, such as integration by parts or by substi-
tutions should be used to complete the answers.)

3. Consider

$$x'(t) = k(t)x(t) + f(t), \quad x(t_0) = x_0,$$

for some continuous functions $k(t)$ and $f(t)$.

(a) Solve the equation when $f = 0$.

(b) Review the derivation of the variation of parameters formula by
assuming that $C(t)e^{\int_{t_0}^{t} k(s)ds}$ is a solution and then find the formula for $C(t)$.

(c) Let

$$y(t) = e^{\int_{t_0}^{t} k(s)ds} \left[x_0 + \int_{t_0}^{t} e^{-\int_{t_0}^{s} k(h)dh} f(s)ds \right].$$

Find $y'(t)$ in terms of $y(t)$ and $f(t)$.

4. Show that

   (a) $T(t) = e^{kt}$ satisfies the "semigroup property" (S1) and (S2).

   (b) $U(t, s) = e^{\int_s^t k(h)\,dh}$ satisfies the "evolution system property" (E1) and (E2).

5. Draw the direction field for

   (a) $x' = tx$.

   (b) $x' = -x^2$.

   (c) $x_1' = x_1^2,\ x_2' = x_2^2$.

   (d) $x_1' = tx_1 x_2,\ x_2' = -x_2^2$.

6. Sketch the function $f(x) = x^2 + x - 2$. Then sketch, on the x-axis, the flows of the solutions of $x' = f(x)$. Finally sketch the solutions in the (t, x)-plane.

7. Consider the logistic equation in Example 1.2.1.

   (a) Determine the concavity of the solution $x(t)$ when $x(t) > C$ by using Figure 1.12.

   (b) Check that $x_1(t) = 0,\ x_2(t) = C,\ t \geq 0$, are both constant solutions. Then solve for other solutions.

   (c) For the solution given by (2.11), determine the concavity by using the second derivative.

8. Verify that the logistic equation (2.1) can be replaced by

$$x'(t) = rx(t)[1 - x(t)], \tag{3.17}$$

   where $r = aC$.

9. Solve the equation in Example 1.3.10.

10. Show that $x(t) = \cos t,\ y(t) = \sin t$ satisfies $\{x'(t) = -y(t),\ y'(t) = x(t)\}$.

11. Consider

$$x'(t) = 2x(t),\ x(0) = 1, \tag{3.18}$$

and define

$$
\begin{aligned}
x_0(t) &= 1, \\
x_1(t) &= 1 + \int_0^t 2x_0(s)ds, \\
x_m(t) &= 1 + \int_0^t 2x_{m-1}(s)ds, \quad m = 2, 3, \cdots.
\end{aligned}
$$

(a) Use an induction to find a formula for $x_m(t)$.

(b) For $t \in \Re$ fixed, find $\lim_{m \to \infty} x_m(t)$.

(c) Define $x(t) = \lim_{m \to \infty} x_m(t)$. Check if $x(t)$ is a solution of Eq. (3.18).

12. Examine the change of the number of critical points for

   (a) $x' = \mu + x^2$,

   (b) $x' = \mu - x - e^{-x}$,

   where μ is regarded as a parameter.

13. Start with any real number $x_0 \in (0, 1)$ and use a calculator or Maple to find $x_1, x_2, \cdots, x_m$ up to $m = 30$ for

   (a) $x_{m+1} = \sin x_m$,

   (b) $x_{m+1} = \cos x_m$,

   (c) $x_{m+1} = 0.5 \sin \pi x_m$,

   (d) $x_{m+1} = 0.76 \sin \pi x_m$,

   (e) $x_{m+1} = 0.92 \sin \pi x_m$,

   (f) $x_{m+1} = 0.94 \sin \pi x_m$,

   (g) $x_{m+1} = 0.98 \sin \pi x_m$.

14. Find a V function for the equation

$$
x'(t) = -t^4 x(t),
$$

such that its derivative in t satisfies $V' \le 0$.

15. Find all the periodic solutions of the scalar differential equation

$$
x' = x[x - 1][x + 1].
$$

Argue why they are all the periodic solutions for the equation.

Chapter 2

Existence and Uniqueness

2.1 Introduction

After reading Chapter 1, you are probably convinced that, besides learning some numerical methods, it is important and beneficial to study some qualitative properties of differential equations. This will help you apply the knowledge of differential equations in your future studies and careers. In fact, in most applications of differential equations in sciences, it is more important to obtain certain qualitative properties rather than solve the differential equations analytically or numerically.

We start with a description of general differential equations that will be used for the rest of the book. From the examples we have encountered in Chapter 1, we see that in applications, a differential equation models the rate of change with respect to time of certain quantity or quantities. Thus we use $t \in \Re = (-\infty, \infty)$ to denote the time variable, and use $x = x(t)$ to denote the quantity or quantities at time t.

For example, in Chapter 1, we use $x(t)$ in Example 1.1.1 to denote the population of a university; in Example 1.2.3 of Lotka-Volterra competition equation, there are two populations $x_1(t)$ and $x_2(t)$, then we use $x(t) = [x_1(t), x_2(t)]^T$ to denote the two populations (here T means the transpose, so $x(t)$ is a 2×1 vector in $\Re^2$).

In general, we let x be an $n \times 1$ vector in $\Re^n$, denoted by

$$x = \begin{bmatrix} x_1 \\ x_2 \\ \cdots \\ x_n \end{bmatrix} = [x_1, x_2, \cdots, x_n]^T, \quad x_i \in \Re, \ i = 1, 2, \cdots, n.$$

The distance of x to the origin 0 in $\Re^n$ (also called the **norm**) is defined to be

$$|x| = \sum_{i=1}^{n} |x_i|. \tag{1.1}$$

When $n = 1$, $|\cdot|$ is just the absolute value. (If y is a complex number, then we also use $|y|$ to denote the distance from y to the origin $(0,0)$.)

Note that $|x| = \sum_{i=1}^{n} |x_i|$ is equivalent to another commonly used definition of distance,

$$r(x) = \sqrt{x_1^2 + x_2^2 + \cdots + x_n^2}, \tag{1.2}$$

because

$$(\frac{1}{\sqrt{n}})|x| \leq r(x) \leq |x|, \tag{1.3}$$

(see an exercise). We will use $|x| = \sum_{i=1}^{n} |x_i|$ in most places in this book because it is easier to work with, but all results done with $|x|$ are also true if $|x|$ is replaced by $r(x) = \sqrt{x_1^2 + x_2^2 + \cdots + x_n^2}$.

For an $n \times n$ matrix $A = [a_{ij}]$, we define

$$|A| = \sum_{i,j=1}^{n} |a_{ij}|.$$

It follows (see an exercise) that, for $n \times n$ matrices A, B, and an $n \times 1$ vector x,

$$|A + B| \leq |A| + |B|, \ |AB| \leq |A||B|, \ |Ax| \leq |A||x|. \tag{1.4}$$

If we consider the formats of the differential equations in Chapter 1, we find that a first-order differential equation is presented in such a way that the left-hand side is $x'(t)$ (or x') where $x(t)$ is a value or a vector, and the right-hand side is a function $f(t, x)$ where the position of x is replaced by $x(t)$. For example, in Example 1.2.2, $f(t, x) = k(t)x + \sin(tx)$, and we write $x(t)$

for x, thus the differential equation is $x'(t) = k(t)x(t) + \sin(tx(t))$. Also, Eq. (1.16) in Chapter 1 gives an example of a differential equation in a matrix form.

In general, when $f(t, x)$ is **linear in** x for $x \in \Re^n$, that is, when $f(t, x) = f(t, x_1, x_2, \cdots, x_n)$ is given by

$$
f(t, x) = \begin{bmatrix} a_{11}(t)x_1 + a_{12}(t)x_2 + \cdots + a_{1n}(t)x_n + f_1(t) \\ a_{21}(t)x_1 + a_{22}(t)x_2 + \cdots + a_{2n}(t)x_n + f_2(t) \\ \cdots \\ a_{n1}(t)x_1 + a_{n2}(t)x_2 + \cdots + a_{nn}(t)x_n + f_n(t) \end{bmatrix} \tag{1.5}
$$

for some functions $a_{ij}(t)$ and $f_i(t)$ in $\Re$, we will rewrite $f(t, x)$ using matrix and vector notations as

$$
\begin{aligned}
f(t, x) &= \begin{bmatrix} a_{11}(t) & a_{12}(t) & . & . & a_{1n}(t) \\ a_{21}(t) & a_{22}(t) & . & . & a_{2n}(t) \\ . & & . & . & . \\ . & & . & . & . \\ a_{n1}(t) & a_{n2}(t) & . & . & a_{nn}(t) \end{bmatrix} \begin{bmatrix} x_1 \\ x_2 \\ . \\ . \\ x_n \end{bmatrix} + \begin{bmatrix} f_1(t) \\ f_2(t) \\ . \\ . \\ f_n(t) \end{bmatrix} \\
&= A(t)x + f(t), \tag{1.6}
\end{aligned}
$$

where $A(t) = [a_{ij}(t)]$ is an $n \times n$ matrix function and $f(t) = [f_1(t), \cdots, f_n(t)]^T$ is an $n \times 1$ vector function formed from $a_{ij}(t)$ and $f_i(t)$ in (1.5).

Therefore, in general, we will consider a function $f(t, x)$ that is defined on a **domain** (a connected open set) D in the $(n + 1)$-dimensional (t, x)-space $\Re \times \Re^n$, with its function values in $\Re^n$, $n \geq 1$, see **Figure 2.1**. That is,

$$
f : D \subset \Re \times \Re^n \longrightarrow \Re^n.
$$

For example, in Chapter 1, $f(t, x) = kx + f(t)$ in Eq. (1.5) is from $\Re \times \Re$ into $\Re$; in Eq. (2.4) of Lotka-Volterra competition equation, $f(t, x) = f(t, x_1, x_2) = [\beta_1 x_1(K_1 - x_1 - \mu_1 x_2), \ \beta_2 x_2(K_2 - x_2 - \mu_2 x_1)]^T$ is from $\Re \times \Re^2$ into $\Re^2$.

Next, for a function $x(t) = [x_1(t), x_2(t), \cdots, x_n(t)]^T$ from $\Re$ to $\Re^n$, we define

$$
x'(t) = [x_1'(t), x_2'(t), \cdots, x_n'(t)]^T,
$$

and

$$
\int_a^b x(t)dt = [\int_a^b x_1(t)dt, \ \int_a^b x_2(t)dt, \cdots, \int_a^b x_n(t)dt]^T.
$$

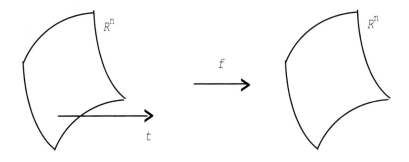

Figure 2.1: $f : D \subset \Re \times \Re^n \to \Re^n$

Now, we are ready to define the general first-order **ordinary differential equation in** $\Re^n$, $n \geq 1$, that we will use in the rest of the book. It is given as

$$x'(t) = f(t, x(t)), \quad x(t_0) = x_0, \tag{1.7}$$

on a domain $D \subset \Re \times \Re^n$, where $(t_0, x_0) \in D$ and $f(t, x)$ is a function from $D \subset \Re \times \Re^n$ into $\Re^n$. We state the following definition concerning its solutions.

Definition 2.1.1 *A function $x(t)$ is said to be a* **solution** *of Eq. (1.7) on an interval I if $t_0 \in I$, $(t, x(t)) \in D$ for $t \in I$, and $x(t)$ is differentiable for $t \in I$ and satisfies Eq. (1.7) for $t \in I$. See* **Figure 2.2**.

Notice that here, as we also did for the examples in Chapter 1, we have made a **fundamental assumption** that when modeling using a differential equation, the time rate, given by $x'(t)$, depends only on the current status, given by $f(t, x(t))$. Note also that Eq. (1.7) is written as $x' = f(t, x)$ sometimes, and in general, solutions of Eq. (1.7) may exist on both sides of t_0.

This chapter is organized as follows: In Section 2, we study existence and uniqueness of solutions, that is, we examine under what conditions a differential equation has solutions and how the solutions are uniquely determined, without solving the differential equation analytically. We follow the idea of an example in Chapter 1 where a sequence of functions is defined using the Picard approximations. To verify for general cases that the

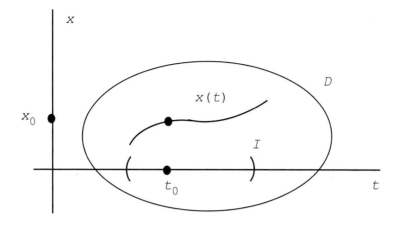

Figure 2.2: A solution $x(t)$ of Eq. (1.7) on I

---

sequence converges to a unique solution, a condition called the **Lipschitz condition** is utilized. In Section 3, we show under certain conditions that solutions are continuous and differentiable with respect to initial data and parameters. In Section 4, we determine structures of the maximal intervals of existence for solutions, and study properties of solutions with respect to the maximal intervals of existence. In Section 5 (which may be optional), we introduce the Fixed Point Method. We use the contraction mapping principle to derive existence and uniqueness if a local Lipschitz condition is satisfied. Then, when a local Lipschitz condition is not assumed, we use Schauder's second fixed point theorem to obtain existence of solutions, in which case, uniqueness is not guaranteed.

Exercises 2.1

1. Take a square and use an induction to show that

$$\frac{\sum_{i=1}^n |x_i|}{\sqrt{n}} \leq \sqrt{x_1^2 + x_2^2 + \cdots + x_n^2} \leq \sum_{i=1}^n |x_i|, \quad x_i \in \Re.$$

2. Show that $|\int_a^b f(t)dt| \leq \int_a^b |f(t)|dt$, where $a \leq b$ and $f : \Re \to \Re^n$.

3. Let A and B be 2×2 matrices.

(a) Find $|AB|$ and $|A||B|$.

(b) Show the inequalities in (2.2) when $n = 2$.

4. Show the inequalities in (2.2) for any $n \geq 1$.

2.2   Existence and Uniqueness

In this section, we will find some conditions that guarantee existence and uniqueness of solutions for Eq. (1.7). First, we note that a very effective way to attack the **differential equation** (1.7) is to convert it to an equivalent **integral equation**, since an integral equation is more convenient to work with, as we will soon see. We present this **equivalency** as follows.

Lemma 2.2.1 *Let $f(t, x)$ be a continuous function on a domain $D \subset \Re \times \Re^n$ and let $x(t)$ be a continuous function on some interval I containing t_0 such that $(t, x(t)) \in D$ for $t \in I$. Then $x(t)$ is a solution of Eq. (1.7) on the interval I if and only if*

$$x(t) = x_0 + \int_{t_0}^{t} f(s, x(s))ds, \quad t \in I. \tag{2.1}$$

Proof. If $x(t)$ is a solution of Eq. (1.7), then integrating Eq. (1.7) from t_0 to t yields (2.1).

On the other hand, if $x(t)$ satisfies (2.1), then, as $f(t, x(t))$ is continuous in t, the right-hand side of (2.1) is differentiable in t, thus $x(t)$ given by (2.1) is differentiable. Now, $x(t_0) = x_0$, and, taking a derivative of t in (2.1), we obtain

$$\frac{d}{dt}x(t) = \frac{d}{dt}\int_{t_0}^{t} f(s, x(s))ds = f(t, x(t)), \quad t \in I, \tag{2.2}$$

where the fundamental theorem of calculus is used. Therefore, $x(t)$ is a solution of Eq. (1.7) on the interval I. This completes the proof.   ♠

Based on Lemma 2.2.1, you should have some idea why we utilize the **Picard approximations**, given by

$$\begin{cases} x_0(t) & = & x_0, \\ x_1(t) & = & x_0 + \int_{t_0}^{t} f(s, x_0(s))ds, \\ x_m(t) & = & x_0 + \int_{t_0}^{t} f(s, x_{m-1}(s))ds, \quad m = 2, 3, \cdots. \end{cases} \tag{2.3}$$

The reason is that the approximation schemes suggest that if $x_m(t)$ converges to some function $x(t)$, and if the limit as $m \to \infty$ can be taken inside the integral $\int_{t_0}^{t} f(s, x_{m-1}(s))ds$ and also inside the function f, then, from (2.3), we will obtain

$$x(t) = x_0 + \int_{t_0}^{t} f(s, x(s))ds,$$

hence $x(t)$ will be a solution of Eq. (1.7), using Lemma 2.2.1.

This is exactly what happened in Example 1.3.9 in Section 3 of Chapter 1, where, for

$$x'(t) = x(t), \quad x(0) = 1, \tag{2.4}$$

one has $f(t, x) = x$, and then the Picard approximations yield

$$\begin{cases} x_0(t) = 1, \\ x_1(t) = 1 + \int_0^t f(s, x_0(s))ds = 1 + \int_0^t x_0(s)ds = 1 + \int_0^t ds = 1 + t, \\ x_2(t) = 1 + \int_0^t f(s, x_1(s))ds = 1 + \int_0^t x_1(s)ds = 1 + \int_0^t (1 + s)ds, \\ x_m(t) = 1 + \int_0^t f(s, x_{m-1}(s))ds = 1 + \int_0^t x_{m-1}(s)ds, \quad m = 3, 4, \cdots \end{cases} \tag{2.5}$$

and an induction shows that $x_m(t) = 1 + t + \frac{t^2}{2} + \cdots + \frac{t^m}{m!}$. Consequently,

$$\lim_{m \to \infty} x_m(t) = \lim_{m \to \infty} [1 + t + \frac{t^2}{2} + \cdots + \frac{t^m}{m!}] = e^t,$$

and taking a limit in (2.5) yields

$$e^t = 1 + \int_0^t e^s ds = x(0) + \int_0^t f(s, e^s)ds,$$

hence $x(t) = e^t$ is the solution of Eq. (2.4).

Next, let's investigate uniqueness of solutions. Recall that in applications, the point (t_0, x_0) (or sometimes just x_0) is called initial condition, or initial point, or initial data. For a given initial point, we only want one solution to pass through it. Such as in Example 1.1.1 in Section 1 of Chapter 1, we only want one solution to be used as the population of that university.

Now, let's look at what will happen if $x(t)$ and $y(t)$ are both solutions of Eq. (1.7). Then, under this assumption, we know from Lemma 2.2.1 that

$$x(t) = x_0 + \int_{t_0}^{t} f(s, x(s))ds, \tag{2.6}$$

$$y(t) = x_0 + \int_{t_0}^{t} f(s, y(s))ds. \tag{2.7}$$

Thus, (assuming $t \geq t_0$)

$$|x(t) - y(t)| = |(x_0 - x_0) + \int_{t_0}^t f(s, x(s)) - f(s, y(s))ds|$$

$$\leq 0 + \int_{t_0}^t |f(s, x(s)) - f(s, y(s))|ds. \qquad (2.8)$$

In a special case when $f(t, x)$ is linear in x, say for example,

$$f(t, x) = kx + f(t) \qquad (2.9)$$

for $x \in \Re$ and some constant $k > 0$, we obtain $f(t, x) - f(t, y) = kx + f(t) - ky - f(t)$, and then

$$|f(t, x) - f(t, y)| \leq k|x - y|, \qquad (2.10)$$

(in fact, we should get equality in (2.10) for this case, but we use inequality for other purposes). Hence, using (2.10), we find that (2.8) becomes

$$|x(t) - y(t)| \leq 0 + \int_{t_0}^t k|x(s) - y(s)|ds, \quad (\text{assuming } t \geq t_0). \qquad (2.11)$$

But on the other hand, this $f(t, x) = kx + f(t)$ in (2.9) is the function on the right-hand side of Eq. (1.5) in Section 1 of Chapter 1, whose solutions are given **uniquely** by the variation of parameters formula (1.11) also in Section 1 of Chapter 1. Therefore, as x and y are now solutions of $x' = f(t, x) = kx + f(t)$, $x(t_0) = x_0$, we find that $x - y$ is a solution of $z' = kz$, $z(t_0) = 0$, hence, from the uniqueness using the variation of parameters formula, $x - y = 0$.

This indicates that for $f(t, x)$ given in (2.9), we should also get $|x(t) - y(t)| = 0$ from (2.11). Therefore, it would help if there is a device that allows us to estimate $|x(t) - y(t)|$ based on (2.11) and conclude that $|x(t) - y(t)| = 0$. The following inequality will provide such a device and help us get exactly what we need.

Lemma 2.2.2 (Gronwall's inequality) *Let $M \geq 0$ be a constant. If $u(t)$ and $v(t)$ are real-valued nonnegative continuous functions such that*

$$u(t) \leq M + \int_{t_0}^t u(s)v(s)ds, \quad t_0 \leq t < T, \quad (T \leq \infty) \qquad (2.12)$$

then

$$u(t) \leq M \exp\left(\int_{t_0}^t v(s)ds\right), \quad t_0 \leq t < T. \qquad (2.13)$$

Proof. First, assume $M > 0$. Then (2.12) implies

$$\frac{u(h)v(h)}{M + \int_{t_0}^h u(s)v(s)ds} \leq v(h), \quad t_0 \leq h < T,$$

and

$$\int_{t_0}^t \frac{u(h)v(h)}{M + \int_{t_0}^h u(s)v(s)ds} dh \leq \int_{t_0}^t v(h)dh, \quad t_0 \leq t < T. \tag{2.14}$$

But

$$\int_{t_0}^t \frac{u(h)v(h)}{M + \int_{t_0}^h u(s)v(s)ds} dh = \ln\left[M + \int_{t_0}^t u(s)v(s)ds\right] - \ln M. \tag{2.15}$$

Thus we obtain, from (2.14) and (2.15),

$$\ln \frac{M + \int_{t_0}^t u(s)v(s)ds}{M} \leq \int_{t_0}^t v(h)dh, \tag{2.16}$$

hence

$$M + \int_{t_0}^t u(s)v(s)ds \leq M \exp\left(\int_{t_0}^t v(s)ds\right). \tag{2.17}$$

Therefore, (2.13) is true using (2.12) and (2.17).

If $M = 0$, then (2.12) becomes

$$u(t) \leq \int_{t_0}^t u(s)v(s)ds, \quad t_0 \leq t < T.$$

Now, for any $m \geq 1$, we have

$$u(t) \leq \frac{1}{m} + \int_{t_0}^t u(s)v(s)ds, \quad t_0 \leq t < T.$$

From what we have just proved,

$$u(t) \leq \frac{1}{m} \exp\left(\int_{t_0}^t v(s)ds\right), \quad t_0 \leq t < T.$$

Thus for any fixed $t \in [t_0, T)$, we can let $m \to \infty$ to conclude $u(t) \leq 0$. This completes the proof. ♠

For t on the left-hand side of t_0, we have a similar result, whose proof is left as an exercise.

Lemma 2.2.2.a *Let $M \geq 0$ be a constant. If $u(t)$ and $v(t)$ are real-valued nonnegative continuous functions such that*

$$u(t) \leq M + \int_t^{t_0} u(s)v(s)ds, \quad S < t \leq t_0, \quad (S \geq -\infty),$$

then

$$u(t) \leq M \exp\left(\int_t^{t_0} v(s)ds\right), \quad S < t \leq t_0. \qquad \spadesuit \qquad (2.18)$$

Now, let's look at (2.11) again and apply Gronwall's inequality there, then we obtain

$$|x(t) - y(t)| \leq 0 e^{k|t-t_0|} = 0, \quad \text{or} \quad x(t) = y(t),$$

which verifies the uniqueness of solutions for Eq. (1.7) when $f(t,x) = kx + f(t)$.

For a general function $f(t,x)$ in Eq. (1.7), we deduce from the above analysis that if $|f(t,x) - f(t,y)|$ can be controlled by $k|x-y|$ for a constant $k > 0$, that is, if the inequality (2.10) holds for a general function $f(t,x)$ in Eq. (1.7), then we still have the uniqueness of solutions for Eq. (1.7). Accordingly, we introduce the following condition, which is formulated based on the inequality (2.10).

Definition 2.2.3 (Lipschitz condition) *A function $f(t,x)$ on a domain $D \subset \Re \times \Re^n$ is said to satisfy a **Lipschitz condition** (also called a **global Lipschitz condition** sometimes) with respect to x on D if there exists a constant $k > 0$ (called a **Lipschitz constant**) such that*

$$|f(t,x) - f(t,y)| \leq k|x-y|, \quad \text{for } (t,x), (t,y) \in D. \qquad (2.19)$$

Example 2.2.4 For a linear (in x) function $f(t,x) = A(t)x + f(t)$, where $A(t)$ is an $n \times n$ matrix function and $f(t)$ is an $n \times 1$ vector function, one has

$$|f(t,x) - f(t,y)| \leq |A(t)||x-y|, \qquad (2.20)$$

thus a Lipschitz condition is satisfied if $|A(t)|$ is bounded by a constant. $\spadesuit$

The next example indicates that many nonlinear (in x) functions also satisfy a Lipschitz condition.

Example 2.2.5 For $f(t, x) = 5 \sin t \cos x$, we have

$$
\begin{aligned}
|f(t, x) - f(t, y)| &= |5 \sin t[\cos x - \cos y]| \\
&\leq |5 \sin t||\cos x - \cos y| \leq 5|\cos x - \cos y| \\
&\leq 5|x - y|,
\end{aligned}
$$

where we have used the **mean value theorem** at the last step. Thus a Lipschitz condition is satisfied. ♠

So far, after analyzing the structure of some linear functions, we have formulated a Lipschitz condition, and it seems that this condition implies uniqueness of solutions. Now, it is natural to ask whether there exists an example where a Lipschitz condition is not satisfied and where uniqueness is violated. The answer is YES, as can be seen from the following example from physics.

Example 2.2.6 (Leaky bucket) Consider a bucket with a hole in the bottom, shown in **Figure 2.3**. When you find that the bucket is empty, can you tell how much water the bucket had initially? The answer is obviously NO, because the bucket could be full initially, or could be half-full initially, or could even be empty initially.

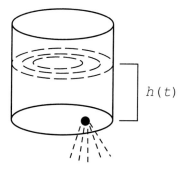

Figure 2.3: A leaky bucket

To analyze the situation, let's use Torricelli's law in physics, which says that the rate at which water drains from the bucket is proportional to the square root of the height of the water remaining inside the bucket. If we denote $h(t)$ the height of the water in the bucket at time t, and assume that the initial height is h_0 at the initial time $t = 0$, then, according to Torricelli's law, there is a positive constant k such that

$$h'(t) = -k\sqrt{h(t)}, \quad h(0) = h_0, \quad t \geq 0, \qquad (2.21)$$

where the negative sign is needed because $h(t)$ is a decreasing function in t.

Note that $\phi(t) = 0$, $t \geq 0$, is certainly a solution of Eq. (2.21) (with zero the initial value). To find out if we also have nonzero solutions, we solve

$$\int \frac{1}{\sqrt{h}} dh = \int -k\, dt,$$

and obtain

$$h(t) = \frac{k^2}{4}\left[t - \frac{2\sqrt{h_0}}{k}\right]^2, \quad 0 \leq t \leq \frac{2\sqrt{h_0}}{k}, \qquad (2.22)$$

where the initial condition $h(0) = h_0$ is used. The formula (2.22) indicates that the height of the water becomes zero or the bucket becomes empty when $t = \frac{2\sqrt{h_0}}{k}$. Thus, for things to be meaningful in physics, the bucket will remain empty or the height will remain zero after $t = \frac{2\sqrt{h_0}}{k}$. Therefore, we define

$$x(t, 0, h_0) = \begin{cases} \frac{k^2}{4}\left[t - \frac{2\sqrt{h_0}}{k}\right]^2, & 0 \leq t \leq \frac{2\sqrt{h_0}}{k}, \\ 0, & t > \frac{2\sqrt{h_0}}{k}. \end{cases} \qquad (2.23)$$

Then, for $x(t) = x(t, 0, h_0)$ defined in (2.23), we can verify that

$$x'(t) = \begin{cases} \frac{k^2}{2}\left[t - \frac{2\sqrt{h_0}}{k}\right] = -k\sqrt{x(t)}, & 0 \leq t \leq \frac{2\sqrt{h_0}}{k}, \\ 0 = -k\sqrt{x(t)}, & t > \frac{2\sqrt{h_0}}{k}. \end{cases} \qquad (2.24)$$

(For the derivative at $t = \frac{2\sqrt{h_0}}{k}$, the left and the right derivatives should be calculated, which is left as an exercise.) That is, for any $h_0 \geq 0$ fixed, $x(t, 0, h_0)$ defined in (2.23) gives the unique solution of Eq. (2.21) for $t \geq 0$, which uniquely determines the height of the water in the bucket at any time $t \geq 0$. See **Figure 2.4**.

In particular, if we use F to denote the height of the water when the bucket is full, then we replace h_0 by F and find that at time

$$t_F = \frac{2\sqrt{F}}{k},$$

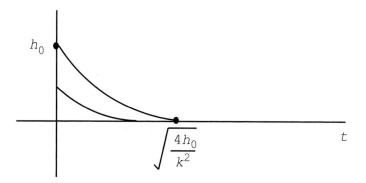

Figure 2.4: The solutions of Eq. (2.21)

the bucket becomes empty. Similarly, for a half-full bucket, we replace h_0 by $\frac{F}{2}$ and find that it takes

$$t_H = \frac{\sqrt{2F}}{k} > \frac{t_F}{2}$$

to drain a half-full bucket to empty. See **Figure 2.5**.

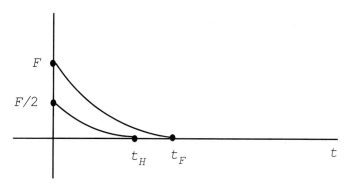

Figure 2.5: The time taken to drain a full or a half-full bucket to empty

In the above, **for any initial data** $(0, h_0)$**, a solution is uniquely determined for** $t \geq 0$**, or when** t **increases from** $t = 0$.

However, in Figure 2.5, if we start with the point $(t_F, 0)$ and look **backward**, then we find that **the solutions on the left-hand side of** $(t_F, 0)$

are **NOT uniquely determined**. In fact, from Figure 2.4, we find that the solutions on the left-hand side of any point on the positive t-axis are not uniquely determined. That is, for Eq. (2.21), we have the **nonuniqueness of the solutions to the left-hand side**. It says, in practice, that if at a time you find that the bucket is empty, then you cannot tell how much water the bucket had initially. The height of the initial water could be any value from zero to F. See Figure 2.5.

Next, let's check the Lipschitz condition. We have $f(t, x) = -k\sqrt{x}$; and if $x \neq 0$ and $y = 0$, then

$$|f(t, x) - f(t, y)| = k|x^{1/2}| = k|x^{-1/2}x| = \frac{k}{|x^{1/2}|}|x - y|, \qquad (2.25)$$

and $\frac{k}{|x^{1/2}|} \to \infty$ as $x \to 0$. Thus, $f(t, x) = -k\sqrt{x}$ does not satisfy a Lipschitz condition on any domain containing a point $(t, 0)$, $t > 0$, see **Figure 2.6**. ♠

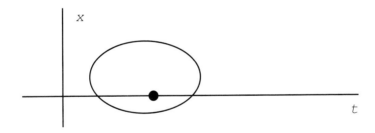

Figure 2.6: A Lipschitz condition is not satisfied on any domain containing a point $(t, 0)$, $t > 0$

Example 2.2.6 indicates that nonuniqueness of solutions may occur if a Lipschitz condition is not satisfied. Therefore, to guarantee uniqueness of solutions, we should assume a Lipschitz condition, which allows us to estimate the difference of a nonlinear function $f(t, x)$, $|f(t, x) - f(t, y)|$, almost like that of a linear function, such that Gronwall's inequality can be used to derive uniqueness of solutions. We also point out that a differential equation not satisfying a Lipschitz condition may or may not have uniqueness.

Now, we are ready to prove the following existence and uniqueness result, utilizing the Picard approximations and the Lipschitz condition.

Theorem 2.2.7 (Picard's local existence and uniqueness theorem)
*Assume that $f(t, x)$ is continuous on a domain $D \subset \Re \times \Re^n$ and satisfies
a Lipschitz condition with respect to x on D. Let $(t_0, x_0) \in D$. Then there
exist positive constants a and b such that the region*

$$R = \{(t, x) : |t - t_0| \leq a, \ |x - x_0| \leq b\} \tag{2.26}$$

is in D. Moreover, if we define

$$r = \min\{a, \ \frac{b}{M}\}, \quad \text{where} \ \ M = \max_{(t,x) \in R} |f(t, x)|, \tag{2.27}$$

*then $r > 0$ is finite and on the interval $I = (t_0 - r, t_0 + r)$, Eq. (1.7) has a
unique solution, denoted by $x(t, t_0, x_0)$, passing through (t_0, x_0). See* **Figure
2.7**.

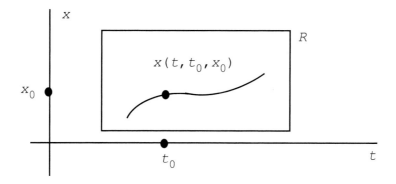

Figure 2.7: The domain R and the solution $x(t, t_0, x_0)$

Note that if the function f is identically zero on R, then $M = 0$. In
this case we can take $r = a$, and then $x(t) = x_0$, $|t - t_0| < a$, is a constant
solution. So that in the following analysis, we assume

$$M = \max_{(t,x) \in R} |f(t, x)| > 0.$$

Before we prove Theorem 2.2.7, let's note that if, for $(t, x) \in D, |f(t, x)| \leq
M$ for some constant $M > 0$, then in the two-dimensional (t, x)-space when
considering $x \in \Re$, we find (see an exercise) that a solution of Eq. (1.7) (if

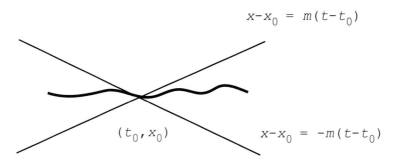

$$x - x_0 = m(t - t_0)$$

$$(t_0, x_0)$$

$$x - x_0 = -m(t - t_0)$$

Figure 2.8: Straight lines $x - x_0 = \pm M(t - t_0)$ and possible solutions

exists) lies between the two straight lines $x - x_0 = \pm M(t - t_0)$ with slopes $\pm M$. See **Figure 2.8**.

Accordingly, we have the two cases for the straight lines $x - x_0 = \pm M(t - t_0)$ to intersect the boundary of the region R, given in **Figure 2.9**, from which we detect that the requirement $r = \min\{a, \frac{b}{M}\}$ is used to make sure that any solution of Eq. (1.7) (if exists) doesn't get outside of the region R. (Otherwise, if a candidate for a solution gets outside of R, it may get outside of the domain D, then the function f is not defined there.)

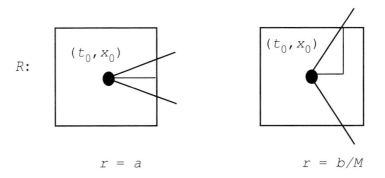

Figure 2.9: Two cases of how the straight lines $x - x_0 = \pm M(t - t_0)$ intersect the boundary of the region R

Proof of Theorem 2.2.7. Note that a domain is (by definition) an open set, thus we can find positive constants a and b such that the region R defined by (2.26) is in D. Since f is a continuous function on the closed and bounded region R, we conclude that

$$M = \max_{(t,x) \in R} |f(t,x)| < \infty,$$

hence r in (2.27) is well defined, and is positive and finite.

We will prove the theorem by showing the following four things:

(A). Based on Lemma 2.2.1, define the Picard approximations as

$$\begin{cases} x_0(t) = x_0, & |t - t_0| \leq r, \\ x_1(t) = x_0 + \int_{t_0}^t f(s, x_0(s))ds, & |t - t_0| \leq r, \\ x_m(t) = x_0 + \int_{t_0}^t f(s, x_{m-1}(s))ds, & |t - t_0| \leq r, \ m = 2, 3, \cdots. \end{cases} \tag{2.28}$$

Then the sequence $\{x_m(t)\}_{m \geq 0}$ is **defined** on $[t_0 - r, t_0 + r]$, and $(t, x_m(t)) \in R$, $m \geq 0$, $t \in I$.

(B). The sequence $\{x_m(t)\}_{m \geq 0}$ **converges** uniformly on $[t_0 - r, t_0 + r]$ to a continuous function, denoted by $x(t)$.

(C). The function $x(t)$ is **a solution** of Eq. (1.7) on I.

(D). The solution $x(t)$ is **unique**. That is, if $y(t)$ is also a solution of Eq. (1.7) on I, then $y(t) = x(t)$, $t \in I$.

(A): We verify that the sequence $\{x_m(t)\}_{m \geq 0}$ is defined on $[t_0 - r, t_0 + r]$ and that $(t, x_m(t)) \in R$ for $m \geq 0$, $t \in I$. To this end, let's prove that the sequence $\{x_m(t)\}_{m \geq 0}$ is defined on $[t_0 - r, t_0 + r]$ and

$$|x_m(t) - x_0| \leq M|t - t_0|, \quad m \geq 0, \ t \in I, \tag{2.29}$$

because if (2.29) is true, then

$$|x_m(t) - x_0| \leq M|t - t_0| \leq Mr \leq b, \quad t \in I, \tag{2.30}$$

hence, by the definition of the region R, we have $(t, x_m(t)) \in R$ for $m \geq 0$, $t \in I$.

We will use an induction to verify that the sequence $\{x_m(t)\}_{m \geq 0}$ is defined on $[t_0 - r, t_0 + r]$ and that (2.29) is true. For $m = 0$, $x_0(t) = x_0$ is defined on $[t_0 - r, t_0 + r]$ and $|x_0(t) - x_0| = |x_0 - x_0| = 0 \leq M|t - t_0|$. Assume that $x_m(t)$ is defined on $[t_0 - r, t_0 + r]$ and (2.29) is true for m. Then, for $t \in [t_0 - r, t_0 + r]$, one has $(t, x_m(t)) \in R$ from (2.29) and (2.30), hence $f(t, x_m(t))$ is defined and satisfies $|f(t, x_m(t))| \leq M$. Thus, from (2.28), $x_{m+1}(t)$ is defined on $[t_0 - r, t_0 + r]$, and

$$|x_{m+1}(t) - x_0| = |\int_{t_0}^{t} f(s, x_m(s))ds| \leq M|t - t_0|,$$

which proves (2.29) for $m + 1$, and therefore completes the induction.

(B): To show that $\{x_m(t)\}_{m \geq 0}$ is uniformly convergent on $[t_0 - r, t_0 + r]$, let's note that

$$
\begin{aligned}
x_m(t) &= x_0(t) + [x_1(t) - x_0(t)] + [x_2(t) - x_1(t)] + \cdots + [x_m(t) - x_{m-1}(t)] \\
&= x_0(t) + \sum_{j=0}^{m-1} [x_{j+1}(t) - x_j(t)]. \quad (2.31)
\end{aligned}
$$

Thus, showing the uniform convergence on $[t_0 - r, t_0 + r]$ of the sequence $\{x_m(t)\}_{m \geq 0}$ is equivalent to showing that the series

$$\sum_{j=0}^{\infty} [x_{j+1}(t) - x_j(t)], \quad t \in [t_0 - r, t_0 + r], \quad (2.32)$$

is uniformly convergent on $[t_0 - r, t_0 + r]$. From (2.28), (2.29), and using the Lipschitz condition (with constant k), we obtain

$$
\begin{aligned}
|x_2(t) - x_1(t)| &= |\int_{t_0}^{t} f(s, x_1(s))ds - \int_{t_0}^{t} f(s, x_0(s))ds| \\
&\leq k|\int_{t_0}^{t} |x_1(s) - x_0|ds| \\
&\leq k|\int_{t_0}^{t} M|s - t_0|ds| \\
&= \frac{kM|t - t_0|^2}{2}, \quad t \in [t_0 - r, t_0 + r]. \quad (2.33)
\end{aligned}
$$

Hence, using an induction again, we can verify (left as an exercise, see (2.33) for a hint) that

$$|x_{j+1}(t) - x_j(t)| \leq \frac{Mk^j|t - t_0|^{j+1}}{(j+1)!}, \quad j \geq 0, \quad t \in [t_0 - r, t_0 + r], \quad (2.34)$$

which implies, for $t \in [t_0 - r, \, t_0 + r]$,

$$
\begin{aligned}
\sum_{j=0}^{\infty} |x_{j+1}(t) - x_j(t)| &\leq \sum_{j=0}^{\infty} \frac{Mk^j |t - t_0|^{j+1}}{(j+1)!} \\
&= \frac{M}{k} \sum_{j=0}^{\infty} \frac{|k(t - t_0)|^{j+1}}{(j+1)!} \\
&\leq \frac{M}{k} \sum_{j=0}^{\infty} \frac{(kr)^{j+1}}{(j+1)!} \\
&= \frac{M}{k} [\exp(kr) - 1] < \infty. \qquad (2.35)
\end{aligned}
$$

Now, we apply the Weierstrass M-test concerning the uniform convergence of general function series from the Appendix, and conclude that the series (2.32) converges uniformly to a continuous function on $[t_0 - r, \, t_0 + r]$. Therefore, $\{x_m(t)\}_{m \geq 0}$ converges uniformly on $[t_0 - r, \, t_0 + r]$ to a continuous function, which we denote by $x(t)$, $t \in [t_0 - r, \, t_0 + r]$.

(Another approach to the uniform convergence is as follows: From

$$
\begin{aligned}
\max_{t \in [t_0 - r, \, t_0 + r]} |x_{j+1}(t) - x_j(t)| &\leq \max_{t \in [t_0 - r, \, t_0 + r]} \left| \int_{t_0}^{t} k |x_j(s) - x_{j-1}(s)| ds \right| \\
&\leq kr \max_{s \in [t_0 - r, \, t_0 + r]} |x_j(s) - x_{j-1}(s)|, \quad (2.36)
\end{aligned}
$$

we have, by an induction,

$$
\max_{t \in [t_0 - r, \, t_0 + r]} |x_{j+1}(t) - x_j(t)| \leq (kr)^j \max_{s \in [t_0 - r, \, t_0 + r]} |x_1(s) - x_0|. \qquad (2.37)
$$

Therefore, if we assume $kr < 1$, then the geometric series convergence results can be applied to derive the uniform convergence of the series (2.32).)

(C): To show that $x(t)$ is a solution of Eq. (1.7), we first verify that $(t, x(t)) \in R$, $t \in I$. To this end, we note, from (2.29),

$$
\begin{aligned}
|x(t) - x_0| &\leq |x(t) - x_m(t)| + |x_m(t) - x_0| \\
&\leq |x(t) - x_m(t)| + M|t - t_0|, \quad m \geq 0, \ t \in I.
\end{aligned}
$$

Thus, for any fixed $t \in I$, since $\{x_m(t)\}_{m \geq 0}$ converges to $x(t)$, we can let $m \to \infty$ to obtain

$$
\begin{aligned}
|x(t) - x_0| &\leq \lim_{m \to \infty} |x(t) - x_m(t)| + M|t - t_0| \\
&= 0 + M|t - t_0| \leq Mr \leq b, \quad t \in I,
\end{aligned}
$$

which implies that $(t, x(t)) \in R$ for $t \in I$, and guarantees that $f(t, x(t))$ is defined for $t \in I$. Next, from the Lipschitz condition (with constant k), we obtain

$$
\begin{aligned}
&\left| \int_{t_0}^t f(s, x_{m-1}(s))ds - \int_{t_0}^t f(s, x(s))ds \right| \\
&\leq k \left| \int_{t_0}^t |x_{m-1}(s) - x(s)|ds \right| \\
&\leq k|t - t_0| \max_{s \in [t_0 - r, t_0 + r]} |x_{m-1}(s) - x(s)|, \quad t \in [t_0 - r, t_0 + r].
\end{aligned}
$$

Then, as $\{x_m(t)\}_{m \geq 0}$ converges uniformly on $[t_0 - r, t_0 + r]$ to $x(t)$, we derive, for $t \in I$,

$$
\lim_{m \to \infty} \left| \int_{t_0}^t f(s, x_{m-1}(s))ds - \int_{t_0}^t f(s, x(s))ds \right| = 0. \tag{2.38}
$$

Now, from (2.38) and the fact that $\{x_m(t)\}_{m \geq 0}$ converges to $x(t)$, we can take a limit as $m \to \infty$ in (2.28) to obtain

$$
x(t) = x_0 + \int_{t_0}^t f(s, x(s))ds, \quad t \in I,
$$

hence $x(t)$ is a solution of Eq. (1.7), using Lemma 2.2.1.

(D): If $y(t)$ is also a solution of Eq. (1.7) on I, then from Lemma 2.2.1,

$$
y(t) = x_0 + \int_{t_0}^t f(s, y(s))ds, \quad t \in I. \tag{2.39}
$$

Now, for $t \in [t_0, t_0 + r)$, the Lipschitz condition (with constant k) yields

$$
\begin{aligned}
|y(t) - x(t)| &\leq \int_{t_0}^t |f(s, y(s)) - f(s, x(s))|ds \\
&\leq k \int_{t_0}^t |y(s) - x(s)|ds, \tag{2.40}
\end{aligned}
$$

therefore we can apply Gronwall's inequality to conclude that

$$
|y(t) - x(t)| = 0, \quad t \in [t_0, t_0 + r).
$$

The same conclusion can also be proved in the same way for $t \in (t_0 - r, t_0]$ by using Gronwall's inequality for $t \leq t_0$. Therefore, uniqueness holds.

The above four steps in (A)–(D) complete the proof.   ♠

Note that $x(t, t_0, x_0)$ is used to denote the solution of Eq. (1.7) such that

$$x(t_0, t_0, x_0) = x_0.$$

Here, t_0 and x_0 should also be regarded as variables that are subject to change, thus $x(t, t_0, x_0)$ should be regarded as a function in the three variables (t, t_0, x_0). Therefore, (t_0, x_0) may be replaced by different points, and in some cases, when a point $(\tau, \overline{x}_0)$ is in the domain of Eq. (1.7), we may say that "$x(t) = x(t, \tau, \overline{x}_0)$ is a solution of Eq. (1.7)," which means $x'(t) = f(t, x(t))$ and

$$x(\tau, \tau, \overline{x}_0) = \overline{x}_0.$$

In the proof of Theorem 2.2.7, we find that the requirement $r = \min\{a, \frac{b}{M}\}$ is used to guarantee that $(t, x_m(t))$ and $(t, x(t))$ are all in $R \subset D$ for $t \in I = (t_0 - r, t_0 + r)$, thus $f(t, x_m(t))$ and $f(t, x(t))$ are defined for $t \in I$. Note that r could be very small, that is the reason why Theorem 2.2.7 is called a **local** existence and uniqueness theorem. In Section 4 of this chapter, we will look at the size of the interval on which $x(t)$ is a solution.

In the next example, let's examine the role $r = \min\{a, \frac{b}{M}\}$ plays in ensuring the existence of a solution on $(t_0 - r, t_0 + r)$.

Example 2.2.8 Consider the scalar differential equation,

$$x'(t) = x^2(t), \quad x(0) = 1. \tag{2.41}$$

Using separation of variables, the solution is given by

$$x(t) = (1 - t)^{-1},$$

hence $x(t)$ is defined on $(-\infty, 1)$, and $\lim_{t \to 1^-} x(t) = \infty$, in which sense, we say that for Eq. (2.41), solutions **blow up** at finite times, see **Figure 2.10**.

Now, $f(t, x) = x^2$, and

$$|f(t, x) - f(t, y)| = |x^2 - y^2| = |x + y||x - y|,$$

thus f satisfies a Lipschitz condition on any bounded domain $D \subset \Re \times \Re$. Accordingly, we can select, for example, the region R (see **Figure 2.11**) to be

$$R = \{(t, x) : |t - 0| \le 2, \ |x - 1| \le b\}, \quad (b > 0).$$

For this region R, we have $M = \max_{(t,x) \in R} |f(t, x)| = \max_{(t,x) \in R} |x^2| = (1 + b)^2$, and Theorem 2.2.7 implies that the solution is defined on $(-r, r)$, where

$$r = \min\{a, \frac{b}{M}\} = \min\{2, \frac{b}{(1 + b)^2}\}.$$

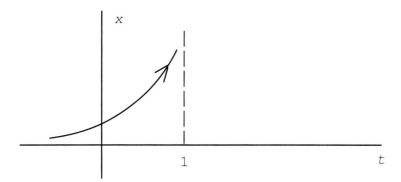

Figure 2.10: The solution $x(t) = (1 - t)^{-1}$ blows up at the time $t = 1$

Now, if $r > 1$, then we will be in trouble, because the solution $x(t) = (1 - t)^{-1}$ with $t_0 = 0$ is not defined at $t = 1$; that is, we must need $r < 1$ in this case. To find out how big $\frac{b}{(1+b)^2}$ could get as a function in $b > 0$, we can use the optimization technique from calculus and set the first derivative to be zero,

$$\frac{d}{db} \frac{b}{(1+b)^2} = \frac{1 - b^2}{(1+b)^4} = 0,$$

and obtain $b = 1$. Then we can use the first or the second derivative test to conclude (see an exercise) that

$$\max_{b>0} \frac{b}{(1+b)^2} = \frac{b}{(1+b)^2}\Big|_{b=1} = \frac{1}{4}.$$

Therefore, we can rest easy, knowing that

$$r = \min\{2, \frac{b}{(1+b)^2}\} \leq \frac{1}{4},$$

hence the interval of existence determined by Theorem 2.2.7 is given as

$$(t_0 - r, t_0 + r) = (-r, r) \subset (-\frac{1}{4}, \frac{1}{4}),$$

which does not include $t = 1$.   ♠

Many related results emerge based on the proof of Theorem 2.2.7. For example, we find that the (global) Lipschitz condition can be weakened as follows, which can be applied directly to linear differential equations.

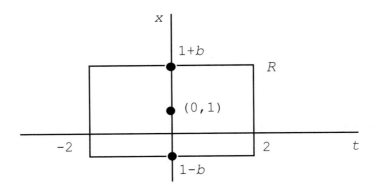

Figure 2.11: The region R

Definition 2.2.9 *A function $f(t, x)$ on a domain $D \subset \Re \times \Re^n$ is said to satisfy a* **weak Lipschitz condition** *with respect to x on D if there exists a nonnegative continuous function $k(t)$ such that*

$$|f(t, x) - f(t, y)| \leq k(t)|x - y|, \quad \text{for } (t, x), (t, y) \in D. \tag{2.42}$$

Example 2.2.10 Let's look at Example 2.2.4 again, where the linear function is given as $f(t, x) = A(t)x + f(t)$ with $A(t)$ an $n \times n$ matrix function and $f(t)$ an $n \times 1$ vector function. Now,

$$|f(t, x) - f(t, y)| \leq |A(t)||x - y|,$$

thus a weak Lipschitz condition is satisfied if $A(t)$ is continuous. ♠

The (global) Lipschitz condition can also be weakened to a local one.

Definition 2.2.11 *A function $f(t, x)$ on a domain $D \subset \Re \times \Re^n$ is said to satisfy a* **local Lipschitz condition** *with respect to x on D if for any $(t_1, x_1) \in D$, there exists a domain D_1 such that $(t_1, x_1) \in D_1 \subset D$ and that $f(t, x)$ satisfies a Lipschitz condition with respect to x on D_1. That is, there exists a positive constant k_1 such that*

$$|f(t, x) - f(t, y)| \leq k_1|x - y| \text{ for } (t, x), (t, y) \in D_1. \tag{2.43}$$

Note that the weak Lipschitz condition implies the local Lipschitz condition, thus the weak Lipschitz condition is a notion falls between the local

Lipschitz condition and the (global) Lipschitz condition, (see an exercise). It will be seen in Theorem 2.4.4 and Remark 2.4.12 that the weak Lipschitz condition is very useful in some situations.

We can modify the proof of Theorem 2.2.7 so as to obtain the following result, whose proof is left as an exercise.

Theorem 2.2.7.a (Picard's local existence and uniqueness theorem)
Assume that $f(t, x)$ is continuous on a domain D and satisfies a weak Lipschitz or a local Lipschitz condition with respect to x on D. Then the same results as in Theorem 2.2.7 hold. ♠

If we examine the Lipschitz condition when $x \in \Re$, then we find that the inequality (2.10) can be changed to

$$|\frac{f(t,x) - f(t,y)}{x - y}| \leq k,$$

which suggests that the Lipschitz condition is related to the partial derivative with respect to x. In this regard, we present a result that guarantees a local, a weak, or a global Lipschitz condition using partial derivatives, whose proof is left as an exercise. Recall that for $f : D \subset \Re \times \Re^n \to \Re^n$, we can write, from multivariable calculus, $f(t, x) = [f_1(t, x), f_2(t, x), \cdots, f_n(t, x)]^T$, where $f_i \in \Re$, then $\frac{\partial f(t,x)}{\partial x}$ is defined to be the $n \times n$ matrix

$$\frac{\partial f(t,x)}{\partial x} = [\frac{\partial f_i(t,x)}{\partial x_j}].$$

Theorem 2.2.12 *Assume the domain $D \subset \Re \times \Re^n$ is such that if (t, x_1), $(t, x_2) \in D$, then $(t, \lambda x_1 + (1 - \lambda)x_2) \in D$ for $0 \leq \lambda \leq 1$.*

(a). *If $\frac{\partial f(t,x)}{\partial x}$ exists and is continuous, then $f(t, x)$ satisfies a local Lipschitz condition with respect to x on D.*

(b). *If $|\frac{\partial f(t,x)}{\partial x}| \leq k(t)$ for some nonnegative continuous function $k(t)$, then $f(t, x)$ satisfies a weak Lipschitz condition with respect to x on D.*

(c). *If $|\frac{\partial f(t,x)}{\partial x}| \leq k$ for some positive constant k, then $f(t, x)$ satisfies a (global) Lipschitz condition with respect to x on D.* ♠

Theorem 2.2.12 indicates that certain conditions on the partial derivatives of f with respect to x imply certain Lipschitz conditions. In an exercise, you are asked to find an example such that a (global) Lipschitz condition is satisfied but the partial derivative of f with respect to x does not exist. Therefore, "having partial derivatives" is a stronger notion than "satisfying Lipschitz conditions."

Next, let's use Theorem 2.2.7.a and Theorem 2.2.12 to check the existence and uniqueness for the nonlinear differential equations we have encountered in Chapter 1.

Example 2.2.13 For the logistic equation, we have $f(t, x) = ax[C - x]$, and

$$\frac{\partial f(t, x)}{\partial x} = \frac{d}{dx} ax[C - x] = aC - 2ax,$$

therefore, a local Lipschitz condition is satisfied for any domain in $\Re \times \Re$, and a (global) Lipschitz condition is satisfied for any bounded domain in $\Re \times \Re$. Thus the existence and uniqueness is guaranteed. ♠

Example 2.2.14 For $x'(t) = k(t)x(t) + \sin(tx(t))$, we have $f(t, x) = k(t)x + \sin(tx)$, and

$$|\frac{\partial f(t, x)}{\partial x}| = |k(t) + t\cos(tx)| \le |k(t)| + |t|,$$

thus a weak Lipschitz condition is satisfied for any domain in $\Re \times \Re$ when $k(t)$ is continuous. ♠

Example 2.2.15 In the Lotka-Volterra competition equation, we have

$$f(t, x) = [\beta_1 x_1(K_1 - x_1 - \mu_1 x_2), \ \beta_2 x_2(K_2 - x_2 - \mu_2 x_1)]^T = [f_1(t, x), \ f_2(t, x)]^T,$$

and

$$\frac{\partial f(t, x)}{\partial x} = \begin{bmatrix} \frac{\partial f_1(t, x)}{\partial x_1} & \frac{\partial f_1(t, x)}{\partial x_2} \\ \frac{\partial f_2(t, x)}{\partial x_1} & \frac{\partial f_2(t, x)}{\partial x_2} \end{bmatrix}$$
$$= \begin{bmatrix} \beta_1(K_1 - \mu_1 x_2) - 2\beta_1 x_1 & -\beta_1 \mu_1 x_1 \\ -\beta_2 \mu_2 x_2 & \beta_2(K_2 - \mu_2 x_1) - 2\beta_2 x_2 \end{bmatrix}. \quad (2.44)$$

Therefore, a local Lipschitz condition is satisfied for any domain in $\Re \times \Re^2$, and a (global) Lipschitz condition is satisfied for any bounded domain in $\Re \times \Re^2$. ♠

Example 2.2.16 For the equation of a simple pendulum given by $\theta''(t) + k\theta'(t) + q\sin\theta(t) = 0$, we let $x_1 = \theta$, $x_2 = \theta'$, then we obtain the system

$$\begin{cases} x_1' & = & x_2, \\ x_2' & = & -kx_2 - q\sin x_1. \end{cases} \qquad (2.45)$$

Similar to Example 2.2.15, one can verify (see an exercise) that in this case, a (global) Lipschitz condition is satisfied for any domain in $\Re \times \Re^2$. ♦

Note here that for Examples 2.2.14–2.2.16, we **cannot solve** or **write out** the solutions explicitly, but we **can prove** that solutions exist and are unique.

We have now proved existence and uniqueness of solutions under some Lipschitz conditions, and we also know from Example 2.2.6 that uniqueness of solutions is not guaranteed without assuming a Lipschitz condition. The remaining question is: **What about existence of solutions?** The following result ensures that we can still obtain existence of solutions even without a Lipschitz condition, by using **Euler's method** in calculus for approximating solutions of differential equations with piecewisely straight lines. Of course, in this case, uniqueness is not expected.

Therefore, let's recall a few details of Euler's method in calculus. To solve a scalar differential equation

$$x'(t) = f(t, x(t)), \quad x(t_0) = x_0, \quad t \in [t_0, b], \qquad (2.46)$$

we divide $[t_0, b]$ into m equal subintervals $[t_0, t_1], [t_1, t_2], \cdots, [t_{m-1}, b]$. On the first interval, we require Eq. (2.46) to be satisfied at (t_0, x_0), that is, $x'(t_0) = f(t_0, x_0)$, then use $f(t_0, x_0)$ as the slope and (t_0, x_0) as the point so as to obtain the straight line

$$x = x_0 + f(t_0, x_0)(t - t_0), \quad t \in [t_0, t_1], \qquad (2.47)$$

see **Figure 2.12**.

On the second interval $[t_1, t_2]$, evaluate the straight line (2.47) at t_1 to obtain $x_1 = x_0 + f(t_0, x_0)(t_1 - t_0)$. We require Eq. (2.46) to be satisfied at (t_1, x_1), that is, $x'(t_1) = f(t_1, x_1)$, then use $f(t_1, x_1)$ as the slope and (t_1, x_1) as the point so as to obtain the straight line

$$x = x_1 + f(t_1, x_1)(t - t_1), \quad t \in [t_1, t_2].$$

This way, we get, on $[t_0, b]$, a function z (for zigzag) consisting of piecewisely straight lines, see Figure 2.12. This constitutes Euler's method of

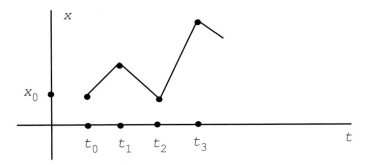

Figure 2.12: Using piecewisely straight lines to approximate a solution

approximating a solution of differential equation (2.46) using piecewisely straight lines.

When m gets larger and larger, the corresponding function z satisfies Eq. (2.46) at more and more left-end points of those small subintervals in $[t_0, b]$, (at this stage, the derivative for function z at the left end of an interval is the right-hand side derivative, since typically z has a "corner point" there). Thus, when $m \to \infty$, the limit function of z should satisfy Eq. (2.46) "everywhere" in $[t_0, b]$. Or, the function z should approach a solution of Eq. (2.46) on $[t_0, b]$. In the following, we will explore this idea and use it to derive solutions for general differential equations in $\Re^n$, $n \geq 1$.

Theorem 2.2.17 (Existence without uniqueness) *Assume that $f(t, x)$ is continuous on a domain $D \subset \Re \times \Re^n$ and let $(t_0, x_0) \in D$. Then there exist positive constants a and b such that*

$$R = \left\{ (t, x) : |t - t_0| \leq a, \ |x - x_0| \leq b \right\} \subset D, \qquad (2.48)$$

and Eq. (1.7) has a solution $x(t)$ on $I = (t_0 - r, t_0 + r)$, where $r = \min\{a, \frac{b}{M}\}$ with $M = \max_{(t,x) \in R} |f(t, x)|$.

Proof. Since (t_0, x_0) is in domain D, the a, b, R, M, and r specified above can be defined. Next, we only consider interval $[t_0, t_0 + r]$ and define a sequence of functions on $[t_0, t_0 + r]$, which will be used to approximate a solution. The case for the left-hand side of t_0 is similar.

For each $k = 1, 2, \cdots$, divide $[t_0, t_0 + r]$ into k equal subintervals $[t_0, t_1]$, $[t_1, t_2], \cdots, [t_{k-1}, t_k] = [t_{k-1}, t_0 + r]$, with the distance of every subinterval given by $\frac{r}{k}$. Define a continuous function $x_k(t)$ on $[t_0, t_0 + r]$ that is piecewisely "straight lines" as follows:

On $[t_0, t_1]$, define $x_k(t)$ to be the straight line with the point (t_0, x_0) and the "slope" $f(t_0, x_0)$. That is,

$$x_k(t) = x_0 + f(t_0, x_0)(t - t_0), \quad t \in [t_0, t_1].$$

On $[t_1, t_2]$, define $x_k(t)$ to be the straight line with the point $(t_1, x_k(t_1))$ and the "slope" $f(t_1, x_k(t_1))$. That is,

$$x_k(t) = x_k(t_1) + f(t_1, x_k(t_1))(t - t_1), \quad t \in [t_1, t_2].$$

This way, we define a continuous function $x_k(t)$ that is piecewisely straight lines on $[t_0, t_0 + r]$.

We verify that

$$(t, x_k(t)) \in R \ \text{ for } \ t \in [t_0, t_0 + r]. \tag{2.49}$$

Consider $t \in [t_0, t_1]$. Since $(t_0, x_0) \in R$, one has $|f(t_0, x_0)| \leq M$. Thus,

$$|x_k(t) - x_0| \leq |f(t_0, x_0)(t - t_0)| \leq M(\frac{r}{k}) \leq \frac{b}{k} \leq b,$$

hence (2.49) is true for $t \in [t_0, t_1]$.

Assume that (2.49) is true for $t \in [t_0, t_j]$, where $1 \leq j \leq k - 1$. Then $|f(t_i, x_k(t_i))| \leq M$, $i = 1, 2, \cdots, j$; and for $t \in [t_j, t_{j+1}]$, one has

$$
\begin{aligned}
|x_k(t) - x_0| &\leq |x_k(t) - x_k(t_j)| + |x_k(t_j) - x_k(t_{j-1})| + \cdots + |x_k(t_1) - x_k(t_0)| \\
&\leq |f(t_j, x_k(t_j))(t - t_j)| + \cdots + |f(t_0, x_0)(t_1 - t_0)| \\
&\leq M(\frac{r}{k}) + M(\frac{r}{k}) + \cdots + M(\frac{r}{k}) \\
&= (j+1)M(\frac{r}{k}) \leq kM(\frac{r}{k}) = rM \leq b, \tag{2.50}
\end{aligned}
$$

thus, (2.49) is proved using an induction.

Similarly, we can verify (see an exercise) that

$$|x_k(\alpha) - x_k(\beta)| \leq M|\alpha - \beta|, \quad \alpha, \ \beta \in [t_0, t_0 + r]. \tag{2.51}$$

Therefore, the sequence $\{x_k(t)\}$ is uniformly bounded and equi-continuous on $[t_0, t_0 + r]$, hence from Arzela-Ascoli's theorem (see the Appendix), this sequence has a subsequence, which we denote by $\{x_k(t)\}$ again, that converges uniformly on $[t_0, t_0 + r]$ to a continuous function $x(t)$.

Due to the construction of $x_k(t)$, we know that $x'_k(t)$ exists on the interval $[t_0, t_0 + r]$ except at $t_1, t_2, \cdots, t_{k-1}$. Let $t \in [t_0, t_0 + r]$, then $t \in [t_i, t_{i+1}]$ for some $i \in \{0, 1, 2, \cdots, k-1\}$. Now,

$$
\begin{aligned}
\int_{t_0}^{t} x'_k(s)ds &= \int_{t_0}^{t_1} x'_k(s)ds + \cdots + \int_{t_i}^{t} x'_k(s)ds \\
&= [x_k(t_1) - x_k(t_0)] + [x_k(t_2) - x_k(t_1)] + \cdots + [x_k(t) - x_k(t_i)] \\
&= x_k(t) - x_k(t_0) = x_k(t) - x_0. \quad (2.52)
\end{aligned}
$$

Thus, (2.52) is true for $t \in [t_0, t_0 + r]$, and we can write (2.52) as

$$
\begin{aligned}
x_k(t) &= x_0 + \int_{t_0}^{t} x'_k(s)ds \\
&= x_0 + \int_{t_0}^{t} [f(s, x_k(s)) + \delta_k(s)]ds, \quad t \in [t_0, t_0 + r], \quad (2.53)
\end{aligned}
$$

where $\delta_k(t)$ is defined for $t \in [t_0, t_0 + r]$ as

$$
\delta_k(t) = \begin{cases} x'_k(t) - f(t, x_k(t)), & \text{if } t \neq t_i, \ i = 0, 1, \cdots, k-1, \\ 0, & \text{if } t = t_i, \ i = 0, 1, \cdots, k-1. \end{cases} \quad (2.54)
$$

Next, we verify that

$$
\delta_k(t) \to 0 \ \text{uniformly on} \ [t_0, t_0 + r] \ \text{as} \ k \to \infty. \quad (2.55)
$$

Let $t \in [t_0, t_0 + r]$, then $t \in [t_i, t_{i+1}]$ for some $i \in \{0, 1, \cdots, k-1\}$. Now,

$$
|\delta_k(t)| \leq |x'_k(t) - f(t, x_k(t))| \leq |f(t_i, x_k(t_i)) - f(t, x_k(t))|,
$$

and we have $|t_i - t| \leq \frac{r}{k}$ and

$$
|x_k(t_i)) - x_k(t)| \leq |f(t_i, x_k(t_i))|(\frac{r}{k}) \leq M(\frac{r}{k}).
$$

Thus, as $f(t, x)$ is uniformly continuous on the bounded and closed set R and $\frac{r}{k}$ is very small when k is large, we conclude that (2.55) is true. Now, taking a limit as $k \to \infty$ in (2.53) and using the fact that the sequence $\{x_k(t)\}$ converges to x uniformly and f is uniformly continuous on the region R, we derive

$$
x(t) = x_0 + \int_{t_0}^{t} f(s, x(s))ds, \quad t \in [t_0, t_0 + r], \quad (2.56)
$$

therefore, $x(t)$ is a solution of Eq. (1.7) using Lemma 2.2.1. This completes
the proof.   ♠

   You probably didn't see this proof in your calculus class, but now you
know why Euler's method of approximating solutions of differential equations
using piecewisely straight lines works.
   We point out that the method of combining Lemma 2.2.1 and Picard's
approximations used in the proof of Theorem 2.2.7 is very useful in many
studies. It not only provides a framework for some numerical approximations
but also leads to the definition of a mapping when using a functional analysis
approach. See, for example, the related study in Section 2.5 in this chapter.

Exercises 2.2

1. Rewrite the following differential equations as equivalent equations
   without derivatives.

   (a) $x'(t) = \sin t \cos 3t + x^6(t)$, $x(0) = 4$.

   (b) $x''(t) = t^4 \cos 3x(t) + x^6(t)$, $x(1) = 4$, $x'(1) = 3$.

2. Show that $x''(t) = f(t, x(t))$, $x(t_0) = x_0$, $x'(t_0) = x_1$ is equivalent to
   $x(t) = x_0 + x_1(t - t_0) + \int_{t_0}^{t} (t - s)f(s, x(s))ds$.

3. Establish the existence and uniqueness for $x''(t) = f(t, x(t))$, $x(t_0) = x_0$, $x'(t_0) = x_1$.

4. Prove Lemma 2.2.2.a: Gronwall's inequality for $t \leq t_0$.

5. Determine all continuous functions f such that $0 \leq f(t) \leq \int_0^t f(s)ds$,
   $0 \leq t < \infty$.

6. If $f(t)$ and $g(t)$ are nonnegative, continuous, and $f(t) \leq C_1 + C_2(t - t_0) + \int_{t_0}^{t} f(s)g(s)ds$ for some positive constants C_1, C_2, then find an
   inequality for $f(t)$.

7. If $x(t_0) = x_0 \in \Re$ and for all t, $|x'(t)| \leq M$ for some constant $M > 0$,
   then show that for any given $T > 0$, the curve of $x(t)$ cannot escape
   the region determined by the straight lines $x - x_0 = \pm M(t - t_0)$ for
   $t \in [t_0 - T, t_0 + T]$. (A picture may be useful in visualizing the problem.)

8. In the proof of Theorem 2.2.7, use an induction to determine the inequality concerning $|x_{j+1}(t) - x_j(t)|$, $j \geq 0$.

9. Draw pictures and explain why $r = \min\{a, \frac{b}{M}\}$ is needed in the proof of Theorem 2.2.7.

10. (a). Find $\max_{b>0} \frac{b}{(1+b)^2}$ in Example 2.2.8.
    (b). Repeat Example 2.2.8 with $f(t, x) = x^3$.

11. Show that the (global) Lipschitz condition implies the weak Lipschitz condition, and that the weak Lipschitz condition implies the local Lipschitz condition. Find examples to demonstrate that the local Lipschitz condition does not imply the weak Lipschitz condition, and that the weak Lipschitz condition does not imply the (global) Lipschitz condition.

12. Prove Theorem 2.2.7.a.

13. (a). Prove Theorem 2.2.12 for $n = 1$.
    (b). Prove Theorem 2.2.12 for any $n \geq 1$.

14. Find an example such that a global Lipschitz condition is satisfied but the partial derivative of f with respect to x does not exist.

15. Let $f(t, x) = x^\alpha$ for $\alpha \in \Re$. Determine the conditions on α such that f satisfies a local Lipschitz condition.

16. Determine whether the following functions on their domains satisfy a local Lipschitz, a weak Lipschitz, or a (global) Lipschitz condition.

    (a) $f(t, x) = t^6 + \sin 4x$, $D = \{(t, x) : (t, x) \in \Re \times \Re\}$.
    (b) $f(t, x) = t^6 \sin 4x$, $D = \{(t, x) : (t, x) \in \Re \times \Re\}$.
    (c) $f(t, x) = t^6 |x|$, $D = \{(t, x) : |t| < 3, \ |x| < \infty\}$.
    (d) $f(t, x) = t^6 \sqrt{x}$, $D = \{(t, x) : |t| < 3, \ x > 0\}$.

17. Find the left and the right derivatives at $t = 1$ for $x_2(t)$ in Example 2.2.6.

18. Prove that the function $f(x) = 0$, $x \leq 0$; $f(x) = x^{1/3}$, $x > 0$, does not satisfy a local Lipschitz condition.

19. Use the idea of Example 2.2.6 to construct another example for which a local Lipschitz condition and uniqueness do not hold.

20. Assume that existence and uniqueness is satisfied for the scalar equation $x'(t) = f(t, x(t))$ on $\Re \times \Re$. If $x_1 < x_2$, show that $x(t, t_0, x_1) < x(t, t_0, x_2)$ for $t \geq t_0$.

21. Assume that existence and uniqueness is satisfied for equation $x'(t) = f(t, x(t))$ on $\Re \times \Re^n$ with $f(t, 0) = 0$. If x is a solution and if $x(t_0) = 0$ for some $t_0 \in \Re$, then show that $x(t) = 0$ for all $t \in \Re$.

22. Rewrite the equation of a simple pendulum $\theta''(t) + k\theta'(t) + q \sin \theta(t) = 0$ into a first-order system and then check if some Lipschitz conditions are satisfied.

23. Verify (2.51).

24. Provide the details for getting (2.55) and (2.56).

2.3   Dependence on Initial Data and Parameters

From the previous section, we learned that under some Lipschitz conditions, Picard's local existence and uniqueness theorem 2.2.7 implies that for any $(t_0, x_0) \in D$, there is an $r > 0$, such that Eq. (1.7) has a unique solution denoted by $x(t, t_0, x_0)$ on the interval $I = (t_0 - r, t_0 + r)$, and $x(t, t_0, x_0)$ passes through (t_0, x_0), that is, the solution $x(t, t_0, x_0)$ is such that

$$x(t_0, t_0, x_0) = x_0. \tag{3.1}$$

Here, (t_0, x_0) (or sometimes just x_0) denotes "initial data" or "initial condition" in applications. In some experiments in physics, initial condition is associated with initial measurements, such as putting one gallon of acid initially to create certain reactions in some experiments, which inevitably involves some errors. In mathematics, this indicates that the initial data (t_0, x_0) is subject to some change, which is the reason why we mentioned in the previous section that $x(t, t_0, x_0)$ should be regarded as a function in the three variables (t, t_0, x_0). It was also mentioned there that when we say "$x(t) = x(t, \tau, \overline{x}_0)$ is a solution of Eq. (1.7)," we mean $x'(t) = f(t, x(t))$ and

$$x(\tau, \tau, \overline{x}_0) = \overline{x}_0.$$

Now, the question is: **As (t_0, x_0) changes, what are the changes it will bring to the corresponding solutions?** That is, we need to study properties of solutions with respect to the variation in initial data. We will eventually verify that the solution $x(t, t_0, x_0)$ is a continuous and continuously differentiable function in the three variables (t, t_0, x_0). However, the notations may get a little complicated if we go to the three variables (t, t_0, x_0) right away. Thus we will start with the dependence of solutions with respect to the change in x_0 only, which will be helpful in understanding other cases.

Before we proceed, let's look at the following example.

Example 2.3.1 Consider the scalar differential equation,

$$x' = x^2, \quad x(0) = x_0 > 0, \quad t \geq 0. \tag{3.2}$$

The solution is given by

$$x(t) = \frac{x_0}{1 - x_0 t}, \tag{3.3}$$

therefore, it is defined on interval

$$[0, \frac{1}{x_0}). \tag{3.4}$$

If x_0 changes arbitrarily, say for example $x_0 = n \to \infty$, then the size of the corresponding interval in (3.4), now $\frac{1}{n}$, will go to zero, see **Figure 2.13**.

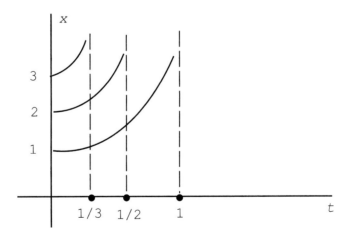

Figure 2.13: The solutions of Eq. (3.2) with $x(0) = 1, 2, 3, \cdots$

Hence, the corresponding solutions for $x_0 = 1, 2, \cdots$ cannot exist on a common interval with a positive size. However, if x_0 changes slightly, say for example $x_0 \in [1,2]$, then $\frac{1}{x_0} \geq \frac{1}{2}$, therefore, the corresponding solutions now exist on a common interval $[0, \frac{1}{2}]$. ♠

The following theorem verifies the result in Example 2.3.1 for general cases. That is, when x_0 changes slightly, the corresponding solutions exist on a common interval.

Theorem 2.3.2 *Assume that $f(t,x)$ is continuous on a domain $D \subset \Re \times \Re^n$ and satisfies a Lipschitz condition (or weak or local Lipschitz condition) with respect to x on D. Let $(t_0, \overline{x}_0) \in D$. Then there exist constants $d > 0$ and $r > 0$ such that for any $x_0 \in B_d(\overline{x}_0) = \{x \in \Re^n : |x - \overline{x}_0| \leq d\}$ (called the ball centered at $\overline{x}_0$ with radius d), Eq. (1.7) has a unique solution $x(t, t_0, x_0)$ on $I = (t_0 - r, t_0 + r)$, that is, the corresponding solutions exist on the common interval I. See* **Figure 2.14**.

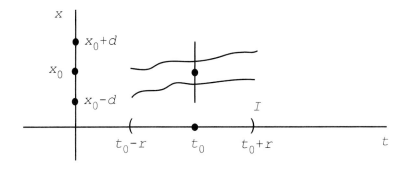

Figure 2.14: The solutions exist on a common interval

Proof. First, as the domain D is an open set, there exist positive constants a, b, and d such that

$$R^* = \{(t,x) : |t - t_0| \leq a, \ |x - \overline{x}_0| \leq b + d\} \subset D. \tag{3.5}$$

For any $x_0 \in B_d(\overline{x}_0)$, since $|x - \overline{x}_0| \leq |x - x_0| + |x_0 - \overline{x}_0| \leq |x - x_0| + d$, we find that

$$\begin{aligned} R' &= \{(t,x) : |t - t_0| \leq a, \ |x - x_0| \leq b\} \\ &\subset \{(t,x) : |t - t_0| \leq a, \ |x - \overline{x}_0| \leq b + d\} = R^* \subset D. \end{aligned} \tag{3.6}$$

That is, when x_0 changes in $B_d(\overline{x}_0)$, the set $R' = \{(t, x) : |t - t_0| \leq a,$ $|x - x_0| \leq b\}$ is always in R^*, see **Figure 2.15**.

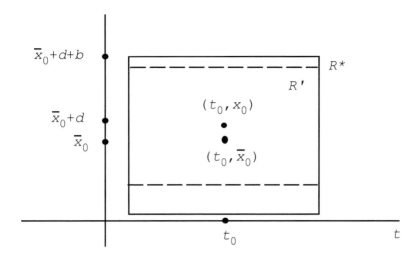

Figure 2.15: Regions R^* and R'

Now, define $M = \max_{(t,x)\in R^*} |f(t, x)|$. Then for any $x_0 \in B_d(\overline{x}_0)$, the set R' can be treated as the set R in the proof of Theorem 2.2.7, so we get a unique solution $x(t, t_0, x_0)$ on $I = (t_0 - r, t_0 + r)$, where $r = \min\{a, \frac{b}{M}\}$ is independent of $x_0 \in B_d(\overline{x}_0)$. Therefore, the corresponding solutions exist on the common interval $I = (t_0 - r, t_0 + r)$. ♠

Theorem 2.3.2 also says that there is a correspondence, or a **mapping**, from a **point** $x_0 \in B_d(\overline{x}_0)$ to its corresponding **solution** $x(t, t_0, x_0)$ on $I = (t_0 - r, t_0 + r)$. That is, there exists $r > 0$ such that for x_0 in a closed and bounded set, the mapping

$$x_0 \longrightarrow x(t, t_0, x_0), \quad t \in I = (t_0 - r, t_0 + r), \tag{3.7}$$

is well defined, which maps a **point** to a **function**, see **Figure 2.16**.

For example, in Example 2.3.1, for $x_0 \in [1, 2]$, the mapping is given by

$$x_0 \longrightarrow x(t, t_0, x_0) = \frac{x_0}{1 - x_0 t}, \quad t \in [0, \frac{1}{2}]. \tag{3.8}$$

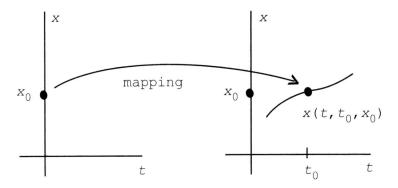

Figure 2.16: A mapping from a point x_0 to a function $x(t, t_0, x_0)$

The formula of the solution in (3.8) indicates that the mapping in (3.8) is continuous with respect to x_0 uniformly for $t \in [0, \frac{1}{2}]$, which we explain and prove for general cases next.

Theorem 2.3.3 *Assume that $f(t, x)$ is continuous on a domain $D \subset \Re \times \Re^n$ and satisfies a Lipschitz condition (or weak or local Lipschitz condition) with respect to x on D. Let $(t_0, \overline{x}_0) \in D$. Then there exist constants $d > 0$ and $r > 0$ such that the mapping described above in (3.7) is continuous. That is, the solution $x(t, t_0, x_0)$ is continuous with respect to $x_0 \in B_d(\overline{x}_0)$ in the following sense: If $y, x_0 \in B_d(\overline{x}_0)$, then $x(t, t_0, y)$ and $x(t, t_0, x_0)$ are solutions of Eq. (1.7), and*

$$\lim_{y \to x_0} x(t, t_0, y) = x(t, t_0, x_0) \tag{3.9}$$

*uniformly for $t \in I = (t_0 - r, t_0 + r)$. See **Figure 2.17**.*

Proof. Let $y, x_0 \in B_d(\overline{x}_0)$, then from Theorem 2.3.2 and Lemma 2.2.1, $x(t, t_0, y)$ and $x(t, t_0, x_0)$ are solutions of Eq. (1.7) and are given by

$$x(t, t_0, y) = y + \int_{t_0}^t f(s, x(s, t_0, y)) ds, \quad t \in I, \tag{3.10}$$

$$x(t, t_0, x_0) = x_0 + \int_{t_0}^t f(s, x(s, t_0, x_0)) ds, \quad t \in I. \tag{3.11}$$

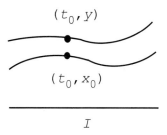

Figure 2.17: Continuity of the mapping with respect to x_0

Next, we assume a Lipschitz condition (with constant k) and $t \geq t_0$, (other cases are similar). Then we get

$$|x(t, t_0, y) - x(t, t_0, x_0)| \leq |y - x_0| + \int_{t_0}^{t} |f(s, x(s, t_0, y)) - f(s, x(s, t_0, x_0))| ds$$

$$\leq |y - x_0| + k \int_{t_0}^{t} |x(s, t_0, y) - x(s, t_0, x_0)| ds, \quad (3.12)$$

hence, Gronwall's inequality implies

$$|x(t, t_0, y) - x(t, t_0, x_0)| \leq |y - x_0| e^{k(t-t_0)} \leq |y - x_0| e^{kr}, \quad t \in I, \quad (3.13)$$

which implies that (3.9) is true uniformly for $t \in I = (t_0 - r, t_0 + r)$. This completes the proof. ♠

The above is the **continuous dependence of solutions with respect to x_0**. Next, let's look at the continuous dependence of solutions with respect to (t_0, x_0), that is, we will now allow t_0 to change. Using the same idea as in the proof of Theorem 2.3.2, we can verify that when (t_0, x_0) changes a little, the corresponding solutions exist on intervals of the same length.

Theorem 2.3.4 *Assume that $f(t, x)$ is continuous on a domain $D \subset \Re \times \Re^n$ and satisfies a Lipschitz condition (or weak or local Lipschitz condition) with respect to x on D. Let $(\bar{t}_0, \bar{x}_0) \in D$. Then there exist constants $c > 0$, $d > 0$, and $r > 0$ such that for any $t_0 \in [\bar{t}_0 - c, \bar{t}_0 + c]$ and any $x_0 \in B_d(\bar{x}_0)$, Eq. (1.7) has a unique solution $x(t, t_0, x_0)$ on $I = (t_0 - r, t_0 + r)$, that is, the corresponding solutions exist on the intervals of the same length $2r$. See* **Figure 2.18**.

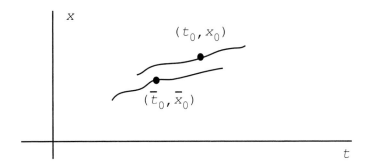

Figure 2.18: The solutions exist on the intervals of the same length $2r$

Proof. Similar to the proof of Theorem 2.3.2, there exist positive constants a, b, c, and d such that

$$R_1^* = \{(t, x) : |t - \bar{t}_0| \leq a + c, \ |x - \bar{x}_0| \leq b + d\} \subset D. \tag{3.14}$$

Now, for any $t_0 \in [\bar{t}_0 - c, \bar{t}_0 + c]$ and any $x_0 \in B_d(\bar{x}_0)$, the set

$$\begin{aligned} R' &= \{(t, x) : |t - t_0| \leq a, \ |x - x_0| \leq b\} \\ &\subset \{(t, x) : |t - \bar{t}_0| \leq a + c, \ |x - \bar{x}_0| \leq b + d\} = R_1^* \subset D. \end{aligned} \tag{3.15}$$

That is, when t_0 changes in $[\bar{t}_0 - c, \bar{t}_0 + c]$ and x_0 changes in $B_d(\bar{x}_0)$, the set $R' = \{(t, x) : |t - t_0| \leq a, \ |x - x_0| \leq b\}$ is always in R_1^*, see **Figure 2.19**.

The rest of the proof is the same as in Theorem 2.3.2, hence the proof is complete.   ♠

Notice in Theorem 2.3.4 that when t_0 varies near $\bar{t}_0$, say for example $t_0 = \bar{t}_0 \pm \delta$ for a small $\delta > 0$, the corresponding solutions exist on $(t_0 - r, t_0 + r) = (\bar{t}_0 \pm \delta - r, \bar{t}_0 \pm \delta + r)$. The sizes of these intervals are the same, $2r$, but these intervals are different. Now, if we select $\delta < r$, then all those solutions exist on a common interval

$$(\bar{t}_0 + \delta - r, \bar{t}_0 - \delta + r) = \left(\bar{t}_0 - (r - \delta), \bar{t}_0 + (r - \delta)\right), \tag{3.16}$$

because $(\bar{t}_0 + \delta - r, \bar{t}_0 - \delta + r) \subset (\bar{t}_0 \pm \delta - r, \bar{t}_0 \pm \delta + r)$, see Figure 2.18. The next result says that on this common interval, the solutions are continuous with respect to (t_0, x_0).

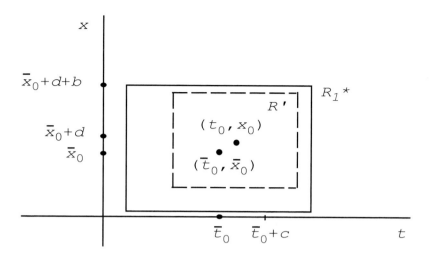

Figure 2.19: Regions R_1^* and R'

Theorem 2.3.5 *Assume that $f(t, x)$ is continuous on a domain $D \subset \Re \times \Re^n$ and satisfies a Lipschitz condition (or weak or local Lipschitz condition) with respect to x on D. Let $(\bar{t}_0, \bar{x}_0) \in D$. Then there exist constants $\delta > 0$, $\delta' > 0$, and $d > 0$ such that the solutions of Eq. (1.7) with $(t_0, x_0) \in [\bar{t}_0 - \delta, \bar{t}_0 + \delta] \times B_d(\bar{x}_0)$ exist on a common interval $I = (\bar{t}_0 - \delta', \bar{t}_0 + \delta')$. And the solution $x(t, t_0, x_0)$ is continuous with respect to $(t_0, x_0) \in [\bar{t}_0 - \delta, \bar{t}_0 + \delta] \times B_d(\bar{x}_0)$ in the following sense: If (τ, y), $(t_0, x_0) \in [\bar{t}_0 - \delta, \bar{t}_0 + \delta] \times B_d(\bar{x}_0)$, then*

$$\lim_{(\tau, y) \to (t_0, x_0)} x(t, \tau, y) = x(t, t_0, x_0) \tag{3.17}$$

*uniformly for $t \in I = (\bar{t}_0 - \delta', \bar{t}_0 + \delta')$. See **Figure 2.20**.*

Proof. Let the positive constants c, d, and r be from Theorem 2.3.4 and let $0 < \delta < \min\{r, c\}$ and define $\delta' = r - \delta > 0$. Then from (3.16), the solutions of Eq. (1.7) with $(t_0, x_0) \in [\bar{t}_0 - \delta, \bar{t}_0 + \delta] \times B_d(\bar{x}_0)$ exist on the common interval $I = (\bar{t}_0 - \delta', \bar{t}_0 + \delta')$.

Let (τ, y), $(t_0, x_0) \in [\bar{t}_0 - \delta, \bar{t}_0 + \delta] \times B_d(\bar{x}_0)$. To prove the continuity, we note from Lemma 2.2.1 that solutions $x(t, \tau, y)$ and $x(t, t_0, x_0)$ are given by

$$x(t, \tau, y) = y + \int_\tau^t f(s, x(s, \tau, y)) ds, \quad t \in I, \tag{3.18}$$

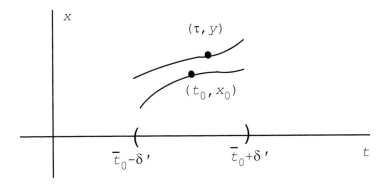

Figure 2.20: Solutions are continuous with respect to (t_0, x_0)

$$x(t, t_0, x_0) = x_0 + \int_{t_0}^{t} f(s, x(s, t_0, x_0)) ds, \quad t \in I. \qquad (3.19)$$

Next, we assume a Lipschitz condition (with constant k) and $t \geq t_0 \geq \tau$, (other cases are similar). From the proof of Theorem 2.3.4, solutions $x(t, \tau, y)$ and $x(t, t_0, x_0)$ are in a closed and bounded set R_1^* defined in (3.14), hence there is a $M > 0$ such that $|f(t, x(t, \tau, y))| \leq M$, $t \in I$. Then we get, for $t \in I$,

$$|x(t, \tau, y) - x(t, t_0, x_0)| \leq |y - x_0| + \int_{\tau}^{t_0} |f(s, x(s, \tau, y))| ds$$

$$+ \int_{t_0}^{t} |f(s, x(s, \tau, y)) - f(s, x(s, t_0, x_0))| ds$$

$$\leq |y - x_0| + M|t_0 - \tau| + k \int_{t_0}^{t} |x(s, \tau, y) - x(s, t_0, x_0)| ds,$$

hence, Gronwall's inequality implies

$$|x(t, \tau, y) - x(t, t_0, x_0)| \leq \left(|y - x_0| + M|t_0 - \tau| \right) e^{k(t - t_0)}$$

$$\leq \left(|y - x_0| + M|t_0 - \tau| \right) e^{k2r}, \quad t \in I,$$

which implies that (3.17) is true uniformly for $t \in I = (\bar{t}_0 - \delta', \bar{t}_0 + \delta')$. This completes the proof. ♠

Finally, since a solution $x(t, t_0, x_0)$ is also continuous in t, and, according to Theorem 2.3.5, is continuous in (t_0, x_0) uniformly for t, we have the following result stating that $x(t, t_0, x_0)$ is a continuous function in its three variables (t, t_0, x_0), the proof is left as an exercise.

Theorem 2.3.6 *Assume that $f(t, x)$ is continuous on a domain $D \subset \Re \times \Re^n$ and satisfies a Lipschitz condition (or weak or local Lipschitz condition) with respect to x on D. Let $(\bar{t}_0, \overline{x}_0) \in D$. Then there exist constants $\delta > 0$, $\delta' > 0$, and $d > 0$ such that the solution $x(t, t_0, x_0)$ of Eq. (1.7) is a continuous function in its variables (t, t_0, x_0) in $(\bar{t}_0 - \delta', \bar{t}_0 + \delta') \times [\bar{t}_0 - \delta, \bar{t}_0 + \delta] \times B_d(\overline{x}_0)$. That is, if $(t_1, t_0, x_0), (t_2, \tau, y) \in (\bar{t}_0 - \delta', \bar{t}_0 + \delta') \times [\bar{t}_0 - \delta, \bar{t}_0 + \delta] \times B_d(\overline{x}_0)$, then*

$$\lim_{(t_2, \tau, y) \to (t_1, t_0, x_0)} x(t_2, \tau, y) = x(t_1, t_0, x_0). \qquad \spadesuit \qquad (3.20)$$

In applications, the process of collecting initial values involve some errors. For example, when weighing one gallon of water, there is no guarantee that the measurement is exactly one gallon every time. That is, in applications, real data always have some inherent uncertainty and initial values taken from real data are never known precisely. Theorems 2.3.3, 2.3.5, and 2.3.6 indicate that if the initial error is small, or if the situation changes a little, then the corresponding solution changes a little, or the change is continuous. Differential equations have this good property so they can be used to model real situations effectively. This continuity property is studied here for t on a finite interval. Later, we will study a stronger version of this property, called **stability**, which is basically the same "continuity" property but for t on $[t_0, \infty)$.

Next, we look at the **differential equations with parameter**,

$$x'(t) = f(t, x(t), \mu), \quad x(t_0) = x_0, \qquad (3.21)$$

where μ is a parameter in a domain $D_0 \subset \Re^k$, $k \geq 1$. We have already seen these type of differential equations. For example, in Chapter 1, the logistic equation $x' = ax[C - x]$ can be regarded as a differential equation with parameters a and C; the Lotka-Volterra competition equation can be regarded as a differential equation with parameters $\beta_1, \mu_1, \beta_2, \mu_2$; and the motion of a simple pendulum can be regarded as a differential equation with parameters k and q.

Differential equations with parameters are very important in applications, especially in control theory and bifurcation. For example, for equation $x' = \mu - x^2$ where μ is a parameter (regarded as a control parameter of the system), we have seen in Example 1.3.2 of Section 3 in Chapter 1 that when

μ is varied, critical points (solutions of $x' = 0$) can be created and destroyed, thus bifurcations will occur.

In Eq. (3.21), for each fixed $\mu \in D_0$, it is assumed that $f(t, x, \mu)$ satisfies the same conditions as in Picard's existence and uniqueness theorem 2.2.7. In other words, when μ is fixed, Eq. (3.21) is treated as a regular differential equation without a parameter. Thus, for each (t_0, x_0, μ), we immediately have the existence and uniqueness of a solution

$$x(t) = x(t, t_0, x_0, \mu)$$

for Eq. (3.21). Now, from the results of Theorem 2.3.3, we expect the solution $x(t) = x(t, t_0, x_0, \mu)$ to be continuous with respect to (x_0, μ) in the sense described in Theorem 2.3.3. For example, for the logistic equation $x' = ax[C - x]$ where a and C are regarded as parameters, the solution

$$x(t) = \frac{Cx_0}{x_0 + [C - x_0]e^{-aC(t-t_0)}}, \quad t \geq t_0, \tag{3.22}$$

given in Chapter 1 is certainly continuous in a and C. Let's look at one more example.

Example 2.3.7 Similar to Example 2.3.1, let's consider the scalar differential equation,

$$x' = \mu x^2, \quad x(0) = x_0 > 0, \quad \mu \in (0, \infty), \quad t \geq 0. \tag{3.23}$$

For any fixed $\mu \in (0, \infty)$, the solution is given by

$$x(t) = \frac{x_0}{1 - \mu x_0 t}, \tag{3.24}$$

therefore, it is defined on interval

$$[0, \frac{1}{\mu x_0}). \tag{3.25}$$

If x_0 or μ changes arbitrarily, then similar to Example 2.3.1, the corresponding solutions cannot exist on a common interval. However, if x_0 and μ change slightly, say for example $x_0 \in [1, 2]$ and $\mu \in [3, 4]$, then $\frac{1}{\mu x_0} \geq \frac{1}{4 \times 2} = \frac{1}{8}$. Therefore, the corresponding solutions now exist on a common interval $[0, \frac{1}{8}]$, and (3.24) indicates that in this case the solution is continuous with respect to (x_0, μ) uniformly for $t \in [0, \frac{1}{8}]$. ♠

In general, we have the following result.

Theorem 2.3.8 *Assume that $f(t, x, \mu)$ is continuous on a domain $D \times D_0 \subset \Re \times \Re^n \times \Re^k$ (D is from Theorem 2.2.7) and satisfies a Lipschitz condition (or weak or local Lipschitz condition) with respect to x on $D \times D_0$, where the Lipschitz constant k is independent of μ, that is,*

$$|f(t, x, \mu) - f(t, y, \mu)| \le k|x - y|$$

for all $\mu \in D_0$. Let $(t_0, \overline{x}_0) \in D$ and let $D_{00} \subset D_0$ be any closed and bounded set. Then there exist constants $d > 0$ and $r > 0$ such that for any $x_0 \in B_d(\overline{x}_0) = \{x \in \Re^n : |x - \overline{x}_0| \le d\}$ and any $\mu \in D_{00}$, Eq. (3.21) has a unique solution $x(t, t_0, x_0, \mu)$ on $I = (t_0 - r, t_0 + r)$. Moreover, the solution $x(t, t_0, x_0, \mu)$ is continuous with respect to $(x_0, \mu) \in B_d(\overline{x}_0) \times D_{00}$ in the following sense: if $(y, \lambda), (x_0, \mu) \in B_d(\overline{x}_0) \times D_{00}$, then

$$\lim_{(y, \lambda) \to (x_0, \mu)} x(t, t_0, y, \lambda) = x(t, t_0, x_0, \mu) \tag{3.26}$$

uniformly for $t \in I = (t_0 - r, t_0 + r)$.

Proof. The proof of the first conclusion (existence and uniqueness) is similar to that of Theorem 2.3.2, because now we can find positive constants a, b, and d such that

$$R^\star = \{(t, x, \mu) : |t - t_0| \le a, \ |x - \overline{x}_0| \le b + d, \ \mu \in D_{00}\} \subset D \times D_0.$$

Then we define $M = \max_{(t,x,\mu) \in R^\star} |f(t, x, \mu)|$ to get $I = (t_0 - r, t_0 + r)$ with $r = \min\{a, \frac{b}{M}\}$.

To get the second conclusion (continuity), let $(y, \lambda), (x_0, \mu) \in B_d(\overline{x}_0) \times D_{00}$. Note from Lemma 2.2.1 that solutions $x(t, t_0, y, \lambda)$ and $x(t, t_0, x_0, \mu)$ are given by

$$x(t, t_0, y, \lambda) = y + \int_{t_0}^t f(s, x(s, t_0, y, \lambda), \lambda)ds, \quad t \in I, \tag{3.27}$$

$$x(t, t_0, x_0, \mu) = x_0 + \int_{t_0}^t f(s, x(s, t_0, x_0, \mu), \mu)ds, \quad t \in I. \tag{3.28}$$

In the following, we assume a Lipschitz condition (with constant k) and $t \ge t_0$, because other cases are similar. Then we get, for $t \in I$,

$$|x(t, t_0, y, \lambda) - x(t, t_0, x_0, \mu)|$$
$$\le |y - x_0| + \int_{t_0}^t |f(s, x(s, t_0, y, \lambda), \lambda) - f(s, x(s, t_0, x_0, \mu), \mu)|ds$$

$$\leq |y - x_0| + \int_{t_0}^t |f(s, x(s, t_0, y, \lambda), \lambda) - f(s, x(s, t_0, x_0, \mu), \lambda)$$

$$+ f(s, x(s, t_0, x_0, \mu), \lambda) - f(s, x(s, t_0, x_0, \mu), \mu)| ds$$

$$\leq |y - x_0| + k \int_{t_0}^t |x(s, t_0, y, \lambda) - x(s, t_0, x_0, \mu)| ds$$

$$+ \int_{t_0}^t |f(s, x(s, t_0, x_0, \mu), \lambda) - f(s, x(s, t_0, x_0, \mu), \mu)| ds$$

$$\leq |y - x_0| + k|t - t_0| \max_{s \in I} |x(s, t_0, y, \lambda) - x(s, t_0, x_0, \mu)|$$

$$+ |t - t_0| \max_{s \in I} |f(s, x(s, t_0, x_0, \mu), \lambda) - f(s, x(s, t_0, x_0, \mu), \mu)|$$

$$\leq |y - x_0| + kr \max_{s \in I} |x(s, t_0, y, \lambda) - x(s, t_0, x_0, \mu)|$$

$$+ r \max_{s \in I} |f(s, x(s, t_0, x_0, \mu), \lambda) - f(s, x(s, t_0, x_0, \mu), \mu)|. \tag{3.29}$$

Next, we require r to be small such that $kr \leq \frac{1}{2}$. Then (3.29) becomes

$$\frac{1}{2} \max_{t \in I} |x(t, t_0, y, \lambda) - x(t, t_0, x_0, \mu)| \leq |y - x_0|$$

$$+ r \max_{s \in I} |f(s, x(s, t_0, x_0, \mu), \lambda) - f(s, x(s, t_0, x_0, \mu), \mu)|. \tag{3.30}$$

Now, as $f(s, x(s, t_0, x_0, \mu), \lambda)$ is continuous for (s, λ) in the closed and bounded set $[t_0 - r, t_0 + r] \times D_{00}$, it is uniformly continuous. Therefore, taking a limit as $(y, \lambda) \to (x_0, \mu)$ in (3.30), we get

$$\lim_{(y, \lambda) \to (x_0, \mu)} \frac{1}{2} \max_{t \in I} |x(t, t_0, y, \lambda) - x(t, t_0, x_0, \mu)| \leq \lim_{(y, \lambda) \to (x_0, \mu)} \Big\{ |y - x_0|$$

$$+ r \max_{s \in I} |f(s, x(s, t_0, x_0, \mu), \lambda) - f(s, x(s, t_0, x_0, \mu), \mu)| \Big\} = 0,$$

which completes the proof.   ♠

Another way to see why Theorem 2.3.8 is true is to use the fact that μ is fixed, such that we can rewrite Eq. (3.21) as

$$\begin{cases} x'(t) = f(t, x(t), \mu(t)), \\ \mu'(t) = 0, \\ x(t_0) = x_0, \ \mu(t_0) = \mu. \end{cases} \tag{3.31}$$

That is, we "stack" the parameter μ (in $\Re^k$) to x and come up with a differential equation about (x, μ) in $\Re^{n+k}$ where μ is actually a constant.

Now, Eq. (3.31) is a differential equation in the variables $(x, \mu) \in \Re^{n+k}$ without parameters, so Eq. (3.31) can be regarded as a special case of the general differential equation (1.7). Thus Theorems 2.3.2 and 2.3.3 can be applied to obtain the continuity of solutions of Eq. (3.31) with respect to $(x(t_0), \mu(t_0)) = (x_0, \mu)$, which is the same as the continuity of solutions of Eq. (3.21) with respect to (x_0, μ). But, anyway, we proved Theorem 2.3.8 directly in case you want to see a "real proof" dealing with parameters.

The continuity of solutions of Eq. (3.21) with respect to (t, t_0, x_0, μ) is left as an exercise.

Next, we verify that if $f(t, x, \mu)$ is continuously differentiable with respect to (x, μ), then $x(t, t_0, x_0, \mu)$ is also continuously differentiable with respect to (x_0, μ), which is certainly the case for the solutions given in Examples 2.3.1 and 2.3.7. To this end, we introduce the following result which is a useful replacement of the mean value theorem of scalar functions when dealing with vector functions.

Lemma 2.3.9 *Assume $f(t, x) : \Re \times D_1 \subset \Re \times \Re^n \to \Re^n$, and has continuous partial derivatives, where $x = (x^1, x^2, \cdots, x^n)$ and the domain $D_1 \subset \Re^n$ is convex, that is, $x_1, x_2 \in D_1$ implies $\lambda x_1 + (1 - \lambda)x_2 \in D_1$ for $0 \le \lambda \le 1$. Then there exist continuous (vector) functions $f_k(t, x_1, x_2)$, $k = 1, \cdots, n$, on $\Re \times D_1 \times D_1$ such that*

$$f_k(t, x, x) = \frac{\partial f}{\partial x^k}(t, x), \quad f(t, x_2) - f(t, x_1) = \sum_{k=1}^{n} f_k(t, x_1, x_2)(x_2^k - x_1^k). \quad (3.32)$$

The functions f_k are given by

$$f_k(t, x_1, x_2) = \int_0^1 \frac{\partial f}{\partial x^k}(t, \, sx_2 + (1-s)x_1)ds. \quad (3.33)$$

Proof. Now, we can define $F(s) = f(t, \, sx_2 + (1-s)x_1)$ for $s \in [0, 1]$ and $x_1, x_2 \in D_1$. Then

$$F'(s) = \sum_{k=1}^{n}(x_2^k - x_1^k)\frac{\partial f}{\partial x^k}(t, \, sx_2 + (1-s)x_1).$$

Thus we obtain, for f_k defined in (3.33),

$$f(t, x_2) - f(t, x_1) \quad = \quad F(1) - F(0) = \int_0^1 F'(s)ds$$

$$= \sum_{k=1}^{n} (x_2^k - x_1^k) \int_0^1 \frac{\partial f}{\partial x^k}(t, sx_2 + (1-s)x_1)ds$$

$$= \sum_{k=1}^{n} f_k(t, x_1, x_2)(x_2^k - x_1^k),$$

which gives (3.32). This completes the proof.   ♠

Theorem 2.3.10 *Assume that $f(t, x, \mu)$ is continuous on $D \times D_0 \subset \Re \times \Re^n \times \Re^k$ (D is from Theorem 2.2.7) and is continuously differentiable with respect to (x, μ), then the solution $x(t, t_0, x_0, \mu)$ of Eq. (3.21) is continuously differentiable with respect to (x_0, μ) in its interval of existence.*

Proof. We have seen that Eq. (3.21) can be reformulated as Eq. (3.31) without parameters, and Eq. (3.31) can be regarded as a special case of the general Eq. (1.7), thus we only need to prove the results for Eq. (1.7), which in turn implies the corresponding results for the solutions of Eq. (3.21). That is, let $(t_0, \eta) \in D$ and let $x(t, t_0, \eta)$ be the solution of Eq. (1.7) with $x(t_0) = \eta$, then we need to prove that $x(t, t_0, \eta)$ is continuously differentiable with respect to η, or $\frac{\partial x(t, t_0, \eta)}{\partial \eta_i}$ exists and is continuous in η for $i = 1, 2, \cdots, n$.

Notice from Theorem 2.3.2 that there is an $r > 0$ such that for sufficiently small scalar h,

$$x_h(t) = x(t, t_0, \eta + he_i) \tag{3.34}$$

is a solution (passing through $(t_0, \eta + he_i)$) defined on $I = (t_0 - r, t_0 + r)$, where $\{e_i\}$ is the standard unit basis of $\Re^n$. From Theorem 2.3.3,

$$x_h(t) \to x(t) = x(t, t_0, \eta), \quad h \to 0, \tag{3.35}$$

uniformly on the interval I. Define

$$w_h(t) = \frac{x_h(t) - x(t)}{h}, \quad h \neq 0, \tag{3.36}$$

then, to prove the existence and continuity of $\frac{\partial x(t, t_0, \eta)}{\partial \eta_i}$, we need to prove that $\lim_{h \to 0} w_h(t)$ exists and is continuous in η.

Now, using Lemma 2.3.9,

$$w_h'(t) = \frac{1}{h}[x_h(t) - x(t)]' = \frac{1}{h}[f(t, x_h(t)) - f(t, x(t))]$$

$$= \frac{1}{h} \sum_{k=1}^{n} f_k(t, x(t), x_h(t))[x_h^k(t) - x^k(t)] \tag{3.37}$$

$$\overset{\text{def}}{=} A(t; h) \frac{1}{h} [x_h(t) - x(t)]$$

$$= A(t; h) w_h(t), \tag{3.38}$$

where $A(t; h)$ is the matrix from (3.37) such that the kth column is the vector $f_k(t, x(t), x_h(t))$. Thus, for $h \neq 0$, $w_h(t)$ is a solution of the homogeneous linear differential equation with a parameter h,

$$y'(t) = A(t; h)y(t), \quad y(t_0) = e_i. \tag{3.39}$$

For Eq. (3.39), h can be zero, in which case we have, from Lemma 2.3.9,

$$A(t; h)|_{h=0} = A(t; 0) = \frac{\partial f}{\partial x}\left(t, x(t)\right) = \frac{\partial f}{\partial x}\left(t, x(t, t_0, \eta)\right). \tag{3.40}$$

Now, the proof in Theorems 2.2.7 and 2.3.2 indicate that for r and h small, the solutions $x(t) = x(t, t_0, \eta)$ and $x_h(t) = x(t, t_0, \eta + he_i)$ are all inside a closed, bounded, and convex set $R \subset D$. Moreover, $\frac{\partial f(t, x)}{\partial x}$ is continuous and hence is bounded on R, thus the matrix $A(t; h)$ is continuous and bounded for $t \in I = (t_0 - r, t_0 + r)$ when h is small, and $A(t; h) \rightarrow A(t; 0)$ as $h \rightarrow 0$. Therefore, on $(t_0 - r, t_0 + r) \times \Re^n$, Eq. (3.39) satisfies a (global) Lipschitz condition independently of small parameter h. Hence, Theorem 2.3.8 implies that the solutions of Eq. (3.39) are continuous with respect to the parameter h. That is, when the parameter $h \rightarrow 0$, the unique solution $w_h(t)$ of Eq. (3.39) satisfies

$$w_h(t) \rightarrow w(t), \quad h \rightarrow 0, \tag{3.41}$$

where $w(t)$ is the unique solution of Eq. (3.39) with $h = 0$, that is, $w(t)$ is the unique solution of

$$w'(t) = A(t; 0)w(t) = \left[\frac{\partial f}{\partial x}\left(t, x(t, t_0, \eta)\right)\right]w(t), \quad w(t_0) = e_i. \tag{3.42}$$

Therefore, $x(t, t_0, \eta)$ has a partial derivative with respect to η since $\lim_{h \rightarrow 0} w_h(t) = w(t)$ exists. To get the continuity of this derivative, we need to check the continuity of $w(t)$ with respect to η. To this end, we can regard Eq. (3.42) as an equation with the parameter η and apply Theorem 2.3.8 again to obtain the continuity of $w(t)$ with respect to η. This completes the proof. ♠

Corollary 2.3.11 *Assume that $f(t, x)$ is continuous in (t, x) and continuously differentiable in $x \in \Re^n$, then the corresponding solution $x = x(t, t_0, x_0)$ (in its interval of existence) is continuously differentiable in x_0, and the $n \times n$ matrix $J(t, t_0, x_0) = \frac{\partial x(t, t_0, x_0)}{\partial x_0}$ is the solution of*

$$J'(t, t_0, x_0) = \frac{\partial f}{\partial x}(t, x(t, t_0, x_0)) J(t, t_0, x_0), \quad J(t_0, t_0, x_0) = E,$$

where E is the $n \times n$ unit or identity matrix.   ♠

The differentiability of solutions with respect to t_0 is left as an exercise. In fact, if f is analytic, as are the cases in most applications, then the corresponding solutions are also analytic. Use Picard's approximations to prove this and argue that each successive approximation is analytic, and hence a solution, which is the uniform limit of these analytic functions, is also analytic. We leave it to you to find the conditions and formulate and prove the results.

Exercises 2.3

1. Consider

$$\begin{aligned} x'(t) &= F(t, x(t)), \quad x(t_0) = x_0, \quad t \in [t_0, t_0 + T], \\ y'(t) &= G(t, y(t)), \quad y(t_0) = y_0, \quad t \in [t_0, t_0 + T]. \end{aligned}$$

   If $|F(t, w) - G(t, w)| \leq \varepsilon$, and F or G satisfies a Lipschitz condition, then show that

$$|x(t) - y(t)| \leq [|x_0 - y_0| + \varepsilon T] e^{Tk},$$

   where k is the corresponding Lipschitz constant. Roughly speaking, this result says that if $|x_0 - y_0| \to 0$ and $|F - G| \to 0$, then $|x - y| \to 0$.

2. Let $\{f_m(t, x)\}$ be a sequence of continuous functions defined on $R = \{(t, x) : |t - t_0| \leq a, \ |x - x_0| \leq b\}$ for some positive constants a and b, and let $x_m(t)$ be a solution of

$$x'_m(t) = f_m(t, x_m(t)), \quad x_m(t_m) = x_m^0, \quad t \in [t_0 - a, t_0 + a],$$

   where $t_m \to t_0$, $x_m^0 \to x_0$ as $m \to \infty$. If

$$f(t, x) = \lim_{m \to \infty} f_m(t, x)$$

uniformly on R and if

$$x'(t) = f(t, x(t)), \quad x(t_0) = x_0,$$

has a unique solution x on $[t_0 - a, t_0 + a]$, then show that

$$x(t) = \lim_{m \to \infty} x_m(t)$$

uniformly on $[t_0 - a, t_0 + a]$.

3. Prove Theorem 2.3.6.

4. Formulate the conditions and prove the continuity of solutions with respect to (t, t_0, x_0, μ).

5. Formulate the conditions and prove the differentiability of solutions with respect to t_0.

6. Formulate the conditions and prove the analyticity of solutions when f is analytic.

7. Can you prove the continuity and differentiability of solutions with respect to the variables using Picard's approximations?

8. Prove Corollary 2.3.11.

2.4   Maximal Interval of Existence

From Theorem 2.2.7, we know that under some Lipschitz conditions, the differential Eq. (1.7) has a unique solution $x(t) = x(t, t_0, x_0)$ on an interval $I = (t_0 - r, t_0 + r)$ for some $r > 0$. However, $x(t)$ may exist on an interval that is larger than $I = (t_0 - r, t_0 + r)$, or $x(t)$ may even exist on the whole t-interval $(-\infty, \infty)$.

For example, for the differential equation $x'(t) = x^2(t)$, $x(0) = 1$, in Example 2.2.8, using the region R selected there, we derive $r \leq \frac{1}{4}$, thus, $I = (t_0 - r, t_0 + r) \subset (-\frac{1}{4}, \frac{1}{4})$. However, the solution is given by $x(t) = (1 - t)^{-1}$ and hence is defined on the interval $(-\infty, 1)$. The solutions of the linear differential equation $x'(t) = k(t)x(t) + f(t)$ given in Chapter 1 exist on $(-\infty, \infty)$.

Therefore, the question here is: What is **the largest interval** on which $x(t)$ is a unique solution? To answer this question, we first define what do we mean by "the largest interval."

Definition 2.4.1 *For a differential equation (such as Eq. (1.7)), if $x(t)$ is a unique solution defined on an interval q, and if there is no interval p such that $q \subset p$, $q \neq p$, and $x(t)$ is also a unique solution of the same differential equation on p, then q is called* **the maximal interval of existence** *of $x(t)$.*

In other words, q is the largest or maximal interval of existence of a solution x if there is no interval of existence of x that **properly contains** q. In applications in biology, chemistry, and physics, the variable t in a differential equation means the "time," and "time goes **on**." So in this section we sometimes only consider the maximal interval of existence of a solution to the right-hand side of an "initial time," such as t_0 in Eq. (1.7). The same results can be obtained to the left-hand side as well.

In order to find the maximal intervals of existence, let's look at the following situation: If $x(t)$ is a solution of an equation which is defined on a domain D, and if $(t, x(t))$ has a limit that is **inside** the domain D when t approaches a finite value, then this limit of $(t, x(t))$ can be regarded as an initial point from which another solution $u(t)$ can be derived. Now, we may **glue** the two solutions $x(t)$ and $u(t)$ such that $x(t)$ can be extended. Accordingly, we have the following result.

Theorem 2.4.2 (Extension) *Assume that $f(t, x)$ is continuous on a domain $D \subset \Re \times \Re^n$ and satisfies a Lipschitz condition (or weak or local Lipschitz condition) with respect to x on D. Let $(t_0, x_0) \in D$ and let $x(t) = x(t, t_0, x_0)$ be the unique solution of Eq. (1.7) on an interval $[t_0, d)$ with $d < \infty$. If $\lim_{t \to d^-} x(t)$ exists with $(d, x_d) \in D$, where $x_d = \lim_{t \to d^-} x(t)$, then there exists a $\delta > 0$ such that $x(t)$ can be extended to become a unique solution of Eq. (1.7) on $[t_0, d + \delta)$. That is, there is a function $y(t)$ defined on $[t_0, d + \delta)$ such that $y(t)$ is the unique solution of Eq. (1.7) on $[t_0, d + \delta)$, and $y(t) = x(t)$ for $t \in [t_0, d)$. See* **Figure 2.21**.

Proof. Since $(d, x_d) \in D$, we can treat (d, x_d) as an initial point in Theorem 2.2.7 to conclude that there exists a $\delta > 0$ and a unique solution $u(t)$ on $[d, d + \delta)$ with $u(d) = x_d$. Now we "glue $x(t)$ and $u(t)$ together" (see Figure 2.21) by defining

$$y(t) = \begin{cases} x(t), & t \in [t_0, d), \\ u(t), & t \in [d, d + \delta). \end{cases} \tag{4.1}$$

For $t \in [t_0, d)$, $y'(t) = x'(t) = f(t, x(t)) = f(t, y(t))$; for $t \in (d, d + \delta)$, $y'(t) = u'(t) = f(t, u(t)) = f(t, y(t))$. Next, let's check the derivative of

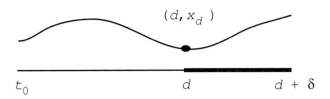

Figure 2.21: Glue $x(t)$ and $u(t)$ together

$y(t)$ at $t = d$. To do this, we observe that

$$
\begin{aligned}
\text{for } t \in [t_0, d) : \quad y(t) &= x(t) = x_0 + \int_{t_0}^{t} f(s, x(s)) ds \\
&= [x_0 + \int_{t_0}^{d} f(s, x(s)) ds] + \int_{d}^{t} f(s, x(s)) ds \\
&= \lim_{t \to d^-} x(t) + \int_{d}^{t} f(s, x(s)) ds \\
&= x_d + \int_{d}^{t} f(s, x(s)) ds \\
&= y(d) + \int_{d}^{t} f(s, y(s)) ds, \quad\quad (4.2) \\
\text{for } t \in (d, d + \delta) : \quad y(t) &= u(t) = x_d + \int_{d}^{t} f(s, u(s)) ds \\
&= y(d) + \int_{d}^{t} f(s, y(s)) ds, \quad\quad (4.3)
\end{aligned}
$$

which, by using the mean value theorem for integration, implies that the left and the right derivatives of $y(t)$ at $t = d$ are both given by $f(d, y(d))$ (see an exercise). Thus we have verified that $y(t)$ is a solution of Eq. (1.7) on $[t_0, d + \delta)$. The uniqueness of $y(t)$ can be shown in the same way as in Theorem 2.2.7. ♠

Theorem 2.4.2 indicates that to find the maximal intervals of existence, we only need to look at open intervals, because closed intervals can be extended. Now we are ready to prove that the maximal intervals of existence do exist.

Theorem 2.4.3 *Assume that $f(t, x)$ is continuous on a domain $D \subset \Re \times \Re^n$ and satisfies a Lipschitz condition (or weak or local Lipschitz condition) with respect to x on D. Let $(t_0, x_0) \in D$ and let $x(t) = x(t, t_0, x_0)$ be the unique solution of Eq. (1.7). Then there exist constants $\alpha \geq -\infty$ and $\beta \leq \infty$ such that $t_0 \in (\alpha, \beta)$ and (α, β) is the maximal interval of existence of $x(t)$. Moreover, α and β depend continuously on (t_0, x_0).*

Proof. From Theorem 2.2.7, we know that there exists an $r > 0$ such that Eq. (1.7) has a unique solution $x(t) = x(t, t_0, x_0)$ on $(t_0 - r, t_0 + r)$. Hence the following collection of intervals

$$K = \{I = (a, b) : t_0 \in I, x(t, t_0, x_0) \text{ is a unique solution of Eq.}(1.7) \text{ on } I\}, (4.4)$$

is nonempty (here, every element in K is an interval). In (4.4), we only use open intervals because from Theorem 2.4.2, closed intervals can be extended. Let's consider the union of these intervals in $\Re$,

$$I^* = \bigcup_{I \in K} I, \tag{4.5}$$

given in **Figure 2.22**.

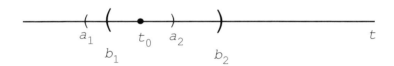

Figure 2.22: The union of the intervals in (4.4)

From the construction of I^*, we find that there is no such interval I' that $I^* \subset I'$, $I' \neq I^*$, and $x(t)$ is a unique solution of Eq. (1.7) on I'. Therefore, all we need is to verify that $x(t) = x(t, t_0, x_0)$ is a unique solution of Eq. (1.7) on I^* and I^* has the form of (α, β) for some constants $\alpha \geq -\infty$ and $\beta \leq \infty$, that is, I^* must be an open interval.

First, $t_0 \in I^*$ from (4.4)–(4.5). And we will only consider the structure of I^* on the right-hand side of t_0 and leave the structure on the left-hand side as an exercise. We need to consider two cases.

The first case is that there exist intervals $I_m = (a_m, b_m) \in K$ such that $\lim_{m \to \infty} b_m = \infty$. Then for any $t^0 \in [t_0, \infty)$, there is an N such that

$t^0 \in [t_0, b_N)$, and $x(t)$ is a unique solution of Eq. (1.7) on $[t_0, b_N)$. Therefore, $x(t)$ is a solution of Eq. (1.7) on $[t_0, \infty)$. If $y(t)$ is another solution of Eq. (1.7) on $[t_0, \infty)$, then on any $[t_0, b_n)$, we must have $x(t) = y(t)$ since $x(t)$ is the only solution on $[t_0, b_n)$. Thus, $x(t)$ is a unique solution of Eq. (1.7) on $[t_0, \infty)$, in which case, we define $\beta = \infty$.

The second case is that all the right-end values of the intervals in K are bounded by a finite value, that is,

$$b^* = \sup_{(a,b)\in K} b < \infty. \tag{4.6}$$

Now, there exist intervals $I_m = (a_m, b_m) \in K$ such that $\lim_{m\to\infty} b_m = b^*$. Similar to the case of $\lim_{m\to\infty} b_m = \infty$ discussed above, we conclude that $x(t)$ is a unique solution of Eq. (1.7) on $[t_0, b^*)$. Now, we define $\beta = b^*$ and claim that $[t_0, \beta)$ is the maximal interval of existence on the right-hand side of t_0, since $x(t)$ cannot exist on any interval that is beyond b^* in view of (4.6). The continuity of α and β on (t_0, x_0) is left as an exercise. This completes the proof. ♠

Theorem 2.4.2 also indicates that a solution can be extended further if there is no "barrier." In fact, utilizing the notion of a weak Lipschitz condition, we obtain the following "global existence" result for differential equations defined on $\Re \times \Re^n$, which can be applied directly to linear differential equations (see Example 2.2.10).

Theorem 2.4.4 (Global existence) *Assume that $f(t, x)$ is continuous on $\Re \times \Re^n$ and satisfies a weak Lipschitz condition with respect to x on $\Re \times \Re^n$. Let $x(t) = x(t, t_0, x_0)$ be the unique solution of Eq. (1.7) on its maximal interval of existence (α, β). Then $\alpha = -\infty$, $\beta = \infty$.*

Proof. We only prove $\beta = \infty$, since the case for α is similar. We will use the method of **proof by contradiction**, that is, we assume $\beta < \infty$ and then derive a contradiction.

Suppose $\beta < \infty$. We first show that $x(t)$ is bounded for $t \in [t_0, \beta)$. From Lemma 2.2.1 and a weak Lipschitz condition,

$$|x(t) - x_0| = |\int_{t_0}^t f(s, x(s))ds|$$

$$\leq |\int_{t_0}^t [f(s, x(s)) - f(s, x_0)]ds| + |\int_{t_0}^t f(s, x_0)ds|$$

$$\leq \int_{t_0}^t k(s)|x(s) - x_0|ds + |\int_{t_0}^t f(s, x_0)ds|, \quad t \in [t_0, \beta), \tag{4.7}$$

(where $k(t)$ is from a weak Lipschitz condition.) Since x_0 is fixed, there exists a constant $M > 0$ such that

$$k(s) \leq M, \quad |f(s, x_0)| \leq M, \quad s \in [t_0, \beta]. \tag{4.8}$$

Thus (4.7) becomes

$$|x(t) - x_0| \leq M(\beta - t_0) + M \int_{t_0}^{t} |x(s) - x_0| ds, \quad t \in [t_0, \beta), \tag{4.9}$$

hence, Gronwall's inequality implies that

$$|x(t) - x_0| \leq M(\beta - t_0) \exp\left(M(\beta - t_0)\right), \quad t \in [t_0, \beta), \tag{4.10}$$

therefore, $x(t)$ is bounded for $t \in [t_0, \beta)$.

Next, we show that $\lim_{t \to \beta^-} x(t)$ exists. The boundedness of $x(t)$ for $t \in [t_0, \beta)$ from (4.10) implies that the closure

$$\overline{\{(t, x(t)) : t \in [t_0, \beta)\}} \tag{4.11}$$

is bounded in $\Re \times \Re^n$. Therefore, since a continuous function on a bounded and closed set is bounded, $f(t, x(t))$ is bounded for $t \in [t_0, \beta)$. That is, there exists a constant $M_1 > 0$ such that

$$\max_{t \in [t_0, \beta)} |f(t, x(t))| \leq M_1. \tag{4.12}$$

Then, using Lemma 2.2.1, we see that for $t_0 \leq t_1 \leq t_2 < \beta$,

$$\begin{aligned} |x(t_2) - x(t_1)| &= \int_{t_1}^{t_2} |f(s, x(s))| ds \\ &\leq M_1(t_2 - t_1), \quad t_0 \leq t_1 \leq t_2 < \beta. \end{aligned} \tag{4.13}$$

This property says that for any $t_m \to \beta^-$, $x(t_m)$ is a Cauchy sequence (consult a reference book if needed, such as Goldberg [1976]). Therefore, $\lim_{t \to \beta^-} x(t)$ exists, and we denote it by x_β.

Now, (β, x_β) is a finite point in $\Re \times \Re^n$, hence the result of Theorem 2.4.2 can be applied to extend the solution $x(t)$ to the right-hand side of β, which contradicts the assumption that (α, β) is the maximal interval of existence of $x(t)$. Thus we must have $\beta = \infty$. This completes the proof. ♠

From the proof of Theorem 2.4.4, we see that the most important step is to show (4.12), that is, $f(t, x(t))$ is bounded. Therefore, we also derive the following result, whose proof is left as an exercise.

Theorem 2.4.4.a (Global existence) *Assume that $f(t, x)$ is continuous on $\Re \times \Re^n$ and satisfies a local Lipschitz condition with respect to x on $\Re \times \Re^n$, and that for some constant $M > 0$, $|f(t, x)| \leq M$, $(t, x) \in \Re \times \Re^n$. Let $x(t) = x(t, t_0, x_0)$ be the unique solution of Eq. (1.7) on its maximal interval of existence (α, β). Then $\alpha = -\infty$, $\beta = \infty$.* ♠

Theorem 2.4.4.a says that if the weak Lipschitz condition in Theorem 2.4.4 is reduced to a local Lipschitz condition, but an additional assumption that f is bounded is assumed, then we still have the global existence. Next, note that the solution of $x' = x^2$, $x(0) = 1$, exists only on $(-\infty, 1)$, and $f(t, x) = x^2$ satisfies only a local Lipschitz condition and is unbounded on $\Re \times \Re$. Therefore, it indicates that the weak Lipschitz condition in Theorem 2.4.4 cannot be weakened to a local Lipschitz condition, and that the boundedness assumption in Theorem 2.4.4.a cannot be removed.

Now, let's look at Example 2.2.10 again, where $x' = A(t)x + f(t)$ is an equation in $\Re^n$. We assume that $A(t)$ and $f(t)$ are continuous on $\Re$ so that the equation is defined on $\Re \times \Re^n$. Using Theorem 2.4.4, we immediately obtain that solutions exist globally on $\Re$. However, Theorem 2.4.4.a does not apply well in this case, because now $f(t, x) = A(t)x + f(t)$ may not be bounded on $\Re \times \Re^n$.

Theorems 2.4.4 and 2.4.4.a deal with differential equations defined on $\Re \times \Re^n$. However, we note from the proof of Theorem 2.4.4 that we have actually shown the following interesting results that are useful when dealing with differential equations defined on a general domain $D \subset \Re \times \Re^n$.

Proposition 2.4.5 *Assume that $f(t, x)$ is continuous on a domain $D \subset \Re \times \Re^n$ and satisfies a weak Lipschitz condition with respect to x on D. Let $(t_0, x_0) \in D$ and let $x(t) = x(t, t_0, x_0)$ be the unique solution of Eq. (1.7) on an interval $[t_0, b)$ with $b < \infty$. Then the following are true:*

(P1). If $k(t)$ (from a weak Lipschitz condition) and $f(t, x_0)$ are bounded for $t \in [t_0, b)$, then $x(t)$ is bounded for $t \in [t_0, b)$.

(P2). If $\overline{\{(t, x(t)) : t \in [t_0, b)\}}$ is in D and is bounded, or if $f(t, x(t))$ is bounded for $t \in [t_0, b)$, then $\lim_{t \to b^-} x(t)$ exists. ♠

A condition that guarantees Proposition 2.4.5(P2) is given below.

Proposition 2.4.6 *Assume that $f(t, x)$ is continuous on $\overline{D}$, the closure of D. Let $y(t)$ be a function on a finite interval I such that $(t, y(t)) \in D$ for $t \in I$. If $y(t)$ is bounded for $t \in I$, then $f(t, y(t))$ is bounded for $t \in I$.*

Proof. In this case, when $y(t)$ is bounded on a finite interval I, $\overline{\{(t, y(t)) : t \in I\}}$ is a closed and bounded set in $\overline{D}$. Hence $f(t, x)$ is bounded on $\overline{\{(t, y(t)) : t \in I\}}$, which implies that $f(t, y(t))$ is bounded for $t \in I$. ♠

To determine the structure of the maximal interval of existence (α, β) for a general domain $D \subset \Re \times \Re^n$, and also to determine the behavior of solutions as $t \to \beta^-$, we look at the following examples.

Example 2.4.7 We have seen that $\beta = \infty$ for linear differential equations defined on $\Re$. Next, for $x' = x^2$, $x(0) = 1$, defined on $\Re \times \Re$, we have $(\alpha, \beta) = (-\infty, 1)$. That is, $\beta = 1 < \infty$. In this case, when $t \to \beta^- = 1^-$, one has $\lim_{t \to \beta^-} |x(t)| = \lim_{t \to 1^-} (1 - t)^{-1} = \infty$, see **Figure 2.23**. ♠

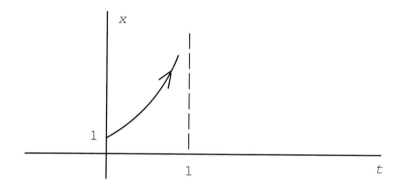

Figure 2.23: A case where $\lim_{t \to \beta^-} |x(t)| = \infty$

Example 2.4.8 Consider $x' = x^2, x(0) = 1$, defined on $D = \Re \times (-2, 2)$. The solution is given by $x(t) = (1 - t)^{-1}$, thus if we solve $x(t) = (1 - t)^{-1} = 2$, we get $t = \frac{1}{2}$ and conclude that the maximal interval of existence is $(\alpha, \beta) = (-\infty, \frac{1}{2})$, and

$$\lim_{t \to \beta^-} |x(t)| = \lim_{t \to \frac{1}{2}^-} (1 - t)^{-1} = 2.$$

Now, observe that $(\frac{1}{2}, 2)$ is on the boundary of the domain $D = \Re \times (-2, 2)$, see **Figure 2.24**. ♠

We claim that the three cases described in Examples 2.4.7 and 2.4.8 are all the possible cases. That is, we can prove the following result.

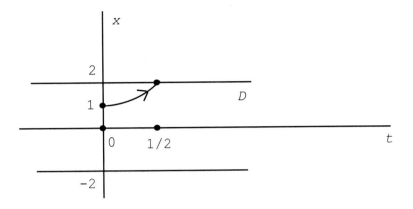

Figure 2.24: A case where $\lim_{t \to \beta^-}(t, x(t))$ is a boundary point

Theorem 2.4.9 *Assume that $f(t,x)$ is continuous on a domain $D \subset \Re \times \Re^n$ and satisfies a weak Lipschitz condition with respect to x on D, and assume further that $f(t,x)$ is continuous on $\overline{D}$ (or assume that for every finite interval I, $f(t, y(t))$ is defined and bounded on I for every $y(t)$ that is defined and bounded on I, see Propositions 2.4.5 and 2.4.6). Let $x(t) = x(t, t_0, x_0)$ be the unique solution of Eq. (1.7) on its maximal interval of existence (α, β). Then one of the following three cases (given in* **Figure 2.25***) must happen:*

(C1). $\beta = \infty$.

(C2). $\beta < \infty$, *and there exists $t_m \to \beta^-$ as $m \to \infty$ such that $\lim_{m \to \infty} |x(t_m)| = \infty$.*

(C3). $\beta < \infty$, *and $\lim_{t \to \beta^-}(t, x(t))$ exists as a finite point in $\Re \times \Re^n$, and this point lies on the boundary of D.*

Proof. If (C1) and (C2) do not happen, then $\beta < \infty$ and $x(t)$ remains bounded on $[t_0, \beta)$. From the assumption and Proposition 2.4.6, $f(t, x(t))$ is bounded on $[t_0, \beta)$. Using Proposition 2.4.5(P2), we see that $\lim_{t \to \beta^-} x(t)$ exists. Therefore, $\lim_{t \to \beta^-}(t, x(t))$ exists as a finite point in $\Re \times \Re^n$. This point must lie on the boundary of D, because otherwise Theorem 2.4.2 (Extension) can be applied to extend the solution further to the right-hand side of β, which contradicts the definition of β. This completes the proof. ♠

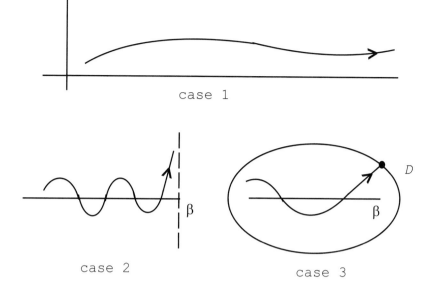

Figure 2.25: Three cases for the maximal intervals of existence

---

Corollary 2.4.10 *Assume that $f(t, x)$ is continuous on $\Re \times \Re^n$ and satisfies a weak Lipschitz condition with respect to x on $\Re \times \Re^n$. If the unique solution of Eq. (1.7) is bounded on any finite t interval, then it is defined on $\Re$.* ♠

From Proposition 2.4.5(P2) and the proof in Theorem 2.4.9, we find that when f is bounded, further results can be obtained, which we state below.

Corollary 2.4.11 *Assume that $f(t, x)$ is continuous on a domain $D \subset \Re \times \Re^n$ and satisfies a local Lipschitz condition with respect to x on D, and assume further that $f(t, x)$ is bounded on D. Let $x(t) = x(t, t_0, x_0)$ be the unique solution of Eq. (1.7) on its maximal interval of existence (α, β). Then one of the following two cases must happen:*

(D1). $\beta = \infty$.

(D2). $\beta < \infty$, and $\lim_{t \to \beta^-} (t, x(t))$ exists as a finite point in $\Re \times \Re^n$, and this point lies on the boundary of D. ♠

Remark 2.4.12 Note that $f(t,x) = x^2$ in Example 2.4.7 satisfies a local Lipschitz condition, thus Example 2.4.7 indicates that a local Lipschitz condition cannot guarantee the global existence of solutions on $(-\infty, \infty)$. Thus, to get the global existence on $(-\infty, \infty)$, many texts, such as Coddington and Levinson [1955], Perko [1991], and Cronin [1994], assumed that either the function f satisfies a (global) Lipschitz condition or f is bounded on $\Re \times \Re^n$. When this assumption is checked for the simplest case, that is, the linear differential equation $x' = A(t)x + f(t)$, the boundedness of $A(t)x$ does not apply in $\Re \times \Re^n$ (because x is unbounded in $\Re^n$); and the (global) Lipschitz condition would require that $A(t)$ to be bounded on $\Re$, which is too strong and restrictive. Therefore, Theorem 2.4.4 of this book gives a better result for the global existence on $(-\infty, \infty)$ by utilizing the notion of a weak Lipschitz condition, and this condition can be applied directly to the linear differential equations $x' = A(t)x + f(t)$ in $\Re^n$. ♠

Note that in our later studies, such as stability, boundedness, and periodicity for general differential equations, we will examine the "long term" behavior of solutions and require that the solutions exist on $[t_0, \infty)$ in those studies.

Exercises 2.4

1. In the proof of Theorem 2.4.2, find the left and the right derivatives of $y(t)$ at $t = d$.

2. For $x \in \Re$, determine the maximal interval of existence for

   (a) $x' = x \sin t$, $x(0) = a$.

   (b) $x' = x^2$, $x(t_0) = a$.

   (c) $x' = x^2 - 3$, $x(t_0) = a$.

   (d) $x' = x^3$, $x(t_0) = a$.

   (e) $x' = 1/x$, $x(t_0) = a$.

3. Determine the maximal interval of existence for the system $x_1' = x_1^2$, $x_2' = x_1^2 + x_2$.

4. For $x' = x(x-1)$, $x(0) = x_0 \in \Re$, determine the maximal interval of existence according to different x_0 values.

5. In the proof of Theorem 2.4.3, discuss the cases for the left-end point α.

6. In the proof of Theorem 2.4.3, verify the continuity of α and β on (t_0, x_0).

7. Prove that $\alpha = -\infty$ in Theorem 2.4.4.

8. Prove Theorem 2.4.4.a.

9. Prove Proposition 2.4.5.

10. Prove that the scalar equation

$$x'(t) = \frac{e^{-x^4 \cos^4 t}}{3 + t^6 \sin^2 t}, \quad x(t_0) = x_0,$$

has a unique solution on $\Re$.

11. Let (α, β) be the maximal interval of existence of x of $x' = f(t, x)$. If the arc length of $\{x(t) : 0 \le t < \beta\}$ is finite, then prove that $\lim_{t \to \beta^-} x(t)$ exists.

12. Prove Corollary 2.4.10.

13. Prove Corollary 2.4.11.

2.5   Fixed Point Method

Here, we provide a brief treatment showing that fixed point method can be used to derive existence and uniqueness of solutions if a local Lipschitz condition is satisfied; and, if a local Lipschitz condition is not assumed, then we can still obtain existence of solutions, in which case, uniqueness is not guaranteed. Since some knowledge of Functional Analysis is used, this section may be optional.

Let $f : \Re \times \Re^n \to \Re^n$ be continuous. From Lemma 2.2.1, we know that for some $r > 0$, a continuous function x on $[t_0 - r, t_0 + r]$ is a solution of

$$x'(t) = f(t, x(t)), \quad x(t_0) = x_0, \quad t \in [t_0 - r, t_0 + r], \tag{5.1}$$

if and only if

$$x(t) = x_0 + \int_{t_0}^{t} f(s, x(s))ds, \quad t \in [t_0 - r, t_0 + r]. \tag{5.2}$$

Thus, we are led to the following definition of a **mapping** P, such that if x is a continuous function on $[t_0 - r, t_0 + r]$, then define

$$(Px)(t) = x_0 + \int_{t_0}^t f(s, x(s))ds, \quad t \in [t_0 - r, t_0 + r]. \tag{5.3}$$

We observe that a continuous function x on $[t_0 - r, t_0 + r]$ is a solution of Eq. (5.1) if and only if

$$x(t) = x_0 + \int_{t_0}^t f(s, x(s))ds = (Px)(t), \quad t \in [t_0 - r, t_0 + r], \tag{5.4}$$

or if and only if

$$x = Px. \tag{5.5}$$

Such an x is called a **fixed point** of the mapping P. Thus, we obtain

Lemma 2.5.1 *A continuous function x on $[t_0 - r, t_0 + r]$ is a solution of Eq. (5.1) on $[t_0 - r, t_0 + r]$ if and only if x is a fixed point of the mapping P defined by (5.3).* ♠

Therefore, we are going to utilize two fixed point theorems presented in the Appendix to deal with Eq. (5.1). (See the Appendix for the definitions of the notions used in the following analysis.)

Contraction mapping principle. *Let P be a contraction mapping on a complete metric space X, then there is a unique $x \in X$ with $Px = x$.* ♠

Schauder's second fixed point theorem. *Let X be a nonempty convex subset of a Banach space Y and let $P : X \to X$ be a compact mapping (that is, P is continuous and maps a bounded set into a precompact set). Then P has a fixed point in X.* ♠

First, we apply the contraction mapping principle to derive local existence and uniqueness of solutions for Eq. (5.1) if a local Lipschitz condition is satisfied.

Theorem 2.5.2 (Local existence and uniqueness) *Assume that $f(t, x)$ is continuous on a domain $D \subset \Re \times \Re^n$ and satisfies a local Lipschitz condition with respect to x on D. Let $(t_0, x_0) \in D$. Then there exists a positive constant $r = r(t_0, x_0)$ such that Eq. (5.1) has a unique solution $x(t) = x(t, t_0, x_0)$ on $I = (t_0 - r, t_0 + r)$.*

Proof. For $(t_0,\, x_0) \in D$ given, as $f(t,x)$ satisfies a local Lipschitz condition with respect to x on D, there exists a domain D_1 such that $(t_0,\, x_0) \in D_1 \subset D$ and there exists a positive constant $k_1 = k_1(t_0, x_0)$ such that

$$|f(t,x) - f(t,y)| \le k_1 |x - y| \quad \text{for } (t,x),\, (t,y) \in D_1. \tag{5.6}$$

As D_1 is a domain, there exist positive constants a and b such that

$$R = \Big\{ (t,x) : |t - t_0| \le a,\ |x - x_0| \le b \Big\} \subset D_1, \tag{5.7}$$

see **Figure 2.26**.

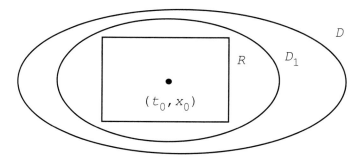

Figure 2.26: A picture for R, D_1 and D

Similar to the proof of Theorem 2.2.7, we let $M = \max_{(t,x) \in R} |f(t,x)|$ and $r = \min\{a, \frac{b}{M}\}$. Now, if a function x satisfies $|x(t) - x_0| \le b$ for $|t - t_0| \le r$, then $(t, x(t)) \in R$, hence

$$\Big| \int_{t_0}^{t} |f(s, x(s))| ds \Big| \le Mr \le b \quad \text{for} \quad |t - t_0| \le r. \tag{5.8}$$

Next, let $C([t_0 - r, t_0 + r], \Re^n)$ be the linear space of all continuous functions from $[t_0 - r, t_0 + r]$ to $\Re^n$ with the sup-norm

$$\|\phi\|_C = \sup_{|t - t_0| \le r} |\phi(t)|,$$

then it can be checked (see an exercise) that $C([t_0 - r, t_0 + r], \Re^n)$ is a Banach space, and that

$$X = \Big\{ x \in C([t_0 - r, t_0 + r], \Re^n) : \quad x(t_0) = x_0,\ |x(t) - x_0| \le b,$$

$$t \in [t_0 - r, t_0 + r] \Big\} \tag{5.9}$$

is a complete metric space with the metric

$$\rho(x, y) = \|x(\cdot) - y(\cdot)\|_C = \sup_{|t-t_0| \le r} |x(t) - y(t)|,$$

(note that X is not a linear space).

Consider the mapping P defined by (5.3) on X, that is, for $x \in X$,

$$(Px)(t) = x_0 + \int_{t_0}^t f(s, x(s))ds, \quad t \in [t_0 - r, t_0 + r]. \tag{5.10}$$

For $x \in X$, we see that Px is in $C([t_0 - r, t_0 + r], \Re^n)$ and $(Px)(t_0) = x_0$. Moreover, from (5.8),

$$|(Px)(t) - x_0| = |\int_{t_0}^t f(s, x(s))ds| \le Mr \le b, \quad t \in [t_0 - r, t_0 + r]. \tag{5.11}$$

Therefore, we have

$$P : X \to X.$$

To apply the contraction mapping principle, note that for $x, y \in X$, one has $(t, x(t)), (t, y(t)) \in R \subset D_1$ for $|t - t_0| \le r$. Thus,

$$
\begin{aligned}
|(Px)(t) - (Py)(t)| &= |\int_{t_0}^t |f(s, x(s)) - f(s, y(s))|ds| \\
&\le k_1 |\int_{t_0}^t |x(s) - y(s)|ds| \\
&\le k_1 r \sup_{|s-t_0| \le r} |x(s) - y(s)| \\
&= k_1 r \rho(x, y), \quad t \in [t_0 - r, t_0 + r].
\end{aligned}
\tag{5.12}
$$

Therefore, if we make an additional assumption that $r_1 = k_1 r < 1$, then we obtain

$$\rho(Px, Py) \le r_1 \rho(x, y), \tag{5.13}$$

hence, P is a contraction mapping on the complete metric space X, thus P has a unique fixed point $x \in X$. That is, $x(t) \in C([t_0 - r, t_0 + r], \Re^n)$ is such that

$$x(t) = x_0 + \int_{t_0}^t f(s, x(s))ds, \quad t \in [t_0 - r, t_0 + r], \tag{5.14}$$

therefore, $x(t)$ gives rise to a solution of Eq. (5.1) on $I = (t_0 - r, t_0 + r)$ using Lemma 2.2.1. Finally, $x(t)$ is the unique solution of Eq. (5.1) since any solution of Eq. (5.1) is a fixed point of the mapping P, and P has a unique fixed point. This completes the proof. ♠

In Theorem 2.5.2, a local Lipschitz condition is assumed such that we can derive existence and uniqueness. Next, we present a result showing that we can still obtain existence of solutions even if a local Lipschitz condition is not satisfied, in which case, uniqueness is not guaranteed.

Theorem 2.5.3 (Existence without uniqueness) *Assume that $f(t, x)$ is continuous on a domain $D \subset \Re \times \Re^n$ and let $(t_0, x_0) \in D$. Then there exist positive constants a and b such that*

$$R = \left\{ (t, x) : |t - t_0| \leq a, \ |x - x_0| \leq b \right\} \subset D, \tag{5.15}$$

and Eq. (5.1) has a solution $x(t)$ on $I = (t_0 - r, t_0 + r)$, where $r = \min\{a, \frac{b}{M}\}$ with $M = \max_{(t,x) \in R} |f(t, x)|$.

Proof. Since (t_0, x_0) is in domain D, the a, b, R, M, and r specified above can be defined. Now, X defined by (5.9) is a nonempty convex subset of the Banach space $C([t_0 - r, t_0 + r], \Re^n)$, and $P : X \to X$, where P is defined by (5.3).

First, we show that $P : X \to X$ is continuous. For $\phi, \psi \in X$,

$$|(P\phi)(t) - (P\psi)(t)| \leq |\int_{t_0}^{t} |f(s, \phi(s)) - f(s, \psi(s))| ds|. \tag{5.16}$$

Now, $f(t, x)$ is uniformly continuous on R, thus for any $\varepsilon > 0$ there is a $\delta = \delta(\varepsilon) > 0$ such that $\|\phi - \psi\|_C = \sup_{|s - t_0| \leq r} |\phi(s) - \psi(s)| \leq \delta$ implies $|f(s, \phi(s)) - f(s, \psi(s))| \leq \frac{\varepsilon}{r}$. Therefore, if $\|\phi - \psi\|_C \leq \delta$, then for $t \in [t_0 - r, t_0 + r]$, (5.16) becomes

$$|(P\phi)(t) - (P\psi)(t)| \leq |\int_{t_0}^{t} |f(s, \phi(s)) - f(s, \psi(s))| ds| \leq (\frac{\varepsilon}{r}) r = \varepsilon.$$

That is, $\|\phi - \psi\|_C \leq \delta$ implies $\|P\phi - P\psi\|_C \leq \varepsilon$, hence P is a continuous mapping on X.

Next, we show that P is a compact mapping. Note that for $\phi \in X$, one has $(t, \phi(t)) \in R$ for $|t - t_0| \leq r$. Thus, for $\phi \in X$ and $t, h \in [t_0 - r, t_0 + r]$,

$$|(P\phi)(t) - (P\phi)(h)| \leq |\int_{h}^{t} |f(s, \phi(s))| ds| \leq M|t - h|. \tag{5.17}$$

Therefore,

$$P : X \to S_0 \subset X,$$

where

$$S_0 = \Big\{ \phi \in C([t_0 - r, t_0 + r], \Re^n) : \ \phi(t_0) = x_0, \ |\phi(t) - \phi(h)| \le M|t - h|,$$

$$|\phi(t) - x_0| \le b, \ t, h \in [t_0 - r, t_0 + r] \Big\}.$$

Now, any sequence in the set S_0 is uniformly bounded and equi-continuous on the interval $[t_0 - r, t_0 + r]$. Hence, from Arzela-Ascoli's theorem (see the Appendix), this sequence has a convergent subsequence, thus the set S_0 is compact and, therefore, P is a compact mapping. By Schauder's second fixed point theorem, P has a fixed point, which gives rise to a solution of Eq. (5.1) on $I = (t_0 - r, t_0 + r)$. This completes the proof. ♠

Note that in the proof of Theorem 2.5.3, we used Schauder's second fixed point theorem, which does not guarantee uniqueness of fixed points. Therefore, uniqueness of solutions in Theorem 2.5.3 is not guaranteed, because solutions are derived from fixed points of the mapping P.

Exercises 2.5

1. Prove that $C([t_0 - r, t_0 + r], \Re^n)$ in the proof of Theorem 2.5.2 is a Banach space.

2. Prove that X defined in (5.9) is a nonempty convex subset of the Banach space $C([t_0 - r, t_0 + r], \Re^n)$.

3. Prove that X defined in (5.9) is a complete metric space with the metric
$$\rho(x, y) = \|x(\cdot) - y(\cdot)\|_C = \sup_{|t - t_0| \le r} |x(t) - y(t)|.$$

4. Prove that the set S_0 in the proof of Theorem 2.5.3 is a compact set.

Chapter 3

Linear Differential Equations

3.1 Introduction

In this chapter, we study linear differential equations, which means that $f(t,x) = f(t, x_1, x_2, \cdots, x_n)$ in the differential equation $x'(t) = f(t, x(t))$ is given by

$$
f(t,x) = \begin{bmatrix} a_{11}(t)x_1 + a_{12}(t)x_2 + \cdots + a_{1n}(t)x_n + f_1(t) \\ a_{21}(t)x_1 + a_{22}(t)x_2 + \cdots + a_{2n}(t)x_n + f_2(t) \\ \cdots \\ a_{n1}(t)x_1 + a_{n2}(t)x_2 + \cdots + a_{nn}(t)x_n + f_n(t) \end{bmatrix}
$$

$$
= \begin{bmatrix} a_{11}(t) & a_{12}(t) & . & . & a_{1n}(t) \\ a_{21}(t) & a_{22}(t) & . & . & a_{2n}(t) \\ . & & . & . & . \\ . & & . & . & . \\ a_{n1}(t) & a_{n2}(t) & . & . & a_{nn}(t) \end{bmatrix} \begin{bmatrix} x_1 \\ x_2 \\ . \\ . \\ x_n \end{bmatrix} + \begin{bmatrix} f_1(t) \\ f_2(t) \\ . \\ . \\ f_n(t) \end{bmatrix}, \quad (1.1)
$$

where $a_{ij}(t)$ and $f_i(t)$ are some real functions. Now, we use matrix notations and let $A(t) = [a_{ij}(t)]_{1 \le i,j \le n}$ be the $n \times n$ matrix function formed from functions $a_{ij}(t)$ and let $f(t) = [f_1(t), f_2(t), \cdots, f_n(t)]^T$ be the $n \times 1$ vector function formed from functions $f_i(t)$. Therefore, the differential equation

$$
x'(t) = f(t, x(t)), \ x(t_0) = x_0,
$$

will take the following form in $\Re^n$:

$$
x'(t) = A(t)x(t) + f(t), \quad x(t_0) = x_0. \quad (1.2)
$$

106

The study of linear differential equations is very important for the following reasons. First, the study provides us with some basic knowledge for understanding general nonlinear differential equations. Second, certain nonlinear differential equations can be written as summations of linear differential equations and some small nonlinear perturbations. Thus, under certain conditions, the qualitative properties of linear differential equations can be used to infer essentially the same qualitative properties for nonlinear differential equations. For example, let's look at the following two cases.

Example 3.1.1 Consider the scalar nonlinear differential equation

$$x' = \sin x. \tag{1.3}$$

As $\sin x = x - \frac{x^3}{3!} + \frac{x^5}{5!} - \cdots = x + O(x^3)$ (where O, called "big O," denotes the terms of the same or higher orders than x^3), Eq. (1.3) becomes

$$x' = x + O(x^3). \tag{1.4}$$

We know from calculus that when $x \approx 0$, we have $\sin x \approx x$ or $\frac{\sin x}{x} \approx 1$, see **Figure 3.1**.

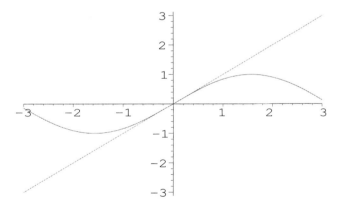

Figure 3.1: $\sin x \approx x$ when $x \approx 0$

Therefore, $O(x^3)$ can be regarded as a small perturbation when $x \approx 0$. We will see later that when $x \approx 0$, Eq. (1.4) or Eq. (1.3) will have essentially the same qualitative properties as the linear differential equation

$$x' = x. \qquad \spadesuit \tag{1.5}$$

Example 3.1.2 For the motion of a simple pendulum given by

$$\theta''(t) + k\theta'(t) + q\sin\theta(t) = 0, \tag{1.6}$$

we let $x_1 = \theta$, $x_2 = \theta'$, then we obtain the system

$$\begin{cases} x_1' & = & x_2, \\ x_2' & = & -kx_2 - q\sin x_1 = -kx_2 - qx_1 + O(x_1^3), \end{cases} \tag{1.7}$$

or in matrix notations,

$$\begin{bmatrix} x_1 \\ x_2 \end{bmatrix}' = \begin{bmatrix} 0 & 1 \\ -q & -k \end{bmatrix}\begin{bmatrix} x_1 \\ x_2 \end{bmatrix} + \begin{bmatrix} 0 \\ O(x_1^3) \end{bmatrix}. \tag{1.8}$$

We will see later that when $[x_1, x_2]^T \approx [0,0]^T$, Eq. (1.8) will have essentially the same qualitative properties as the linear differential equation

$$\begin{bmatrix} x_1 \\ x_2 \end{bmatrix}' = \begin{bmatrix} 0 & 1 \\ -q & -k \end{bmatrix}\begin{bmatrix} x_1 \\ x_2 \end{bmatrix}. \qquad \spadesuit \tag{1.9}$$

We will also verify that the semigroup and evolution system properties, as well as the variation of parameters formulas derived in Chapter 1 for one-dimensional linear differential equations can be derived for linear differential equations in $\Re^n$, $n \geq 1$. Therefore, we can develop an elegant theory that will allow us to completely understand the structure of solutions of linear differential equations in $\Re^n$, $n \geq 1$, and provide further knowledge related to the study of linear algebra.

In Eq. (1.2), if $f(t)$ is not identically zero, then Eq. (1.2) is called a **nonhomogeneous** linear differential equation in $\Re^n$. If, otherwise, $f(t)$ is identically zero, that is, if we have

$$x'(t) = A(t)x(t), \quad x(t_0) = x_0, \tag{1.10}$$

then Eq. (1.10) is called a **homogeneous** linear differential equation in $\Re^n$.

For homogeneous linear differential equations, a very important result is the principle of superposition: A linear combination of solutions is also a solution, see Theorem 3.1.3.

Theorem 3.1.3 (Principle of Superposition) *If $x_1(t)$ and $x_2(t)$ are both solutions of*

$$x'(t) = A(t)x(t) \tag{1.11}$$

on an interval I, and if a and b are constants, then $ax_1(t) + bx_2(t)$ is also a solution of Eq. (1.11) on I.

Proof. Let $x_1(t)$ and $x_2(t)$ be solutions of Eq. (1.11) on I, then

$$
\begin{aligned}
\frac{d}{dt}\Big[ax_1(t) + bx_2(t)\Big] &= ax_1'(t) + bx_2'(t) \\
&= aA(t)x_1(t) + bA(t)x_2(t) \\
&= A(t)\Big[ax_1(t) + bx_2(t)\Big], \quad t \in I, \qquad (1.12)
\end{aligned}
$$

which completes the proof. ♠

Example 3.1.4 Let $x_1(t)$ and $x_2(t)$ be solutions of

$$
x'(t) = 5x(t) + \sin t. \qquad (1.13)
$$

Then

$$
\begin{aligned}
\frac{d}{dt}\Big[x_1(t) + x_2(t)\Big] &= x_1'(t) + x_2'(t) \\
&= 5x_1(t) + \sin t + 5x_2(t) + \sin t \\
&= 5[x_1(t) + x_2(t)] + 2\sin t. \qquad (1.14)
\end{aligned}
$$

Thus $x_1(t) + x_2(t)$ is not a solution of Eq. (1.13). ♠

Therefore, we conclude that the principle of superposition holds only for homogeneous linear differential equations. Some important consequences of the principle of superposition will be seen later.

Note, a scalar linear differential equation of order n can be written as a first-order linear differential equation in $\Re^n$ governed by Eq. (1.2).

Example 3.1.5 Consider the scalar linear differential equation of order n,

$$
x^{(n)}(t) + a_1(t)x^{(n-1)}(t) + \cdots + a_{n-1}(t)x'(t) + a_n(t)x(t) + a_{n+1}(t) = 0.
$$

We can define

$$
x_1(t) = x(t), \ x_2(t) = x'(t), \cdots, x_{n-1}(t) = x^{(n-2)}(t), \ x_n(t) = x^{(n-1)}(t),
$$

and then obtain

$$
\begin{aligned}
x_1'(t) &= x_2(t), \\
x_2'(t) &= x_3(t), \\
&\cdots \\
x_{n-1}'(t) &= x_n(t), \\
x_n'(t) &= x^{(n)}(t) = -a_1(t)x_n(t) - a_2(t)x_{n-1}(t) - \cdots - a_{n+1}(t),
\end{aligned}
$$

which is a first-order linear differential equation in $\Re^n$ of the form of Eq. (1.2). ♠

This chapter is organized as follows: In Section 2, we study general non-homogeneous linear differential equations and obtain the fundamental matrix solutions and verify that they satisfy the "evolution system property." Then we derive the variation of parameters formula using the fundamental matrix solutions and observe what these solutions should look like. In Section 3, we look at equations with constant coefficients and examine detailed structure of solutions in terms of eigenvalues of the leading constant matrix, using the Jordan canonical form theorem. In addition we derive the Putzer algorithm that can be used to actually solve or compute solutions for equations with constant coefficients. This result will appeal to readers interested in computation. In Section 4, we look at equations with periodic coefficients and study Floquet theory, which allows us to transform equations with periodic coefficients into equations with constant coefficients. The results of Section 3 can then be applied to the transformed equations. The concept of the Liapunov exponent is also briefly introduced in Section 4.

Exercises 3.1

1. For the equations

$$\begin{aligned}
x_1'(t) &= 2(\cos t)x_3(t) - x_1(t) + \sin t, \\
x_2'(t) &= e^t x_1(t) - 2x_3(t) + 5tx_2(t) - e^{2t}, \\
x_3'(t) &= x_2(t) - 4t^2 x_1(t) - \cos t + t^2,
\end{aligned}$$

find a matrix $A(t)$ and a vector $f(t)$ so as to rewrite the above equations as $x'(t) = A(t)x(t) + f(t)$.

2. Rewrite $3x'''(t) + 5(\sin t)x''(t) - 8e^t x'(t) - \ln(3t) = 7$ as a system of linear differential equations, and then discuss the existence and uniqueness of solutions.

3. Rewrite

$$\begin{aligned}
3x_1''(t) + 5x_1(t) + 2x_2(t) - 2x_2'(t) + 5 &= 0, \\
4x_1'(t) + x_2(t) - 5x_2'(t) - 6 &= 0,
\end{aligned}$$

as a system of linear differential equations, and then discuss the existence and uniqueness of solutions.

4. Rewrite $x^{(n)}(t) + a_1 x^{(n-1)}(t) + \cdots + a_{n-1} x'(t) + a_n x(t) = 0$ as a system of linear differential equations and verify that the coefficient matrix is

$$A = \begin{bmatrix} 0 & 1 & 0 & . & . & . & 0 \\ 0 & 0 & 1 & . & . & . & 0 \\ . & . & . & . & . & . & . \\ . & . & . & . & . & . & . \\ 0 & 0 & 0 & . & . & . & 1 \\ -a_n & -a_{n-1} & -a_{n-2} & . & . & . & -a_1 \end{bmatrix}.$$

Then verify, using an induction, that $\det[A - \lambda E] = (-1)^n(\lambda^n + a_1\lambda^{n-1} + \cdots + a_{n-1}\lambda + a_n)$, where E is the unit or identity $n \times n$ matrix, and det means determinant.

5. Find the solution of the system $x' = Ax$, $x(t_0) = x_0$, where

$$A = \begin{bmatrix} 1 & -1 & 1 \\ 0 & 2 & 1 \\ 0 & 0 & 2 \end{bmatrix}.$$

6. If $a \in \Re$ is not an eigenvalue of matrix A, then find a vector P such that Pe^{at} is a solution of $x'(t) = Ax(t) + Ce^{at}$.

3.2   General Nonhomogeneous Linear Equations

In Chapter 2, we defined $|A| = \sum_{i,j=1}^n |a_{ij}|$ for an $n \times n$ matrix $A = [a_{ij}]$. Similarly, we make the following definitions.

Definition 3.2.1 *For an $n \times n$ matrix function $A(t) = [a_{ij}(t)]$ on an interval (a, b), define*

$$|A(t)| = \sum_{i,j=1}^n |a_{ij}(t)|, \quad \frac{d}{dt}A(t) = [\frac{d}{dt}a_{ij}(t)], \quad \int_a^b A(t)dt = [\int_a^b a_{ij}(t)dt]. \quad (2.1)$$

It follows, similar to the results in Chapter 2, that for $n \times n$ matrices $A(t), B(t)$, and an $n \times 1$ vector x,

$$\begin{cases} |A(t) + B(t)| & \leq & |A(t)| + |B(t)|, \\ |A(t)B(t)| & \leq & |A(t)||B(t)|, \\ |A(t)x| & \leq & |A(t)||x|. \end{cases} \qquad (2.2)$$

Now, we can directly apply some results in Chapter 2 to obtain the existence and uniqueness for the linear differential equation (1.2).

Theorem 3.2.2 *Assume that $A(t)$ and $f(t)$ are continuous on an interval (a, b). Let $t_0 \in (a, b)$. Then for any $x_0 \in \Re^n$, Eq. (1.2) has a unique solution $x(t) = x(t, t_0, x_0)$, and its maximal interval of existence is (a, b). In particular, if $(a, b) = (-\infty, \infty)$, then the maximal interval of existence is $(-\infty, \infty)$.*

Proof. Now, the right-hand side of Eq. (1.2) is given by $f(t, x) = A(t)x + f(t)$ on $(a, b) \times \Re^n$, so it satisfies a weak Lipschitz condition. By Theorems 2.2.7.a and 2.4.3 in Chapter 2, Eq. (1.2) has a unique solution on its maximal interval of existence, denoted by (α, β). Hence $a \leq \alpha < \beta \leq b$.

If $\beta < b$, then $A(t)$, $f(t)$, and $f(t, x_0) = A(t)x_0 + f(t)$ are all bounded on $[t_0, \beta]$. From Proposition 2.4.5(P1), $x(t)$ is bounded on $[t_0, \beta)$. This implies that $f(t, x(t)) = A(t)x(t) + f(t)$ is bounded on $[t_0, \beta)$. Then Proposition 2.4.5(P2) implies that $\lim_{t \to \beta^-} x(t)$ exists, which we denote by x_β.

Now, the domain for $f(t, x) = A(t)x + f(t)$ in Eq. (1.2) is $D = (a, b) \times \Re^n$, thus $(\beta, x_\beta) \in D$. Then Theorem 2.4.2 (Extension) can be applied to extend $x(t)$ to the right-hand side of β, which is a contradiction. Similarly, $\alpha = a$. This completes the proof.   ♠

Theorem 3.2.2 indicates that for a general linear differential equation, if it is defined on (a, b), then its maximal interval of existence is (a, b); that is, its maximal interval of existence will not reduce from the original interval on which the equation is defined. This is not the case for nonlinear differential equations. For example, equation $x' = x^2$, $x(0) = 1$, is defined on $\Re$, but its maximal interval of existence is $(-\infty, 1) \neq \Re$.

Recall from Chapter 1 that for the scalar homogeneous linear differential equation,

$$x'(t) = k(t)x(t), \quad x(t_0) = 1, \qquad (2.3)$$

its unique solution is given by

$$U_1(t, t_0) = e^{\int_{t_0}^{t} k(h)dh}, \qquad (2.4)$$

and any solution of Eq. (2.3) with $x(t_0) = x_0$ is given by

$$x(t, t_0, x_0) = U_1(t, t_0)x_0. \qquad (2.5)$$

Note here that

$$U_1(t_0, t_0) = 1, \tag{2.6}$$

where 1 is the unit (or identity) element in $\Re$.

In this section, we attempt to generalize the results concerning formulas (2.4), (2.5), and (2.6) to the homogeneous linear differential equation (1.10) in $\Re^n$, $n \geq 1$.

Compare Eq. (2.3) and Eq. (1.10). We find that the counterpart of the **scalar function** $k(t)$ in Eq. (2.3) is now the $n \times n$ **matrix function** $A(t)$ in Eq. (1.10). Thus, the counterpart of "1" in Eq. (2.6) for scalar differential equations will now be the $n \times n$ unit or identity matrix, given by

$$E = \begin{bmatrix} 1 & 0 & . & . & 0 \\ 0 & 1 & . & . & 0 \\ . & . & . & . & . \\ . & . & . & . & . \\ 0 & 0 & . & . & 1 \end{bmatrix} = [e_1, e_2, \cdots, e_n], \tag{2.7}$$

where

$$e_i = \begin{bmatrix} 0 \\ . \\ . \\ . \\ 0 \\ 1_i \\ 0 \\ . \\ . \\ 0 \end{bmatrix} = [0, \cdots, 0, 1_i, 0, \cdots, 0]^T, \quad i = 1, 2, \cdots, n.$$

Now, we assume that $A(t)$ and $f(t)$ are continuous on an interval (a, b), so from Theorem 3.2.2, for $t_0 \in (a, b)$ and for each $i = 1, 2, \cdots, n$,

$$x'(t) = A(t)x(t), \quad x(t_0) = e_i, \tag{2.8}$$

has a unique solution on (a, b), which we denote by $x_{(i)}(t)$. Let's construct a matrix by putting these $x_{(i)}(t)$, $i = 1, 2, \cdots, n$, as the columns. That is, define

$$U(t, t_0) = [x_{(1)}(t), x_{(2)}(t), \cdots, x_{(n)}(t)]. \tag{2.9}$$

Example 3.2.3 Consider $x' = A(t)x$, $x(0) = e_i$, $i = 1, 2$, where

$$A(t) = \begin{bmatrix} t & 1 \\ 0 & 2 \end{bmatrix}.$$

To find $U(t, 0)$, we solve

$$
\begin{aligned}
x'_{(1)}(t) &= A(t)x_{(1)}(t), \quad x_{(1)}(0) = \begin{bmatrix} 1 \\ 0 \end{bmatrix}, \\
x'_{(2)}(t) &= A(t)x_{(2)}(t), \quad x_{(2)}(0) = \begin{bmatrix} 0 \\ 1 \end{bmatrix}.
\end{aligned}
$$

For $x_{(1)}(t)$, we find that the second component is from a linear differential equation with the zero initial value, thus the second component is zero. The first component is now from $x' = tx$, $x(0) = 1$. Therefore,

$$x_{(1)}(t) = \begin{bmatrix} e^{t^2/2} \\ 0 \end{bmatrix}.$$

For $x_{(2)}(t)$, the second component is e^{2t}. The first component is now from $x' = tx + e^{2t}$, $x(0) = 0$. Therefore, using the variation of parameters formula for scalar linear differential equations, we obtain

$$x_{(2)}(t) = \begin{bmatrix} e^{t^2/2} \int_0^t e^{2s-(s^2/2)}ds \\ e^{2t} \end{bmatrix}.$$

Accordingly, the matrix U in this case is given by

$$U(t,0) = \begin{bmatrix} e^{t^2/2} & e^{t^2/2} \int_0^t e^{2s-(s^2/2)}ds \\ 0 & e^{2t} \end{bmatrix}. \qquad \spadesuit \qquad (2.10)$$

The $n \times n$ matrix $U(t, t_0)$ defined by (2.9) has the following properties, which are similar to those of the scalar $U_1(t, t_0) = e^{\int_{t_0}^t k(h)dh}$ for scalar homogeneous linear differential equations.

Theorem 3.2.4 *For $U(t, t_0)$ defined in (2.9) with $t, t_0 \in (a, b)$,*

(U1). $U(t, t_0)$ is the unique matrix solution of $\frac{d}{dt}U(t, t_0) = A(t)U(t, t_0)$,
   $U(t_0, t_0) = E$, $t \in (a, b)$.

(U2). *For any $x_0 \in \Re^n$, the unique solution of Eq. (1.10) is given by $x(t, t_0, x_0)$
$= U(t, t_0)x_0$. That is, $x(t, t_0, x_0)$ is a linear combination of the columns
in $U(t, t_0)$ with the combination coefficients given by the components
of x_0.*

(U3). *For any fixed $t \in (a, b)$, the columns of $U(t, t_0)$ are linearly independent
vectors in $\Re^n$, and hence the matrix $U(t, t_0)$ is invertible.*

Proof. (U1): Since $U(t, t_0) = [x_{(1)}(t), x_{(2)}(t), \cdots, x_{(n)}(t)]$, we have

$$\frac{d}{dt}U(t, t_0) = [x'_{(1)}(t), x'_{(2)}(t), \cdots, x'_{(n)}(t)],$$

and

$$
\begin{aligned}
A(t)U(t, t_0) &= A(t)[x_{(1)}(t), x_{(2)}(t), \cdots, x_{(n)}(t)] \\
&= [A(t)x_{(1)}(t), A(t)x_{(2)}(t), \cdots, A(t)x_{(n)}(t)].
\end{aligned}
$$

Thus we can compare each column to derive the equalities in (U1). The
uniqueness can also be obtained in this way using the uniqueness for Eq.
(2.8).

(**U2**): Since each column in $U(t, t_0)$ is a solution of Eq. (1.11), by the
principle of superposition, we find that $U(t, t_0)x_0$, as a linear combination
of n such solutions, is a solution of Eq. (1.10). Its uniqueness is guaranteed
by Theorem 3.2.2.

(**U3**): Suppose there exists a $t_1 \in (a, b)$ such that the vectors $\{x_{(1)}(t_1),
x_{(2)}(t_1), \cdots, x_{(n)}(t_1)\}$ are linearly dependent in $\Re^n$. Then there exist con-
stants $c_1, c_2, \cdots, c_n$, not all zero, such that

$$c_1 x_{(1)}(t_1) + c_2 x_{(2)}(t_1) + \cdots + c_n x_{(n)}(t_1) = 0.$$

Now define

$$y(t) = c_1 x_{(1)}(t) + c_2 x_{(2)}(t) + \cdots + c_n x_{(n)}(t), \quad t \in (a, b). \qquad (2.11)$$

By the principle of superposition, we find that $y(t)$ is a solution of

$$y'(t) = A(t)y(t), \quad y(t_1) = 0. \qquad (2.12)$$

But the 0 vector in $\Re^n$ is also a solution of (2.12), thus by the uniqueness
from Theorem 3.2.2, we get $y(t) = 0$, $t \in (a, b)$. In particular, $y(t_0) = 0$,

which implies, from (2.8),

$$
\begin{aligned}
0 = y(t_0) &= c_1 x_{(1)}(t_0) + c_2 x_{(2)}(t_0) + \cdots + c_n x_{(n)}(t_0) \\
&= c_1 e_1 + c_2 e_2 + \cdots + c_n e_n \\
&= [c_1, c_2, \cdots, c_n]^T,
\end{aligned}
\tag{2.13}
$$

that is, $c_i = 0$ for $i = 1, 2, \cdots, n$, which is a contradiction to the assumption that the constants $c_1, c_2, \cdots, c_n$ are not all zero.   ♠

Using the language of linear algebra, we have the following corollary of Theorem 3.2.4. First, let's recall some definitions in linear algebra.

Definition 3.2.5 *Let I be an interval and let $f_i : I \to \Re^n$ be a vector function, $i = 1, 2, \cdots, s$. We say that $f_1(t), f_2(t), \cdots, f_s(t)$ are **linearly independent functions** on I if there do not exist constants $\{a_1, a_2, \cdots, a_s\}$, not all zero, such that*

$$
\sum_{i=1}^{s} a_i f_i(t) \equiv 0, \quad t \in I.
\tag{2.14}
$$

Or equivalently, if (2.14) is true for some constants $\{a_1, a_2, \cdots, a_s\}$, then we must have
$$
a_1 = a_2 = \cdots = a_s = 0.
$$

*Let S be a linear space consisting of some functions defined on I. We say that S is **n-dimensional** if S has a set, called a **basis**, consisting of n linearly independent functions, such that any element of S is a linear combination of the functions in the basis.*

Corollary 3.2.6 *Let Eq. (1.11) be defined on an interval (a, b). Then all solutions of Eq. (1.11) (which are defined on (a, b) by Theorem 3.2.2) form an n-dimensional linear space. Let t_0 be any value in (a, b), then the columns in $U(t, t_0)$ is a basis for this space.*

Proof. From the principle of superposition, all solutions of Eq. (1.11) form a linear space. Let t_0 be any value in (a, b), then the n columns of $U(t, t_0)$ are n solutions of Eq. (1.11) and are linearly independent functions on (a, b) using Theorem 3.2.4(U3) and Definition 3.2.5. Next, let $x(t)$ be any solution of Eq. (1.11) on (a, b). Then $(t_0, x(t_0))$ is well defined and can be regarded as an initial point for $x(t)$. Now, from Theorem 3.2.4(U2), one has $x(t) = U(t, t_0)x(t_0)$. That is, any solution of Eq. (1.11) is a linear combination of the columns in $U(t, t_0)$. Therefore, the columns in $U(t, t_0)$ can be used as a basis. This completes the proof. ♠

Example 3.2.7 Consider Example 3.2.3 again, where $A(t) = \begin{bmatrix} t & 1 \\ 0 & 2 \end{bmatrix}$, and

$$x_{(1)}(t) = \begin{bmatrix} e^{t^2/2} \\ 0 \end{bmatrix}, \quad x_{(2)}(t) = \begin{bmatrix} e^{t^2/2} \int_0^t e^{2s-(s^2/2)} ds \\ e^{2t} \end{bmatrix}.$$

If for some constants a and b, one has

$$a \begin{bmatrix} e^{t^2/2} \\ 0 \end{bmatrix} + b \begin{bmatrix} e^{t^2/2} \int_0^t e^{2s-(s^2/2)} ds \\ e^{2t} \end{bmatrix} = 0,$$

then we must have $b = 0$ from the second component and then $a = 0$ from the first component. Thus $x_{(1)}(t)$ and $x_{(2)}(t)$ are linearly independent. ♠

Now, we make the following definition concerning the matrix $U(t, t_0)$.

Definition 3.2.8 *The matrix $U(t, t_0)$ defined in (2.9) is called the* **fundamental matrix solution** *of Eq. (1.10). That is, $U(t, t_0)$ is the unique matrix solution of $\frac{d}{dt}U(t, t_0) = A(t)U(t, t_0)$, $U(t_0, t_0) = E$.*

Next, knowing that $U(t, t_0)x_0$ is the unique solution of the homogeneous linear differential equation (1.10), we follow the ideas of the variation of parameters formulas derived for scalar linear differential equations, and try to find a solution of the nonhomogeneous linear differential equation (1.2) in $\Re^n, n \geq 1$, of the form $U(t, t_0)C(t)$, where $C(t)$ is an unknown $n \times 1$ vector function to be determined. Then, from Eq. (1.2) and the product rule, we are led to

$$[\frac{d}{dt}U(t, t_0)]C(t) + U(t, t_0)C'(t) = A(t)U(t, t_0)C(t) + f(t).$$

From Theorem 3.2.4(U1), $[\frac{d}{dt}U(t, t_0)]C(t) = A(t)U(t, t_0)C(t)$, hence

$$U(t, t_0)C'(t) = f(t).$$

Using Theorem 3.2.4(U3), $U^{-1}(t, t_0)$ exists for $t \in (a, b)$. Therefore,

$$\begin{aligned} C(t) &= C(t_0) + \int_{t_0}^t U^{-1}(s, t_0)f(s)ds \\ &= x_0 + \int_{t_0}^t U^{-1}(s, t_0)f(s)ds, \end{aligned} \tag{2.15}$$

hence

$$U(t,t_0)C(t) = U(t,t_0)\Big[x_0 + \int_{t_0}^t U^{-1}(s,t_0)f(s)ds\Big]$$

$$= U(t,t_0)x_0 + \int_{t_0}^t U(t,t_0)U^{-1}(s,t_0)f(s)ds. \quad (2.16)$$

This formula is similar to formula (1.13) in Section 1 of Chapter 1 for scalar differential equations, where we know for the scalar $U_1(t,t_0) = e^{\int_{t_0}^t k(h)dh}$ that

$$U_1(t,t_0)U_1^{-1}(s,t_0) = e^{\int_{t_0}^t k(h)dh} e^{-\int_{t_0}^s k(h)dh}$$

$$= e^{\int_s^t k(h)dh} = U_1(t,s).$$

Thus, in order to simplify (2.16) and derive something that looks like formula (1.14) in Section 1 of Chapter 1 involving an evolution system, let's find out what is $U(t,t_0)U^{-1}(s,t_0)$ for differential equations in $\Re^n$. First, note that in a similar way, $U(r,s)$ is defined to be the fundamental matrix solution of $\frac{d}{dr}U(r,s) = A(r)U(r,s)$, $U(s,s) = E$.

Lemma 3.2.9 *Let Eq. (1.10) be defined on (a,b). For the fundamental matrix solutions, the **evolution system property** is satisfied. That is, we have, for any t, t_0, $t_1 \in (a,b)$,*

$$\begin{cases} U(t,t) = E, \\ U(t,t_1)U^{-1}(t_0,t_1) = U(t,t_0), \\ U^{-1}(t_0,t_1) = U(t_1,t_0), \\ U(t,t_1)U(t_1,t_0) = U(t,t_0). \end{cases} \quad (2.17)$$

Proof. $U(t,t) = E$ is from the definition. From Theorem 3.2.4(U2), $U(t,t_0)x_0$ is the unique solution of Eq. (1.10). Next, define $y(t) = U(t,t_1)U^{-1}(t_0,t_1)x_0$. We have $y(t_0) = U(t_0,t_1)U^{-1}(t_0,t_1)x_0 = x_0$, and

$$y'(t) = [\frac{d}{dt}U(t,t_1)][U^{-1}(t_0,t_1)x_0]$$

$$= A(t)U(t,t_1)[U^{-1}(t_0,t_1)x_0]$$

$$= A(t)[U(t,t_1)U^{-1}(t_0,t_1)x_0]$$

$$= A(t)y(t),$$

which imply that $y(t)$ is also a solution of Eq. (1.10). Then the uniqueness for Eq. (1.10) implies $y(t) = [U(t, t_0)]x_0$. That is, we have

$$[U(t, t_1)U^{-1}(t_0, t_1)]x_0 = [U(t, t_0)]x_0. \tag{2.18}$$

But x_0 in (2.18) can be arbitrary, so that by letting $x_0 = e_i$, $i = 1, 2, \cdots, n$, where $\{e_1, e_2, \cdots, e_n\}$ form the standard unit basis of $\Re^n$, we obtain

$$U(t, t_1)U^{-1}(t_0, t_1) = U(t, t_0).$$

The third equality in (2.17) is true by letting $t = t_1$ in the second equality, and the last equality in (2.17) is true from the first three equalities. This completes the proof. ♠

Example 3.2.10 In Example 3.2.3 with $A(t) = \begin{bmatrix} t & 1 \\ 0 & 2 \end{bmatrix}$, we have calculated $U(t, 0)$. Now, the same calculation gives

$$U(t, h) = \begin{bmatrix} e^{(t^2/2)-(h^2/2)} & e^{t^2/2}\int_h^t e^{2(s-h)-(s^2/2)}ds \\ 0 & e^{2(t-h)} \end{bmatrix}.$$

Thus,

$$U^{-1}(t, h) = e^{-(t^2/2)+(h^2/2)-2(t-h)}\begin{bmatrix} e^{2(t-h)} & -e^{t^2/2}\int_h^t e^{2(s-h)-(s^2/2)}ds \\ 0 & e^{(t^2/2)-(h^2/2)} \end{bmatrix}.$$

Therefore,

$$\begin{aligned}
&U(t, t_1)U^{-1}(t_0, t_1) \\
&= \begin{bmatrix} e^{(t^2/2)-(t_1^2/2)} & e^{t^2/2}\int_{t_1}^t e^{2(s-t_1)-(s^2/2)}ds \\ 0 & e^{2(t-t_1)} \end{bmatrix} \\
&\quad \cdot e^{-(t_0^2/2)+(t_1^2/2)-2(t_0-t_1)}\begin{bmatrix} e^{2(t_0-t_1)} & -e^{t_0^2/2}\int_{t_1}^{t_0} e^{2(s-t_1)-(s^2/2)}ds \\ 0 & e^{(t_0^2/2)-(t_1^2/2)} \end{bmatrix} \\
&= \begin{bmatrix} e^{(t^2/2)-(t_0^2/2)} & e^{(t^2/2)-2(t_0-t_1)}\int_{t_0}^t e^{2(s-t_1)-(s^2/2)}ds \\ 0 & e^{2(t-t_0)} \end{bmatrix} \\
&= \begin{bmatrix} e^{(t^2/2)-(t_0^2/2)} & e^{t^2/2}\int_{t_0}^t e^{2(s-t_0)-(s^2/2)}ds \\ 0 & e^{2(t-t_0)} \end{bmatrix} \\
&= U(t, t_0),
\end{aligned}$$

which is the second equality in (2.17). Based on this, other equalities in (2.17) can be checked for this example. ♠

After showing the evolution system property in Lemma 3.2.9, we derive the following variation of parameters formula from (2.16). The formula is similar to formula (1.14) in Section 1 of Chapter 1 for scalar linear differential equations.

Theorem 3.2.11 (Variation of Parameters Formula) *Assume that $A(t)$ and $f(t)$ are continuous on an interval (a,b). Let $t_0 \in (a,b)$ and let $U(t,t_0)$ be the fundamental matrix solution of Eq. (1.10). Then the unique solution of Eq. (1.2) is given by*

$$
\begin{aligned}
x(t) &= U(t,t_0)\Big[x_0 + \int_{t_0}^t U^{-1}(s,t_0)f(s)ds\Big] \\
&= U(t,t_0)x_0 + \int_{t_0}^t U(t,s)f(s)ds, \quad t \in (a,b). \qquad \spadesuit \quad (2.19)
\end{aligned}
$$

This approach to formula (2.19) demonstrates that structures of solutions for all first-order linear differential equations in $\Re^n$, $n \geq 1$, are the same, and are also determined by the fundamental matrix solutions which satisfy the "evolution system property" described in Lemma 3.2.9. The solution formula (2.19) also indicates the "affine" structure of Eq. (1.2), meaning that the solution set is some particular solution of Eq. (1.2) plus any solution of the corresponding homogeneous equation. This is very useful when we study some qualitative properties of solutions for Eq. (1.2), such as stability.

We have one **problem** left, however. That is, for the scalar linear differential equation $x'(t) = k(t)x(t)$, we know from Chapter 1 that the fundamental (scalar) solution $U_1(t,t_0)$ can be expressed as an **exponential**, that is, we have

$$
U_1(t,t_0) = \exp\Big(\int_{t_0}^t k(s)ds\Big). \tag{2.20}
$$

Now, for $n > 1$, we ask the following question:

What does the fundamental matrix solution $U(t,t_0)$ look like for Eq. (1.10) in $\Re^n$?

We may **mimic** the formula (2.20) for scalar differential equations and **guess** that in $\Re^n$ the fundamental matrix solution $U(t,t_0)$ is given by

$$
\exp\Big(\int_{t_0}^t A(s)ds\Big).
$$

To pursue in this direction, we must understand the meaning of putting a matrix into the exponential function. From the Taylor series expansion of

$e^a = 1 + a + \frac{a^2}{2!} + \frac{a^3}{3!} + \cdots$ in $\Re$, we have

Lemma 3.2.12 *Let A be an $n \times n$ matrix. Then the series*

$$E + A + \frac{A^2}{2!} + \frac{A^3}{3!} + \cdots \qquad (2.21)$$

(where E is the $n \times n$ unit or identity matrix) converges to an $n \times n$ matrix.

Proof. Let m be any integer. Since $|A|$ is a finite number, we have,

$$|E| + |A| + \frac{|A|^2}{2!} + \frac{|A|^3}{3!} + \cdots + \frac{|A|^m}{m!} = n + |A| + \frac{|A|^2}{2!} + \frac{|A|^3}{3!} + \cdots + \frac{|A|^m}{m!}$$
$$\leq (n-1) + e^{|A|} < \infty, \qquad (2.22)$$

which implies that the series in (2.21) converges absolutely. This completes the proof. (Cauchy criterion can also be used to prove the result.) ♠

Definition 3.2.13 *Based on Lemma 3.2.12, we define the limit in (2.21) as e^A, also written as $\exp(A)$. That is,*

$$e^A = \exp(A) = \lim_{m \to \infty} \left[E + A + \frac{A^2}{2!} + \frac{A^3}{3!} + \cdots + \frac{A^m}{m!} \right] = \sum_{k=0}^{\infty} \frac{A^k}{k!}. \qquad (2.23)$$

Now we know how to put a matrix into the exponential function, thus $\exp\left(\int_{t_0}^{t} A(s)ds \right)$ makes sense. For scalar linear differential equations, $\exp\left(\int_{t_0}^{t} A(s)ds \right)$ is always a fundamental (scalar) solution according to (2.20). Thus, at this stage, we ask the following question:

For $n > 1$, is $\exp\left(\int_{t_0}^{t} A(s)ds \right)$ always a fundamental matrix solution of Eq. (1.10)?

Unfortunately, the answer is NO, as we shall see from the following example.

Example 3.2.14 Consider $x'(t) = A(t)x(t)$, where

$$A(t) = \begin{bmatrix} t & 1 \\ 0 & 0 \end{bmatrix}.$$

We have

$$\exp\left(\int_0^t A(s)ds\right) = \exp\begin{bmatrix} \frac{t^2}{2} & t \\ 0 & 0 \end{bmatrix}$$

$$= \begin{bmatrix} 1 & 0 \\ 0 & 1 \end{bmatrix} + \begin{bmatrix} \frac{t^2}{2} & t \\ 0 & 0 \end{bmatrix} + \frac{1}{2!}\begin{bmatrix} \frac{t^2}{2} & t \\ 0 & 0 \end{bmatrix}^2 + \cdots$$

$$= \begin{bmatrix} 1 & 0 \\ 0 & 1 \end{bmatrix} + \begin{bmatrix} \frac{t^2}{2} & t \\ 0 & 0 \end{bmatrix} + \frac{1}{2!}\begin{bmatrix} (\frac{t^2}{2})^2 & \frac{t^2}{2}t \\ 0 & 0 \end{bmatrix} + \cdots. \qquad (2.24)$$

Now, the $(1,2)$ position (the first row and the second column) of the matrix (2.24) is

$$t + \frac{1}{2!}\frac{t^2}{2}t + \frac{1}{3!}(\frac{t^2}{2})^2 t + \frac{1}{4!}(\frac{t^2}{2})^3 t + \cdots$$

$$= \frac{2}{t}[\frac{t^2}{2} + \frac{1}{2!}(\frac{t^2}{2})^2 + \frac{1}{3!}(\frac{t^2}{2})^3 + \frac{1}{4!}(\frac{t^2}{2})^4 + \cdots]$$

$$= \frac{2}{t}[e^{t^2/2} - 1], \qquad (2.25)$$

therefore, the matrix (2.24) becomes

$$\exp\left(\int_0^t A(s)ds\right) = \begin{bmatrix} e^{t^2/2} & \frac{2}{t}[e^{t^2/2} - 1] \\ 0 & 1 \end{bmatrix}. \qquad (2.26)$$

If the matrix given in (2.26) is a matrix solution of $x' = A(t)x$, then $g(t) = \frac{2}{t}[e^{t^2/2} - 1]$ on the $(1,2)$ position shall satisfy

$$g'(t) = [t, 1] \times \begin{bmatrix} g(t) \\ 1 \end{bmatrix} = tg(t) + 1. \qquad (2.27)$$

However,

$$g'(t) = \frac{d}{dt}\left(\frac{2}{t}[e^{t^2/2} - 1]\right) = \frac{-2}{t^2}[e^{t^2/2} - 1] + 2e^{t^2/2}, \qquad (2.28)$$

but

$$tg(t) + 1 = t\frac{2}{t}[e^{t^2/2} - 1] + 1 = 2[e^{t^2/2} - 1] + 1 = 2e^{t^2/2} - 1, \qquad (2.29)$$

thus (2.27) is not satisfied, therefore, we conclude that the matrix $\exp\left(\int_0^t A(s)ds\right)$ is not a matrix solution of $x' = A(t)x$. That is, for this $A(t)$,

$$\frac{d}{dt}\exp\left(\int_0^t A(s)ds\right) \neq A(t)\left[\exp\left(\int_0^t A(s)ds\right)\right]. \qquad \spadesuit \qquad (2.30)$$

In view of Example 3.2.14, let's define $C(t) = \exp\left(\int_{t_0}^t A(s)ds\right)$ and find the conditions required for $C(t)$ to be a fundamental matrix solution. Suppose for the moment that $C(t)$ is a fundamental matrix solution, that is, $C'(t) = A(t)C(t)$; also suppose for the moment that we require the chain rule to be held for $C(t) = \exp\left(\int_{t_0}^t A(s)ds\right)$ (in fact, it is a complicated subject, see Example 3.2.17), that is, $C'(t) = C(t)A(t)$. Then we would require $A(t)C(t) = C(t)A(t)$, or require $A(t)\left(\int_{t_0}^t A(s)ds\right) = \left(\int_{t_0}^t A(s)ds\right)A(t)$. We will show that under this condition, $\exp\left(\int_{t_0}^t A(s)ds\right)$ is a fundamental matrix solution.

Theorem 3.2.15 *Let Eq. (1.10) be defined on (a,b). If $A(t)$ commutes with its integral, that is, if*

$$A(t)\left(\int_{t_0}^t A(s)ds\right) = \left(\int_{t_0}^t A(s)ds\right)A(t), \quad t_0,\ t \in (a,b), \tag{2.31}$$

then the fundamental matrix solution $U(t,t_0)$ of Eq. (1.10) is given by

$$U(t,t_0) = \exp\left(\int_{t_0}^t A(s)ds\right). \tag{2.32}$$

In this case, the solution of Eq. (1.2) for $t \in (a,b)$ is given by

$$x(t) = \left[\exp\left(\int_{t_0}^t A(s)ds\right)\right]\left[x_0 + \int_{t_0}^t \{\exp\left(\int_{t_0}^s A(r)dr\right)\}^{-1}f(s)ds\right]$$

$$= \left[\exp\left(\int_{t_0}^t A(s)ds\right)\right]x_0 + \int_{t_0}^t\left[\exp\left(\int_s^t A(r)dr\right)\right]f(s)ds. \tag{2.33}$$

Proof. Define $B(t) = \int_{t_0}^t A(s)ds$. We first use an induction to verify that

$$\frac{d}{dt}[B(t)]^m = mA(t)[B(t)]^{m-1}, \quad m = 1, 2, \cdots. \tag{2.34}$$

Formula (2.34) is true when $m = 1$ because we have $\frac{d}{dt}B(t) = A(t)$ from the definition of $B(t)$. Assume that (2.34) is true for $m = k$. Then for $m = k+1$, we have, from the product rule and the fact that $A(t)$ and $B(t)$ commute,

$$\begin{aligned}\frac{d}{dt}[B(t)]^{k+1} &= \frac{d}{dt}\{[B(t)]^k B(t)\}\\ &= \{\frac{d}{dt}[B(t)]^k\}B(t) + [B(t)]^k\frac{d}{dt}B(t)\\ &= kA(t)[B(t)]^{k-1}B(t) + [B(t)]^k A(t)\\ &= kA(t)[B(t)]^k + A(t)[B(t)]^k = (k+1)A(t)[B(t)]^k, \tag{2.35}\end{aligned}$$

which proves (2.34) for $m = k+1$, and hence completes the induction. Next, note that from the proof of Picard's local existence and uniqueness theorem 2.2.7 in Chapter 2, the solution of Eq. (1.10) is the limit of the sequence

$$\begin{cases} x_0(t) & = x_0, \\ x_1(t) & = x_0 + \int_{t_0}^{t} A(s)x_0(s)ds = \left[E + B(t)\right]x_0, \\ x_{m+1}(t) & = x_0 + \int_{t_0}^{t} A(s)x_m(s)ds, \quad m = 1, 2, \cdots. \end{cases} \qquad (2.36)$$

Now, let's use an induction to verify that $x_m(t)$ defined in (2.36) satisfies

$$x_m(t) = \left[E + B(t) + \frac{1}{2!}B^2(t) + \cdots + \frac{1}{m!}B^m(t)\right]x_0, \quad m = 1, 2, \cdots. \quad (2.37)$$

If this can be verified, then taking a limit as $m \to \infty$ in (2.37), we see that the left-hand side of (2.37) goes to the solution of Eq. (1.10), given by $U(t, t_0)x_0$; and the right-hand side of (2.37) goes to $e^{B(t)}x_0$. Thus,

$$U(t, t_0)x_0 = e^{B(t)}x_0 = e^{\int_{t_0}^{t} A(s)ds}x_0,$$

therefore, (2.32) is proved since x_0 is arbitrary.

To verify (2.37), note that (2.37) is true for $m = 1$ from (2.36). Assume that (2.37) is true for $m = k$. Then for $m = k + 1$, we have, from (2.36) and (2.34) (which also says that $\frac{d}{dt}\frac{[B(t)]^m}{m!} = \frac{A(t)[B(t)]^{m-1}}{(m-1)!}$),

$$\begin{aligned} \frac{d}{dt}x_{k+1}(t) & = A(t)x_k(t) \\ & = A(t)\left[E + B(t) + \frac{1}{2!}B^2(t) + \cdots + \frac{1}{k!}B^k(t)\right]x_0 \\ & = \left[A(t) + A(t)B(t) + \frac{1}{2!}A(t)B^2(t) + \cdots + \frac{1}{k!}A(t)B^k(t)\right]x_0 \\ & = \frac{d}{dt}\left[B(t) + \frac{1}{2!}B^2(t) + \cdots + \frac{1}{(k+1)!}B^{k+1}(t)\right]x_0, \qquad (2.38) \end{aligned}$$

then

$$x_{k+1}(t) = H + \left[B(t) + \frac{1}{2!}B^2(t) + \cdots + \frac{1}{(k+1)!}B^{k+1}(t)\right]x_0,$$

for some constant vector H. Since $B(t_0) = 0$, we have

$$H = x_{k+1}(t_0) = x_0,$$

therefore

$$x_{k+1}(t) = x_0 + \left[B(t) + \frac{1}{2!}B^2(t) + \cdots + \frac{1}{(k+1)!}B^{k+1}(t)\right]x_0$$

$$= \left[E + B(t) + \frac{1}{2!}B^2(t) + \cdots + \frac{1}{(k+1)!}B^{k+1}(t)\right]x_0, \quad (2.39)$$

which verifies (2.37) for $m = k + 1$, and hence completes the induction. Therefore, (2.32) is true. Finally, note that (2.33) is from the variation of parameters formula in Theorem 3.2.11. This completes the proof. ♠

Observe that (2.31) is not true in general. That is, if you randomly pick up a matrix $A(t)$, then chances are that (2.31) does not hold for that $A(t)$. Let's look at the following example.

Example 3.2.16 Let

$$A(t) = \begin{bmatrix} 1 & t \\ t^2 & t^3 \end{bmatrix}.$$

Then

$$A(t)\left(\int_0^t A(s)ds\right) = \begin{bmatrix} 1 & t \\ t^2 & t^3 \end{bmatrix}\begin{bmatrix} t & \frac{t^2}{2} \\ \frac{t^3}{3} & \frac{t^4}{4} \end{bmatrix} = \begin{bmatrix} t + \frac{t^4}{3} & \frac{t^2}{2} + \frac{t^5}{4} \\ t^3 + \frac{t^6}{3} & \frac{t^4}{2} + \frac{t^7}{4} \end{bmatrix},$$

but

$$\left(\int_0^t A(s)ds\right)A(t) = \begin{bmatrix} t & \frac{t^2}{2} \\ \frac{t^3}{3} & \frac{t^4}{4} \end{bmatrix}\begin{bmatrix} 1 & t \\ t^2 & t^3 \end{bmatrix} = \begin{bmatrix} t + \frac{t^4}{2} & t^2 + \frac{t^5}{2} \\ \frac{t^3}{3} + \frac{t^6}{4} & \frac{t^4}{3} + \frac{t^7}{4} \end{bmatrix},$$

therefore, (2.31) does not hold for this $A(t)$. ♠

However, if, for example, $A(t)$ is independent of t, that is, $A(t) = A$ for a constant matrix A, then (2.31) is true, because now,

$$A(t)\left(\int_{t_0}^t A(s)ds\right) = A\left(\int_{t_0}^t Ads\right) = (t - t_0)A^2 = \left(\int_{t_0}^t A(s)ds\right)A(t)$$

since $A(t)$ is independent of t. This case will be treated in the next section.

Recall that the **chain rule** for exponential functions says that if $a(t)$ is a scalar differentiable function, then

$$\frac{d}{dt}e^{a(t)} = e^{a(t)}\frac{d}{dt}a(t) = \left[\frac{d}{dt}a(t)\right]e^{a(t)}.$$

Now, we point out that in general, chain rule does not hold for $\exp\left(\int_{t_0}^t A(s)ds\right)$ when $A(t)$ is an $n \times n$ matrix function with $n > 1$.

Example 3.2.17 (Chain rule doesn't hold for $\exp\left(\int_{t_0}^t A(s)ds\right)$ when $n > 1$). In general, for an $n \times n$ matrix function $A(t)$ with $n > 1$,

$$\frac{d}{dt}\exp\left(\int_{t_0}^t A(s)ds\right) \;\neq\; \left[\exp\left(\int_{t_0}^t A(s)ds\right)\right]A(t), \qquad (2.40)$$

$$\frac{d}{dt}\exp\left(\int_{t_0}^t A(s)ds\right) \;\neq\; A(t)\left[\exp\left(\int_{t_0}^t A(s)ds\right)\right]. \qquad (2.41)$$

To see why, we look at Example 3.2.14 again, where $A(t) = \begin{bmatrix} t & 1 \\ 0 & 0 \end{bmatrix}$,

and $\exp\left(\int_0^t A(s)ds\right) = \begin{bmatrix} e^{t^2/2} & \frac{2}{t}[e^{t^2/2} - 1] \\ 0 & 1 \end{bmatrix}$. The $(1,2)$ position of

$\frac{d}{dt}\exp\left(\int_0^t A(s)ds\right)$ is $\frac{d}{dt}\frac{2}{t}[e^{t^2/2} - 1] = \frac{-2}{t^2}[e^{t^2/2} - 1] + 2e^{t^2/2}$; but the $(1,2)$

position of $\left[\exp\left(\int_0^t A(s)ds\right)\right]A(t)$ is $e^{t^2/2}$, thus we get (2.40) in this case. We also get (2.41) in this case because of (2.30) in Example 3.2.14.   ♠

Next, we close this section by showing the following interesting result concerning a general matrix solution (not necessarily the fundamental one).

Theorem 3.2.18 (Liouville's formula) *Let $X(t)$ be an $n \times n$ matrix function such that $X'(t) = A(t)X(t)$, $t \in (a,b)$. Then*

$$\frac{d}{dt}[\det X(t)] = [tr A(t)][\det X(t)], \quad t \in (a,b), \qquad (2.42)$$

and

$$\det X(t) = [\det X(t_0)]\exp\left(\int_{t_0}^t tr A(s)ds\right), \quad t_0, t \in (a,b), \qquad (2.43)$$

where $tr A(t) = \sum_{i=1}^n a_{ii}(t)$ for the $n \times n$ matrix $A(t) = [a_{ij}(t)]$, and $\det$ denotes determinant.

Proof. Let $X(t) = [x_{ij}(t)]$. From the operations of matrices,

$$[\det X(t)]' = \begin{vmatrix} x'_{11} & x'_{12} & . & . & x'_{1n} \\ x_{21} & x_{22} & . & . & x_{2n} \\ . & . & . & . & . \\ . & . & . & . & . \\ x_{n1} & x_{n2} & . & . & x_{nn} \end{vmatrix} + \cdots + \begin{vmatrix} x_{11} & x_{12} & . & . & x_{1n} \\ x_{21} & x_{22} & . & . & x_{2n} \\ . & . & . & . & . \\ . & . & . & . & . \\ x'_{n1} & x'_{n2} & . & . & x'_{nn} \end{vmatrix}. \qquad (2.44)$$

Since $X(t)$ satisfies $X' = A(t)X(t)$, we have

$$x'_{ij}(t) = \sum_{k=1}^{n} a_{ik}(t)x_{kj}(t).$$

Thus, (2.44) becomes

$$
[\det X(t)]'
$$

$$
= \begin{vmatrix} \sum_{k=1}^{n} a_{1k}x_{k1} & \sum_{k=1}^{n} a_{1k}x_{k2} & \cdot & \cdot & \sum_{k=1}^{n} a_{1k}x_{kn} \\ x_{21} & x_{22} & & \cdot & \cdot & x_{2n} \\ & & & \cdot & \cdot & \cdot \\ \cdot & \cdot & \cdot & \cdot & & \cdot \\ x_{n1} & x_{n2} & \cdot & \cdot & x_{nn} \end{vmatrix} + \cdots
$$

$$
+ \begin{vmatrix} x_{11} & x_{12} & \cdot & \cdot & x_{1n} \\ x_{21} & x_{22} & \cdot & \cdot & x_{2n} \\ \cdot & \cdot & \cdot & \cdot & & \cdot \\ \cdot & \cdot & \cdot & \cdot & & \cdot \\ \sum_{k=1}^{n} a_{nk}x_{k1} & \sum_{k=1}^{n} a_{nk}x_{k2} & \cdot & \cdot & \sum_{k=1}^{n} a_{nk}x_{kn} \end{vmatrix}. \qquad (2.45)
$$

In the first determinant on the right-hand side of (2.45), multiply the second row by a_{12} and subtract it from the first row, multiply the third row by a_{13} and subtract it from the first row, $\cdots$, and multiply the last row by a_{1n} and subtract it from the first row. Then the first determinant on the right-hand side of (2.45) now equals

$$
\begin{vmatrix} a_{11}x_{11} & a_{11}x_{12} & \cdot & \cdot & a_{11}x_{1n} \\ x_{21} & x_{22} & \cdot & \cdot & x_{2n} \\ \cdot & \cdot & \cdot & \cdot & & \cdot \\ \cdot & \cdot & \cdot & \cdot & & \cdot \\ x_{n1} & x_{n2} & \cdot & \cdot & x_{nn} \end{vmatrix} = a_{11} \begin{vmatrix} x_{11} & x_{12} & \cdot & \cdot & x_{1n} \\ x_{21} & x_{22} & \cdot & \cdot & x_{2n} \\ \cdot & \cdot & \cdot & \cdot & & \cdot \\ \cdot & \cdot & \cdot & \cdot & & \cdot \\ x_{n1} & x_{n2} & \cdot & \cdot & x_{nn} \end{vmatrix} = a_{11}[\det X(t)].
$$

Similar operations on other determinants on the right-hand side of (2.45) yield

$$
\begin{aligned}
[\det X(t)]' &= a_{11}(t)[\det X(t)] + a_{22}(t)[\det X(t)] + \cdots + a_{nn}(t)[\det X(t)] \\
&= [tr A(t)][\det X(t)], \qquad (2.46)
\end{aligned}
$$

which is (2.42). To derive (2.43), all we need is to observe that (2.42) implies that $\det X(t)$ is a solution of $w' = [tr A(t)]w$. Solving it, we obtain (2.43). This completes the proof. ♠

Theorem 3.2.18 says that for a general matrix solution $X(t)$ of $X'(t) = A(t)X(t)$ on (a, b), $\det X(t)$ is always given as an exponential function, even though the fundamental matrix solution or $X(t)$ itself may not, in general, be given as an exponential matrix function.

Theorem 3.2.18 also says that the linear independence of the vector functions consisting of the columns of the matrix solution $X(t)$ is determined at any one point in (a, b). In fact, we have the following corollary from Theorem 3.2.18.

Corollary 3.2.19 *Let $X(t)$ be an $n \times n$ matrix function such that $X'(t) = A(t)X(t)$, $t \in (a, b)$. Then the following statements are equivalent.*

1. *There exists a $t_0 \in (a, b)$ such that $\det X(t_0) \neq 0$.*

2. *The vector functions consisting of the columns of $X(t)$ are linearly independent on (a, b).*

3. *For any $t \in (a, b)$, $\det X(t) \neq 0$.*

Proof. We have $(3) \Rightarrow (1)$, and the proof of $(1) \Rightarrow (2)$ is left as an exercise. So we only need to prove $(2) \Rightarrow (3)$.

Assume the statement in (2). If the statement in (3) is not true, then there exists a $t_0 \in (a, b)$ such that $\det X(t_0) = 0$. Let $\{x_{(1)}(t), x_{(2)}(t), \cdots, x_{(n)}(t)\}$ be the n columns of the matrix solution $X(t)$. Then $\det X(t_0) = 0$ implies that the n vectors $\{x_{(1)}(t_0), x_{(2)}(t_0), \cdots, x_{(n)}(t_0)\}$ in $\Re^n$ are linearly dependent. Thus there exist constants $c_1, c_2, \cdots, c_n$, not all zero, such that

$$c_1 x_{(1)}(t_0) + c_2 x_{(2)}(t_0) + \cdots + c_n x_{(n)}(t_0) = 0.$$

Now define

$$y(t) = c_1 x_{(1)}(t) + c_2 x_{(2)}(t) + \cdots + c_n x_{(n)}(t), \quad t \in (a, b). \qquad (2.47)$$

Since each $x_{(i)}(t)$ is a solution of $x'(t) = A(t)x(t)$, by the principle of superposition, we find that $y(t)$ is a solution of

$$y'(t) = A(t)y(t), \quad y(t_0) = 0. \qquad (2.48)$$

But the 0 vector in $\Re^n$ is also a solution of (2.48), hence by the uniqueness from Theorem 3.2.2, we get $y(t) = 0$, $t \in (a, b)$, or

$$c_1 x_{(1)}(t) + c_2 x_{(2)}(t) + \cdots + c_n x_{(n)}(t) = 0, \quad t \in (a, b), \qquad (2.49)$$

which means that the vector functions $\{x_{(1)}(t), x_{(2)}(t), \cdots, x_{(n)}(t)\}$ are linearly dependent on (a, b), a contradiction to the statement (2). This completes the proof. ♠

Consequently, for a matrix solution $X(t)$ on an interval (a, b), $\det X(t)$ is either identically zero on (a, b) or never zero on (a, b), that is, the linear independence of the columns of $X(t)$ on (a, b) is determined at any one point in (a, b). Notice that this is only true for a matrix solution because any column of a matrix solution is a solution. This may not be true for an arbitrary matrix function because for a set of functions, we deduce from Definition 3.2.5 that the linear independence at one point t_0 implies the linear independence on any interval containing t_0, the converse, however, is not true: The linear independence on an interval does not imply the linear independence at any point in that interval, see the following example.

Example 3.2.20 The scalar functions $\sin t$ and $\cos t$ are linearly independent on any t interval. But for **any** fixed t_1, $\sin t_1$ and $\cos t_1$ are linearly dependent numbers in $\Re$. Accordingly, $f_1(t) = [\sin t, \sin t]^T$ and $f_2(t) = [\cos t, \cos t]^T$ are linearly independent vector functions on any t interval, but for any fixed t_1, $[\sin t_1, \sin t_1]^T$ and $[\cos t_1, \cos t_1]^T$ are linearly dependent vectors in $\Re^2$. In other words, $[\sin t, \sin t]^T$ and $[\cos t, \cos t]^T$ are linearly independent vector functions on any t interval, but for any t,

$$\det \begin{bmatrix} \sin t & \cos t \\ \sin t & \cos t \end{bmatrix} = 0. \qquad \spadesuit \qquad (2.50)$$

Exercises 3.2

1. Find 2×2 matrices A and B such that $AB \neq BA$.

2. Randomly pick up a matrix $A(t)$ and check if $A(t)\left(\int_{t_0}^t A(s)ds\right) = \left(\int_{t_0}^t A(s)ds\right)A(t)$.

3. Let $A(t)$ be an $n \times n$ matrix function and let B be an $n \times n$ constant matrix, show that $\int_a^b A(t)Bdt = [\int_a^b A(t)dt]B$ and $\int_a^b BA(t)dt = B[\int_a^b A(t)dt]$, provided the integrals exist.

4. Use the definition of e^A to show that for $A = \begin{bmatrix} \lambda & 0 \\ 0 & \mu \end{bmatrix}$, $e^A = \begin{bmatrix} e^\lambda & 0 \\ 0 & e^\mu \end{bmatrix}$.

5. Find $[e^{tA}]^{-1}$.

6. Let A, B, C be $n \times n$ constant matrices. Show that $e^{tA} C e^{tB}$ satisfies the matrix differential equation $X' = AX + XB$.

7. Find the fundamental matrix solution $U(t, t_0)$ for

(a) $A(t) = \begin{bmatrix} t & 1 \\ 0 & 2 \end{bmatrix}$;   (b) $A(t) = \begin{bmatrix} 1 & -2 & 0 \\ 0 & 2 & 3 \\ 0 & 0 & -2 \end{bmatrix}$.

8. Find solutions for the equation $x' = Ax + f(t)$, where

$$A = \begin{bmatrix} 2 & 1 \\ 0 & 2 \end{bmatrix}, \quad f(t) = \begin{bmatrix} 0 \\ e^{-t} \end{bmatrix}.$$

9. Find an $n \times n$ $(n > 1)$ matrix $A(t)$ (other than the one in Example 3.2.14) such that $\exp\left(\int_{t_0}^t A(s)ds\right)$ is not a matrix solution of $x' = A(t)x$. (Recall that it is a solution when $n = 1$.)

10. (a) Show that $x(t) \equiv 0$ is the only solution of $3x''(t) + 5(\sin t)x'(t) - e^t x(t) = 0$ with $x(t_0) = 0$, $x'(t_0) = 0$.

    (b) Show that if $x(t)$ is a solution of $3x''(t) + 5(\sin t)x'(t) - e^t x(t) = 0$ such that $x(t_1) = 0$, $x'(t_1) = 0$ for some $t_1 \in \Re$, then $x(t) \equiv 0$ for all $t \in \Re$.

11. Let $\Phi(t)$ be an $n \times n$ matrix function, and write $\Phi(t) = [\Phi_1(t), \Phi_2(t), \cdots, \Phi_n(t)]$ with $\Phi_i(t)$ a column, $i = 1, 2, \cdots, n$. Show that

    (a) $\Phi'(t) = A(t)\Phi(t)$ if and only if $\Phi_i'(t) = A(t)\Phi_i(t)$, $i = 1, 2, \cdots, n$.

    (b) $\Phi'(t) = A(t)\Phi(t)$ if and only if $\Phi(t) = U(t, t_0)\Phi(t_0)$ for some t_0, where $U(t, t_0)$ is the fundamental matrix solution. In particular, if $\Phi(t_0)$ is nonsingular for some t_0, then $\Phi(t)$ is nonsingular for every t.

12. Let $A(t)$ be continuous. Show that if $x'(t) = A(t)x(t)$, $x(t_0) = x_0$, has a unique solution, then $x'(t) = A(t)x(t) + f(t)$, $x(t_0) = x_0$, also has a unique solution.

13. Let $\phi_i(t)$, $i = 1, 2, \cdots, n$, be n solutions of $x^{(n)}(t) + a_1(t)x^{(n-1)}(t) + \cdots + a_{n-1}(t)x' + a_n(t)x(t) = 0$, where $a_i(t)$, $i = 1, 2, \cdots, n$, are continuous on an interval I, and let c_i, $i = 1, 2, \cdots, n$, be constants. Define $X(t) = c_1\phi_1(t) + c_2\phi_2(t) + \cdots + c_n\phi_n(t)$, and

$$\Phi(t) = c_1 \begin{bmatrix} \phi_1(t) \\ \phi_1'(t) \\ . \\ . \\ . \\ \phi_1^{(n-1)}(t) \end{bmatrix} + c_2 \begin{bmatrix} \phi_2(t) \\ \phi_2'(t) \\ . \\ . \\ . \\ \phi_2^{(n-1)}(t) \end{bmatrix} + \cdots + c_n \begin{bmatrix} \phi_n(t) \\ \phi_n'(t) \\ . \\ . \\ . \\ \phi_n^{(n-1)}(t) \end{bmatrix}.$$

Prove that the following statements are equivalent.

(a) There exists $c \in I$ such that $X(c) = 0$.

(b) $X(t) = 0$ for all $t \in I$.

(c) There exists $c \in I$ such that $\Phi(c) = 0$.

(d) $\Phi(t) = 0$ for all $t \in I$.

14. Let $\phi_i(t)$, $i = 1, 2, \cdots, n$, be n solutions of $x^{(n)}(t) + a_1(t)x^{(n-1)}(t) + \cdots + a_{n-1}(t)x' + a_n(t)x(t) = 0$, where $a_i(t)$, $i = 1, 2, \cdots, n$, are continuous on an interval I, and define the **Wronskian** for ϕ_i, $i = 1, 2, \cdots, n$, as the determinant

$$W(\phi_1, \phi_2, \cdots, \phi_n)(t) = \det \begin{bmatrix} \phi_1(t) & . & . & . & \phi_n(t) \\ \phi_1'(t) & . & . & . & \phi_n'(t) \\ . & & . & . & . \\ . & . & . & & . \\ \phi_1^{(n-1)}(t) & . & . & . & \phi_n^{(n-1)}(t) \end{bmatrix}.$$

Prove that the following statements are equivalent.

(a) There exists $t_0 \in I$ such that $W(\phi_1, \phi_2, \cdots, \phi_n)(t_0) \neq 0$.

(b) $\phi_1, \phi_2, \cdots, \phi_n$ are linearly independent on the interval I (there do not exist constants c_i, $i = 1, 2, \cdots, n$, not all zero, such that $c_1\phi_1(t) + c_2\phi_2(t) + \cdots + c_n\phi_n(t) = 0$ for all $t \in I$).

(c) For any $t \in I$, $W(\phi_1, \phi_2, \cdots, \phi_n)(t) \neq 0$.

15. Let I be any interval containing 0. Prove that $\phi_1(t) = t^5$, $\phi_2(t) = t^7$ cannot be solutions of $x'' + a_1(t)x' + a_2(t)x = 0$ with $a_i(t)$ continuous on I.

16. For $x'(t) = A(t)x(t)$, prove that for the fundamental matrix solution $U(t, t_0)$,

$$\det U(t, t_0) = \exp\left(\int_{t_0}^{t} tr\, A(s)ds\right).$$

17. Verify, using Definition 3.2.5, that for a set of vector functions, linear independence at one point t_0 implies linear independence on any interval containing t_0.

18. Prove that $(1) \Rightarrow (2)$ in Corollary 3.2.19.

19. Let $X(t)$ be an $n \times n$ matrix function such that $X'(t) = A(t)X(t)$, $t \in (a, b)$. Prove that the following statements are equivalent.

  (a) There exists $t_0 \in (a, b)$ such that $\det X(t_0) = 0$.

  (b) The vector functions consisting of the columns of $X(t)$ are linearly dependent on (a, b).

  (c) For any $t \in (a, b)$, $\det X(t) = 0$.

3.3   Linear Equations with Constant Coefficients

In this section, we assume that $A(t)$ is independent of t, that is,

$$A(t) = A$$

is a constant matrix for t in an interval (a, b). Now, the matrix A commutes with its integral, so Theorem 3.2.15 can be applied to express the fundamental matrix solution as an exponential,

$$U(t, t_0) = \exp\left(\int_{t_0}^{t} A(s)ds\right) = e^{(t-t_0)A}. \tag{3.1}$$

To simplify the notation, we assume that $t_0 = 0$ in the rest of this section, hence Eq. (1.2) now becomes

$$x'(t) = Ax(t) + f(t), \quad x(0) = x_0, \tag{3.2}$$

with the unique solution given by

$$
\begin{aligned}
x(t) &= e^{tA}\left[x_0 + \int_0^t \{e^{sA}\}^{-1}f(s)ds\right] \\
&= e^{tA}x_0 + \int_0^t e^{(t-s)A}f(s)ds.
\end{aligned}
\tag{3.3}
$$

The corresponding homogeneous linear differential equation becomes

$$
x'(t) = Ax(t), \quad x(0) = x_0,
\tag{3.4}
$$

with the unique solution given by

$$
x(t) = e^{tA}x_0.
\tag{3.5}
$$

In the following, we take advantage of the fact that the matrix A is constant, and analyze the properties of the fundamental matrix solution e^{tA} so as to get a better understanding of the structure of solutions of Eq. (3.4) and Eq. (3.2).

We will do two things here. First, examine the theoretical structure of e^{tA}, which will provide us with a basis for long-term behavior of solutions of Eq. (3.2). Second, derive an algorithm that can be used to actually compute e^{tA} and solve Eq. (3.2).

To begin, we observe that if A is a diagonal matrix or a triangular matrix, then the solution of Eq. (3.4) can be easily computed. For example, if

$$
A = diag(\lambda_1, \lambda_2, \cdots, \lambda_n) =
\begin{bmatrix}
\lambda_1 & 0 & 0 & \cdots & 0 & 0 \\
0 & \lambda_2 & 0 & \cdots & 0 & 0 \\
0 & 0 & \lambda_3 & \cdots & 0 & 0 \\
\cdot & \cdot & & & 0 & 0 \\
\cdot & \cdot & & & 0 & 0 \\
\cdot & \cdot & & & 0 & 0 \\
0 & 0 & 0 & \cdots & \lambda_{n-1} & 0 \\
0 & 0 & 0 & \cdots & 0 & \lambda_n
\end{bmatrix},
$$

then an inspection shows that the fundamental matrix solution e^{tA} is given by

$$e^{tA} = diag(e^{\lambda_1 t}, e^{\lambda_2 t}, \cdots, e^{\lambda_n t}) = \begin{bmatrix} e^{\lambda_1 t} & 0 & 0 & . & . & . & 0 & 0 \\ 0 & e^{\lambda_2 t} & 0 & . & . & . & 0 & 0 \\ 0 & 0 & e^{\lambda_3 t} & . & . & . & 0 & 0 \\ . & . & . & . & . & . & 0 & 0 \\ . & . & . & . & . & . & 0 & 0 \\ . & . & . & . & . & . & 0 & 0 \\ 0 & 0 & 0 & . & . & . & e^{\lambda_{n-1} t} & 0 \\ 0 & 0 & 0 & . & . & . & 0 & e^{\lambda_n t} \end{bmatrix}.$$

Hence the solution of Eq. (3.4) is given by

$$x(t) = [x_{01}e^{\lambda_1 t}, \; x_{02}e^{\lambda_2 t}, \cdots, x_{0n}e^{\lambda_n t}]^T,$$

where $[x_{01}, x_{02}, \cdots, x_{0n}]^T = x_0$. Another way to look at this situation is that in this case, we actually have n uncoupled homogeneous scalar linear differential equations, and each has the form of $y' = \lambda_i y$, $y(0) = x_{0i}$, thus the solution is given by $x_{0i}e^{\lambda_i t}$.

If the matrix A is, say for example, lower triangular,

$$A = \begin{bmatrix} a & 0 & 0 & . & . & . & 0 & 0 \\ b & c & 0 & . & . & . & 0 & 0 \\ d & e & f & . & . & . & 0 & 0 \\ . & . & . & . & . & . & 0 & 0 \\ . & . & . & . & . & . & 0 & 0 \\ . & . & . & . & . & . & 0 & 0 \\ . & . & . & . & . & . & . & 0 \\ . & . & . & . & . & . & . & . \end{bmatrix},$$

then we can solve $x_1(t)$ first from $\{x_1'(t) = ax_1(t), \; x_1(0) = x_{01}\}$ to get $x_1(t) = x_{01}e^{at}$. For $x_2(t)$, we have $x_2'(t) = cx_2(t) + bx_1(t) = cx_2(t) + bx_{01}e^{at}$. Then we can compute $x_2(t)$ using the variation of parameters formula (3.3) for scalar equations. Continuing in this way, we can compute $x_3(t), \cdots, x_n(t)$. For an upper triangular matrix, we can compute $x_n(t)$ first, and then $x_{n-1}(t), \cdots, x_1(t)$.

For a general square matrix A, there is a result saying that A can be transformed into a square matrix that is **almost** diagonal. To understand this, let's recall something from linear algebra.

Definition 3.3.1 *Let B be an $n \times n$ constant matrix. The roots (may be complex valued) of*

$$\det(B - \lambda E) = 0 \qquad (3.6)$$

*(where E is the $n \times n$ unit or identity matrix) are called the **eigenvalues** of B. An eigenvalue λ_0 is of multiplicity m if m is the largest integer such that $\det(B - \lambda E) = (\lambda - \lambda_0)^m P(\lambda)$ for some polynomial $P(\lambda)$. A vector $x \neq 0$ (may be complex valued) satisfies $(B - \lambda E)x = 0$ is called an **eigenvector** corresponding to the eigenvalue λ.*

Definition 3.3.2 *Two $n \times n$ matrices A and B are said to be **similar** if there exists a nonsingular $n \times n$ matrix C such that $C^{-1}AC = B$.*

Theorem 3.3.3 *If the $n \times n$ matrices A and B are similar, then A and B have the same eigenvalues.*

Proof. See the Appendix. ♠

Note, for a triangular matrix, the eigenvalues are the numbers on the (main) diagonal. In general, we have the following result which transforms any square matrix into an almost diagonal matrix. The proof can be found in some texts on linear algebra, such as Halmos [1958].

Theorem 3.3.4 (Jordan canonical form theorem) *For any $n \times n$ constant matrix A (may be complex valued), there exists a nonsingular constant matrix P (may be complex valued) such that*

$$P^{-1}AP = J, \qquad (3.7)$$

here the matrix J (may be complex valued) has the form

$$J = \begin{bmatrix} J_1 & 0 & 0 & . & . & . & 0 & 0 \\ 0 & J_2 & 0 & . & . & . & 0 & 0 \\ 0 & 0 & J_3 & . & . & . & 0 & 0 \\ . & . & . & . & . & . & 0 & 0 \\ . & . & . & . & . & . & 0 & 0 \\ . & . & . & . & . & . & 0 & 0 \\ 0 & 0 & 0 & . & . & . & J_{s-1} & 0 \\ 0 & 0 & 0 & . & . & . & 0 & J_s \end{bmatrix}, \qquad (3.8)$$

where the zero matrices 0 *may not be square matrices, but each* J_i, $i =$ 1, 2, $\cdots$, *s*, *is a square matrix, and has the form*

$$J_i = \begin{bmatrix} \lambda_i & 1 & 0 & . & . & . & 0 & 0 \\ 0 & \lambda_i & 1 & . & . & . & 0 & 0 \\ 0 & 0 & \lambda_i & . & . & . & 0 & 0 \\ . & . & . & . & . & . & 0 & 0 \\ . & . & . & . & . & . & 0 & 0 \\ . & . & . & . & . & . & 1 & 0 \\ 0 & 0 & 0 & . & . & . & \lambda_i & 1 \\ 0 & 0 & 0 & . & . & . & 0 & \lambda_i \end{bmatrix}, \tag{3.9}$$

and all the eigenvalues of A *are given by* λ_i, $i = 1, 2, \cdots, s$, *(not necessarily distinct). Matrix* J_i *may be a* 1×1 *matrix. Except for the order in which the matrices* J_i, $i = 1, 2, \cdots, s$, *appear on the diagonal, the matrix* J *is unique.*
♠

The most important aspect of Theorem 3.3.4 is that the matrix A is similar to the "almost" diagonal matrix J, whose special format can be exploited to help us understand the structure of e^{tA}.

However, we point out that the weakness of this theorem is that the matrix P in the theorem is very hard to construct in general. Only in some special cases, such as when we have n linearly independent eigenvectors for the $n \times n$ matrix A, is P easy to construct. We verify this in the following.

Proposition 3.3.5 *If an* $n \times n$ *matrix* A *has* n *linearly independent eigenvectors* $v_1, v_2, \cdots, v_n$, *then the matrix* P *in the Jordan transformation can be taken as* $P = [v_1, v_2, \cdots, v_n]$.

Proof. For the eigenvectors $v_1, v_2, \cdots, v_n$, let $\lambda_1, \lambda_2, \cdots, \lambda_n$ be the corresponding eigenvalues (may not be distinct). Then

$$\begin{aligned} P^{-1}AP &= P^{-1}A[v_1, v_2, \cdots, v_n] = P^{-1}[Av_1, Av_2, \cdots, Av_n] \\ &= P^{-1}[\lambda_1 v_1, \lambda_2 v_2, \cdots, \lambda_n v_n] = [\lambda_1 P^{-1}v_1, \lambda_2 P^{-1}v_2, \cdots, \lambda_n P^{-1}v_n] \\ &= [\lambda_1 e_1, \lambda_2 e_2, \cdots, \lambda_n e_n] = diag(\lambda_1, \lambda_2, \cdots, \lambda_n), \end{aligned}$$

where $[e_1, e_2, \cdots, e_n]$ is the $n \times n$ unit matrix. This completes the proof. ♠

The following result from linear algebra can be used to check linear independency of eigenvectors.

Proposition 3.3.6 *If the eigenvalues $\lambda_1, \lambda_2, \cdots, \lambda_n$ of an $n \times n$ matrix A are distinct, then the corresponding eigenvectors $v_1, v_2, \cdots, v_n$ are linearly independent.*

Proof. See the Appendix. ♠

From Propositions 3.3.5 and 3.3.6, we have

Corollary 3.3.7 *If an $n \times n$ matrix A has n distinct eigenvalues, then the matrix P in the Jordan transformation can be taken as $P = [v_1, v_2, \cdots, v_n]$, where $v_1, v_2, \cdots, v_n$ are the corresponding eigenvectors.* ♠

Now we use some examples to illustrate how to transform a matrix A into J.

Example 3.3.8 Consider the matrix

$$A = \begin{bmatrix} 1 & 6 \\ 5 & 2 \end{bmatrix}.$$

The eigenvalues of A are $\lambda_1 = -4$ and $\lambda_2 = 7$, and eigenvectors are $v_1 = [6, -5]^T$ and $v_2 = [1, 1]^T$, respectively. Then we can form $P = [v_1, v_2]$ to obtain

$$P^{-1}AP = \frac{1}{11} \begin{bmatrix} 1 & -1 \\ 5 & 6 \end{bmatrix} \begin{bmatrix} 1 & 6 \\ 5 & 2 \end{bmatrix} \begin{bmatrix} 6 & 1 \\ -5 & 1 \end{bmatrix} = \begin{bmatrix} -4 & 0 \\ 0 & 7 \end{bmatrix},$$

in which case, $J = \begin{bmatrix} J_1 & 0 \\ 0 & J_2 \end{bmatrix}$, where $J_1 = -4$ and $J_2 = 7$ are 1×1 matrices. ♠

Example 3.3.9 Consider the matrix

$$A = \begin{bmatrix} 7 & -2 \\ 2 & 3 \end{bmatrix}.$$

In this case, 5 is the only eigenvalue with multiplicity 2, and $u_1 = [1, 1]^T$ is an eigenvector. Now, we try $u_2 = [a, b]^T$, $a \neq b$, so that u_1 and u_2 are linearly independent. We can play with

$$P = [u_1, u_2] = \begin{bmatrix} 1 & a \\ 1 & b \end{bmatrix}, \quad (a \neq b)$$

and then find that we can choose $a = \frac{1}{2}$, $b = 0$ to obtain

$$P^{-1}AP = -2 \begin{bmatrix} 0 & -\frac{1}{2} \\ -1 & 1 \end{bmatrix} \begin{bmatrix} 7 & -2 \\ 2 & 3 \end{bmatrix} \begin{bmatrix} 1 & \frac{1}{2} \\ 1 & 0 \end{bmatrix} = \begin{bmatrix} 5 & 1 \\ 0 & 5 \end{bmatrix},$$

in which case, $J = J_1 = \begin{bmatrix} 5 & 1 \\ 0 & 5 \end{bmatrix}$. ♠

Next, let's take advantage of the special form of J and use it to simplify e^{tA}. From $P^{-1}AP = J$ in Theorem 3.3.4, we have $A = PJP^{-1}$. Then,

$$\begin{aligned} e^{tA} &= e^{tPJP^{-1}} \\ &= E + tPJP^{-1} + \frac{[tPJP^{-1}]^2}{2!} + \frac{[tPJP^{-1}]^3}{3!} + \cdots \\ &= E + tPJP^{-1} + \frac{P[tJ]^2P^{-1}}{2!} + \frac{P[tJ]^3P^{-1}}{3!} + \cdots \\ &= P\{E + tJ + \frac{[tJ]^2}{2!} + \frac{[tJ]^3}{3!} + \cdots\}P^{-1} \\ &= Pe^{tJ}P^{-1}. \end{aligned} \qquad (3.10)$$

In order to understand the structure of the fundamental matrix solution e^{tA}, we must understand the structure of e^{tJ}. From (3.8), we have

$$e^{tJ} = E + tJ + \frac{[tJ]^2}{2!} + \frac{[tJ]^3}{3!} + \cdots$$

$$= E + t \begin{bmatrix} J_1 & 0 & 0 & \cdot & \cdot & \cdot & 0 & 0 \\ 0 & J_2 & 0 & \cdot & \cdot & \cdot & 0 & 0 \\ 0 & 0 & J_3 & \cdot & \cdot & \cdot & 0 & 0 \\ \cdot & \cdot & \cdot & \cdot & \cdot & \cdot & 0 & 0 \\ \cdot & \cdot & \cdot & \cdot & \cdot & \cdot & 0 & 0 \\ \cdot & \cdot & \cdot & \cdot & \cdot & \cdot & 0 & 0 \\ 0 & 0 & 0 & \cdot & \cdot & \cdot & J_{s-1} & 0 \\ 0 & 0 & 0 & \cdot & \cdot & \cdot & 0 & J_s \end{bmatrix} + \frac{t^2}{2!}J^2 + \cdots$$

$$= E + \begin{bmatrix} tJ_1 & 0 & 0 & \cdot & \cdot & \cdot & 0 & 0 \\ 0 & tJ_2 & 0 & \cdot & \cdot & \cdot & 0 & 0 \\ 0 & 0 & tJ_3 & \cdot & \cdot & \cdot & 0 & 0 \\ \cdot & \cdot & \cdot & \cdot & \cdot & \cdot & 0 & 0 \\ \cdot & \cdot & \cdot & \cdot & \cdot & \cdot & 0 & 0 \\ \cdot & \cdot & \cdot & \cdot & \cdot & \cdot & 0 & 0 \\ 0 & 0 & 0 & \cdot & \cdot & \cdot & tJ_{s-1} & 0 \\ 0 & 0 & 0 & \cdot & \cdot & \cdot & 0 & tJ_s \end{bmatrix}$$

$$+ \begin{bmatrix} \frac{t^2}{2!}J_1^2 & 0 & 0 & . & . & . & 0 & 0 \\ 0 & \frac{t^2}{2!}J_2^2 & 0 & . & . & . & 0 & 0 \\ 0 & 0 & \frac{t^2}{2!}J_3^2 & . & . & . & 0 & 0 \\ . & . & . & . & . & . & 0 & 0 \\ . & . & . & . & . & . & 0 & 0 \\ . & . & . & . & . & . & 0 & 0 \\ 0 & 0 & 0 & . & . & . & \frac{t^2}{2!}J_{s-1}^2 & 0 \\ 0 & 0 & 0 & . & . & . & 0 & \frac{t^2}{2!}J_s^2 \end{bmatrix} + \cdots$$

$$= \begin{bmatrix} e^{tJ_1} & 0 & 0 & . & . & . & 0 & 0 \\ 0 & e^{tJ_2} & 0 & . & . & . & 0 & 0 \\ 0 & 0 & e^{tJ_3} & . & . & . & 0 & 0 \\ . & . & . & . & . & . & 0 & 0 \\ . & . & . & . & . & . & 0 & 0 \\ . & . & . & . & . & . & 0 & 0 \\ 0 & 0 & 0 & . & . & . & e^{tJ_{s-1}} & 0 \\ 0 & 0 & 0 & . & . & . & 0 & e^{tJ_s} \end{bmatrix}. \tag{3.11}$$

Hence, we must understand e^{tJ_i}, $i = 1, 2, \cdots, s$, in order to understand the fundamental matrix solution e^{tA}. Since J_i is given by (3.9), we have

$$e^{tJ_i} = e^{t(\lambda_i E + N_i)}, \tag{3.12}$$

where E is the unit or identity matrix of the same size of the matrix J_i, and

$$N_i = \begin{bmatrix} 0 & 1 & 0 & . & . & . & 0 & 0 \\ 0 & 0 & 1 & . & . & . & 0 & 0 \\ 0 & 0 & 0 & . & . & . & 0 & 0 \\ . & . & . & . & . & . & 0 & 0 \\ . & . & . & . & . & . & 0 & 0 \\ . & . & . & . & . & . & 1 & 0 \\ 0 & 0 & 0 & . & . & . & 0 & 1 \\ 0 & 0 & 0 & . & . & . & 0 & 0 \end{bmatrix}. \tag{3.13}$$

Now, we need the following result to help us simplify $e^{t(\lambda_i E + N_i)}$. Although the result looks like something in linear algebra, the idea from differential equations can be used to prove it.

Lemma 3.3.10 *If the constant square matrices B and C commute, that is, $BC = CB$, then*

$$Ce^{tB} = e^{tB}C, \quad e^{tB+tC} = e^{tB}e^{tC}, \quad t \in \Re. \tag{3.14}$$

Proof. From Theorem 3.2.15, it follows that if Q is a constant square matrix, then e^{tQ} is a fundamental matrix solution that satisfies

$$(e^{tQ})' = Qe^{tQ}. \tag{3.15}$$

Let $Z(t) = Ce^{tB} - e^{tB}C$. Then,

$$Z'(t) = CBe^{tB} - Be^{tB}C = B[Ce^{tB} - e^{tB}C] = BZ(t), \quad Z(0) = C - C = 0.$$

Hence from uniqueness, $Z(t) = 0$, $t \in \Re$. This shows that

$$Ce^{tB} = e^{tB}C, \quad t \in \Re. \tag{3.16}$$

Next, let $W(t) = e^{t(B+C)}$ and $Y(t) = e^{tB}e^{tC}$. Then we have $W(0) = E$, $Y(0) = E$. Now, treat $(B+C)$ as Q in (3.15), we obtain

$$W'(t) = \frac{d}{dt}e^{t(B+C)} = (B+C)e^{t(B+C)} = (B+C)W(t). \tag{3.17}$$

Also, from (3.16) and the product rule, we have

$$\begin{aligned} Y'(t) &= Be^{tB}e^{tC} + e^{tB}Ce^{tC} \\ &= Be^{tB}e^{tC} + Ce^{tB}e^{tC} = (B+C)e^{tB}e^{tC} \\ &= (B+C)Y(t). \end{aligned} \tag{3.18}$$

Therefore, by applying uniqueness to (3.17) and (3.18) with $W(0) = E$ and $Y(0) = E$, we obtain $W(t) = Y(t)$, that is, $e^{tB+tC} = e^{tB}e^{tC}$. This completes the proof. ♠

Now, we can apply Lemma 3.3.10 to verify that

$$e^{tJ_i} = e^{t(\lambda_i E + N_i)} = e^{t\lambda_i E}e^{tN_i} = (e^{\lambda_i t}E)e^{tN_i} = (e^{\lambda_i t})e^{tN_i}. \tag{3.19}$$

Therefore, according to (3.10), (3.11), and (3.19), we must examine the structure of e^{tN_i} in order to understand the fundamental matrix solution e^{tA}. Assume that the size of N_i is $m_i \times m_i$, and observe that N_i is nilpotent, that is, for $h \geq m_i$, $N_i^h = 0$, (see an exercise). Therefore,

$$e^{tN_i} = E + tN_i + \frac{1}{2!}[tN_i]^2 + \frac{1}{3!}[tN_i]^3 + \cdots + \frac{1}{(m_i - 1)!}[tN_i]^{m_i - 1}$$

$$= E + \begin{bmatrix} 0 & t & 0 & . & . & . & 0 & 0 \\ 0 & 0 & t & . & . & . & 0 & 0 \\ 0 & 0 & 0 & . & . & . & 0 & 0 \\ . & & . & . & . & . & 0 & 0 \\ . & . & . & . & . & . & 0 & 0 \\ . & . & . & . & . & . & t & 0 \\ 0 & 0 & 0 & . & . & . & 0 & t \\ 0 & 0 & 0 & . & . & . & 0 & 0 \end{bmatrix} + \frac{1}{2!} \begin{bmatrix} 0 & t & 0 & . & . & . & 0 & 0 \\ 0 & 0 & t & . & . & . & 0 & 0 \\ 0 & 0 & 0 & . & . & . & 0 & 0 \\ . & . & . & . & . & . & 0 & 0 \\ . & . & . & . & . & . & 0 & 0 \\ . & . & . & . & . & . & t & 0 \\ 0 & 0 & 0 & . & . & . & 0 & t \\ 0 & 0 & 0 & . & . & . & 0 & 0 \end{bmatrix}^2$$

$$+ \cdots + \frac{1}{(m_i - 1)!} \begin{bmatrix} 0 & t & 0 & . & . & . & 0 & 0 \\ 0 & 0 & t & . & . & . & 0 & 0 \\ 0 & 0 & 0 & . & . & . & 0 & 0 \\ . & . & . & . & . & . & 0 & 0 \\ . & . & . & . & . & . & 0 & 0 \\ . & . & . & . & . & . & t & 0 \\ 0 & 0 & 0 & . & . & . & 0 & t \\ 0 & 0 & 0 & . & . & . & 0 & 0 \end{bmatrix}^{m_i - 1}$$

$$= \begin{bmatrix} 1 & t & \frac{t^2}{2!} & \frac{t^3}{3!} & . & . & . & \frac{t^{m_i-1}}{(m_i-1)!} \\ 0 & 1 & t & \frac{t^2}{2!} & . & . & . & . \\ 0 & 0 & 1 & . & . & . & . & . \\ . & . & . & . & . & . & . & . \\ . & . & . & . & . & . & . & . \\ . & . & . & . & . & \frac{t^2}{2!} & & \frac{t^3}{3!} \\ . & . & . & . & . & & t & \frac{t^2}{2!} \\ 0 & 0 & 0 & . & . & . & 1 & t \\ 0 & 0 & 0 & . & . & . & 0 & 1 \end{bmatrix}_{m_i \times m_i} \tag{3.20}$$

Therefore, we have, from (3.19) and (3.20),

$$e^{tJ_i} = (e^{\lambda_i t}) \begin{bmatrix} 1 & t & \frac{t^2}{2!} & \frac{t^3}{3!} & . & . & . & \frac{t^{m_i-1}}{(m_i-1)!} \\ 0 & 1 & t & \frac{t^2}{2!} & . & . & . & . \\ 0 & 0 & 1 & . & . & . & . & . \\ . & . & . & . & . & . & . & . \\ . & . & . & . & . & . & . & . \\ . & . & . & . & . & \frac{t^2}{2!} & & \frac{t^3}{3!} \\ . & . & . & . & . & & t & \frac{t^2}{2!} \\ 0 & 0 & 0 & . & . & . & 1 & t \\ 0 & 0 & 0 & . & . & . & 0 & 1 \end{bmatrix}_{m_i \times m_i} \tag{3.21}$$

Formulas (3.10), (3.11), and (3.21) together give **a clear picture of the fundamental matrix solution** e^{tA}:

> e^{tA} looks like, subject to a transformation $e^{tA} = Pe^{tJ}P^{-1}$, a matrix with some square matrices on the diagonal and with other parts being zero; and for any square matrix on the diagonal, each entry is either 0, or a product of $e^{\lambda_i t}$ and a polynomial in t, where λ_i is an eigenvalue of the matrix A. Here P, J may be complex matrices, but e^{tA} is a real matrix since A in this chapter is real.

With this structure, we find that long-term property of e^{tA} is determined by the eigenvalues of matrix A. In fact, we have the following result concerning long-term property of e^{tA}.

Theorem 3.3.11 *Let λ be a complex number and denote $R(\lambda)$ the real part of λ. Then,*

(A). $|e^{tA}| \leq C_0$, $0 \leq t < \infty$, *for some positive constant C_0 if and only if the following is true: for each eigenvalue λ of the matrix A, either $R(\lambda) < 0$, or $R(\lambda) = 0$ but in this case λ appears only in matrices J_i (in the Jordan canonical form for A) such that J_i is a 1×1 matrix.*

(B). $\lim_{t \to \infty} |e^{tA}| = 0$ *if and only if each eigenvalue of the matrix A has a negative real part.*

(C). $\lim_{t \to \infty} |e^{tA}| = \infty$ *if and only if either there is an eigenvalue λ of the matrix A with $R(\lambda) = 0$ and λ appears in a matrix J_i that is at least 2×2, or there is an eigenvalue of the matrix A with a positive real part.*

Proof. Note first that $|e^{tA}| \leq |P||e^{tJ}||P^{-1}|$ and $|e^{tJ}| \leq |P^{-1}||e^{tA}||P|$ with $|P|$ and $|P^{-1}|$ fixed constants, then from (3.11) and (3.21), we only need to consider each $|e^{tJ_i}|$, $i = 1, 2, \cdots, s$. Observe that

$$|e^{\lambda_i t}| = e^{R(\lambda_i)t}, \quad t \in \Re,$$

then each $|e^{tJ_i}|$ is given as

$$|e^{tJ_i}| = e^{R(\lambda_i)t} M_i(t), \quad t \geq 0,$$

where $M_i(t)$ is a polynomial in t of order less than n, and $M_i(t) \geq 1$ for $t \geq 0$ because the size of e^{tJ_i} is at least 1×1.

If $R(\lambda_i) < 0$, then l'Hôpital's rule can be applied to verify that

$$|e^{tJ_i}| = e^{R(\lambda_i)t} M_i(t) \to 0, \ t \to \infty.$$

If $R(\lambda_i) > 0$, then, since $|M_i(t)| \geq 1$, we have

$$|e^{tJ_i}| = e^{R(\lambda_i)t} M_i(t) \to \infty, \ t \to \infty.$$

If $R(\lambda_i) = 0$ and if the size of J_i is 1×1, then

$$|e^{tJ_i}| = e^{R(\lambda_i)t} M_i(t) = e^{0t} 1 = 1.$$

Otherwise, if $R(\lambda_i) = 0$ and if the size of J_i is at least 2×2, then

$$|e^{tJ_i}| = e^{R(\lambda_i)t} M_i(t) \geq e^{0t}(2+t) = (2+t) \to \infty, \ t \to \infty.$$

Now, the results in (A), (B), and (C) will follow from the cases analyzed above for $R(\lambda_i)$ (see an exercise), hence the proof is complete. ♠

Translating Theorem 3.3.11 in terms of the corresponding solutions of Eq. (3.4), we have

Theorem 3.3.12 *Consider Eq. (3.4). Then,*

(A). Every solution of Eq. (3.4) satisfies $|x(t, 0, x_0)| \leq C_1|x_0|$, $0 \leq t < \infty$, for some positive constant C_1 if and only if the following is true: for each eigenvalue λ of the matrix A, either $R(\lambda) < 0$, or $R(\lambda) = 0$ but in this case λ appears only in matrices J_i such that J_i is a 1×1 matrix.

(B). Every solution of Eq. (3.4) satisfies $\lim_{t \to \infty} |x(t, 0, x_0)| = 0$ if and only if each eigenvalue of the matrix A has a negative real part.

(C). There is a solution x of Eq. (3.4) with $\lim_{t \to \infty} |x(t, 0, x_0)| = \infty$ if and only if either there is an eigenvalue λ of the matrix A with $R(\lambda) = 0$ and λ appears in a matrix J_i that is at least 2×2, or there is an eigenvalue of the matrix A with a positive real part.

Proof. Observe that the solutions of Eq. (3.4) are given by $x(t, 0, x_0) = e^{tA}x_0$, so we can use Theorem 3.3.11 to prove the result. For (A) and (B), we can choose $x_0 = e_i = [0, \cdots, 0, 1_i, 0, \cdots, 0]^T$, $i = 1, 2, \cdots, n$, and use Theorem 3.3.11. For (C): If Eq. (3.4) has a solution that goes to $\pm\infty$, then e^{tA} has an entry that goes to $\pm\infty$ (otherwise, no solutions can go to

$\pm\infty$). Thus, $|e^{tA}|$ goes to ∞, hence Theorem 3.3.11 can be applied. On the other hand, if $|e^{tA}|$ goes to ∞, then e^{tA} has an entry, say for example $q(t)$ on the position $(1,2)$ that goes to $\pm\infty$. In this case, we choose $x_0 = e_2 = [0,1,0,\cdots,0]^T$, such that the corresponding solution satisfies $|x(t,0,x_0)| = |e^{tA}e_2| \geq |q(t)| \to \infty$, $t \to \infty$. This completes the proof. ♠

Example 3.3.13 Consider $x'(t) = Ax(t)$ with

$$A = \begin{bmatrix} 4 & -1 & 12 \\ 37 & -8 & 76 \\ 0 & 0 & -4 \end{bmatrix}.$$

The eigenvalues are given by -4, $-2 \pm i$ ($i = \sqrt{-1}$). Since the real part of each eigenvalue is negative, every solution goes to zero. ♠

With the help of the variation of parameters formula (3.3), results in Theorem 3.3.12 will allow us to determine long-term behavior of the solutions of Eq. (3.2) by examining eigenvalues of the matrix A. For example, in Chapter 1 we noted that eigenvalues can be used to determine stabilities for scalar linear differential equations. We will give a detailed analysis in this direction for linear differential equations in $\Re^n, n \geq 1$, when we study stability in Chapter 5.

The detailed structures of e^{tA}, e^{tJ}, and e^{tJ_i} analyzed above also show a very interesting result concerning the relationship between eigenvalues of e^{tA} and eigenvalues of A, which is the version for finite dimensional spaces of the "spectral mapping theorem" in functional analysis.

Theorem 3.3.14 (Spectral mapping theorem) *Let A be an $n \times n$ matrix with the eigenvalues λ_i, $i = 1, 2, \cdots, n$, (not necessarily distinct). Then all eigenvalues of e^A are given by e^{λ_i}, $i = 1, 2, \cdots, n$. That is, if we define $\sigma(B)$ to be the set of all eigenvalues of matrix B, then*

$$\sigma(e^A) = e^{\sigma(A)}.$$

Moreover, for any fixed real number τ,

$$\sigma(e^{\tau A}) = e^{\tau\sigma(A)}. \qquad ♠ \qquad (3.22)$$

In the above, we analyzed the theoretical structure of e^{tA}, and found that most properties of e^{tA}, especially long-term properties, can be determined by using the eigenvalues of matrix A. These results are summarized in

Theorem 3.3.11, and their applications to differential equations $x' = Ax$ are summarized in Theorem 3.3.12.

Next, we take a different approach to e^{tA}. That is, we will provide an **algorithm** that can be used to actually **calculate** e^{tA}. For a polynomial in λ given by

$$P(\lambda) = \lambda^m + a_1 \lambda^{m-1} + \cdots + a_{m-1} \lambda + a_m,$$

we can define, for a square matrix A,

$$P(A) = A^m + a_1 A^{m-1} + \cdots + a_{m-1} A + a_m E,$$

which is also a square matrix, where E is the unit or identity matrix of the same size of the matrix A. The following is a result from linear algebra.

Theorem 3.3.15 (Cayley-Hamilton) *Let A be an $n \times n$ matrix and let $P(\lambda) = \det(\lambda E - A)$ be the characteristic polynomial of the matrix A, then $P(A) = 0$ (the 0 matrix).*

Proof. See the Appendix. ♠

With this preparation, we can verify the following algorithm.

Theorem 3.3.16 (Putzer algorithm (1966)) *Let A be an $n \times n$ matrix with the eigenvalues $\lambda_1, \lambda_2, \cdots, \lambda_n$ (not necessarily distinct). Then*

$$e^{tA} = \sum_{j=0}^{n-1} u_{j+1}(t) P_j, \tag{3.23}$$

where

$$P_0 = E, \quad P_j = \prod_{k=1}^{j} (A - \lambda_k E), \quad j = 1, 2, \cdots, n, \tag{3.24}$$

and $\{u_1(t), \cdots, u_n(t)\}$ is the solution of the lower triangular system

$$
\begin{aligned}
u_1'(t) &= \lambda_1 u_1(t), \quad u_1(0) = 1, \\
u_j'(t) &= \lambda_j u_j(t) + u_{j-1}(t), \quad u_j(0) = 0, \quad j = 2, 3, \cdots, n.
\end{aligned} \tag{3.25}
$$

Proof. Define $Y(t) = \sum_{j=0}^{n-1} u_{j+1}(t) P_j$. If we can show that $Y'(t) = AY(t)$, $Y(0) = E$, then by uniqueness, $e^{tA} = Y(t)$, proving the theorem. Now,

$$Y(0) = \sum_{j=0}^{n-1} u_{j+1}(0) P_j = u_1(0) P_0 = E, \tag{3.26}$$

and

$$Y'(t) \quad = \quad \sum_{j=0}^{n-1} u'_{j+1}(t)P_j$$

$$= \quad \sum_{j=0}^{n-1} \Big[\lambda_{j+1} u_{j+1}(t) + u_j(t) \Big] P_j, \tag{3.27}$$

where we define $u_0(t) = 0$. Thus, as $P_{j+1} = (A - \lambda_{j+1}E)P_j$, one has

$$Y'(t) - \lambda_n Y(t) = \sum_{j=0}^{n-1} \Big[\lambda_{j+1} u_{j+1}(t) + u_j(t) \Big] P_j - \lambda_n \sum_{j=0}^{n-1} u_{j+1}(t)P_j$$

$$= \sum_{j=0}^{n-1} [\lambda_{j+1} - \lambda_n] u_{j+1}(t)P_j + \sum_{j=1}^{n-1} u_j(t)P_j$$

$$= \sum_{j=0}^{n-1} [\lambda_{j+1} - \lambda_n] u_{j+1}(t)P_j + \sum_{j=0}^{n-2} u_{j+1}(t)P_{j+1}$$

$$= \sum_{j=0}^{n-2} [\lambda_{j+1} - \lambda_n] u_{j+1}(t)P_j + \sum_{j=0}^{n-2} u_{j+1}(t) \Big[(A - \lambda_{j+1}E)P_j \Big]$$

$$= \sum_{j=0}^{n-2} \Big[\lambda_{j+1}E - \lambda_n E + (A - \lambda_{j+1}E) \Big] u_{j+1}(t)P_j$$

$$= \sum_{j=0}^{n-2} \Big[A - \lambda_n E \Big] u_{j+1}(t)P_j$$

$$= \Big[A - \lambda_n E \Big] \sum_{j=0}^{n-2} u_{j+1}(t)P_j$$

$$= \Big[A - \lambda_n E \Big] \Big[\sum_{j=0}^{n-1} u_{j+1}(t)P_j - u_n(t)P_{n-1} \Big]$$

$$= \Big[A - \lambda_n E \Big] \Big[Y(t) - u_n(t)P_{n-1} \Big]$$

$$= \Big[A - \lambda_n E \Big] Y(t) - \Big[A - \lambda_n E \Big] u_n(t)P_{n-1}$$

$$= \Big[A - \lambda_n E \Big] Y(t) - u_n(t)P_n. \tag{3.28}$$

The characteristic equation for the matrix A can be written in factored form as

$$P(\lambda) = (\lambda - \lambda_1)(\lambda - \lambda_2) \cdots (\lambda - \lambda_n), \tag{3.29}$$

then from the Cayley-Hamilton theorem 3.3.15,

$$P_n = (A - \lambda_1 E)(A - \lambda_2 E) \cdots (A - \lambda_n E) = P(A) = 0. \qquad (3.30)$$

Therefore, (3.28) becomes $Y'(t) = AY(t)$, hence we have completed the proof.   ♠

The Putzer algorithm says that to compute e^{tA}, all we need is to solve $u_1, u_2, \cdots, u_n$ successively from the lower triangular system, starting with u_1, then $u_2, \cdots$, then u_n, which can be done using the variation of parameters formula for scalar linear differential equations.

Let's use the following example to illustrate how to apply the Putzer algorithm.

Example 3.3.17 Consider

$$A = \begin{bmatrix} 3 & -1 & 0 \\ 1 & 1 & 0 \\ -1 & 2 & 1 \end{bmatrix}.$$

The characteristic equation is $\det(\lambda E - A) = (\lambda^2 - 4\lambda + 4)(\lambda - 1) = 0$, so the eigenvalues are given by $\lambda_1 = 2$, $\lambda_2 = 2$, $\lambda_3 = 1$. Now, $P_0 = E$,

$$P_1 = (A - \lambda_1 E) = \begin{bmatrix} 1 & -1 & 0 \\ 1 & -1 & 0 \\ -1 & 2 & -1 \end{bmatrix}, \quad P_2 = P_1(A - \lambda_2 E) = \begin{bmatrix} 0 & 0 & 0 \\ 0 & 0 & 0 \\ 2 & -3 & 1 \end{bmatrix}.$$

Next, we compute $u_i(t)$, $i = 1, 2, 3$. For $u_1(t)$, $u_1'(t) = \lambda_1 u_1(t) = 2u_1(t)$ and $u_1(0) = 1$, so $u_1(t) = e^{2t}$. For $u_2(t)$, $u_2'(t) = \lambda_2 u_2(t) + u_1(t) = 2u_2(t) + e^{2t}$ and $u_2(0) = 0$, so $u_2(t) = te^{2t}$. For $u_3(t)$, $u_3'(t) = \lambda_3 u_3(t) + u_2(t) = u_3(t) + te^{2t}$ and $u_3(0) = 0$, so $u_3(t) = te^{2t} - e^{2t} + e^t$. Therefore,

$$\begin{aligned}
e^{tA} &= u_1(t)P_0 + u_2(t)P_1 + u_3(t)P_2 \\
&= e^{2t}E + te^{2t}\begin{bmatrix} 1 & -1 & 0 \\ 1 & -1 & 0 \\ -1 & 2 & -1 \end{bmatrix} + \left(te^{2t} - e^{2t} + e^t\right)\begin{bmatrix} 0 & 0 & 0 \\ 0 & 0 & 0 \\ 2 & -3 & 1 \end{bmatrix} \\
&= \begin{bmatrix} te^{2t} + e^{2t} & -te^{2t} & 0 \\ te^{2t} & e^{2t} - te^{2t} & 0 \\ te^{2t} - 2e^{2t} + 2e^t & 3e^{2t} - te^{2t} - 3e^t & e^t \end{bmatrix}. \qquad (3.31)
\end{aligned}$$

We check by taking a derivative in t that the right-hand side of (3.31) is the unique matrix solution of $X' = AX$, $X(0) = E$; therefore, it is the same as e^{tA}. That is, e^{tA} can be computed as in (3.31).   ♠

Next, we introduce another method to compute e^{tA}:

$$
\begin{aligned}
e^{tA} &= e^{(A-\lambda E)t + \lambda Et} = e^{(A-\lambda E)t} e^{\lambda Et} \\
&= e^{(A-\lambda E)t} e^{\lambda t} \\
&= \Big[\sum_{m=0}^{\infty} \frac{(A-\lambda E)^m}{m!} t^m \Big] e^{\lambda t}.
\end{aligned}
\tag{3.32}
$$

Formula (3.32) is useful sometimes, especially when the $n \times n$ matrix A has n linearly independent vectors v_i, $i = 1, 2, \cdots, n$, with λ_i, $i = 1, 2, \cdots, n$, the corresponding eigenvalues. Because in that case, $(A - \lambda_i E)v_i = 0$, thus $e^{tA}v_i = e^{\lambda_i t}v_i$. Now, any $x_0 \in \Re^n$ is a linear combination of v_i, $x_0 = \sum_{i=1}^{n} c_i v_i$ for some constants c_i, therefore,

$$
e^{tA} x_0 = e^{tA} \sum_{i=1}^{n} c_i v_i = \sum_{i=1}^{n} c_i e^{tA} v_i = \sum_{i=1}^{n} c_i e^{\lambda_i t} v_i.
$$

Exercises 3.3

1. Let D and P be $n \times n$ matrices and let P be nonsingular. Show that $P e^D P^{-1} = e^{P D P^{-1}}$.

2. Let $A = \begin{bmatrix} 0 & 1 & 0 \\ 0 & 0 & 1 \\ 0 & 0 & 0 \end{bmatrix}$. Compute A^m, $m = 2, 3, \cdots$.

3. Let N_i be the nilpotent matrix given by (3.13) with its size $m_i \times m_i$. Show that $N_i^k = 0$ for $k \geq m_i$.

4. Let N be an $n \times n$ nilpotent matrix and let $A = diag(1, a, a^2, \cdots, a^{n-1})$. Prove that $A^{-1} N A = aN$.

5. Transform the following matrices into the Jordan canonical forms:

$$
A_1 = \begin{bmatrix} 1 & 2 \\ 5 & 4 \end{bmatrix}; \quad A_2 = \begin{bmatrix} 1 & 1 \\ -1 & 3 \end{bmatrix}.
$$

6. Find e^{tA} for (a) $\quad A = \begin{bmatrix} 2 & 1 \\ 0 & 2 \end{bmatrix};\quad$ (b) $\quad A = \begin{bmatrix} 3 & 1 & 0 \\ 0 & 3 & 1 \\ 0 & 0 & 3 \end{bmatrix};$

$\quad$ (c) $\quad A = \begin{bmatrix} 4 & 1 & 0 & 0 & 0 \\ 0 & 4 & 0 & 0 & 0 \\ 0 & 0 & 5 & 0 & 0 \\ 0 & 0 & 0 & 6 & 1 \\ 0 & 0 & 0 & 0 & 6 \end{bmatrix}.$

7. Find e^{tA} for $A = \begin{bmatrix} 1 & 0 & 0 \\ 0 & 2 & -1 \\ 0 & 0 & 2 \end{bmatrix}.$

8. Prove that if x is an eigenvector of the matrix A corresponding to the eigenvalue λ, then x is also an eigenvector of the matrix e^A corresponding to the eigenvalue e^λ.

9. Let $a(t) \in \Re$ be differentiable and let A be an $n \times n$ matrix.

$\quad$ (a) Use the definitions of derivative and exponential for a matrix to show that $\frac{d}{dt} e^{a(t)A} = a'(t) A e^{a(t)A}$.

$\quad$ (b) Show that $e^{(\ln t)A}$, $t > 0$, is a solution of $x'(t) = (\frac{1}{t}) A x(t)$.

$\quad$ (c) Solve $x'(t) = (\frac{1}{t}) A x(t) + f(t)$, $x(1) = 2$, $t \geq 1$.

10. Review l'Hôpital's rule and find $\lim_{t \to \infty} e^{-2t} t^3$.

11. Complete the proof of Theorem 3.3.11 by verifying the results in (A), (B), and (C).

12. Prove the spectral mapping theorem 3.3.14.

13. In Example 3.3.17, take a derivative in t of the matrix e^{tA} computed there, then compare the derivative with $A e^{tA}$.

14. $\quad$ (a) Use the Putzer algorithm to compute e^{tA} for

$$(i).\ A = \begin{bmatrix} 1 & -1 \\ 3 & 5 \end{bmatrix}, \quad (ii).\ A = \begin{bmatrix} 1 & -1 & 1 \\ 3 & 5 & 2 \\ 0 & 0 & 3 \end{bmatrix}.$$

$\quad$ (b) Use formula (3.32) to compute $e^{tA} x_0$ for matrix A given in (i) and (ii) above.

3.4 Periodic Coefficients and Floquet Theory

From the previous section, we see that for linear differential equations $x'(t) = A(t)x(t)$, the simplest case is when $A(t) = A$ is independent of t, because then the fundamental matrix solution, given by

$$U(t, t_0) = e^{(t-t_0)A}, \tag{4.1}$$

is completely understood using the eigenvalues of matrix A.

The next simplest case for linear differential equations where we can still get a good understanding of the structure of solutions is given by linear **periodic** differential equations. That is, in the differential equation

$$x'(t) = A(t)x(t), \quad x(t_0) = x_0, \tag{4.2}$$

there exists a constant $T > 0$, called a **period**, such that $A(t)$ is T-periodic, written as

$$A(t) = A(t + T), \quad t \in \Re.$$

We remark that $A(t) = A(t + T)$ means that every entry in the matrix $A(t)$ is T-periodic, and we choose $T > 0$ to be the smallest period.

When $A(t)$ is periodic, $A(t)$ may not commute with its integral, thus the fundamental matrix solution $U(t, t_0)$ of Eq. (4.2) may not be given by an exponential $\exp\left(\int_{t_0}^t A(s)ds\right)$. However, we will show for periodic equations that $U(t, t_0)$ is represented by a formula that looks **almost** like (4.1), therefore we can still understand $U(t, t_0)$ in a better way.

In order to find a good approach to treat periodic linear differential equations in $\Re^n, n \geq 1$, let's first analyze a scalar periodic linear differential equation.

Example 3.4.1 Consider the scalar periodic linear differential equation,

$$x'(t) = a(t)x(t), \quad a(t + T) = a(t). \tag{4.3}$$

Now, $U(t, 0) = e^{\int_0^t a(s)ds}$. Let C be a scalar such that

$$U(T, 0) = e^{\int_0^T a(s)ds} = e^{TC}, \quad \text{or } C = \frac{1}{T}\int_0^T a(s)ds,$$

and define

$$P(t) = U(t, 0)e^{-tC} = e^{\int_0^t a(s)ds}e^{-tC},$$

and

$$y(t) = P^{-1}(t)x(t),$$

where $x(t)$ is a solution of (4.3). Then we have

$$U(t,0) = P(t)e^{tC}, \tag{4.4}$$

and

$$
\begin{aligned}
y'(t) &= -P^{-2}(t)P'(t)x(t) + P^{-1}(t)x'(t) \\
&= -P^{-2}(t)P(t)[a(t) - C]x(t) + P^{-1}(t)a(t)x(t) = P^{-1}(t)Cx(t) \\
&= Cy(t). \qquad\qquad\qquad\qquad\qquad\qquad\qquad \spadesuit \tag{4.5}
\end{aligned}
$$

From Example 3.4.1, we find that the fundamental (scalar) solution is given by $U(t,0) = P(t)e^{tC}$, which is almost an exponential function; and $y(t) = P^{-1}(t)x(t)$ transforms the periodic Eq. (4.3) into Eq. (4.5), an equation with a constant coefficient. In order to carry these to linear periodic equations in $\Re^n, n \geq 1$, we introduce a result that defines the natural logarithmic function for a nonsingular matrix. Recall that for a complex number $B \neq 0$, $C = \ln B$ is well defined, and $B = e^C$. The following result is similar for matrices.

Lemma 3.4.2 *If B is an $n \times n$ constant nonsingular matrix (may be complex valued), then there exists an $n \times n$ constant matrix C (may be complex valued), such that $B = e^C$.*

Proof. Applying the Jordan canonical form theorem 3.3.4 to the matrix B, we obtain matrices P and J such that $B = PJP^{-1}$ and J is given by (3.8) with J_i on the diagonal given by (3.9). If we can prove that J is given by an exponential, say for example $J = e^D$, then B must be given by an exponential, because $J = e^D$ implies

$$B = PJP^{-1} = Pe^D P^{-1} = e^{PDP^{-1}}, \tag{4.6}$$

hence we can let $C = PDP^{-1}$ to finish the proof.

To prove that J is given by an exponential, note from (3.8) that we only need to verify that each J_i can be written as an exponential. Since B is nonsingular, $\det(B - 0E) = \det B \neq 0$, thus each eigenvalue λ_i of B (or of J since they are similar) is nonzero. Then we can write

$$J_i = \lambda_i E + N_i = \lambda_i(E + \frac{N_i}{\lambda_i}), \tag{4.7}$$

where the nilpotent matrix N_i is given by (3.13), with its size $m_i \times m_i$.

Recall that for $x \in \Re$ and $|x| < 1$, the series

$$\ln(1+x) = x - \frac{x^2}{2} + \frac{x^3}{3} - \frac{x^4}{4} + \cdots \tag{4.8}$$

converges, hence we can define

$$\ln(E + \frac{N_i}{\lambda_i}) = \frac{N_i}{\lambda_i} - \frac{(\frac{N_i}{\lambda_i})^2}{2} + \frac{(\frac{N_i}{\lambda_i})^3}{3} - \frac{(\frac{N_i}{\lambda_i})^4}{4} + \cdots + (-1)^{m_i}\frac{(\frac{N_i}{\lambda_i})^{m_i-1}}{(m_i-1)} + \cdots$$

$$= \frac{N_i}{\lambda_i} - \frac{(\frac{N_i}{\lambda_i})^2}{2} + \frac{(\frac{N_i}{\lambda_i})^3}{3} - \frac{(\frac{N_i}{\lambda_i})^4}{4} + \cdots + (-1)^{m_i}\frac{(\frac{N_i}{\lambda_i})^{m_i-1}}{(m_i-1)}, \tag{4.9}$$

which has finite number of terms because $N_i^h = 0$ for $h \geq m_i$. Next, for $x \in \Re$ and $|x| < 1$,

$$1 + x = e^{\ln(1+x)} = 1 + (x - \frac{x^2}{2} + \cdots) + \frac{1}{2!}(x - \frac{x^2}{2} + \cdots)^2 + \cdots, \tag{4.10}$$

thus, for any $k \geq 2$, the coefficient of x^k in the expansion on the right-hand side of (4.10) is zero. The same is true if we replace $\ln(1+x)$ by the finite summation $\ln(E + \frac{N_i}{\lambda_i})$. That is, from (4.10), we have

$$E + \frac{N_i}{\lambda_i} = E + \left[\frac{N_i}{\lambda_i} - \frac{(\frac{N_i}{\lambda_i})^2}{2} + \frac{(\frac{N_i}{\lambda_i})^3}{3} - \frac{(\frac{N_i}{\lambda_i})^4}{4} + \cdots + (-1)^{m_i}\frac{(\frac{N_i}{\lambda_i})^{m_i-1}}{(m_i-1)}\right]$$

$$+ \frac{1}{2!}\left[\frac{N_i}{\lambda_i} - \frac{(\frac{N_i}{\lambda_i})^2}{2} + \frac{(\frac{N_i}{\lambda_i})^3}{3} - \frac{(\frac{N_i}{\lambda_i})^4}{4} + \cdots + (-1)^{m_i}\frac{(\frac{N_i}{\lambda_i})^{m_i-1}}{(m_i-1)}\right]^2$$

$$+ \frac{1}{3!}\left[\frac{N_i}{\lambda_i} - \frac{(\frac{N_i}{\lambda_i})^2}{2} + \frac{(\frac{N_i}{\lambda_i})^3}{3} - \frac{(\frac{N_i}{\lambda_i})^4}{4} + \cdots + (-1)^{m_i}\frac{(\frac{N_i}{\lambda_i})^{m_i-1}}{(m_i-1)}\right]^3 + \cdots. \tag{4.11}$$

Now, from (4.9) and (4.11), we obtain

$$e^{\ln(E+\frac{N_i}{\lambda_i})} = E + \ln(E + \frac{N_i}{\lambda_i}) + \frac{1}{2!}\left[\ln(E + \frac{N_i}{\lambda_i})\right]^2 + \cdots$$

$$= E + \left[\frac{N_i}{\lambda_i} - \frac{(\frac{N_i}{\lambda_i})^2}{2} + \frac{(\frac{N_i}{\lambda_i})^3}{3} - \frac{(\frac{N_i}{\lambda_i})^4}{4} + \cdots + (-1)^{m_i}\frac{(\frac{N_i}{\lambda_i})^{m_i-1}}{(m_i-1)}\right]$$

$$+ \frac{1}{2!}\left[\frac{N_i}{\lambda_i} - \frac{(\frac{N_i}{\lambda_i})^2}{2} + \frac{(\frac{N_i}{\lambda_i})^3}{3} - \frac{(\frac{N_i}{\lambda_i})^4}{4} + \cdots + (-1)^{m_i}\frac{(\frac{N_i}{\lambda_i})^{m_i-1}}{(m_i-1)}\right]^2 + \cdots$$

$$= E + \frac{N_i}{\lambda_i}. \tag{4.12}$$

Therefore, from (4.12) and Lemma 3.3.10, we can rewrite (4.7) as

$$
\begin{aligned}
J_i &= \lambda_i \left(E + \frac{N_i}{\lambda_i} \right) = \lambda_i e^{\ln\left(E + \frac{N_i}{\lambda_i}\right)} = e^{\ln \lambda_i} e^{\ln\left(E + \frac{N_i}{\lambda_i}\right)} = [e^{\ln \lambda_i} E] e^{\ln\left(E + \frac{N_i}{\lambda_i}\right)} \\
&= [e^{(\ln \lambda_i)E}] e^{\ln\left(E + \frac{N_i}{\lambda_i}\right)} = e^{(\ln \lambda_i)E + \ln\left(E + \frac{N_i}{\lambda_i}\right)},
\end{aligned}
\tag{4.13}
$$

which verifies that J_i can be written as an exponential, hence we have completed the proof. ♠

With the help of Lemma 3.4.2, we will be able to transform the fundamental matrix solution $U(t, t_0)$ of a linear periodic differential equation into a simpler form that is "almost" an exponential matrix function. This study is called Floquet theory, which will eventually allow us to transform Eq. (4.2) with $A(t)$ periodic into a linear differential equation with constant coefficients.

Theorem 3.4.3 (Floquet theory) *Let $A(t) = A(t+T)$ for some constant $T > 0$ and let $U(t, t_0)$ be the fundamental matrix solution of Eq. (4.2). Then there exists a constant matrix C and a nonsingular, continuous, T-periodic matrix function $P(t)$, such that*

$$
U(t, t_0) = P(t)e^{tC},
\tag{4.14}
$$

or

$$
U(t, t_0) = \overline{P}(t)e^{(t-t_0)C},
\tag{4.15}
$$

where $\overline{P}(t) = P(t)e^{t_0 C}$ is T-periodic and $\overline{P}(t_0) = E$.

Proof. Define $Y(t) = U(t + T, t_0)$. Then, since $A(t)$ is T-periodic,

$$
\begin{aligned}
Y'(t) &= U'(t + T, t_0) = A(t + T)U(t + T, t_0) = A(t)U(t + T, t_0) \\
&= A(t)Y(t), \quad Y(t_0) = U(t_0 + T, t_0).
\end{aligned}
\tag{4.16}
$$

However, $X(t) = U(t, t_0)U(t_0 + T, t_0)$ is also a solution of (4.16), then by uniqueness, $Y(t) = X(t)$, or

$$
U(t + T, t_0) = U(t, t_0)U(t_0 + T, t_0).
\tag{4.17}
$$

From Lemma 3.4.2, there exists a matrix C such that

$$
U(t_0 + T, t_0) = e^{TC}.
\tag{4.18}
$$

Now, define
$$P(t) = U(t, t_0)e^{-tC}.$$

Since $U(t, t_0)$ is nonsingular and e^{-tC} has the inverse e^{tC}, $P(t)$ is non-singular, and we have
$$U(t, t_0) = P(t)e^{tC}.$$

Also, from (4.17) and (4.18),

$$
\begin{aligned}
P(t + T) &= U(t + T, t_0)e^{-(t+T)C} = U(t, t_0)U(t_0 + T, t_0)e^{-(t+T)C} \\
&= U(t, t_0)e^{TC}e^{-(t+T)C} = U(t, t_0)e^{-tC} = P(t), \qquad (4.19)
\end{aligned}
$$

hence $P(t)$ is T-periodic, and then the proof is complete.   ♠

Next, let's look at how to apply the Floquet theory to transform Eq. (4.2) with $A(t)$ periodic into a linear differential equation with constant coefficients.

Theorem 3.4.4 *Let $A(t)$ be T-periodic and let C and $P(t)$ be given in Theorem 3.4.3, and define $y(t) = P^{-1}(t)x(t)$ with $x(t) = x(t, t_0, x_0)$ the unique solution of Eq. (4.2). Then $y(t)$ satisfies the linear differential equation*

$$y'(t) = Cy(t), \quad y(t_0) = P^{-1}(t_0)x_0. \qquad (4.20)$$

That is, $y(t) = P^{-1}(t)x(t)$ transforms the solution x of a linear periodic differential equation to the solution y of a linear differential equation with constant coefficients.

Proof. We have $x(t) = P(t)y(t)$ and $x' = A(t)x$, then

$$A(t)x(t) = x'(t) = P'(t)y(t) + P(t)y'(t),$$

hence

$$
\begin{aligned}
y'(t) &= P^{-1}(t)[A(t)x(t) - P'(t)y(t)] = P^{-1}(t)[A(t)P(t)y(t) - P'(t)y(t)] \\
&= P^{-1}(t)[A(t)P(t) - P'(t)]y(t).
\end{aligned}
$$

Next, since $P(t) = U(t, t_0)e^{-tC}$ from Theorem 3.4.3, we have

$$
\begin{aligned}
P'(t) &= U'(t, t_0)e^{-tC} - U(t, t_0)e^{-tC}C \\
&= A(t)U(t, t_0)e^{-tC} - U(t, t_0)e^{-tC}C = A(t)P(t) - P(t)C.
\end{aligned}
$$

Therefore,

$$\begin{cases} y'(t) & = & P^{-1}(t)[A(t)P(t) - P'(t)]y(t) = P^{-1}(t)[P(t)C]y(t) = Cy(t), \\ y(t_0) & = & P^{-1}(t_0)x_0, \end{cases}$$

which completes the proof. ♠

Notice from the Floquet theory that $\overline{P}(t)$ in (4.15) is T-periodic and $\overline{P}(t_0) = E$, then

$$U(T + t_0, t_0) = \overline{P}(T + t_0)e^{TC} = \overline{P}(t_0)e^{TC} = e^{TC}.$$

Thus, due to the relationship between the matrices e^{TC} and C, and its applications in the related studies, we define the following.

Definition 3.4.5 *Consider Eq. (4.2) with $A(t)$ of period T and let C be the constant matrix from the Floquet theory in Theorem 3.4.3. The eigenvalues of matrix e^{TC} are called the **characteristic multipliers** of $A(t)$ (or of Eq. (4.2)). The eigenvalues of matrix C are called the **characteristic exponents** (or **Floquet exponents**) of $A(t)$ (or of Eq. (4.2)).*

From the spectral mapping theorem 3.3.14, if λ is a characteristic exponent (or Floquet exponents), then $e^{T\lambda}$ is a characteristic multiplier.

Now, since the matrix C in Eq. (4.20) is a constant matrix, the results in the previous section, such as the Jordan canonical form theorem, and Theorems 3.3.11 and 3.3.12 concerning long-term behavior of solutions, can be applied to Eq. (4.20). The difference between the solutions $x(t)$ of Eq. (4.2) and $y(t)$ of Eq. (4.20) is just a continuous periodic (hence bounded) transformation $x(t) = P(t)y(t)$; therefore, when $A(t)$ is periodic, long-term behavior of solutions of Eq. (4.20) will determine long-term behavior of solutions of Eq. (4.2). We state these results below and leave the proof as an exercise.

Theorem 3.4.6 *Consider Eq. (4.2) with $A(t)$ periodic, and let the matrix C be given in Theorem 3.4.3.*

(A). The following statements are equivalent:

(1). There is a constant $C_1 > 0$ such that every solution of Eq. (4.2) satisfies $|x(t, t_0, x_0)| \leq C_1|x_0|$, $t_0 \leq t < \infty$;

(2). *There is a constant $C_2 > 0$ such that every solution of Eq. (4.20)*
     satisfies $|y(t, t_0, y_0)| \le C_2 |y_0|$, $t_0 \le t < \infty$;

(3). *For each characteristic exponent λ (eigenvalue of matrix C), ei-*
     ther $R(\lambda) < 0$, or $R(\lambda) = 0$ but in this case λ appears only in
     matrices J_i (Jordan canonical form for C) such that J_i is a 1×1
     matrix;

(4). *For each characteristic multiplier η (eigenvalue of matrix e^{TC}),*
     either $|\eta| < 1$, or $|\eta| = 1$ but in this case η appears only in
     matrices J_i (Jordan canonical form for e^{TC}) such that J_i is a
     1×1 matrix.

(B). *The following statements are equivalent:*

(1). *Every solution of Eq. (4.2) satisfies $\lim_{t \to \infty} |x(t, t_0, x_0)| = 0$;*

(2). *Every solution of Eq. (4.20) satisfies $\lim_{t \to \infty} |y(t, t_0, y_0)| = 0$;*

(3). *Each characteristic exponent (eigenvalue of matrix C) has a neg-*
     ative real part;

(4). *Each characteristic multiplier η (eigenvalue of matrix e^{TC}) satis-*
     fies $|\eta| < 1$.

(C). *The following statements are equivalent:*

(1). *There is a solution x of Eq. (4.2) with $\lim_{t \to \infty} |x(t, t_0, x_0)| = \infty$;*

(2). *There is a solution y of Eq. (4.20) with $\lim_{t \to \infty} |y(t, t_0, y_0)| = \infty$;*

(3). *Either there is a characteristic exponent λ (eigenvalue of matrix*
     C) with $R(\lambda) = 0$ and λ appears in a matrix J_i that is at least
     2×2, or there is a characteristic exponent with a positive real
     part;

(4). *Either there is a characteristic multiplier η (eigenvalue of matrix*
     e^{TC}) with $|\eta| = 1$ and η appears in a matrix J_i that is at least
     2×2, or there is a characteristic multiplier η with $|\eta| > 1$.   ♠

The relationship of the results in Theorem 3.4.6 and some long-term
properties, such as stability or boundedness of solutions, will be given later.

Remark 3.4.7 Theorem 3.4.6 is stated for the matrix e^{TC}, but it can also
be stated for matrix $e^{\tau C}$ with τ being any fixed positive number. In fact,
there is a change of variables that transforms Eq. (4.2) with $A(t)$ of period

T into an equation of any period (see an exercise). For example, let $t = Ts$ and define $y(s) = x(Ts)$. Then in variable s,

$$
\begin{aligned}
y'(s) &= x'(Ts)T = A(Ts)x(Ts)T = [TA(Ts)]y(s) \\
&= \overline{A}(s)y(s), \quad\quad (4.21)
\end{aligned}
$$

where $\overline{A}(s)$ is defined in (4.21), and

$$
\overline{A}(s+1) = TA(T(s+1)) = TA(Ts+T) = TA(Ts) = \overline{A}(s), \quad (4.22)
$$

thus $\overline{A}(s)$ is periodic with period 1. That is, $y(s) = x(Ts)$ transforms the T-periodic equation (4.2) to a 1-periodic equation (4.21). ♠

Looking at Theorem 3.4.6, it seems that the study of linear periodic equations is as easy as the study of linear equations with constant coefficients. However, finding the periodic transformation matrix $P(t)$ of Theorem 3.4.3 (Floquet theory) is difficult because the fundamental matrix solution $U(t, t_0)$ of Eq. (4.2) would also be needed. The next example indicates that, unlike linear differential equations with constant coefficients where eigenvalues determine properties of solutions, eigenvalues of $A(t)$ cannot determine the properties of solutions for $x'(t) = A(t)x(t)$, even when $A(t)$ is periodic.

Example 3.4.8 (Marcus and Yamabe (1960)) Consider the periodic equation $x' = A(t)x$, where

$$
A(t) = \begin{bmatrix} -1 + \frac{3}{2}\cos^2 t & 1 - \frac{3}{2}\sin t \cos t \\ -1 - \frac{3}{2}\sin t \cos t & -1 + \frac{3}{2}\sin^2 t \end{bmatrix}, \quad t \in \Re.
$$

The eigenvalues of $A(t)$ are given by $\frac{-1 \pm \sqrt{7}i}{4}$, thus the real parts are negative. However, it can be checked that (see an exercise)

$$
x(t) = e^{t/2}\begin{bmatrix} -\cos t \\ \sin t \end{bmatrix}
$$

is a solution. Accordingly, there are $t_m \to \infty$ as $m \to \infty$ such that $|x(t_m)| \to \infty$, $m \to \infty$. ♠

However, in some situations, the results in this section can be applied to derive good results. One such case is Hill's equation, given in the next example.

Example 3.4.9 (Hill's equation) Let $\phi(t)$ be real and continuous. Consider the scalar linear periodic differential equation, called Hill's equation,

$$y''(t) + \phi(t)y(t) = 0, \quad \phi(t+T) = \phi(t), \quad t \in \Re, \tag{4.23}$$

introduced by Hill [1886] in his study of the motion of the lunar perigee which he described as a function of the mean motions of the sun and moon. If we define $x = [y, y']^T$, then Eq. (4.23) becomes $x'(t) = A(t)x(t)$, where

$$A(t) = \begin{bmatrix} 0 & 1 \\ -\phi(t) & 0 \end{bmatrix}. \tag{4.24}$$

Let $U(t,0)$ be the fundamental matrix solution, then $U(T,0) = e^{TC}$, where C is from the Floquet theory. Now, $tr A(t) = 0$. Thus, we have, using Liouville's formula in Theorem 3.2.18,

$$\det e^{TC} = \det U(T,0) = [\det U(0,0)] \exp \left(\int_0^T tr A(s)ds \right) = [\det E]e^0 = 1.$$

Let η_1 and η_2 be two characteristic multipliers of $A(t)$ (eigenvalues of matrix e^{TC}). Then using the characteristic polynomial of e^{TC}, we obtain

$$\det[e^{TC} - \eta E] = (\eta - \eta_1)(\eta - \eta_2),$$

which implies that

$$\eta_1 \eta_2 = \det e^{TC} = 1. \tag{4.25}$$

Due to (4.25), we cannot have $|\eta_1| < 1$ and $|\eta_2| < 1$, thus the conditions in Theorem 3.4.6(B) are not satisfied. Therefore, Eq. (4.23) has solutions that do not go to zero. This is what we can say for Hill's equation right now. Other long-term properties, such as stability or boundedness of solutions, will be examined later after we make the corresponding definitions. For example, we will restate "Eq. (4.23) has solutions that do not go to zero" as "the zero solution of Eq. (4.23) is not asymptotically stable." (Now that we have made all necessary preparations, you can go directly to Chapter 5 to read the stability definitions and see the analysis for Hill's equation.) ♠

We know that the real parts of eigenvalues of matrix C in Eq. (4.20) (or characteristic exponents) determine long-term properties of solutions of Eq. (4.20) and Eq. (4.2). Next, let's find some relationship between the real parts of eigenvalues of matrix C and solutions of Eq. (4.2).

For a (real) scalar autonomous linear differential equation

$$x'(t) = ax(t), \quad x(0) = x_0, \tag{4.26}$$

the transformed equation using the Floquet theory is itself, or $C = a$. In this case, the solution of Eq. (4.26) is given by $x(t, 0, x_0) = x_0 e^{at}$. Thus, for $x_0 \neq 0$,

$$\frac{1}{t} \ln \left(\frac{|x(t, 0, x_0)|}{|x_0|} \right) = \frac{1}{t} \ln \left(\frac{|e^{at} x_0|}{|x_0|} \right) = \frac{1}{t} \ln \left(e^{at} \right) = \frac{1}{t}(at) = a = C. \tag{4.27}$$

For a (real) scalar T-periodic linear differential equation

$$x'(t) = a(t)x(t), \quad x(0) = x_0, \tag{4.28}$$

we know from Example 3.4.1 that $C = \frac{1}{T} \int_0^T a(s)ds$ in the transformed equation using the Floquet theory. In this case, the solution of Eq. (4.28) is given by $x(t, 0, x_0) = x_0 e^{\int_0^t a(s)ds}$. Thus, for $x_0 \neq 0$, a calculation (see an exercise) shows that

$$
\begin{aligned}
\lim_{t \to \infty} \frac{1}{t} \ln \left(\frac{|x(t, 0, x_0)|}{|x_0|} \right) &= \lim_{t \to \infty} \frac{1}{t} \ln \left(\frac{|e^{\int_0^t a(s)ds} x_0|}{|x_0|} \right) = \lim_{t \to \infty} \frac{1}{t} \int_0^t a(s)ds \\
&= \frac{1}{T} \int_0^T a(s)ds = C. \tag{4.29}
\end{aligned}
$$

The formula in (4.29) is related to something called the "Liapunov exponent," which is a very important concept in the study of differential equations. See Guckenheimer and Holmes [1986], Liapunov [1892], and Wiggins [1990] for related studies. For periodic linear differential equation (4.2), the Liapunov exponent is defined as follows.

Definition 3.4.10 *Let $U(t, 0)$ be the fundamental matrix solution of Eq. (4.2). For a nonzero vector v in $\Re^n$, the* **Liapunov exponent** *of Eq. (4.2) with respect to v is*

$$\xi(v) = \limsup_{t \to \infty} \frac{1}{t} \ln \left(\frac{|U(t, 0)v|}{|v|} \right). \tag{4.30}$$

In view of (4.29), we have

Theorem 3.4.11 *If β is a characteristic exponent (Floquet exponent) of Eq. (4.2), then the real part of β is a Liapunov exponent of Eq. (4.2).*

Proof. From the Floquet theory,

$$U(t,0) = P(t)e^{tC},$$

where P and C are from Theorem 3.4.3 and $P(t)$ is T-periodic. Since (by definition) β is an eigenvalue of matrix C, using the spectral mapping theorem 3.3.14, $e^{\beta T}$ is an eigenvalue of e^{TC}. Thus there is a $v \neq 0$ in $\Re^n$ such that $e^{TC}v = e^{\beta T}v$. For $t > 0$ large, there is a positive integer m and a number τ such that $t = mT + \tau$, $0 \leq \tau < T$, and $t \to \infty$ if and only if $m \to \infty$. Now, for $\beta = a + bi$,

$$
\begin{aligned}
\xi(v) &= \limsup_{t\to\infty} \frac{1}{t}\ln\left(\frac{|U(t,0)v|}{|v|}\right) \\[2mm]
&= \limsup_{m\to\infty} \frac{1}{T}\left(\frac{mT}{mT+\tau}\right)\frac{1}{m}\ln\left(\frac{|U(mT+\tau,0)v|}{|v|}\right) \\[2mm]
&= \limsup_{m\to\infty} \frac{1}{T}\frac{1}{m}\ln\left(\frac{|P(mT+\tau)e^{(mT+\tau)C}v|}{|v|}\right) \\[2mm]
&= \limsup_{m\to\infty} \frac{1}{T}\frac{1}{m}\ln\left(\frac{|P(\tau)e^{\tau C}[e^{TC}]^m v|}{|v|}\right) \\[2mm]
&= \limsup_{m\to\infty} \frac{1}{T}\frac{1}{m}\ln\left(\frac{|P(\tau)e^{\tau C}e^{m\beta T}v|}{|v|}\right) \\[2mm]
&= \limsup_{m\to\infty} \frac{1}{T}\left[\frac{1}{m}\ln|e^{m\beta T}| + \frac{1}{m}\ln\left(\frac{|P(\tau)e^{\tau C}v|}{|v|}\right)\right] \\[2mm]
&= \limsup_{m\to\infty} \frac{1}{T}\left[\frac{1}{m}\ln e^{maT} + \frac{1}{m}\ln\left(\frac{|P(\tau)e^{\tau C}v|}{|v|}\right)\right] \\[2mm]
&= \limsup_{m\to\infty} \frac{1}{T}\left[\frac{1}{m}maT + \frac{1}{m}\ln\left(\frac{|P(\tau)e^{\tau C}v|}{|v|}\right)\right] \\[2mm]
&= \frac{1}{T}[aT + 0] = a = R(\beta),
\end{aligned}
\tag{4.31}
$$

which completes the proof.   ♠

For Eq. (4.2), if Liapunov exponents are positive, then according to Theorems 3.4.6 and 3.4.11, Eq. (4.2) has solutions x such that $|x| \to \infty$ exponentially fast as $t \to \infty$, no matter how small their initial values are, or how close their initial points are to the zero solution. This can be characterized as "long-term sensitive dependence on the initial conditions," in the sense that solutions started from a small neighborhood separate exponentially fast. This should not be confused with the continuous dependence of solutions with respect to initial conditions on finite intervals.

In some related study of nonlinear differential equations, similar results hold. That is, positive Liapunov exponents indicates that the system will demonstrate some complex and strange behavior, or even **chaos**, for which, one feature used to describe the chaotic behavior is the "sensitive dependence on the initial conditions." See Chapter 7 for some discussions, and also, Guckenheimer and Holmes [1986], Liapunov [1892], and Wiggins [1990] for further details.

Exercises 3.4

1. Verify all steps in equality (4.13).

2. Let $A(t) = A(t + T)$ for some constant $T > 0$. Show that a solution $x(t)$, $x(t_0) \neq 0$, of $x' = A(t)x$ has the property $x(t + T) = kx(t)$ for some constant k if and only if k is an eigenvalue of e^{TC} with $x(t_0)$ the corresponding eigenvector, where the matrix C is from Theorem 3.4.3.

3. Let $A(t) = A(t + T)$ for some constant $T > 0$. Show that $x' = A(t)x$ has a nonzero T-periodic solution if and only if 1 is an eigenvalue of e^{TC}.

4. Let $A(t) = A(t+T)$ for some constant $T > 0$ and let $A(t)$ be odd, that is, $A(-t) = -A(t)$. Use the Floquet theory to show that the fundamental matrix solution $U(t, 0)$ of $x' = A(t)x$ is even and T-periodic.

5. Let $A(t)$ be T-periodic and assume that $A(t)$ commutes with its integral. Prove that matrix C in Theorem 3.4.3 can be taken as

$$C = \frac{1}{T} \int_0^T A(s)ds.$$

6. Examine long-term property of solutions of the scalar periodic equation $x'(t) = a(t)x(t)$, where $a(t)$ is T-periodic.

7. Let $a(t)$ be a T-periodic scalar function. Prove that $e^{\int_0^t a(s)ds} e^{-tC}$ is T-periodic.

8. Prove Theorem 3.4.6.

9. Let τ be any positive number. Verify that there is a change of variables that transforms Eq. (4.2) with $A(t)$ of period T into an equation of period τ.

10. In Example 3.4.8, verify that $\frac{-1\pm\sqrt{7}i}{4}$ are the eigenvalues and $x(t) =$
$e^{t/2}\begin{bmatrix} -\cos t \\ \sin t \end{bmatrix}$ is a solution, and there are $t_m \to \infty$ as $m \to \infty$ such
that $|x(t_m)| \to \infty$, $m \to \infty$.

11. Let $a(t)$ and $\phi(t)$ be real, continuous, and periodic with period T.
Consider the scalar equation,

$$y''(t) - a(t)y'(t) + \phi(t)y(t) = 0, \quad t \in \Re. \tag{4.32}$$

Let η_1 and η_2 be two characteristic multipliers of $A(t)$ (after formulat-
ing the equation as a system). Then show that

$$\eta_1\eta_2 = e^{\int_0^T a(s)ds}. \tag{4.33}$$

Can you derive any long-term properties of solutions from $\eta_1\eta_2 = e^{\int_0^T a(s)ds}$? Look at some special cases.

12. Let $\phi(t)$ be real, continuous, and periodic with period π. Consider the
scalar equation,

$$y''(t) - (\cos^2 t)y'(t) + \phi(t)y(t) = 0, \quad t \in \Re. \tag{4.34}$$

Show that there is a solution that goes to ∞ as $t \to \infty$.

13. Calculate the Liapunov exponent directly for the scalar T-periodic
differential equation $x'(t) = a(t)x(t)$.

14. Calculate the Liapunov exponents for the system $x'(t) = ax(t)$, $y'(t) = by(t)$, where a and b are constants.

Chapter 4

Autonomous Differential Equations in $\Re^2$

4.1 Introduction

In this chapter, we study autonomous differential equations, that is, $f(t,x)$ in the differential equation $x'(t) = f(t, x(t))$ is independent of t, or $f(t,x) = f(x)$ for $(t,x) \in \Re \times \Re^n$. Hence we have

$$x'(t) = f(x(t)), \quad \text{or} \quad x' = f(x). \tag{1.1}$$

For example, $x' = \sin x$ and $x' = Ax$, for a constant matrix A, are autonomous differential equations.

We assume that Eq. (1.1) is defined on $\Re \times \Re^n$. For autonomous differential equations, a weak Lipschitz condition is the same as a (global) Lipschitz condition (see an exercise). Thus, if $f(x)$ satisfies a (global) Lipschitz condition, or if $f(x)$ is bounded and satisfies a local Lipschitz condition, then the two global existence theorems in Section 4 of Chapter 2 can be applied to guarantee that solutions of Eq. (1.1) exist on $\Re$.

However, we need to take care of other cases where solutions may not exist on $\Re$. For example, the solution of the differential equation $x' = x^2$, $x(0) = 1$, is given by $x(t) = (1-t)^{-1}$ and exists only on $(-\infty, 1)$, in which case the two global existence theorems in Section 2.4 do not apply because $f(x) = x^2$ is unbounded and does not satisfy a (global) Lipschitz condition for x in $\Re$. For those cases, we will use an idea of Vinograd (see

Nemytskii and Stepanov [1960]), in which the variable t of an autonomous differential equation (whose solutions do not exist on $\Re$) is **rescaled** so that the equation will have the same solution curves in $\Re^n$ (not in $\Re \times \Re^n$) as another autonomous differential equation whose solutions exist on $\Re$. See the details below.

Lemma 4.1.1 *Let $f : \Re^n \to \Re^n$ and $g : \Re^n \to (0, \infty)$ be continuous, then Eq. (1.1) and $y' = f(y)g(y)$ have the same solution curves in $\Re^n$.*

Proof. Let $x(t) = x(t, t_0, x_0)$ be a solution of Eq. (1.1) on an interval (a, b) containing t_0. Define $\tau(t) = \int_{t_0}^{t} \frac{dh}{g(x(h))}$ on (a, b). Since $g > 0$, $\tau(t)$ is a strictly increasing function in t, hence $\tau(t)$ maps the interval (a, b) to an interval (c, d) in a one-to-one fashion. Now define $y(s)$ on (c, d) such that $y(s) = x(t)$ if $s = \tau(t)$, see **Figure 4.1**.

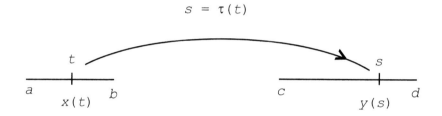

Figure 4.1: Define $y(s) = x(t)$ if $s = \tau(t)$

From $y(s) = x(t)$ (when $s = \tau(t)$) we find that a point on $x(t)$ is also on $y(s)$ and vice versa, thus $x(t)$ and $y(s)$ define the same curve in $\Re^n$. Next,

$$
\begin{aligned}
f(y(s)) &= f(x(t)) = \frac{d}{dt}x(t) = \frac{d}{dt}y(s) = \frac{dy(s)}{ds}\frac{ds}{dt} = \frac{dy(s)}{ds}\tau'(t) \\
&= \frac{dy(s)}{ds}\frac{1}{g(x(t))} = \frac{dy(s)}{ds}\frac{1}{g(y(s))},
\end{aligned} \tag{1.2}
$$

thus

$$
\frac{dy(s)}{ds} = f(y(s))g(y(s)), \tag{1.3}
$$

therefore, y is a solution of

$$
y' = f(y)g(y). \tag{1.4}
$$

On the other hand, if we start with a solution y of Eq. (1.4), then we can regard fg as f and $\frac{1}{g}$ as g, thus the same arguments as above show that there is a solution x of $x' = [f(x)g(x)][\frac{1}{g(x)}] = f(x)$ such that y and x define the same curve in $\Re^n$. This completes the proof. ♠

Lemma 4.1.2 *Assume that $f : \Re^n \to \Re^n$ satisfies a local Lipschitz condition. Then $F(y) = f(y)\frac{1}{1+|f(y)|}$ also satisfies a local Lipschitz condition, and solutions of $y' = F(y)$ exist on $\Re$. Moreover, Eq. (1.1) and $y' = F(y)$ have the same solution curves in $\Re^n$.*

Proof. The proof of the local Lipschitz condition for $F(y) = f(y)\frac{1}{1+|f(y)|}$ is left as an exercise. Next, as $|F(y)| < 1$, the second global existence theorem in Section 4 of Chapter 2 can be used to guarantee that solutions of $y' = F(y)$ exist on $\Re$. Finally, Lemma 4.1.1 implies that Eq. (1.1) and $y' = F(y)$ have the same solution curves in $\Re^n$. This completes the proof. ♠

The proofs in Lemmas 4.1.1 and 4.1.2 indicate that these results can only be applied to autonomous differential equations to guarantee that if $f(x)$ satisfies a local Lipschitz condition, then as long as solution curves in $\Re^n$ are concerned, Eq. (1.1) is equivalent to an autonomous differential equation whose solutions exist on $\Re$.

Now, let's explain the results in Lemmas 4.1.1 and 4.1.2 for the scalar autonomous differential equation $x' = x^2$, $x(0) = 1$, whose solution $x(t) = (1-t)^{-1}$ exists only on $(-\infty, 1)$.

Example 4.1.3 For the scalar autonomous differential equation

$$x' = x^2, \quad x(0) = 1, \tag{1.5}$$

one has $f(x) = x^2$, $t_0 = 0$, and $x(t) = (1-t)^{-1}$ is the solution that exists on $(-\infty, 1)$. From Lemma 4.1.2, the function g of Lemma 4.1.1 is given by

$$g(y) = \frac{1}{1 + |f(y)|} = \frac{1}{1 + y^2},$$

thus the rescaling function s in the proof of Lemma 4.1.1 is given as

$$s = \tau(t) = \int_0^t \frac{dh}{g(x(h))} = \int_0^t [1 + x^2(h)]dh \tag{1.6}$$

$$= \int_0^t [1 + \frac{1}{(1-h)^2}]dh = (h + \frac{1}{1-h})\Big|_0^t = t + \frac{1}{1-t} - 1. \tag{1.7}$$

From (1.6) we find that $s = 0$ implies $t = 0$. We also know from Lemmas 4.1.1 and 4.1.2 that $y(s) = x(t)$ (when $s = \tau(t)$) is the solution of the autonomous differential equation

$$y' = f(y)\frac{1}{1 + |f(y)|} = \frac{y^2}{1 + y^2}, \quad y(0) = x(0) = 1. \tag{1.8}$$

Using (1.7), we can verify (see an exercise) that when t takes values in $(-\infty, 1)$ for the solution $x(t) = (1 - t)^{-1}$, the variable s takes values in $(-\infty, \infty)$ for the corresponding solution $y(s)$. Therefore, $y(s)$ exists on $\Re$.

In fact, in this case we have another way to see why $y(s)$ exists on $\Re$, that is, we can actually solve $y(s)$ from Eq. (1.8) using separation of variables, and obtain (see an exercise)

$$y(s) = \frac{1}{2}[s + \sqrt{s^2 + 4}], \tag{1.9}$$

which certainly exists on $\Re$.

Finally, note that the solution curves in $\Re$ (or the ranges in this case) for $x(t) = (1 - t)^{-1}$ with $t \in (-\infty, 1)$ and for $y(s) = \frac{1}{2}[s + \sqrt{s^2 + 4}]$ with $s \in (-\infty, \infty)$ are the same: $(0, \infty)$, (see an exercise).

The above indicate that Eq. (1.5) and Eq. (1.8) have the same solution curves in $\Re$, and the solution of Eq. (1.8) exists on $\Re$. This explains that the situation for equation $x' = x^2$, $x(0) = 1$, can be handled very well using Lemmas 4.1.1 and 4.1.2. ♠

Based on Lemmas 4.1.1 and 4.1.2, we assume in the rest of this chapter that $f : \Re^n \to \Re^n$ satisfies a local Lipschitz condition, and, if necessary, an appropriate rescaling of the time variable has been made, such that the solutions of Eq. (1.1) exist on $\Re$. We point out that sometimes a rescaling may modify certain properties, such as stability.

To begin the study of Eq. (1.1), let's look at a special feature of autonomous differential equations. Assume that $x(t)$ is a solution of Eq. (1.1), and consider $y(t) = x(t + c)$ for any constant $c \in \Re$. Then, since $f(x)$ is independent of t, $y(t)$ satisfies

$$y'(t) = x'(t + c) = f(x(t + c)) = f(y(t)), \tag{1.10}$$

that is, $y(t) = x(t + c)$ is also a solution of Eq. (1.1). We point out that this is not true for nonautonomous differential equations, as can be seen from the following example.

Example 4.1.4 Let $x(t)$ be a nonzero solution of the scalar nonautonomous differential equation

$$x'(t) = tx(t), \tag{1.11}$$

which can be solved using separation of variables. Consider $y(t) = x(t+c)$ for $c \neq 0$, then

$$y'(t) = x'(t+c) = (t+c)x(t+c) = tx(t+c) + cx(t+c) = ty(t) + cx(t+c),$$

thus $y(t) = x(t+c)$ is not a solution of Eq. (1.11). ♠

Next, let's continue with the idea of adding a constant to the variable of a solution for autonomous differential equations. Assume that $x(t) = x(t, t_0, x_0)$ is a solution of

$$x'(t) = f(x(t)), \quad \text{or} \quad x' = f(x), \quad x(t_0) = x_0. \tag{1.12}$$

Then, similar to the above, we see that $y(t) = x(t + t_0, t_0, x_0)$ satisfies

$$y'(t) = x'(t + t_0) = f(x(t + t_0)) = f(y(t)), \quad y(0) = x(t_0) = x_0. \tag{1.13}$$

On the other hand, if $y(t)$ is a solution of $y'(t) = f(y(t))$, $y(0) = x_0$, then $x(t) = y(t - t_0)$ satisfies

$$x'(t) = y'(t - t_0) = f(y(t - t_0)) = f(x(t)), \quad x(t_0) = y(0) = x_0. \tag{1.14}$$

That is, for autonomous differential equations, a solution starting at t_0 can be **shifted** to become a solution starting at 0, and vice versa, see **Figure 4.2**.

Accordingly, we will simplify the notations by assuming $t_0 = 0$ in Eq. (1.12), hence we will consider the following autonomous differential equation in this chapter,

$$x'(t) = f(x(t)), \quad \text{or} \quad x' = f(x), \quad x(0) = x_0. \tag{1.15}$$

The following is another point of view at shifting solutions around. Let $x(t, x_0)$ be the unique solution of Eq. (1.15) defined on $\Re$ with $x(0) = x_0$. Recall that x_0 in Eq. (1.15) should be regarded as a variable and that when we say $x(t, \bar{x}_0)$ is a solution of Eq. (1.15) we mean $x'(t, \bar{x}_0) = f(x(t, \bar{x}_0))$ and $x(0, \bar{x}_0) = \bar{x}_0$. Then, uniqueness and continuous dependence on initial values of solutions imply the following (see an exercise).

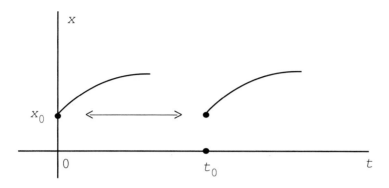

Figure 4.2: For autonomous equations, a solution can be shifted

(i). $x(0, x_0) = x_0$,

(ii). $x(t + s, x_0) = x(t, x(s, x_0))$, $t, s \in \Re$,

(iii). $x(-t, x(t, x_0)) = x_0$, $t \in \Re$,

(iv). $x(t, x_0)$ is continuous in (t, x_0).

Conditions (i)–(iv) are related to the concept of a dynamical system, therefore, we say that "the solutions of Eq. (1.15) define a **dynamical system**." The geometric interpretation of condition (ii) of a dynamical system is that if we start with a point x_0 and go a distance of s, and then use $x(s, x_0)$ as a point and go a distance of t, the result is the same as starting at x_0 and going a distance of $t + s$, see **Figure 4.3**.

The study of autonomous differential equations is a link between the study of linear differential equations and general nonlinear differential equations, especially when we consider stability and boundedness of solutions. To gain further insight and to understand geometric aspects of solutions and dynamical systems of autonomous differential equations, we look at autonomous differential equations in $\Re^2$. That is, in the rest of this chapter we assume $f(x) = [P(x), Q(x)]^T : \Re^2 \to \Re^2$, and consider the autonomous differential equation

$$\begin{cases} x_1'(t) &= P(x_1(t), x_2(t)), \\ x_2'(t) &= Q(x_1(t), x_2(t)), \quad x_1, \ x_2, \ t \in \Re, \end{cases} \qquad (1.16)$$

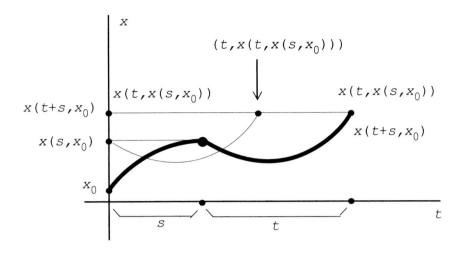

Figure 4.3: Geometric view of condition (ii) of a dynamical system

where the functions $P(x_1, x_2)$ and $Q(x_1, x_2)$ are continuous and satisfy a local Lipschitz condition, and we assume that for any given point (x_1^0, x_2^0) in the (x_1, x_2) plane, Eq. (1.16) has a unique solution $(x_1(t), x_2(t))$ defined on $\Re$ with

$$(x_1(0), x_2(0)) = (x_1^0, x_2^0).$$

Now, as t varies, a solution $(x_1(t), x_2(t))$ of Eq. (1.16) describes a curve in the (x_1, x_2) plane. We call this curve a **trajectory** or an **orbit**, and call the (x_1, x_2) plane the **phase plane**, and call a picture with trajectories of Eq. (1.16) a **phase portrait**, see **Figure 4.4**.

Note that the notion of "a solution" and "a trajectory" are different, see the following example.

Example 4.1.5 Let $\alpha, \beta \in [0, \frac{\pi}{2}]$ be different. Then $(\sin(t + \alpha), \cos(t + \alpha))$ and $(\sin(t + \beta), \cos(t + \beta))$ are different solutions of the system $\{x_1' = x_2, x_2' = -x_1\}$. However, after the variable t varies in a large interval, say $[0, 2\pi]$ or $(-\infty, \infty)$, they represent the same trajectory: the unit circle $x_1^2 + x_2^2 = 1$, see **Figure 4.5**. ♠

Example 4.1.5 indicates that if existence and uniqueness of solutions is assumed, then for a given point in the (x_1, x_2) plane, there is a unique

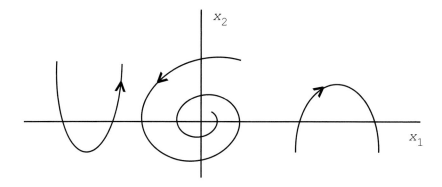

Figure 4.4: A phase portrait in the phase plane (x_1, x_2) with some trajectories

---

solution passing through the point at a specified time, but different solutions could pass through that given point at different times. For example, $(\sin t, \cos t)$ passes through $(0,1)$ at $t = 0$, and $(\sin(t + \frac{\pi}{2}), \cos(t + \frac{\pi}{2}))$ passes through $(0,1)$ at $t = \frac{3\pi}{2}$. Example 4.1.5 also indicates that for a given point in the (x_1, x_2) plane, exactly one trajectory will pass through it, which will be proved below. But first, when we say "a trajectory," we mean a curve drawn in a phase plane after the variable t varies in a sufficiently large interval, typically $[0, \infty)$ or $(-\infty, \infty)$. (Otherwise, we may get only a segment of a whole trajectory. For example, for $(\sin(t), \cos(t))$, if we only consider $t \in [0, \pi]$, then we get a half circle, but the trajectory is a whole circle.)

Theorem 4.1.6 *Assume the existence and uniqueness of solutions for Eq. (1.16). Then through any point in the phase plane* $\Re^2$ *passes exactly one trajectory of Eq. (1.16).*

Proof. If $(x_1(t), y_1(t))$, and $(x_2(t), y_2(t))$ are two trajectories of Eq. (1.16) that pass through a common point in $\Re^2$ at the time t_1 and t_2 respectively, that is,

$$(x_1(t_1), y_1(t_1)) = (x_2(t_2), y_2(t_2)), \tag{1.17}$$

then we prove that the trajectories $(x_1(t), y_1(t))$ and $(x_2(t), y_2(t))$ coincide.

If $t_1 = t_2$, then by uniqueness of solutions, $(x_1(t), y_1(t)) = (x_2(t), y_2(t))$, the two trajectories coincide. If $t_1 \neq t_2$, then, as Eq. (1.16) is autonomous,

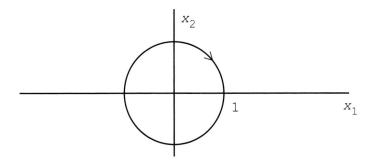

Figure 4.5: The unit circle $x_1^2 + x_2^2 = 1$ is a trajectory for $\{x_1' = x_2,\ x_2' = -x_1\}$

---

an argument similar to the "shifting" in (1.13) shows that

$$\begin{cases} x(t) = x_1(t + t_1 - t_2), \\ y(t) = y_1(t + t_1 - t_2), \end{cases} \qquad (1.18)$$

is also a solution of Eq. (1.16), with

$$(x(t_2),\ y(t_2)) = (x_1(t_1),\ y_1(t_1)) = (x_2(t_2),\ y_2(t_2)). \qquad (1.19)$$

Now, applying uniqueness of solutions again, we obtain $(x(t),\ y(t)) = (x_2(t),\ y_2(t))$, or

$$(x_1(t + t_1 - t_2),\ y_1(t + t_1 - t_2)) = (x_2(t),\ y_2(t)). \qquad (1.20)$$

According to (1.20), a point passed by the trajectory $(x_2,\ y_2)$ at t will be passed by the trajectory $(x_1,\ y_1)$ at $t + t_1 - t_2$. On the other hand, if a point is passed by the trajectory $(x_1,\ y_1)$ at s, then from $s = t + t_1 - t_2$, it follows that the point will be passed by the trajectory $(x_2,\ y_2)$ at $s - t_1 + t_2$. Thus, after the variable t varies in a sufficiently large interval, or $(-\infty,\ \infty)$, the two trajectories coincide. This completes the proof. ♠

Consequently, for any two trajectories in the phase plane $\Re^2$, either they are identical, or they have no common points.

Next, let's recall from calculus that the points in the domain of a scalar function where the first derivative being zero are called critical points, or extreme points, because they are "critically" or "extremely" important in understanding geometric properties of a function. For example, the maximum or minimum values may occur there. Similarly, we make the following definition.

Definition 4.1.7 *In Eq. (1.16), if a point $(x_1^c,\ x_2^c)$ is such that*

$$P(x_1^c,\ x_2^c) = Q(x_1^c,\ x_2^c) = 0, \tag{1.21}$$

*then $(x_1^c,\ x_2^c)$ is called a **critical point** (or an **equilibrium point**) of Eq. (1.16). A point is called a **regular point** of Eq. (1.16) if it is not a critical point of Eq. (1.16).*

Example 4.1.8 For the motion of a simple pendulum,

$$\theta''(t) + k\theta'(t) + q\sin\theta(t) = 0, \tag{1.22}$$

where $k \geq 0$, $q > 0$ are constants, if we let $x_1 = \theta$ and $x_2 = \theta'$, then it is equivalent to

$$\begin{cases} x_1' &=& x_2, \\ x_2' &=& -kx_2 - q\sin x_1. \end{cases} \tag{1.23}$$

Now the critical points for Eq. (1.23) are from $x_2 = 0$ and $kx_2 + q\sin x_1 = 0$, hence they are given by $(x_1, x_2) = (n\pi,\ 0)$, $n = 0, \pm 1,\ \pm 2, \cdots$. ♠

For a regular point $(x_1^r,\ x_2^r) \in \Re^2$, the constant vector $[x_1^r,\ x_2^r]^T$ is not a constant solution of Eq. (1.16); and when a trajectory passes through $(x_1^r,\ x_2^r)$, the direction vector (or slope vector) of the trajectory at $(x_1^r,\ x_2^r)$ is given by the vector function $[P, Q]^T$ evaluated at $(x_1^r,\ x_2^r)$. That is, the direction vector at $(x_1^r,\ x_2^r)$ is given by

$$[P(x_1^r,\ x_2^r),\ Q(x_1^r,\ x_2^r)]^T, \tag{1.24}$$

which is a nonzero vector. Thus by continuity, if another point $(\overline{x},\ \overline{y})$ is very close to $(x_1^r,\ x_2^r)$, then, for a trajectory passing through $(\overline{x},\ \overline{y})$, the direction vector at $(\overline{x},\ \overline{y})$, given by $[P(\overline{x},\ \overline{y}),\ Q(\overline{x},\ \overline{y})]^T$, is very close to $[P(x_1^r,\ x_2^r),\ Q(x_1^r,\ x_2^r)]^T$ and is nonzero. The geometric interpretation is that near a regular point, trajectories **move in essentially the same direction**. Therefore, the local geometrical and qualitative properties for regular points are very clear, see **Figure 4.6**.

However, for a critical point $(x_1^c,\ x_2^c) \in \Re^2$, the constant vector $[x_1^c,\ x_2^c]^T$ is a constant solution of Eq. (1.16). The geometric interpretation is that a

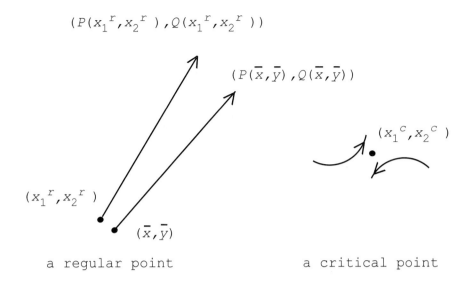

Figure 4.6: Local geometry for a regular point and a critical point

critical point (regarded as a constant solution) is the only trajectory passing through itself; and the direction vector at a critical point is zero, or **directionless**. Thus, trajectories near a critical point could **move in all different directions**, see Figure 4.6. Since this makes the local geometrical and qualitative properties of critical points really complicated, the rest of this chapter is devoted to analyzing them.

We first look at what kinds of trajectories are possible for Eq. (1.16) in $\Re^2$. For example, you may wonder if figures that look like "4," "6," "8," or "0" are possible trajectories for Eq. (1.16). Accordingly, let's recall that the curves that look like a "0" are called **simple closed curves**. For the curves that look like an "8," even though they are closed curves, they are not "simple" closed curves. The following result describes all possible trajectories of Eq. (1.16) in $\Re^2$.

Theorem 4.1.9 *Let $(x_1(t), x_2(t))$ be a solution of Eq. (1.16) for $t \in \Re$. If there exist t_1 and t_2 with $t_1 < t_2$ such that $(x_1(t_1), x_2(t_1)) = (x_1(t_2), x_2(t_2))$, then only the following two cases can occur:*

(i). The solution is a constant solution, that is, a critical point, or

(ii). The solution is periodic with a least positive period.

Therefore, the possible trajectories for Eq. (1.16) in $\Re^2$ are (see **Figure 4.7**):

(a). Critical points, or

(b). Simple closed curves, also called **periodic orbits***, or*

(c). Nonintersecting curves.

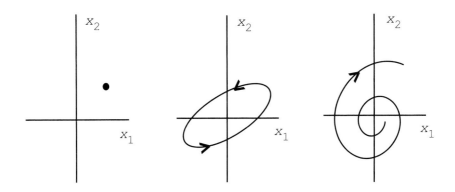

Figure 4.7: Three possible trajectories for Eq. (1.16) in $\Re^2$

Proof. Under the assumption and using the fact that Eq. (1.16) is autonomous, we have

$$(x_1(t),\, x_2(t)) = (x_1(t + t_2 - t_1),\, x_2(t + t_2 - t_1)), \qquad (1.25)$$

because they are all solutions of Eq. (1.16) and pass through the same point when $t = t_1$. This implies that $(x_1(t),\, x_2(t))$ is $(t_2 - t_1)$-periodic. Hence, we can define

$$\bar{p} = \liminf\{\, p :\ 0 < p \leq t_2 - t_1,\ (x_1(t),\, x_2(t)) \text{ is } p\text{--periodic}\}. \qquad (1.26)$$

Consequently, $0 \leq \bar{p} \leq t_2 - t_1$, and there is a sequence $p_m \in [\bar{p},\, t_2 - t_1]$ with $\lim_{m \to \infty} p_m = \bar{p}$, such that $(x_1(t),\, x_2(t))$ is p_m-periodic, see **Figure 4.8**.

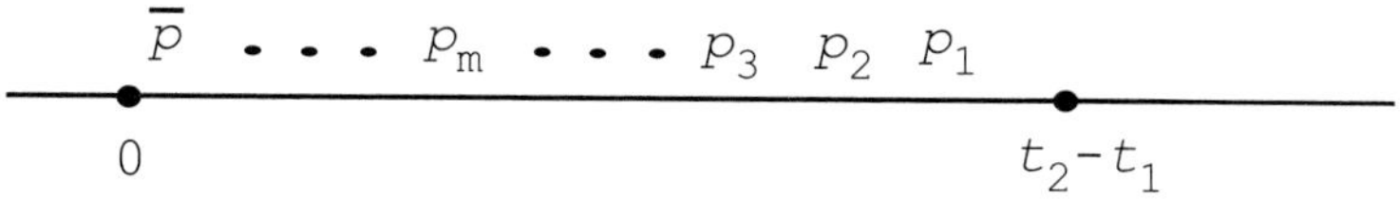

$$\overline{p} \quad \cdots \quad p_m \quad \cdots \quad p_3 \ p_2 \ p_1$$

$$0 \qquad\qquad\qquad\qquad\qquad t_2 - t_1$$

Figure 4.8: $0 \leq \overline{p} \leq t_2 - t_1$ and $\lim_{m \to \infty} p_m = \overline{p}$

If $\overline{p} = 0$, then let's prove that for any $t \in \Re$, $(x_1(t), x_2(t)) = (x_1(0), x_2(0))$, which implies that $(x_1(t), x_2(t))$ is a critical point. We assume $t > 0$ because the case of $t < 0$ is similar. Now, for $p_m > 0$, we get an integer $k_m \geq 0$ such that $t = k_m p_m + t_m$, with $0 \leq t_m < p_m$ and $0 \leq \lim_{m \to \infty} t_m \leq \lim_{m \to \infty} p_m = \overline{p} = 0$. Thus, as the solution $(x_1(\cdot), x_2(\cdot))$ is p_m-periodic,

$$(x_1(t), x_2(t)) = (x_1(k_m p_m + t_m), x_2(k_m p_m + t_m)) = (x_1(t_m), x_2(t_m)),$$

therefore, using continuity of solutions, it follows that

$$
\begin{aligned}
(x_1(t), x_2(t)) &= \lim_{m \to \infty} (x_1(t), x_2(t)) \\
&= \lim_{m \to \infty} (x_1(t_m), x_2(t_m)) = (x_1(0), x_2(0)),
\end{aligned}
$$

that is, the solution $(x_1(t), x_2(t))$ is a critical point in this case.

If $\overline{p} > 0$, let's prove that the solution $(x_1(t), x_2(t))$ is a simple closed curve. Applying continuity and periodicity again, we obtain

$$
\begin{aligned}
(x_1(t), x_2(t)) &= \lim_{m \to \infty} (x_1(t), x_2(t)) = \lim_{m \to \infty} (x_1(t + p_m), x_2(t + p_m)) \\
&= (x_1(t + \overline{p}), x_2(t + \overline{p})), \tag{1.27}
\end{aligned}
$$

which implies that $\overline{p}$ is the least positive period for the solution. Now, when the solution $(x_1(t), x_2(t))$ starts from $t = 0$ at $(x_1(0), x_2(0)) \in \Re^2$, the $\overline{p} > 0$ is the first time such that $(x_1(0), x_2(0)) = (x_1(\overline{p}), x_2(\overline{p}))$, (otherwise, if there is a $0 < p_0 < \overline{p}$, such that $(x_1(0), x_2(0)) = (x_1(p_0), x_2(p_0))$, then uniqueness implies that $(x_1(t), x_2(t)) = (x_1(t + p_0), x_2(t + p_0))$, $t \in \Re$; thus $p_0 < \overline{p}$ is also a positive period, which contradicts that $\overline{p}$ is the least positive period for the solution). After $t = \overline{p}$, the solution will repeat since it is $\overline{p}$-periodic. Therefore, the trajectory of the solution is now a simple closed curve. This completes the proof. ♠

Accordingly, we find that figures that look like "4" or "6" are not possible trajectories because they are intersecting curves. Curves that look like an

"8" are not possible trajectories either because even though they are closed curves, they are not simple closed curves. Curves that look like a "0" are possible trajectories because they are simple closed curves. The following is also true, whose proof is left as an exercise.

Corollary 4.1.10 *Let $x(t) = x(t, x_0)$ be the unique solution of Eq. (1.16) on $\Re$ with $x(0) = x_0$. If x_0 is not a critical point, then there exist $t_1 > 0$ and $t_2 < 0$ such that $x(t_1) \neq x_0$ and $x(t_2) \neq x_0$.*   ♠

Next, if $P(x_1, x_2)$ and $Q(x_1, x_2)$ have continuous partial derivatives, and have the Taylor expansion

$$\begin{cases} P(x_1, x_2) = ax_1 + bx_2 + \varepsilon_1(x_1, x_2), \\ Q(x_1, x_2) = cx_1 + dx_2 + \varepsilon_2(x_1, x_2), \end{cases} \tag{1.28}$$

at $(x_1, x_2) = (0, 0)$ for some functions $\varepsilon_1(x_1, x_2)$ and $\varepsilon_2(x_1, x_2)$, where

$$a = \frac{\partial P}{\partial x_1}(0,0), \ \ b = \frac{\partial P}{\partial x_2}(0,0), \ \ c = \frac{\partial Q}{\partial x_1}(0,0), \ \ d = \frac{\partial Q}{\partial x_2}(0,0), \tag{1.29}$$

then Eq. (1.16) becomes

$$\begin{aligned} \begin{bmatrix} x_1(t) \\ x_2(t) \end{bmatrix}' &= \begin{bmatrix} ax_1(t) + bx_2(t) + \varepsilon_1(x_1(t), x_2(t)) \\ cx_1(t) + dx_2(t) + \varepsilon_2(x_1(t), x_2(t)) \end{bmatrix} \\ &= \begin{bmatrix} a & b \\ c & d \end{bmatrix} \begin{bmatrix} x_1(t) \\ x_2(t) \end{bmatrix} + \begin{bmatrix} \varepsilon_1(x_1(t), x_2(t)) \\ \varepsilon_2(x_1(t), x_2(t)) \end{bmatrix}. \end{aligned} \tag{1.30}$$

Example 4.1.11 Consider

$$\begin{cases} x_1' &= \sin x_2, \\ x_2' &= x_1 e^{x_2}. \end{cases} \tag{1.31}$$

Since $\sin x_2 = x_2 - \frac{x_2^3}{3!} + \frac{x_2^5}{5!} - \cdots$ and $x_1 e^{x_2} = x_1 + x_1 x_2 + \frac{x_1 x_2^2}{2!} + \cdots$, the equation can be written as

$$\begin{bmatrix} x_1(t) \\ x_2(t) \end{bmatrix}' = \begin{bmatrix} 0 & 1 \\ 1 & 0 \end{bmatrix} \begin{bmatrix} x_1(t) \\ x_2(t) \end{bmatrix} + \begin{bmatrix} \varepsilon_1(x_1(t), x_2(t)) \\ \varepsilon_2(x_1(t), x_2(t)) \end{bmatrix}, \tag{1.32}$$

where $\varepsilon_1(x_1, x_2) = -\frac{x_2^3}{3!} + \frac{x_2^5}{5!} - \cdots$ and $\varepsilon_2(x_1, x_2) = x_1 x_2 + \frac{x_1 x_2^2}{2!} + \cdots$.   ♠

Definition 4.1.12 *With constants a, b, c, d given by (1.29), the linear differential equation*

$$\left[\begin{array}{c} x_1(t) \\ x_2(t) \end{array}\right]' = \left[\begin{array}{cc} a & b \\ c & d \end{array}\right]\left[\begin{array}{c} x_1(t) \\ x_2(t) \end{array}\right] \tag{1.33}$$

is called the **linearization** *of Eq. (1.30) (or of Eq. (1.16)) at $(0,0)$.*

Under certain conditions, we can regard Eq. (1.30) as a perturbation of Eq. (1.33). Now it intuitively makes sense to expect that if $\varepsilon_1(x_1, x_2)$ and $\varepsilon_2(x_1, x_2)$ are sufficiently small, then the behavior of Eq. (1.30) should be similar to that of Eq. (1.33). It can be shown that this intuition is essentially, but not completely, correct.

This chapter is organized as follows: In Section 2, we provide a complete analysis for Eq. (1.33) and draw all phase portraits in $\Re^2$ for the different cases according to eigenvalues of the coefficient matrix. We also introduce some terminology, including stability, according to the properties revealed, which leads us to detailed study of the same subject later for general differential equations in $\Re^n, n \geq 1$. In Section 3, we examine the conditions which ensure that solutions of Eq. (1.30) and Eq. (1.33) have essentially the same local geometric and qualitative properties near the origin. In Section 4, we apply the results to analyze an equation of a simple pendulum. In Section 5, we generalize the ideas of a simple pendulum and study the Hamiltonian systems and gradient systems.

Exercises 4.1

1. Verify for autonomous differential equations that a weak Lipschitz condition is the same as a (global) Lipschitz condition.

2. Verify that $f(x) = x^2$ is unbounded and does not satisfy a (global) Lipschitz condition for x in $\Re$.

3. In Example 4.1.3, use (1.7) to verify that when t takes values in $(-\infty, 1)$ for the solution $x(t) = (1-t)^{-1}$, the variable s takes values in $(-\infty, \infty)$ for the corresponding solution $y(s)$.

4. Assume that $f : \Re^n \to \Re^n$ satisfies a local Lipschitz condition. Prove that $F(x) = f(x)\frac{1}{1+|f(x)|}$ also satisfies a local Lipschitz condition.

5. Solve $y' = \frac{y^2}{1+y^2}$, $y(0) = 1$.

6. Verify that the range of $y(s)$ in (1.9) is $(0, \infty)$.

7. Let $x(t)$ be a solution of $x'(t) = x^2 \sin t + x^3$. Is $x(t+2)$ also a solution? Next, find k such that $x(t + k)$ is also a solution.

8. Let $T > 0$ be given. Find conditions on $f(t, x)$ in $x'(t) = f(t, x(t))$ such that if $x(t)$ is a solution, then so is $y(t) = x(t + T)$.

9. Show that the solutions of Eq. (1.15) form a dynamical system.

10. Show that a point (x_1^c, x_2^c) is a critical point of Eq. (1.16) if and only if $(x_1(t), x_2(t)) = (x_1^c, x_2^c)$, $t \in \Re$, is a constant solution.

11. Prove that a trajectory started from a regular point may not reach a critical point in finite time.

12. Find critical points of the system

$$\begin{cases} x' &=& x - 2x^2 - xy, \\ y' &=& -4y + 2xy. \end{cases}$$

13. Show that $x(t) = \cos t$, $y(t) = \sin t$ satisfy

$$\begin{cases} x'(t) &=& -y(t), \\ y'(t) &=& x(t). \end{cases}$$

Then draw $(x(t), y(t)) = (\cos t, \sin t)$ in the (x, y) plane for $t \in [0, \pi]$, $t \in [0, 2\pi]$, $t \in [0, \infty)$, and $t \in (-\infty, \infty)$.

14. Prove Corollary 4.1.10.

4.2   Linear Autonomous Equations in $\Re^2$

In this section, we consider

$$\begin{bmatrix} x_1(t) \\ x_2(t) \end{bmatrix}' = \begin{bmatrix} a & b \\ c & d \end{bmatrix} \begin{bmatrix} x_1(t) \\ x_2(t) \end{bmatrix} = A \begin{bmatrix} x_1(t) \\ x_2(t) \end{bmatrix}, \tag{2.1}$$

where a, b, c, d are constants in $\Re$. Consequently, the origin $(0, 0)$ is a constant solution, or a critical point.

Recall that if P is a 2×2 real matrix and v is a vector in $\Re^2$, then Pv will transform (rotate and/or magnify or shrink) v to become another vector in $\Re^2$. Accordingly, we will find a real matrix P, and use it to transform Eq. (2.1) into a simpler equation for which the trajectories can be drawn easily. If this can be done, then the geometrical and qualitative properties of the transformed simpler equations can be used to indicate the same geometrical and qualitative properties of the original Eq. (2.1), because the trajectories of the two equations differ only by a rotation and/or a magnification or shrinking in $\Re^2$.

Note that the Jordan canonical form theorem in Chapter 3 cannot be applied here, because the transformation matrix P used there may be complex valued. However, we are dealing with 2×2 matrices here, so we can afford to provide a complete analysis and then find ways to transform Eq. (2.1) into simpler forms. First, we recall some general results (with no restrictions on the size of matrices) from linear algebra.

Lemma 4.2.1 *Let A be an $n \times n$ matrix. Then,*

(a). *If λ is an eigenvalue of the matrix A and v is the corresponding eigenvector (they may be complex valued), then $e^{\lambda t}v$ is a solution of $x'(t) = Ax(t)$.*

(b). *If A, u, v are real and if $u(t) + iv(t)$ ($i = \sqrt{-1}$) is a solution of $x'(t) = Ax(t)$, then $u(t)$ and $v(t)$ are both solutions of $x'(t) = Ax(t)$.* ♠

The proof is left as an exercise. With this preparation, we are able to transform Eq. (2.1) into simpler forms.

Theorem 4.2.2 *Let A be a real 2×2 constant matrix in Eq. (2.1). Then there is a real 2×2 constant nonsingular matrix P, such that $[x, y]^T = P^{-1}[x_1, x_2]^T$ transforms Eq. (2.1) into*

$$\left[\begin{array}{c} x(t) \\ y(t) \end{array} \right]' = \left[P^{-1}AP \right] \left[\begin{array}{c} x(t) \\ y(t) \end{array} \right] = B \left[\begin{array}{c} x(t) \\ y(t) \end{array} \right], \qquad (2.2)$$

where $B = P^{-1}AP$ is one of the following three real matrices

$$(I). \ \left[\begin{array}{cc} \lambda & 0 \\ 0 & \mu \end{array} \right], \ (\lambda \neq \mu), \quad (II). \ \left[\begin{array}{cc} \lambda & \star \\ 0 & \lambda \end{array} \right], \quad (III). \ \left[\begin{array}{cc} \alpha & \beta \\ -\beta & \alpha \end{array} \right], \qquad (2.3)$$

where $\star = 0$ or 1. Moreover, (I) occurs when A has two distinct real eigenvalues (or equivalently $(a-d)^2 + 4bc > 0$); (II) occurs when A has a double (or repeated) real eigenvalue (or equivalently $(a-d)^2 + 4bc = 0$); and (III) occurs when A has complex conjugate eigenvalues (or equivalently $(a-d)^2 + 4bc < 0$).

Proof. We prove the theorem case by case.

Case (*I*): The matrix A has two real distinct eigenvalues λ and μ. Now we can let u and v be the real eigenvectors corresponding to λ and μ respectively. From the results in Chapter 3, we let $P = [u, v]$ such that P is nonsingular, and

$$\begin{aligned}
P^{-1}AP &= P^{-1}A[u, v] = P^{-1}[\lambda u, \mu v] \\
&= [\lambda P^{-1}u, \mu P^{-1}v] = [\lambda e_1, \mu e_2] = diag(\lambda, \mu). \quad (2.4)
\end{aligned}$$

Case (*II*): The matrix A has a double real eigenvalue λ. Let u be the corresponding real eigenvector, and let v be any real vector such that $P_1 = [u, v]$ is nonsingular. Now,

$$\begin{aligned}
P_1^{-1}AP_1 &= P_1^{-1}A[u, v] = P_1^{-1}[\lambda u, Av] = [\lambda P_1^{-1}u, P_1^{-1}Av] \\
&= [\lambda e_1, P_1^{-1}Av] = \begin{bmatrix} \lambda & g \\ 0 & h \end{bmatrix}, \quad (2.5)
\end{aligned}$$

for some real numbers g and h. Since similar matrices have the same eigenvalues, we get $h = \lambda$. If $g = 0$, then we can take $P = P_1$, such that

$$P^{-1}AP = \begin{bmatrix} \lambda & 0 \\ 0 & \lambda \end{bmatrix}. \quad (2.6)$$

If $g \neq 0$, let $P_2 = \begin{bmatrix} 1 & 1 \\ 0 & \frac{1}{g} \end{bmatrix}$, then

$$P_2^{-1}\begin{bmatrix} \lambda & g \\ 0 & \lambda \end{bmatrix}P_2 = \begin{bmatrix} 1 & -g \\ 0 & g \end{bmatrix}\begin{bmatrix} \lambda & g \\ 0 & \lambda \end{bmatrix}\begin{bmatrix} 1 & 1 \\ 0 & \frac{1}{g} \end{bmatrix} = \begin{bmatrix} \lambda & 1 \\ 0 & \lambda \end{bmatrix}, \quad (2.7)$$

hence, we can take $P = P_1 P_2$, such that

$$P^{-1}AP = P_2^{-1}P_1^{-1}AP_1P_2 = P_2^{-1}\begin{bmatrix} \lambda & g \\ 0 & \lambda \end{bmatrix}P_2 = \begin{bmatrix} \lambda & 1 \\ 0 & \lambda \end{bmatrix}. \quad (2.8)$$

Case (*III*): The matrix A has complex conjugate eigenvalues $\lambda_1 = \alpha + i\beta$, $\lambda_2 = \alpha - i\beta$, $\beta \neq 0$. Let $u + iv$ be an eigenvector of λ_1 with u and v being real, then $A(u + iv) = (\alpha + i\beta)(u + iv)$. Equating the real and imaginary parts, we get

$$Au = \alpha u - \beta v, \quad Av = \beta u + \alpha v. \quad (2.9)$$

Hence, as $\beta \neq 0$, it follows from (2.9) that $u = 0$ if and only if $v = 0$. But the eigenvector $u + iv$ is nonzero, thus $u \neq 0$ and $v \neq 0$. That is, real or pure imaginary vectors cannot be eigenvectors for the eigenvalue $\alpha + i\beta$.

Moreover, u and v are linearly independent. Suppose this is not the case; then u and v are linearly dependent, and there exist real constants c_1 and c_2, both nonzero in this case, such that $c_1 u + c_2 v = 0$. Then $A(u + iv) = (\alpha + i\beta)(u + iv)$ becomes

$$A\left(1 - \frac{c_1}{c_2}i\right)u = (\alpha + i\beta)\left(1 - \frac{c_1}{c_2}i\right)u, \tag{2.10}$$

which implies $Au = (\alpha + i\beta)u$, hence u is a real eigenvector for the eigenvalue $\alpha + i\beta$. This contradicts what we have just proved that $\alpha + i\beta$ has no real eigenvectors. Thus u and v are linearly independent.

Now, $P = [u, v]$ is nonsingular, and we have, from (2.9),

$$P^{-1}AP = P^{-1}[Au, Av] = P^{-1}[\alpha u - \beta v, \ \beta u + \alpha v]$$

$$= [\alpha e_1 - \beta e_2, \ \beta e_1 + \alpha e_2] = \begin{bmatrix} \alpha & \beta \\ -\beta & \alpha \end{bmatrix}. \tag{2.11}$$

This completes the transformation of the matrix A into the three simpler forms in (2.3). Next, from $[x, y]^T = P^{-1}[x_1, x_2]^T$ and Eq. (2.1), we have

$$\begin{bmatrix} x(t) \\ y(t) \end{bmatrix}' = P^{-1}\begin{bmatrix} x_1(t) \\ x_2(t) \end{bmatrix}' = P^{-1}A\begin{bmatrix} x_1(t) \\ x_2(t) \end{bmatrix} = \begin{bmatrix} P^{-1}AP \end{bmatrix}\begin{bmatrix} x(t) \\ y(t) \end{bmatrix}. \tag{2.12}$$

Finally, observe that the characteristic equation for matrix A is

$$\lambda^2 - (a + d)\lambda + ad - bc = 0,$$

thus the distributions of eigenvalues are determined by

$$(a + d)^2 - 4(ad - bc) = (a - d)^2 + 4bc.$$

This completes the proof. ♠

Next, we will draw all phase portraits according to the transformed simpler equation (2.2) in the (x, y) phase plane, with $B = P^{-1}AP$ given by one of the three forms in (2.3). The phase portraits for the original equation (2.1) in the (x_1, x_2) phase plane will differ from them only by some rotations and/or a magnification or shrinking of the trajectories in the (x, y) phase plane.

We first single out a special case when $\det(A) = 0$, that is, when the zero is an eigenvalue of the matrix A. Since the transformed forms in (2.3) have the same eigenvalues as the matrix A, this may happen when $\lambda = 0$ or $\mu = 0$ in Case (I), or when $\lambda = 0$ in Case (II).

If $\lambda = 0$ and $\mu \neq 0$ in Case (I), then the solutions of Eq. (2.2) are $\{x(t) = x_0, \; y(t) = y_0 e^{\mu t}\}$. In particular, any point on the x-axis (when $y_0 = 0$) is a critical point of Eq. (2.2). The phase portrait is given in **Figure 4.9**, with arrows indicating the directions of increasing time t.

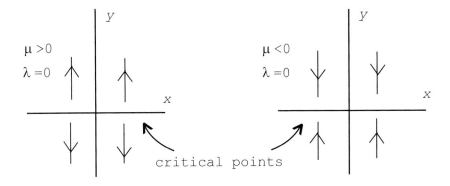

Figure 4.9: Phase portrait for $\lambda = 0$ and $\mu \neq 0$ in Case (I) when $\det(A) = 0$

If now we interchange x and y axes, or rotate the phase portrait for $\{\lambda = 0, \; \mu \neq 0\}$ in Figure 4.9 by 90^0, then we get the phase portrait for $\{\lambda \neq 0, \; \mu = 0\}$ in Case (I).

The phase portraits for Case (II) with $\lambda = 0$ can be drawn in a similar way, see an exercise.

In the following, we assume $\det(A) \neq 0$, that is, the zero is not an eigenvalue of the matrix A, or equivalently, $(0,0)$ is the only critical point of Eq. (2.2). Based on Theorem 4.1.9, we know that in this case, besides the only critical point at the origin, other trajectories are periodic orbits or nonintersecting curves. We will draw phase portraits for Eq. (2.2) in the (x, y) phase plane case by case, according to different forms of $B = P^{-1}AP$ in (2.3).

Case (Ia) : $\begin{bmatrix} \lambda & 0 \\ 0 & \mu \end{bmatrix}$, $(\lambda \neq \mu)$, two distinct real eigenvalues, $\lambda\mu > 0$.

In this case, the fundamental matrix solution of Eq. (2.2) is $\begin{bmatrix} e^{\lambda t} & 0 \\ 0 & e^{\mu t} \end{bmatrix}$, and solutions of Eq. (2.2) are $\{x(t) = x_0 e^{\lambda t},\ y(t) = y_0 e^{\mu t}\}$, where the two eigenvalues λ and μ have the same sign. If $\lambda < \mu < 0$, then, as t approaches ∞, $(x(t), y(t))$ approaches $(0, 0)$, and the slope of the trajectory, $\frac{y(t)}{x(t)} = \frac{y_0}{x_0} e^{(\mu - \lambda)t}$ when $x_0 y_0 \neq 0$, approaches $\pm \infty$ depending on the location of (x_0, y_0). The phase portrait is shown in **Figure 4.10**.

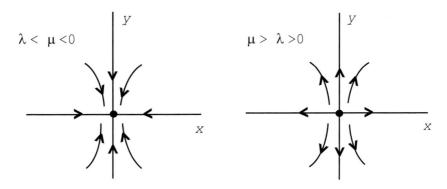

Figure 4.10: Case $(Ia), \lambda \mu > 0$. The origin is an improper node

---

In this case, all nonzero solutions go to the constant solution $[0, 0]^T$ as t approaches ∞; or the constant solution $[0, 0]^T$ **attracts** other solutions. In this sense, we say that the constant solution $[0, 0]^T$ is **asymptotically stable**, or we say that $[0, 0]^T$ is a **sink** or an **attractor**. (A detailed study of stability will be given later.) Now, we call the origin $(0, 0)$ an **improper node**, in the sense that trajectories approach and leave the origin in just two directions. If $\mu < \lambda < 0$, the phase portrait will be rotated by 90^0.

If $\lambda > \mu > 0$, the phase portrait is the same as $\{\lambda < \mu < 0\}$ but with the arrows reversed (consider $t \to -\infty$ near the origin). In this case, the origin $(0, 0)$ is an **unstable node**, or a **repeller**, or a **source**. If $\mu > \lambda > 0$, the phase portrait will be rotated by 90^0.

Case $(Ib):$ $\begin{bmatrix} \lambda & 0 \\ 0 & \mu \end{bmatrix}$, $(\lambda \neq \mu)$, two distinct real eigenvalues, $\lambda \mu < 0$.

In this case, solutions of Eq. (2.2) are $\{x(t) = x_0 e^{\lambda t},\ y(t) = y_0 e^{\mu t}\}$, where the two eigenvalues λ and μ have the opposite signs. Now, as $t \to \infty$, one component may go to 0 while the other goes to $\pm \infty$. For example, if $\mu < 0 < \lambda$, then $|y(t)| = 0$ or $|y(t)|$ goes to 0, and $|x(t)| = 0$ or $|x(t)|$ goes to ∞. The phase portrait is shown in **Figure 4.11**.

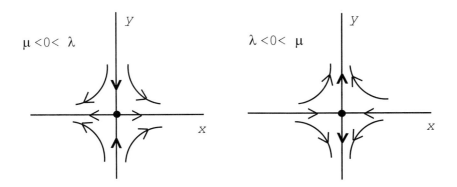

Figure 4.11: Case (Ib), $\lambda\mu < 0$. The origin is a saddle point

---

In this case, we say that the origin $(0,0)$ is a **saddle point**, and the constant solution $[0,0]^T$ is **unstable**. The same is true for the case $\lambda < 0 < \mu$, whose phase portrait is obtained by reversing the arrows.

Case (IIa): $\begin{bmatrix} \lambda & 0 \\ 0 & \lambda \end{bmatrix}$, **double real eigenvalue** $\lambda \neq 0$.

Now, the fundamental matrix solution of Eq. (2.2) is $\begin{bmatrix} e^{\lambda t} & 0 \\ 0 & e^{\lambda t} \end{bmatrix}$, and solutions of Eq. (2.2) are $\{x(t) = x_0 e^{\lambda t}, \ y(t) = y_0 e^{\lambda t}\}$. The trajectories are straight lines $x = 0$, $y = 0$, and $y = \frac{y_0}{x_0}x$, $x_0 \neq 0$. The phase portrait is shown in **Figure 4.12**.

The origin $(0,0)$ is called a **proper node**, in the sense that trajectories approach or leave the origin in all directions. The constant solution $[0,0]^T$ is asymptotically stable (**an attractor**) if $\lambda < 0$ and unstable (**a repeller**) if $\lambda > 0$.

Case (IIb): $\begin{bmatrix} \lambda & 1 \\ 0 & \lambda \end{bmatrix}$, **double real eigenvalue** $\lambda \neq 0$.

Now, the fundamental matrix solution of Eq. (2.2) is $\begin{bmatrix} e^{\lambda t} & te^{\lambda t} \\ 0 & e^{\lambda t} \end{bmatrix}$, and solutions of Eq. (2.2) are $\{x(t) = x_0 e^{\lambda t} + y_0 t e^{\lambda t}, \ y(t) = y_0 e^{\lambda t}\}$. The trajectories as $t \to \infty$ go to $(0,0)$ if $\lambda < 0$, and go far away from $(0,0)$ if $\lambda > 0$. Moreover, the slope $\frac{y(t)}{x(t)} = \frac{y_0}{x_0 + y_0 t} \to 0$, $t \to \infty$. The phase portrait is shown in **Figure 4.13**.

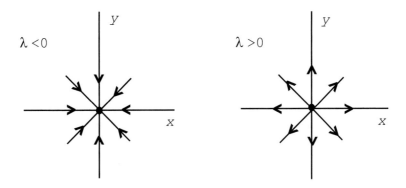

Figure 4.12: Case (IIa), double eigenvalue $\lambda \neq 0$. The origin is a proper node

---

The origin $(0,0)$ now is also called an improper node, and the constant solution $[0,0]^T$ is asymptotically stable if $\lambda < 0$ and unstable if $\lambda > 0$. Note that now an initial point in the second quadrant $(x_0 < 0,\ y_0 > 0)$ will be taken into the first quadrant after $t > -\frac{x_0}{y_0}$ (because then $x(t) = e^{\lambda t}(x_0 + y_0 t) > 0$). Similarly, an initial point in the fourth quadrant will be taken into the third quadrant.

Case $(IIIa)$: $\begin{bmatrix} \alpha & \beta \\ -\beta & \alpha \end{bmatrix}$, **two complex conjugate eigenvalues**
$\alpha \pm i\beta,\ \alpha \neq 0,\ \beta \neq 0.$

In this case, we can check that $[1, i]^T$ is an eigenvector corresponding to the eigenvalue $\alpha + i\beta$. Then from Lemma 4.2.1(a), we know that $e^{\alpha t + i\beta t}[1, i]^T$ is a solution of Eq. (2.2). Now,

$$
e^{\alpha t + i\beta t} \begin{bmatrix} 1 \\ i \end{bmatrix} = e^{\alpha t}(\cos \beta t + i \sin \beta t) \begin{bmatrix} 1 \\ i \end{bmatrix} = \begin{bmatrix} e^{\alpha t}\cos \beta t + i e^{\alpha t} \sin \beta t \\ -e^{\alpha t}\sin \beta t + i e^{\alpha t} \cos \beta t \end{bmatrix}
$$

$$
= \begin{bmatrix} e^{\alpha t}\cos \beta t \\ -e^{\alpha t}\sin \beta t \end{bmatrix} + i \begin{bmatrix} e^{\alpha t}\sin \beta t \\ e^{\alpha t}\cos \beta t \end{bmatrix}, \tag{2.13}
$$

hence from Lemma 4.2.1(b), the real and the imaginary parts in (2.13) are both solutions of Eq. (2.2). Note that at $t = 0$, the real part in (2.13) is $[1, 0]^T = e_1$ and the imaginary part in (2.13) is $[0, 1]^T = e_2$. Thus, according

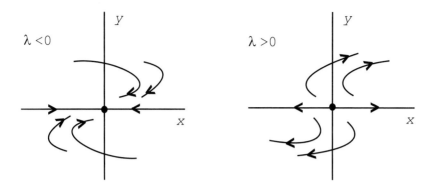

Figure 4.13: Case (*IIb*), double eigenvalue $\lambda \neq 0$. The origin is an improper node

to the study in Chapter 3, the matrix

$$\begin{bmatrix} e^{\alpha t}\cos\beta t & e^{\alpha t}\sin\beta t \\ -e^{\alpha t}\sin\beta t & e^{\alpha t}\cos\beta t \end{bmatrix} \tag{2.14}$$

is the fundamental matrix solution of Eq. (2.2) in this case. Therefore, the solution of Eq. (2.2) with the initial point (x_0, y_0) is given by

$$\begin{bmatrix} x(t) \\ y(t) \end{bmatrix} = \begin{bmatrix} e^{\alpha t}\cos\beta t & e^{\alpha t}\sin\beta t \\ -e^{\alpha t}\sin\beta t & e^{\alpha t}\cos\beta t \end{bmatrix} \begin{bmatrix} x_0 \\ y_0 \end{bmatrix}, \tag{2.15}$$

hence, we obtain

$$x^2(t) + y^2(t) = e^{2\alpha t}(x_0^2 + y_0^2). \tag{2.16}$$

Now, note from (2.15) that when $t \to \infty$, $x(t)$ and $y(t)$ will change signs according to $\sin\beta t$ and $\cos\beta t$, or the trajectory $(x(t), y(t))$ will traverse every quadrant infinitely many times as $t \to \infty$. Thus, if $\alpha < 0$, then from (2.15) and (2.16), we find that the trajectory will spiral to the origin as $t \to \infty$. The trajectory will spiral out from the origin if $\alpha > 0$. The phase portrait is shown in **Figure 4.14**.

In this case, the origin is called a **spiral point** (or a **focus**), according to the fashion a trajectory approaches or leaves the origin. The constant solution $[0, 0]^T$ is asymptotically stable if $\alpha < 0$ or unstable if $\alpha > 0$.

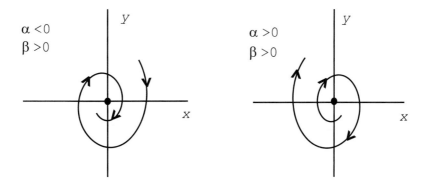

Figure 4.14: Case $(IIIa)$. The origin is a spiral point

Case $(IIIb)$: $\begin{bmatrix} 0 & \beta \\ -\beta & 0 \end{bmatrix}$, **two complex conjugate eigenvalues**

$\pm i\beta$, $\beta \neq 0$.

In this case, all we need is to let $\alpha = 0$ in (2.15)–(2.16) to obtain

$$x^2(t) + y^2(t) = x_0^2 + y_0^2, \tag{2.17}$$

such that the trajectories are circles (periodic orbits). Now, any given trajectory (circle) will not approach the origin, thus the constant solution $[0,0]^T$ is not asymptotically stable. However, if an initial point (x_0, y_0) is near the origin, then the corresponding trajectory will stay near the origin. In this sense, we say that the constant solution $[0,0]^T$ is **stable**. The origin is now called a **center**. The phase portrait is shown in **Figure 4.15**.

We have now completed the analysis and phase portraits for Eq. (2.2) with $B = P^{-1}AP$ given by the three forms in (2.3). This detailed analysis also reveals a **Distribution Diagram** given in **Figure 4.16**, which can be used to determine the nature of the origin very easily by simply looking at the coefficient matrix A of Eq. (2.1).

Corollary 4.2.3 *For Eq. (2.1), let $p = tr A = a + d$, $q = \det A = ad - bc \neq 0$. Then on the (p,q) plane, one has the Distribution Diagram in Figure 4.16 for the origin of Eq. (2.1). That is, one has*

1. *If $q < 0$, then the origin is a saddle point.*

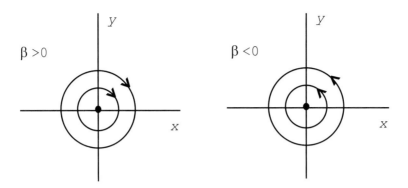

Figure 4.15: Case $(IIIb)$. The origin is a center

2. *If $q > 0$ and $p = 0$, then the origin is a center.*

3. *If $q > 0$, $p > 0$, and $q = \frac{1}{4}p^2$, then the origin is an unstable proper or improper node.*

4. *If $q > 0$, $p > 0$, and $q < \frac{1}{4}p^2$, then the origin is an unstable improper node.*

5. *If $q > 0$, $p > 0$, and $q > \frac{1}{4}p^2$, then the origin is an unstable spiral point.*

6. *If $q > 0$, $p < 0$, and $q = \frac{1}{4}p^2$, then the origin is a stable proper or improper node.*

7. *If $q > 0$, $p < 0$, and $q < \frac{1}{4}p^2$, then the origin is a stable improper node.*

8. *If $q > 0$, $p < 0$, and $q > \frac{1}{4}p^2$, then the origin is a stable spiral point.*

Proof. Let λ_1 and λ_2 be the two eigenvalues of the matrix A. Compare

$$\det(A - \lambda E) = \lambda^2 - (a + d)\lambda + ad - bc = (\lambda - \lambda_1)(\lambda - \lambda_2),$$

we find that p and q can be replaced by

$$p = \lambda_1 + \lambda_2, \quad q = \lambda_1 \lambda_2.$$

Note that the matrix A is similar to the matrices given in (2.3), and similar matrices have the same eigenvalues, thus λ_1 and λ_2 are also eigenvalues

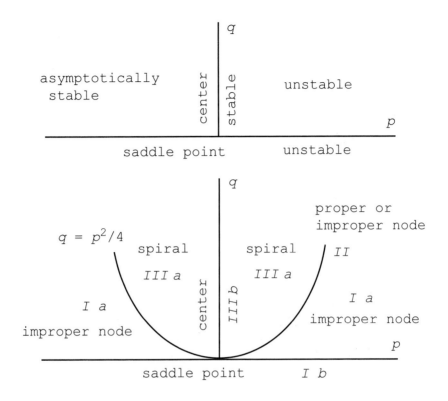

Figure 4.16: Distribution Diagram for the origin of Eq. (2.1)

---

of the matrices given in (2.3). Hence all we need next is to examine each of the above cases from Case (Ia) to Case $(IIIb)$.

In Case (Ia), $p = \lambda_1 + \lambda_2 = \lambda + \mu$, $q = \lambda_1 \lambda_2 = \lambda\mu$ and $\lambda \neq \mu$. Thus, $q = \lambda\mu > 0$ and $q < \frac{1}{4}p^2$. If $p < 0$, which is the same as $\lambda < 0$ and $\mu < 0$ since $\lambda\mu > 0$, then the origin is asymptotically stable and is an improper node. If $p > 0$, which is the same as $\lambda > 0$ and $\mu > 0$, then the origin is unstable and is also an improper node. In Case (Ib), $q = \lambda\mu < 0$ and the origin is a saddle point. In Case (IIa), $p = 2\lambda$, $q = \lambda^2$. Thus, $q = \frac{1}{4}p^2$ and the origin is a proper node, which is asymptotically stable when $p = 2\lambda < 0$ or unstable when $p = 2\lambda > 0$. In Case $(IIIb)$, $q = (i\beta)(-i\beta) = \beta^2 > 0$ and $p = 0 + 0 = 0$, and the origin is a center. Other cases are left as an exercise. This completes the proof. ♠

Figure 4.16 can also be regarded as a "bifurcation diagram" because when the parameters p and q change in the (p, q) plane, the stability of the origin of Eq. (2.1) will change among four types: asymptotically stable, stable, unstable, and half stable (a saddle point).

Next, we point out that the phase portraits of the original differential equation (2.1) differ from the above phase portraits using the three forms in (2.3) only by some rotations and/or a magnification or shrinking of the trajectories. To see this, let's use some examples to demonstrate how to use phase portraits of the transformed equation (2.2) to draw phase portraits of the original equation (2.1). In doing so, we need to know the transformation matrix P. From the proof of Theorem 4.2.2, we have

Corollary 4.2.4 *The matrix P in the transformations in Theorem 4.2.2 can be constructed as follows:*

(C1). If the matrix A has two real distinct eigenvalues, then $P = [u,\, v]$, where u and v are the corresponding real eigenvectors.

(C2). If the matrix A has a double real eigenvalue, then let u be the corresponding real eigenvector, and let v be any real vector such that $P_1 = [u,\, v]$ is nonsingular. If $g = 0$ in (2.5), take $P = P_1$; otherwise take $P = P_1 P_2$ with P_2 given by (2.7).

(C3). If the matrix A has complex conjugate eigenvalues, then $P = [u,\, v]$, where $u + iv$ is an eigenvector with u and v being real vectors.   ♠

Example 4.2.5 Consider the matrix

$$A = \begin{bmatrix} -1 & -2 \\ -2 & -1 \end{bmatrix}. \tag{2.18}$$

The eigenvalues of A are 1 and -3. Then the procedures described in Corollary 4.2.4 can be used to find the corresponding real eigenvectors and then construct

$$P = \begin{bmatrix} 1 & 1 \\ -1 & 1 \end{bmatrix}, \tag{2.19}$$

such that

$$P^{-1}AP = \begin{bmatrix} 1 & 0 \\ 0 & -3 \end{bmatrix} = B. \tag{2.20}$$

For the equation with the coefficient matrix B, the origin $(0,0)$ is a saddle point in the (x, y) plane. The phase portrait for the original equation with the coefficient matrix A in the (x_1, x_2) plane is determined by $[x_1, x_2]^T = P[x, y]^T$, or by

$$Pe_1 = \begin{bmatrix} 1 & 1 \\ -1 & 1 \end{bmatrix} \begin{bmatrix} 1 \\ 0 \end{bmatrix} = \begin{bmatrix} 1 \\ -1 \end{bmatrix}, \quad Pe_2 = \begin{bmatrix} 1 & 1 \\ -1 & 1 \end{bmatrix} \begin{bmatrix} 0 \\ 1 \end{bmatrix} = \begin{bmatrix} 1 \\ 1 \end{bmatrix}.$$

Now, P takes e_1 in the (x, y) plane into $[1, -1]^T$ in the (x_1, x_2) plane, and takes e_2 in the (x, y) plane into $[1, 1]^T$ in the (x_1, x_2) plane. Thus we obtain the following phase portraits in **Figure 4.17** for the trajectories of the transformed equation in the (x, y) plane and the original equation in the (x_1, x_2) plane, which differ by a rotation. ♠

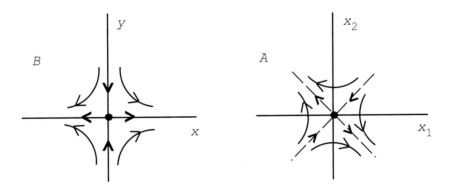

Figure 4.17: Trajectories for the original and the transformed equations differ by a rotation

---

Example 4.2.6 Consider the matrix

$$A = \begin{bmatrix} 0 & 2 \\ -\frac{1}{2} & 0 \end{bmatrix}. \tag{2.21}$$

The eigenvalues of A are $\pm i$, and $\begin{bmatrix} -2i \\ 1 \end{bmatrix}$ is an eigenvector for the eigenvalue i. Then, the procedures described in Corollary 4.2.4 can be used to construct

$$P = \begin{bmatrix} 0 & -2 \\ 1 & 0 \end{bmatrix}, \tag{2.22}$$

such that

$$P^{-1}AP = \begin{bmatrix} 0 & 1 \\ -1 & 0 \end{bmatrix} = B. \tag{2.23}$$

For the equation with the coefficient matrix B, the origin $(0,0)$ is a center in the (x, y) plane; and

$$Pe_1 = \begin{bmatrix} 0 & -2 \\ 1 & 0 \end{bmatrix} \begin{bmatrix} 1 \\ 0 \end{bmatrix} = \begin{bmatrix} 0 \\ 1 \end{bmatrix}, \ Pe_2 = \begin{bmatrix} 0 & -2 \\ 1 & 0 \end{bmatrix} \begin{bmatrix} 0 \\ 1 \end{bmatrix} = \begin{bmatrix} -2 \\ 0 \end{bmatrix}. \tag{2.24}$$

Accordingly, we have the following phase portraits in **Figure 4.18** for the trajectories of the transformed equation in the (x, y) plane and the original equation in the (x_1, x_2) plane, which differ by a magnification. ♠

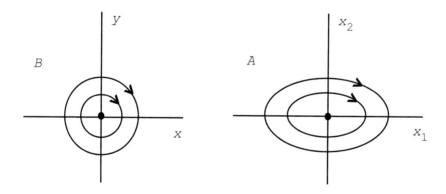

Figure 4.18: Trajectories for the original and the transformed equations differ by a magnification

Examples for other cases can be given in a similar way to conclude that Eq. (2.1) and Eq. (2.2) have essentially the same phase portraits, subject to some transformations (rotations and/or a magnification or shrinking). In this sense, we say that the phase portraits of Eq. (2.1) have the same geometrical and qualitative properties as those of Eq. (2.2) described above. Therefore, we have now completed all phase portraits and the descriptions of the geometric and qualitative properties of the autonomous linear differential equation (2.1) in $\Re^2$.

Exercises 4.2

1. Prove Lemma 4.2.1.

2. Assume that the matrix A has a negative eigenvalue. Prove that $x' = Ax$ has at least one nonzero solution that goes to zero as $t \to \infty$.

3. Draw the phase portraits for Case (II) with $\lambda = 0$. Also find all critical points.

4. Show that if $\det(A) \neq 0$, then the zero is not an eigenvalue of the matrix A, hence $(0,0)$ is the only critical point of $x'(t) = Ax(t)$.

5. Verify that $\alpha + i\beta$ is an eigenvalue of the matrix $\begin{bmatrix} \alpha & \beta \\ -\beta & \alpha \end{bmatrix}$, then find the corresponding eigenvectors.

6. Complete the proof of Corollary 4.2.3.

7. For the following matrices, determine if the origin for $x' = Ax$ is a saddle, a node, a spiral, or a center. Also determine stabilities of each spiral or node.

   (a) $A = \begin{bmatrix} 1 & 2 \\ 3 & 4 \end{bmatrix}$; (b) $A = \begin{bmatrix} 0 & -1 \\ 2 & 0 \end{bmatrix}$; (c) $A = \begin{bmatrix} 3 & 1 \\ 1 & 3 \end{bmatrix}$;

   (d) $A = \begin{bmatrix} 1 & -1 \\ 2 & 3 \end{bmatrix}$.

8. Sketch phase portraits and solve $x' = Ax$ for

   (a) $A = \begin{bmatrix} 1 & 3 \\ 3 & 1 \end{bmatrix}$; (b) $A = \begin{bmatrix} 1 & 3 \\ 0 & -2 \end{bmatrix}$; (c) $A = \begin{bmatrix} 1 & -3 \\ 1 & 1 \end{bmatrix}$;

   (d) $A = \begin{bmatrix} 1 & -3 \\ 0 & 1 \end{bmatrix}$.

9. Solve the linear differential equation $x' = Ax + f(t)$, $x(0) = x_0$, where

$$A = \begin{bmatrix} 1 & -1 \\ 1 & 1 \end{bmatrix}, \quad f(t) = \begin{bmatrix} e^t \\ 2e^t \end{bmatrix}, \quad x_0 = \begin{bmatrix} 1 \\ 0 \end{bmatrix}.$$

4.3   Perturbations on Linear Equations in $\Re^2$

From the study of phase portraits of Eq. (2.2) in the previous section, we observe two important things. First, the signs of the real parts of eigenvalues determine the destinations of trajectories as $t \to \infty$; second, the function

$$V(t) = x_1^2(t) + x_2^2(t), \tag{3.1}$$

where $(x_1(t), x_2(t))$ is a solution, can be used as a good measurement of how close the trajectory $(x_1(t), x_2(t))$ is to the origin $(0,0)$.

Consequently, we expect that if a small perturbation given by $\varepsilon_1(x_1, x_2)$ and $\varepsilon_2(x_1, x_2)$ is applied to the linear differential equation

$$\begin{bmatrix} x_1(t) \\ x_2(t) \end{bmatrix}' = \begin{bmatrix} a & b \\ c & d \end{bmatrix} \begin{bmatrix} x_1(t) \\ x_2(t) \end{bmatrix} = A \begin{bmatrix} x_1(t) \\ x_2(t) \end{bmatrix} \tag{3.2}$$

to come up with the differential equation

$$\begin{bmatrix} x_1(t) \\ x_2(t) \end{bmatrix}' = \begin{bmatrix} a & b \\ c & d \end{bmatrix} \begin{bmatrix} x_1(t) \\ x_2(t) \end{bmatrix} + \begin{bmatrix} \varepsilon_1(x_1(t), x_2(t)) \\ \varepsilon_2(x_1(t), x_2(t)) \end{bmatrix}, \tag{3.3}$$

then, in most cases, the geometrical and qualitative properties of Eq. (3.2) near the origin $(0,0)$ may not be changed under this small perturbation. In other words, we expect in most cases that Eq. (3.2) and Eq. (3.3) have essentially the same geometrical and qualitative properties near the origin when ε_1 and ε_2 are small.

To explain this, we first think of ε_i as being linear, so that Eq. (3.3) takes the form of $x' = Bx$ with the matrix B close to A. Assume that the real parts of eigenvalues of the matrix A are nonzero, then the real parts of eigenvalues of the matrix B will have the same signs as those of A, hence Eq. (3.2) and Eq. (3.3) have the same geometrical and qualitative properties near the origin in this case. This can also be seen from the Distribution Diagram in Figure 4.16, where, if the real parts of eigenvalues of A are nonzero, then the corresponding point in the (p, q) plane of the Distribution Diagram is either **inside** the first or the second quadrant, or **below** the p-axis, (remember that we assumed $\det A \neq 0$, or $q \neq 0$). Thus, after a small change to that point which resulted in the matrix B, the new point should still stay in the corresponding region.

However, in some special cases, perturbations on Eq. (3.2) may cause some significant changes. For example, if the matrix A has two complex conjugate eigenvalues with zero real parts, then the origin for Eq. (3.2) is a center, and hence the trajectories are periodic orbits enclosing the origin. Now, after a linear perturbation, the periodicity (which is delicate) may be destroyed, or the periodic orbits may be sufficiently distorted to make them spiral. Thus in this case, the origin may not be a center after a perturbation. The reason is that the real parts of eigenvalues are zero for a center, and those zero real parts are so easily changed to nonzero after a perturbation. Hence a center may be changed after a perturbation. This can also be seen from the Distribution Diagram in Figure 4.16, where a center corresponds to a point on the positive q-axis. After a perturbation, this point may move *inside* the first or the second quadrant and hence becomes a spiral point.

Based on the above remarks, we can detect the following from the Distribution Diagram in Figure 4.16. After a small linear perturbation, a saddle point, a spiral point, and an improper node for a point below the graph of $q = \frac{1}{4}p^2$ will not change; however, a center for a point on the positive q-axis, and a proper or improper node for a point on the graph of $q = \frac{1}{4}p^2$ may change.

When ε_i are nonlinear, we expect the same things to happen. For example, when the trajectories of Eq. (3.2) spiral to the origin, then in Eq. (3.3), the small $\varepsilon_i(x_1, x_2)$ terms should not significantly change the direction vector (or slope vector), thus the trajectories of Eq. (3.3) should also spiral to the origin. To measure the smallness of $\varepsilon_i(x_1, x_2)$ in Eq. (3.3), we make the following hypothesis.

(H). In Eq. (3.3), assume that $\det A = ad - bc \neq 0$ and that

$$\lim_{r \to 0} \frac{\varepsilon_i(x_1, x_2)}{r} = 0, \quad r = \sqrt{x_1^2 + x_2^2}, \quad i = 1, 2. \tag{3.4}$$

This hypothesis guarantees that the perturbations $\varepsilon_i(x_1, x_2)$ tend to zero faster than the linear terms in Eq. (3.3) and that (see an exercise) the origin is an isolated critical point for Eq. (3.3) (that is, there exists a circle around the origin in which the origin is the only critical point). Under this hypothesis, we have the following results, matching what are true for linear perturbations and what we expect for nonlinear perturbations.

Theorem 4.3.1 *Let the Hypothesis (H) be satisfied and assume the existence and uniqueness of solutions for Eq. (3.3). Then the following statements are true near the origin* $(0,0)$.

(a). *If the origin is a spiral point of Eq. (3.2), then it is a spiral point of Eq. (3.3) (meaning roughly that trajectories starting near the origin spiral into or out from the origin).*

(b). *If the origin is a saddle point of Eq. (3.2), then it is a saddle point of Eq. (3.3) (meaning roughly that there are two trajectories approaching the origin along opposite directions, and all other trajectories close to either of them and to the origin tend away from them).*

(c). *If the origin is an improper node of Eq. (3.2), then it is an improper node of Eq. (3.3) (meaning roughly that trajectories approach or leave the origin along finite directions).*

(d). *If the origin is a proper node of Eq. (3.2), then, under a stronger condition that* $\lim_{r\to 0} \frac{\varepsilon_i(x_1,x_2)}{r^{1+a}} = 0$ *for some* $a > 0$, *it is a proper node of Eq. (3.3) (meaning roughly that trajectories approach or leave the origin along all directions).*

(e). *If the origin is a center of Eq. (3.2), then it is either a center of Eq. (3.3) (meaning roughly that in any small neighborhood of the origin there are at least countably many periodic orbits enclosing the origin) or a spiral point of Eq. (3.3).*

(f). *If the origin is a node (proper or improper) of Eq. (3.2), and if* ε_i, $i = 1, 2$, *have continuous second partial derivatives, then it is a node of Eq. (3.3).*   ♠

A complete proof of these results is extensive, since each individual case needs a different treatment, which is not within the scope of this book. However, reference books, such as Andronov, et al. [1973], Coddington and Levinson [1955], and Perko [1991] offer some detailed analysis. When we study stability in Chapter 5, Theorem 5.3.3 will provide a partial proof of some results in Theorem 4.3.1, such as why stable spiral points and stable nodes for linear differential equations in $\Re^2$ are preserved under the small perturbations satisfying the Hypothesis (H). See also Remark 5.3.4. In addition, some similar results for differential equations in $\Re^n$, called Stable Manifolds and Hartman-Grobman theorems, will be given in Chapter 8.

However, we will prove one special case here, since the proof introduces the very useful **Liapunov method**. We will discuss this in detail when we study stability in Chapter 5.

Proof of Theorem 4.3.1(a): Assume that the origin $(0,0)$ is a spiral point of Eq. (3.2). Applying the transformation given in Theorem 4.2.2 if necessary, we may assume that the matrix A in Eq. (3.3) has been transformed, hence we consider

$$\begin{bmatrix} x_1(t) \\ x_2(t) \end{bmatrix}' = \begin{bmatrix} \alpha & \beta \\ -\beta & \alpha \end{bmatrix} \begin{bmatrix} x_1(t) \\ x_2(t) \end{bmatrix} + \begin{bmatrix} \varepsilon_1(x_1(t), x_2(t)) \\ \varepsilon_2(x_1(t), x_2(t)) \end{bmatrix}, \quad (\alpha\beta \neq 0), \quad (3.5)$$

where $\alpha < 0$, (the case for $\alpha > 0$ is similar). Let $[x_1(t), x_2(t)]^T$ be a solution of Eq. (3.5), whose existence and uniqueness is guaranteed (even though we may not be able to find a formula for it). Define

$$V(t) = \frac{1}{2}[x_1^2(t) + x_2^2(t)]. \quad (3.6)$$

Then, without solving Eq. (3.5), we obtain the following by plugging in Eq. (3.5),

$$\begin{aligned} \frac{d}{dt}V(t) &= x_1(t)x_1'(t) + x_2(t)x_2'(t) \\ &= x_1(t)[\alpha x_1(t) + \beta x_2(t) + \varepsilon_1(x_1(t), x_2(t))] \\ &\quad + x_2(t)[-\beta x_1(t) + \alpha x_2(t) + \varepsilon_2(x_1(t), x_2(t))] \\ &= \alpha[x_1^2 + x_2^2] + [x_1\varepsilon_1(x_1, x_2) + x_2\varepsilon_2(x_1, x_2)]. \quad (3.7) \end{aligned}$$

Under the Hypothesis (H), we find that

$$\begin{aligned} \left| \frac{x_1\varepsilon_1(x_1, x_2) + x_2\varepsilon_2(x_1, x_2)}{x_1^2 + x_2^2} \right| &= \left| \frac{x_1}{r}\frac{\varepsilon_1(x_1, x_2)}{r} + \frac{x_2}{r}\frac{\varepsilon_2(x_1, x_2)}{r} \right| \\ &\leq \left| \frac{\varepsilon_1(x_1, x_2)}{r} \right| + \left| \frac{\varepsilon_2(x_1, x_2)}{r} \right| \longrightarrow 0, \quad \text{as } r = \sqrt{x_1^2 + x_2^2} \to 0. \quad (3.8) \end{aligned}$$

Therefore, there is a $\delta > 0$ such that $r \leq \delta$ implies

$$x_1\varepsilon_1(x_1, x_2) + x_2\varepsilon_2(x_1, x_2) \leq \frac{-\alpha}{2}[x_1^2 + x_2^2]. \quad (3.9)$$

Thus, (3.7) and (3.9) imply

$$\frac{d}{dt}V(t) \leq \frac{\alpha}{2}[x_1^2 + x_2^2] \leq 0, \quad \text{if } r = \sqrt{x_1^2 + x_2^2} \leq \delta. \quad (3.10)$$

Now, let the initial point of the solution $[x_1(t), x_2(t)]^T$ start near the origin, that is, assume $r(0) = \sqrt{x_1^2(0) + x_2^2(0)} < \delta$. Then (3.10) implies that $V'(t) \le 0$ for $t \ge 0$ and small, which implies that $r(t) = \sqrt{x_1^2(t) + x_2^2(t)} < \delta$ for $t \ge 0$ and small. Consequently, we obtain

$$r(t) = \sqrt{x_1^2(t) + x_2^2(t)} \le \delta, \quad t \ge 0, \tag{3.11}$$

therefore, we find that (3.10) becomes

$$V'(t) \le \frac{\alpha}{2}[x_1^2(t) + x_2^2(t)] = \alpha V(t), \quad t \ge 0. \tag{3.12}$$

Now, applying a **differential inequality** (see an exercise), we obtain

$$\frac{1}{2}[x_1^2(t) + x_2^2(t)] = V(t) \le V(0)e^{\alpha t} \longrightarrow 0, \quad t \to \infty, \quad (\alpha < 0) \tag{3.13}$$

which implies that the trajectory $[x_1(t), x_2(t)]^T$ approaches the origin as $t \to \infty$.

To determine the angles of the trajectory in order to check whether it spirals, let's change from Euclidean coordinates to polar coordinates. That is, let

$$x_1(t) = r(t)\cos\theta(t), \quad x_2(t) = r(t)\sin\theta(t). \tag{3.14}$$

Then, by taking a derivative in t on both sides of

$$\tan\theta(t) = \frac{x_2(t)}{x_1(t)},$$

we obtain

$$
\begin{aligned}
\theta'(t) &= \frac{x_2'(t)x_1(t) - x_2(t)x_1'(t)}{r^2(t)} \\
&= \frac{-\beta[x_1^2(t) + x_2^2(t)] + [x_1(t)\varepsilon_2(x_1, x_2) - x_2(t)\varepsilon_1(x_1, x_2)]}{r^2(t)} \\
&= -\beta + \frac{x_1(t)\varepsilon_2(x_1, x_2) - x_2(t)\varepsilon_1(x_1, x_2)}{r^2(t)}.
\end{aligned}
\tag{3.15}
$$

Since we have shown in (3.13) that $[x_1^2(t) + x_2^2(t)] \to 0$, $t \to \infty$, then we can verify, similar to (3.8), that

$$\frac{x_1(t)\varepsilon_2(x_1(t), x_2(t)) - x_2(t)\varepsilon_1(x_1(t), x_2(t))}{r^2(t)} \longrightarrow 0, \quad t \to \infty.$$

Hence,

$$\lim_{t\to\infty} \theta'(t) = -\beta \neq 0, \tag{3.16}$$

then

$$\lim_{t\to\infty} \theta(t) = \infty \quad \text{or} \quad \lim_{t\to\infty} \theta(t) = -\infty. \tag{3.17}$$

Therefore, the trajectory $[x_1(t), x_2(t)]^T$ will spiral to the origin in this case, or the origin is a spiral point of Eq. (3.3). ♠

In the above proof, we obtain some properties of solutions without solving the differential equation; instead, we take a derivative of the function $V(t) = \frac{1}{2}[x_1^2(t) + x_2^2(t)]$ by plugging in the differential equation and then use $V'(t)$ to tell us how small $V(t)$ is or how close a solution is to the origin. This will be a major point in understanding the treatment of stability for general nonlinear differential equations in Chapter 5.

The above proof also shows that for some nonlinear differential equations, the technique of changing from Euclidean coordinates to polar coordinates is very useful in determining phase portraits. We will apply this technique in the following examples in order to determine phase portraits for equations with perturbations.

Example 4.3.2 (A spiral point to a spiral point) Consider the system

$$x' = y - x[1 - x^2 - y^2], \quad y' = -x - y[1 - x^2 - y^2]. \tag{3.18}$$

The linearization of Eq. (3.18) is

$$x'(t) = -x(t) + y(t), \quad y'(t) = -x(t) - y(t), \tag{3.19}$$

hence the origin is a stable spiral point for the linear system using the Distribution Diagram in Figure 4.16. Now, for Eq. (3.18), taking a derivative in t on both sides of $r^2(t) = x^2(t) + y^2(t)$ and $\tan\theta(t) = \frac{y(t)}{x(t)}$, we obtain

$$
\begin{aligned}
r'(t) &= \frac{x(t)x'(t) + y(t)y'(t)}{r(t)} = -r(t)[1 - r^2(t)], \\
\theta'(t) &= \frac{x(t)y'(t) - y(t)x'(t)}{r^2(t)} = -1.
\end{aligned}
$$

The initial point should start near the origin, so we assume $r(0) < 1$. Then $r'(t) \leq 0$, therefore $r(t) < 1$ for $t \geq 0$. Now, using partial fractions, we get

$$\int \frac{dr}{r(1-r^2)} = \int \frac{dr}{r} + \frac{1}{2}\int \frac{dr}{1-r} - \frac{1}{2}\int \frac{dr}{1+r} = \ln \frac{r}{\sqrt{1-r^2}}, \qquad (3.20)$$

hence the solutions are given by

$$r(t) = \frac{1}{\sqrt{1+ce^{2t}}}, \quad c > 0, \quad \theta(t) = -t + d, \quad t \geq 0, \qquad (3.21)$$

where c and d are constants, and the solutions spiral into the origin as $t \to \infty$. Therefore, the origin is a spiral point for Eq. (3.18). ♠

Example 4.3.3 (A center to a spiral point) Consider the system

$$x' = -y - x\sqrt{x^2 + y^2}, \quad y' = x - y\sqrt{x^2 + y^2}. \qquad (3.22)$$

The linearization of Eq. (3.22) is

$$x'(t) = -y(t), \quad y'(t) = x(t), \qquad (3.23)$$

hence the origin is a center for the linear system. Now, for Eq. (3.22), the polar system is

$$r'(t) = \frac{x(t)x'(t) + y(t)y'(t)}{r(t)} = -r^2(t),$$

$$\theta'(t) = \frac{x(t)y'(t) - y(t)x'(t)}{r^2(t)} = 1,$$

then the solutions are given by

$$r(t) = \frac{1}{t+c}, \quad c > 0, \quad \theta(t) = t + d, \quad t \geq 0. \qquad (3.24)$$

The solutions spiral into the origin as $t \to \infty$, therefore the origin is a spiral point for Eq. (3.22). ♠

Example 4.3.4 (A center to a spiral point) Consider the system

$$x' = -y + x(x^2 + y^2), \quad y' = x + y(x^2 + y^2). \qquad (3.25)$$

The linearization of Eq. (3.25) is

$$x'(t) = -y(t), \quad y'(t) = x(t), \tag{3.26}$$

hence the origin is a center for the linear system. Now, for Eq. (3.25), the polar system is

$$r'(t) \quad = \quad \frac{x(t)x'(t) + y(t)y'(t)}{r(t)} = r^3(t),$$

$$\theta'(t) \quad = \quad \frac{x(t)y'(t) - y(t)x'(t)}{r^2(t)} = 1,$$

then the solutions are given by

$$r(t) = \frac{1}{\sqrt{c - 2t}}, \quad c > 0, \quad \theta(t) = t + d, \quad t \geq 0. \tag{3.27}$$

Therefore, the solutions spiral out of the origin as $t \to \infty$, and hence the origin is a spiral point for Eq. (3.25). ♠

Example 4.3.5 (A center to a center) Consider the system

$$\begin{cases} x' = -y + x[x^2 + y^2]\sin\dfrac{\pi}{\sqrt{x^2+y^2}}, \\ y' = x + y[x^2 + y^2]\sin\dfrac{\pi}{\sqrt{x^2+y^2}}, \end{cases} \tag{3.28}$$

for $(x, y) \neq (0, 0)$, and define the right-hand side of Eq. (3.28) at $(0, 0)$ as $f(0, 0) = (0, 0)$. The linearization of Eq. (3.28) is

$$x'(t) = -y(t), \quad y'(t) = x(t), \tag{3.29}$$

hence the origin is a center for the linear system. Now, for Eq. (3.28), the polar system is

$$r'(t) \quad = \quad \frac{x(t)x'(t) + y(t)y'(t)}{r(t)} = r^3(t)\sin\frac{\pi}{r},$$

$$\theta'(t) \quad = \quad \frac{x(t)y'(t) - y(t)x'(t)}{r^2(t)} = 1.$$

For $n = 1, 2, \cdots$, the circles given by

$$r(t) = \frac{1}{n}, \quad \theta(t) = t + d, \quad t \geq 0, \tag{3.30}$$

are periodic orbits of Eq. (3.28) enclosing the origin, and their diameters tend to zero. Therefore, the origin is a center for Eq. (3.28). ♠

Example 4.3.6 (A center to a center) Consider the system

$$\begin{cases} x' = y - xy^2, \\ y' = -x + x^2y. \end{cases} \tag{3.31}$$

The linearization of Eq. (3.31) is

$$x'(t) = y(t), \quad y'(t) = -x(t), \tag{3.32}$$

hence the origin is a center for the linear system. Now, for Eq. (3.31), the polar system is

$$r'(t) = \frac{x(t)x'(t) + y(t)y'(t)}{r(t)} = 0,$$

$$\theta'(t) = \frac{x(t)y'(t) - y(t)x'(t)}{r^2(t)} = xy - 1.$$

Thus for x and y sufficiently small, say for example $|xy| < \frac{1}{2}$, one has

$$r(t) = \text{constant}, \quad \theta'(t) \le -\frac{1}{2}, \quad t \ge 0, \tag{3.33}$$

hence, there are infinitely many periodic orbits of Eq. (3.31) enclosing the origin with their diameters tend to zero. Therefore, the origin is a center for Eq. (3.31). ♠

Example 4.3.7 (A proper node to a spiral point) Consider the system

$$\begin{cases} x' = -x - \frac{y}{\ln\sqrt{x^2+y^2}}, \\ y' = -y + \frac{x}{\ln\sqrt{x^2+y^2}}, \end{cases} \tag{3.34}$$

for $(x, y) \ne (0, 0)$, and define the right-hand side of Eq. (3.34) at $(0, 0)$ as $f(0, 0) = (0, 0)$. The linearization of Eq. (3.34) is

$$x'(t) = -x(t), \quad y'(t) = -y(t), \tag{3.35}$$

hence the origin is a proper node for the linear system. Now, for Eq. (3.34), the polar system is

$$r'(t) = \frac{x(t)x'(t) + y(t)y'(t)}{r(t)} = -r,$$

$$\theta'(t) = \frac{x(t)y'(t) - y(t)x'(t)}{r^2(t)} = \frac{1}{\ln r}.$$

Hence $r(t) = r_0 e^{-t}$ and, plugging it into the equation about θ, we solve $\theta(t) = \theta_0 + \ln|\ln r_0| - \ln|\ln r_0 - t|$. Therefore, $r(t) \to 0$ and $\theta(t) \to -\infty$ as $t \to \infty$. Accordingly, the origin is a spiral point for Eq. (3.34). ♠

Next, let's look at the Distribution Diagram in Figure 4.16 again. We see that a proper or improper node for a point on the graph of $q = \frac{1}{4}p^2$ may change after a perturbation, but the stability will preserve. However, if a center for a point on the positive q-axis is changed after a perturbation to become a spiral point, then the stability will also change. Thus, a change due to perturbation to a center is more significant than a change due to perturbation to a proper or improper node. Accordingly, there is a great deal of research in determining when a center will be preserved after some perturbations. The results in this area are generally complicated. However, for the second-degree polynomial systems

$$\begin{cases} x' = -y + a_{20}x^2 + a_{11}xy + a_{02}y^2, \\ y' = x + b_{20}x^2 + b_{11}xy + b_{02}y^2, \end{cases} \tag{3.36}$$

Li [1982] obtained the following useful and easy-to-use criterion.

Theorem 4.3.8 *Consider Eq. (3.36), where the origin of the linearization is a center. Define*

$$\begin{cases} W_1 &= A\alpha - B\beta, \\ W_2 &= [\beta(5A - \beta) + \alpha(5B - \alpha)]\gamma, \\ W_3 &= (A\beta + B\alpha)\gamma\delta, \end{cases} \tag{3.37}$$

where

$$\begin{cases} A = a_{20} + a_{02}, \quad B = b_{20} + b_{02}, \\ \alpha = a_{11} + 2b_{02}, \quad \beta = b_{11} + 2a_{20}, \\ \gamma = b_{20}A^3 - (a_{20} - b_{11})A^2B + (b_{02} - a_{11})AB^2 - a_{02}B^3, \\ \delta = a_{02}^2 + b_{20}^2 + a_{02}A + b_{20}B. \end{cases} \tag{3.38}$$

Then the origin of Eq. (3.36) is a center if and only if $W_1 = W_2 = W_3 = 0$. ♠

Let's see how to apply Theorem 4.3.8 to the second-degree polynomial systems.

Example 4.3.9 (A center is preserved) Consider the system

$$x' = -y - xy, \quad y' = x + x^2. \tag{3.39}$$

The origin is a center for the linearized linear system. Now, we have $a_{20} = a_{02} = 0, a_{11} = -1, b_{20} = 1, b_{11} = b_{02} = 0$. Thus $A = 0, B = 1, \alpha = -1, \beta = 0, \gamma = 0$, and then $W_1 = W_2 = W_3 = 0$. According to Theorem 4.3.8, the origin of Eq. (3.39) is a center, or a center is preserved.

Next, let's check this result by using the derivatives of $r^2(t) = x^2(t) + y^2(t)$ and $\tan \theta(t) = \frac{y(t)}{x(t)}$. Then

$$
\begin{aligned}
r'(t) &= \frac{xx' + yy'}{r} = \frac{-xy - x^2y + xy + x^2y}{r} = 0, \\
\theta'(t) &= \frac{xy' - yx'}{r^2} = \frac{x^2 + x^3 + y^2 + xy^2}{r^2} = 1 + x > 0
\end{aligned}
$$

for $x > -1$. Thus the trajectories of Eq. (3.39) on the right-hand side of $x = -1$ are circles enclosing the origin, and hence the origin is a center for Eq. (3.39).   ♠

Example 4.3.10 (A center is not preserved) Consider the system

$$x' = -y + xy + y^2, \quad y' = x. \tag{3.40}$$

The origin is a center for the linearized linear system. Now, we have $a_{20} = 0, a_{11} = 1, a_{02} = 1, b_{20} = b_{11} = b_{02} = 0$. Thus $A = 1, B = 0, \alpha = 1, \beta = 0$, and then $W_1 = 1$. According to Theorem 4.3.8, the origin of Eq. (3.40) is not a center, or a center is not preserved. In an exercise, you are asked to use a computer-generated curve to verify this.   ♠

Exercises 4.3

1. Show that condition (3.4) and $\det A = ad - bc \neq 0$ guarantee that the origin is an isolated critical point for Eq. (3.3) (that is, there exists a circle around the origin in which the origin is the only critical point).

2. If $\lim_{t \to \infty} \theta'(t)$ exists and is nonzero and finite, then prove that $\lim_{t \to \infty} \theta(t) = \infty$ or $\lim_{t \to \infty} \theta(t) = -\infty$.

3. Show the following **differential inequality** for $x \in \Re$: If $x'(t) \leq ax(t)$, $t \geq t_0$, for a constant a, then $x(t) \leq x(t_0)e^{a(t-t_0)}$, $t \geq t_0$.

4. Let b be a constant and let $a(t)$, $c(t)$ be continuous functions on $\Re^+$ such that $a(t) \le d$, $c(t) \le d$ for some negative constant d. Use $V(t) = \frac{1}{2}[x_1^2(t) + x_2^2(t)]$ to prove that all solutions of

$$\begin{cases} x_1'(t) = & a(t)x_1(t) - bx_2(t), \\ x_2'(t) = & bx_1(t) + c(t)x_2(t), \end{cases}$$

go to zero exponentially.

5. Let $a(t)$, b, $c(t)$ be the same as above and let $f_i(t, x_1, x_2)$ be such that $\frac{f_i(t,x_1,x_2)}{\sqrt{x_1^2+x_2^2}} \to 0$ as $x_1^2 + x_2^2 \to 0$, $i = 1, 2$. Use $V(t) = \frac{1}{2}[x_1^2(t) + x_2^2(t)]$ to prove that all solutions of

$$\begin{cases} x_1'(t) = & a(t)x_1(t) - bx_2(t) + f_1(t, x_1, x_2), \\ x_2'(t) = & bx_1(t) + c(t)x_2(t) + f_2(t, x_1, x_2), \end{cases}$$

go to zero exponentially.

6. Find the solutions of

$$\begin{cases} x'(t) = & -y - x(x^2 + y^2 - 1), \\ y'(t) = & x - y(x^2 + y^2 - 1), \end{cases}$$

using polar coordinates. Show that the circle $x^2 + y^2 = 1$ is a trajectory of the equation. Next, if $(x(t), y(t))$ is a solution of the equation, find $\lim_{t \to \infty} [x^2(t) + y^2(t)]$.

7. Rewrite using polar coordinates and solve

(a) $\begin{cases} x' = & ax - by, \\ y' = & bx + ay. \end{cases}$ (b) $\begin{cases} x' = & -y + xy, \\ y' = & x - x^2. \end{cases}$

(c) $\begin{cases} x' = & -y - x^3 - xy^2, \\ y' = & x - y^3 - x^2y. \end{cases}$ (d) $\begin{cases} x' = & -y + x^3 + xy^2, \\ y' = & x + y^3 + x^2y. \end{cases}$

8. Determine if the origin is a center for

(a) $\begin{cases} x' = & -y - xy, \\ y' = & x - y^2. \end{cases}$ (b) $\begin{cases} x' = & -y - 2x^2 + xy - y^2, \\ y' = & x + x^2 - 3xy. \end{cases}$

9. In Example 4.3.10, use a computer-generated curve to verify that the origin is not a center for

$$x' = -y + xy + y^2, \quad y' = x. \tag{3.41}$$

4.4   An Application: A Simple Pendulum

Here, we apply Theorem 4.3.1 to analyze a simple pendulum. The motion of a simple pendulum is given by the following differential equation (see Chapter 1):

$$\theta''(t) + k\theta'(t) + q\sin\theta(t) = 0, \qquad (4.1)$$

where $k \geq 0$, $q > 0$ are constants with k related to a damping or friction term, and $\theta(t)$ measures the angle formed by the pendulum and the vertical downward direction.

To change the second-order equation (4.1) into a first-order system, we let $x_1 = \theta$, $x_2 = \theta'$. Then we obtain the system

$$\begin{cases} x_1'(t) & = & x_2(t), \\ x_2'(t) & = & -kx_2(t) - q\sin x_1(t). \end{cases} \qquad (4.2)$$

The critical points are $(n\pi, 0)$, $n = 0, \pm 1, \pm 2, \cdots$, and we will examine each of them. For the critical point $(0,0)$, note that the Hypothesis (H) in (3.4) is not satisfied if $\sin x_1$ is regarded as $\varepsilon_2(x_1, x_2)$, since $\frac{\sin x_1}{x_1} \to 1$ as $x_1 \to 0$. However, from the Taylor expansion of $\sin x_1$, we know that $x_1 - \sin x_1$ satisfies (3.4). Thus we replace $x_2'(t) = -kx_2(t) - q\sin x_1(t)$ by $x_2'(t) = -qx_1(t) - kx_2(t) + q[x_1(t) - \sin x_1(t)]$, and consider

$$\left[\begin{array}{c} x_1(t) \\ x_2(t) \end{array} \right]' = \left[\begin{array}{cc} 0 & 1 \\ -q & -k \end{array} \right] \left[\begin{array}{c} x_1(t) \\ x_2(t) \end{array} \right] + \left[\begin{array}{c} 0 \\ q[x_1(t) - \sin x_1(t)] \end{array} \right]. \qquad (4.3)$$

The matrix from the linearization, $\left[\begin{array}{cc} 0 & 1 \\ -q & -k \end{array} \right]$, is nonsingular since $q > 0$, and its eigenvalues are given by

$$\frac{-k \pm \sqrt{k^2 - 4q}}{2},$$

thus the phase portraits near $(0,0)$ are determined by $k^2 - 4q$. We first assume $k > 0$. If $k^2 - 4q < 0$ (when the damping is weak), then the origin is a stable spiral point for the linearization. Hence, according to Theorem 4.3.1, the origin is also a stable **spiral point** for the nonlinear system (4.3), or Eq. (4.1). Now, the physical and geometric interpretation is that $(\theta, \theta') = (0,0)$ corresponds to the pendulum hanging vertically downward ($\theta = 0$) with zero velocity ($\theta' = 0$), thus, when the damping is weak, a perturbation will cause

Figure 4.19: The motion of a pendulum with a weak damping

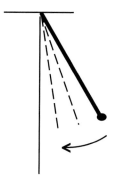

Figure 4.20: The motion of a pendulum with a strong damping

the pendulum to spiral and approach the vertically downward position. See **Figure 4.19**.

If $k^2 - 4q \geq 0$ (when the damping is strong), then the origin for Eq. (4.3) is a stable **node** (see an exercise). The explanation now is that since the damping or friction is very strong, the pendulum will go to the vertically downward position without spirals. See **Figure 4.20**. This case is similar to the simplified linear differential equation $\theta' = -\frac{q}{k}\theta$ from Eq. (4.1) discussed in Section 3 of Chapter 1.

To determine phase portraits for other critical points, let's change variables so as to shift them to the origin $(0,0)$ in order to apply Theorem 4.3.1. For example, for the critical point $(\pi, 0)$ for (θ, θ'), we can change the variable $\theta = \eta + \pi$, such that Eq. (4.1) is transformed to

$$\eta''(t) + k\eta'(t) - q\sin\eta(t) = 0. \tag{4.4}$$

Now, $(0,0)$ is a critical point for (η, η') that corresponds to $(\pi, 0)$ for (θ, θ'). A similar analysis as above (treat $-q$ in Eq. (4.4) as q in Eq. (4.1)) shows that the eigenvalues for the linearization of Eq. (4.4) are

$$\frac{-k \pm \sqrt{k^2 + 4q}}{2}.$$

Thus, the origin is a saddle point for the equation in η, and hence $(\pi, 0)$ is a **saddle** point for the original equation (4.1) in θ. Now, the physical and geometric explanation is that $(\theta, \theta') = (\pi, 0)$ corresponds to the pendulum in a vertically upward position ($\theta = \pi$) with zero velocity ($\theta' = 0$), hence a perturbation will cause the pendulum to move away from that position.

Other critical points can be treated similarly to conclude that $(2n\pi, 0)$, $n = \pm 1, \pm 2, \cdots$, is stable, and $((2n+1)\pi, 0), n = \pm 1, \pm 2, \cdots$, is unstable. A phase portrait in terms of $x_1 = \theta$, $x_2 = \theta'$ with a weak damping is given in **Figure 4.21**.

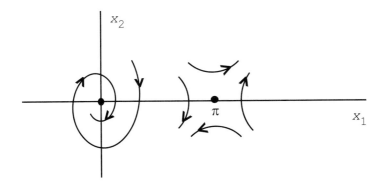

Figure 4.21: A phase portrait of a simple pendulum with a weak damping

---

When there is no damping or friction, that is, when $k = 0$ in Eq. (4.1), the same analysis given above shows that $(2n\pi, 0), n = \pm 1, \pm 2, \cdots$, is a **center**, and $((2n+1)\pi, 0), n = \pm 1, \pm 2, \cdots$, is a **saddle** point. A phase portrait in terms of $x_1 = \theta$, $x_2 = \theta'$ without damping is given in **Figure 4.22**.

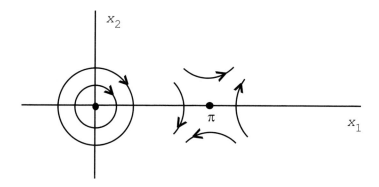

Figure 4.22: A phase portrait of a simple pendulum without damping

---

Exercises 4.4

1. Give more details why at $(0,0)$, $\sin x_1$ cannot be regarded as $\varepsilon_2(x_1, x_2)$ in the Hypothesis (H).

2. For Eq. (4.3), verify that if $k^2 - 4q < 0$, then the origin is a stable spiral point for the linearization.

3. For Eq. (4.3), verify that if $k^2 - 4q \geq 0$, then the origin is a stable node for the linearization. (Provide the details, such as $k = 0$ or $k > 0$, and the cases for proper or improper node.)

4. For the pendulum given by equation (4.1), provide some details for the cases $\theta = 2\pi$, $\theta = 3\pi$, $\theta = -2\pi$, $\theta = -3\pi$.

5. Determine the phase portraits for $\theta''(t) + k\theta'(t) + q \cos \theta(t) = 0$.

6. Determine the phase portraits for $\theta''(t) + k\theta'(t) + q\theta(t) = 0$.

7. Carry out the analysis for Eq. (4.1) with $k > 0$ to the case with $k = 0$.

8. Use separation of variables to solve Eq. (4.1) when $k = 0$.

4.5 Hamiltonian and Gradient Systems

From the analysis of a simple pendulum in the previous section, we know that when there is no damping the critical points are either saddles or centers, and when there is a damping, the critical points are either saddles or nodes (or spiral points).

Next, let's find some relationship between these results and something we did in Chapter 1 concerning the **law of conservation of energy** in physics. For Eq. (4.2) with $k = 0$, which is a special case of Eq. (3.14) in Chapter 1 with $g(u) = q \sin u$ and $x_1 = u$, $x_2 = u'$, the **total energy** of the system is

$$H(x_1, x_2) = \frac{1}{2}x_2^2 + \int_0^{x_1} q \sin s \, ds. \tag{5.1}$$

This is related to a Liapunov function to be discussed in Chapter 5. Now, for a solution $(x_1(t), x_2(t))$ of Eq. (4.2) with $k = 0$, we have

$$\begin{aligned}
\frac{d}{dt}H(x_1(t), x_2(t)) &= x_2 x_2' + q(\sin x_1)x_1' \\
&= x_2(-q \sin x_1) + q(\sin x_1)x_2 = 0. \tag{5.2}
\end{aligned}$$

Thus the **energy is conserved**, and in geometry, this solution lies on a **level curve**

$$H(x_1, x_2) = \text{constant} \tag{5.3}$$

of the function $H(x_1, x_2)$ with the variables x_1 and x_2. See **Figure 4.23** for a demonstration of a level curve.

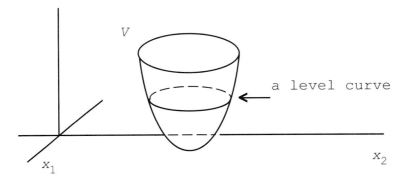

Figure 4.23: A level curve of a function $V(x_1, x_2)$

For $H(x_1, x_2)$ given in (5.1) from Eq. (4.2) with $k = 0$, the projection onto the (x_1, x_2) plane of typical level curves, and the corresponding trajectories are shown in **Figure 4.24**, which explains in geometry that saddles and centers are possible while nodes and spiral points are impossible for Eq. (4.2) with $k = 0$ (without damping).

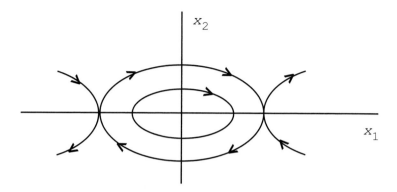

Figure 4.24: Level curves of H in (5.1) and the trajectories

On the other hand, for a solution $(x_1(t), x_2(t))$ of Eq. (4.2) with $k > 0$, we have, for the total energy of the system given in (5.1),

$$
\begin{aligned}
\frac{d}{dt} H(x_1(t), x_2(t)) &= x_2 x_2' + q(\sin x_1)x_1' \\
&= x_2(-kx_2 - q \sin x_1) + q(\sin x_1)x_2 \\
&= -kx_2^2 < 0 \quad \text{if } x_2 \neq 0.
\end{aligned}
\tag{5.4}
$$

Thus the **energy is dissipated**, and in geometry, this solution will not stay on a level curve. Therefore, centers are impossible for Eq. (4.2) with $k > 0$ (with damping).

Based on these analyses, we point out that a simple pendulum, together with its issue of conservativeness, is actually a typical example related to the well-known Hamiltonian systems (also called conservative systems) and gradient systems (also called dissipative systems) that have wide applications in physics and celestial mechanics. Before we define these systems, we observe that for $H(x_1, x_2)$ in (5.1), Eq. (4.2) with $k = 0$ can be written as

$$
\begin{cases}
x_1'(t) &= x_2 = \frac{\partial H}{\partial x_2}, \\
x_2'(t) &= -q \sin x_1 = -\frac{\partial H}{\partial x_1}.
\end{cases}
\tag{5.5}
$$

Definition 4.5.1 *A planar system is called a* **Hamiltonian system** *if there is a differentiable function $H(x_1, x_2) : \Re^2 \to \Re$ such that*

$$\begin{cases} x_1'(t) & = & \frac{\partial H}{\partial x_2}, \\ x_2'(t) & = & -\frac{\partial H}{\partial x_1}. \end{cases} \tag{5.6}$$

The function $H(x_1, x_2)$ is called the **Hamiltonian function** *(or the* **first integral***) of Eq. (5.6) (can be varied up to a constant).*

Hamiltonian systems can be defined for higher dimensional spaces, but we only treat planar systems here. Now, we see that Eq. (4.2) with $k = 0$ is a Hamiltonian system with the Hamiltonian function given in (5.1).

For the Hamiltonian system (5.6), the total energy is always conserved. That is, let $(x_1(t), x_2(t))$ be any solution of Eq. (5.6), we have

$$\begin{aligned} \frac{d}{dt} H(x_1(t), x_2(t)) & = & \frac{\partial H}{\partial x_1} x_1' + \frac{\partial H}{\partial x_2} x_2' \\ & = & \frac{\partial H}{\partial x_1} \frac{\partial H}{\partial x_2} + \frac{\partial H}{\partial x_2} \left(-\frac{\partial H}{\partial x_1} \right) = 0. \end{aligned} \tag{5.7}$$

In this sense, a Hamiltonian function is sometimes called an energy function.

Next, we analyze the critical points of Eq. (5.6), which are the same as the critical points of the function $H(x_1, x_2)$ (defined in calculus as solutions (x_1, x_2) of $\frac{\partial H}{\partial x_1} = \frac{\partial H}{\partial x_2} = 0$). Due to the special structure of Hamiltonian systems, we expect to obtain some specific results. Let (x_1^c, x_2^c) be a critical point of Eq. (5.6), using the change of variables $x = x_1 - x_1^c$, $y = x_2 - x_1^c$ if necessary, we can assume that the origin is a critical point of Eq. (5.6), and we only need to state results for the origin.

Lemma 4.5.2 *Let the origin be a critical point of the Hamiltonian system (5.6). If the function $H(x_1, x_2)$ has a strict local maximum or minimum at the origin, then there is a constant $r > 0$ such that there is no such a point $(x_1^0, x_2^0) \in B_r(0, 0)$ (the ball centered at the origin with radius r) with $(x_1^0, x_2^0) \neq (0, 0)$ and $(x_1(t, x_1^0, x_2^0), x_2(t, x_1^0, x_2^0)) \to (0, 0)$, $|t| \to \infty$. In particular, in this case, the origin is neither a node nor a spiral point for Eq. (5.6).*

Proof. If a point $(x_1^0, x_2^0) \neq (0, 0)$ is such that $(x_1(t, x_1^0, x_2^0), x_2(t, x_1^0, x_2^0)) \to (0, 0)$, $t \to \infty$ (the case $t \to -\infty$ is similar), then from the continuity of the function H and the fact that $H(x_1(t, x_1^0, x_2^0), x_2(t, x_1^0, x_2^0))$ is a constant as the energy is conserved for a Hamiltonian system, we have

$$
\begin{aligned}
H(x_1^0, x_2^0) &= \lim_{t \to \infty} H(x_1^0, x_2^0) \\
&= \lim_{t \to \infty} H(x_1(t, x_1^0, x_2^0), x_2(t, x_1^0, x_2^0)) = H(0, 0), \quad (5.8)
\end{aligned}
$$

that is, H has the same function value at (x_1^0, x_2^0) and at the origin. If such a point (x_1^0, x_2^0) can get arbitrarily close to the origin, then the function $H(x_1, x_2)$ cannot have a strict local maximum or minimum at the origin. Therefore, the conclusion in the lemma is true. ♠

Definition 4.5.3 *A critical point of a system is called* **nondegenerate** *if the matrix of the linearization of the system at that point has no zero eigenvalues.*

Theorem 4.5.4 *Let the function H in the Hamiltonian system (5.6) have continuous second partial derivatives. Then any nondegenerate critical point of Eq. (5.6) is either a center or a saddle. More specifically, if the function H has a strict local maximum or minimum at this point, then it is a center for Eq. (5.6); if the function H has a saddle at this point, then it is a saddle for Eq. (5.6).*

Proof. We assume that this nondegenerate critical point is at the origin. Then

$$
\frac{\partial H}{\partial x_1}(0, 0) = \frac{\partial H}{\partial x_2}(0, 0) = 0, \qquad (5.9)
$$

and the matrix of the linearization of Eq. (5.6) at $(0, 0)$ is given by

$$
A = \begin{bmatrix} \dfrac{\partial^2 H}{\partial x_2 \partial x_1}(0, 0) & \dfrac{\partial^2 H}{\partial x_2^2}(0, 0) \\ -\dfrac{\partial^2 H}{\partial x_1^2}(0, 0) & -\dfrac{\partial^2 H}{\partial x_1 \partial x_2}(0, 0) \end{bmatrix}. \qquad (5.10)
$$

For this matrix A, we have $\operatorname{tr} A = 0$ and

$$
\det A = \frac{\partial^2 H}{\partial x_1^2}(0, 0) \frac{\partial^2 H}{\partial x_2^2}(0, 0) - \left[\frac{\partial^2 H}{\partial x_1 \partial x_2}(0, 0) \right]^2, \qquad (5.11)
$$

and, since the matrix A has no zero eigenvalues, $\det A \neq 0$.

Now, recall from calculus that (5.9) and (5.11) are used in the second derivative test to determine the nature of the critical point $(0, 0)$ of the function H. If $\det A > 0$, which is the same as saying that the function H has a strict local maximum or minimum at $(0,0)$, then from the Distribution Diagram of Corollary 4.2.3, the origin is a center for the linearization with the matrix A. From Theorem 4.3.1, the origin is either a center or a spiral point of Eq. (5.6). However, using Lemma 4.5.2, a spiral point is impossible and the origin is now a center for Eq. (5.6).

If $\det A < 0$, which is the same as saying that the function H has a saddle at $(0,0)$, then from Corollary 4.2.3, the origin is a saddle for the linearization with the matrix A. Applying Theorem 4.3.1, the origin is also a saddle for Eq. (5.6). This completes the proof. ♠

One special Hamiltonian system is the **Newtonian system**

$$x'' = f(x), \tag{5.12}$$

where f is continuously differentiable on $\Re$. Eq. (5.12) includes a simple pendulum without damping, and can be written as a Hamiltonian system

$$x' = y, \quad y' = f(x), \tag{5.13}$$

with the Hamiltonian function given by

$$H(x,y) = \frac{1}{2}y^2 + \left(-\int_0^x f(s)ds\right), \tag{5.14}$$

where $\frac{1}{2}y^2$ is the kinetic energy and $P(x) = -\int_0^x f(s)ds$ is the potential energy. We have the following result, which gives the **phase portrait analysis** based on the potential energy for the Newtonian system (5.12). The proof is left as an exercise.

Theorem 4.5.5 *The critical points of the Newtonian system (5.13) are given by $(x_c, 0)$ where $f(x_c) = 0$. If the potential function $P(x)$ has a strict local minimum at x_c (or when $f'(x_c) < 0$), then $(x_c, 0)$ is a center for Eq. (5.13); if the potential function $P(x)$ has a strict local maximum at x_c (or when $f'(x_c) > 0$), then $(x_c, 0)$ is a saddle for Eq. (5.13).* ♠

Now, we find that Figure 4.22 for a simple pendulum without damping (Eq. (4.1) with $k = 0$) can also be constructed by using Theorem 4.5.5 and the corresponding potential function

$$P(x) = \int_0^x q\sin s\, ds = q(1 - \cos x),$$

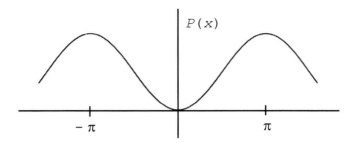

Figure 4.25: Potential function $q(1 - \cos x)$ for Eq. (4.1) with $k = 0$

showing in **Figure 4.25**, which has a strict local minimum at $x_c = 0$ and a strict local maximum at $x_c = \pi$.

Next, we determine the conditions for a given system

$$x_1' = f(x_1, x_2), \quad x_2' = g(x_1, x_2), \tag{5.15}$$

to be a Hamiltonian system. If Eq. (5.15) is a Hamiltonian system, then there is a function H such that $\frac{\partial H}{\partial x_2} = f$ and $-\frac{\partial H}{\partial x_1} = g$. Hence, assuming H has continuous second partial derivatives, we obtain

$$\frac{\partial f}{\partial x_1} = \frac{\partial^2 H}{\partial x_1 \partial x_2} = \frac{\partial^2 H}{\partial x_2 \partial x_1} = -\frac{\partial g}{\partial x_2}. \tag{5.16}$$

In fact, (5.16) is also a sufficient condition for Eq. (5.15) to be a Hamiltonian system.

Lemma 4.5.6 *A given differentiable system (5.15) is a Hamiltonian system if and only if*

$$\frac{\partial f}{\partial x_1} = -\frac{\partial g}{\partial x_2}. \tag{5.17}$$

Proof. We only need to show that if (5.17) is true, then we can construct a Hamiltonian function for Eq. (5.15). Now, solving $H(x_1, x_2)$ from

$$\frac{\partial H}{\partial x_2} = f(x_1, x_2),$$

we obtain

$$H(x_1, x_2) = \int f(x_1, x_2) dx_2 + C(x_1),$$

where the "constant of integration" $C(x_1)$ may depend on x_1 because H is a function of (x_1, x_2). To determine the conditions on $C(x_1)$ so that $H(x_1, x_2)$ becomes a Hamiltonian function of Eq. (5.15), we need

$$\frac{\partial H}{\partial x_1} = \frac{\partial}{\partial x_1} \int f(x_1, x_2)dx_2 + C'(x_1) = -g(x_1, x_2),$$

or

$$C'(x_1) = -g(x_1, x_2) - \frac{\partial}{\partial x_1} \int f(x_1, x_2)dx_2,$$

which can be solved for $C(x_1)$ by taking an integration,

$$C(x_1) = -\int g(x_1, x_2)dx_1 - \int \left(\frac{\partial}{\partial x_1} \int f(x_1, x_2)dx_2 \right)dx_1. \qquad (5.18)$$

Finally, we verify that the right-hand side of (5.18) is really a function of x_1 alone, because under the condition (5.17), we have

$$\frac{\partial}{\partial x_2}\left[\int g(x_1, x_2)dx_1 + \int \left(\frac{\partial}{\partial x_1} \int f(x_1, x_2)dx_2 \right)dx_1 \right]$$

$$= \int \frac{\partial g}{\partial x_2}dx_1 + \int \frac{\partial}{\partial x_2}\left(\frac{\partial}{\partial x_1} \int f(x_1, x_2)dx_2 \right)dx_1$$

$$= \int \frac{\partial g}{\partial x_2}dx_1 + \int \frac{\partial}{\partial x_1}\left(\frac{\partial}{\partial x_2} \int f(x_1, x_2)dx_2 \right)dx_1$$

$$= \int \frac{\partial g}{\partial x_2}dx_1 + \int \frac{\partial f}{\partial x_1}dx_1$$

$$= \int \frac{\partial g}{\partial x_2}dx_1 - \int \frac{\partial g}{\partial x_2}dx_1 = 0.$$

This completes the proof.   ♠

Example 4.5.7 Consider the system

$$\begin{cases} x_1'(t) & = & x_1 + x_2 = f(x_1, x_2), \\ x_2'(t) & = & x_1 + x_2 = g(x_1, x_2). \end{cases} \qquad (5.19)$$

It is not a Hamiltonian system since $\frac{\partial f}{\partial x_1} = 1 \neq -1 = -\frac{\partial g}{\partial x_2}$. Note also that now the function C in the proof of Lemma 4.5.6 is given by

$$-\int g(x_1, x_2)dx_1 - \int \left(\frac{\partial}{\partial x_1} \int f(x_1, x_2)dx_2 \right)dx_1$$

$$= -(\frac{x_1^2}{2} + x_1 x_2) - \int x_2 dx_1 = -(\frac{x_1^2}{2} + 2x_1 x_2),$$

which is not a function of x_1 alone.   ♠

Example 4.5.8 Consider the system

$$\begin{cases} x_1'(t) &= x_2 = f(x_1, x_2), \\ x_2'(t) &= x_1 = g(x_1, x_2). \end{cases} \tag{5.20}$$

It is a Hamiltonian system since $\frac{\partial f}{\partial x_1} = 0 = -\frac{\partial g}{\partial x_2}$. Now the function C in the proof of Lemma 4.5.6 is given by

$$-\int g(x_1, x_2) dx_1 - \int \left(\frac{\partial}{\partial x_1} \int f(x_1, x_2) dx_2 \right) dx_1 = -\frac{x_1^2}{2},$$

which is a function of x_1 alone, and the corresponding Hamiltonian function is given by

$$H(x_1, x_2) = \int f(x_1, x_2) dx_2 + C(x_1) = \frac{x_2^2}{2} - \frac{x_1^2}{2}. \qquad \spadesuit$$

The above are some basic results of Hamiltonian systems, or conservative systems. Next, let's look at a type of dissipative system.

Definition 4.5.9 *A planar system is called a* **gradient system** *if there is a differentiable function* $H(x_1, x_2) : \Re^2 \to \Re$ *such that*

$$\begin{cases} x_1'(t) &= -\frac{\partial H}{\partial x_1}, \\ x_2'(t) &= -\frac{\partial H}{\partial x_2}, \end{cases} \tag{5.21}$$

that is, $[x_1', x_2']^T = -grad\, H$, *where* $grad\, H = [\frac{\partial H}{\partial x_1}, \frac{\partial H}{\partial x_2}]^T$ *is the gradient of the function* H.

Now, if we still regard the function H as an energy function, then for a solution $(x_1(t), x_2(t))$ of Eq. (5.21), we have

$$\begin{aligned} \frac{d}{dt} H(x_1(t), x_2(t)) &= \frac{\partial H}{\partial x_1} x_1' + \frac{\partial H}{\partial x_2} x_2' \\ &= -\left[\left(\frac{\partial H}{\partial x_1} \right)^2 + \left(\frac{\partial H}{\partial x_2} \right)^2 \right] < 0 \end{aligned} \tag{5.22}$$

if $\left(\frac{\partial H}{\partial x_1} \right)^2 + \left(\frac{\partial H}{\partial x_2} \right)^2 \neq 0$. In this sense, gradient systems are dissipative systems. Similar to Hamiltonian systems, we have the following results for gradient systems.

Theorem 4.5.10 *Let the function H in the gradient system (5.21) have continuous second partial derivatives. Then any nondegenerate critical point of Eq. (5.21) is either a node or a saddle. More specifically, if the function H has a strict local maximum or minimum at this point, then it is respectively an unstable or a stable node for Eq. (5.21); if the function H has a saddle at this point, then it is a saddle for Eq. (5.21).*

Proof. Similar to the proof of Theorem 4.5.4, we assume that this nondegenerate critical point is at the origin, and obtain

$$\frac{\partial H}{\partial x_1}(0,0) = \frac{\partial H}{\partial x_2}(0,0) = 0. \tag{5.23}$$

Now, the matrix of the linearization of Eq. (5.21) at $(0,0)$ is given by

$$A = \begin{bmatrix} -\frac{\partial^2 H}{\partial x_1^2}(0,0) & -\frac{\partial^2 H}{\partial x_1 \partial x_2}(0,0) \\ -\frac{\partial^2 H}{\partial x_2 \partial x_1}(0,0) & -\frac{\partial^2 H}{\partial x_2^2}(0,0) \end{bmatrix}, \tag{5.24}$$

hence we have

$$tr A = -\left[\frac{\partial^2 H}{\partial x_1^2}(0,0) + \frac{\partial^2 H}{\partial x_2^2}(0,0)\right],$$

and

$$\det A = \frac{\partial^2 H}{\partial x_1^2}(0,0)\frac{\partial^2 H}{\partial x_2^2}(0,0) - \left[\frac{\partial^2 H}{\partial x_1 \partial x_2}(0,0)\right]^2, \tag{5.25}$$

and, since the matrix A has no zero eigenvalues, $\det A \neq 0$.

If $\det A > 0$, then $\frac{\partial^2 H}{\partial x_1^2}(0,0)$ and $\frac{\partial^2 H}{\partial x_2^2}(0,0)$ must have the same sign. If they are positive, which is the same as saying that the function H has a strict local minimum at $(0,0)$, then $tr A < 0$, and (see an exercise)

$$\det A \leq \frac{1}{4}(tr A)^2. \tag{5.26}$$

Hence, from Corollary 4.2.3, the origin is a stable node for the linearization with the matrix A. From Theorem 4.3.1(f), the origin is also a stable node for Eq. (5.21) (the result concerning stability should be expected and can be verified using Theorem 5.3.3 in Chapter 5). Other cases can be handled similarly (see an exercise). This completes the proof. ♠

Example 4.5.11 Consider the function $H(x_1, x_2) = x_1^2(x_1 - 2)^2 + x_2^2$ and the corresponding gradient system

$$\begin{cases} x_1'(t) & = & -\frac{\partial H}{\partial x_1} = -4x_1(x_1 - 1)(x_1 - 2), \\ x_2'(t) & = & -\frac{\partial H}{\partial x_2} = -2x_2. \end{cases} \tag{5.27}$$

Since H has a strict minimum at $(0, 0)$ and at $(2, 0)$ and a saddle at $(1, 0)$, we know from Theorem 4.5.10 that $(0, 0)$ and $(2, 0)$ are stable nodes for Eq. (5.27) and $(1, 0)$ is a saddle for Eq. (5.27). See **Figure 4.26**. ♠

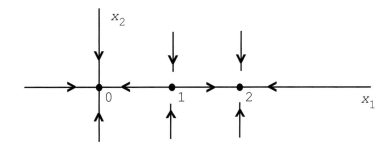

Figure 4.26: The phase portrait for the gradient system (5.27)

Finally, let's look at the relationship between the Hamiltonian system (5.6) and the gradient system (5.21). If they are using the same function H, then they are **orthogonal**, that is, at every point, the trajectories of the two systems are perpendicular. Moreover, centers of Eq. (5.6) correspond to nodes of Eq. (5.21), and saddles of Eq. (5.6) correspond to saddles of Eq. (5.21).

For the related studies of Hamiltonian and gradient systems in higher dimensional spaces and their applications in physics and celestial mechanics, see, for example, Guckenheimer and Holmes [1986] and Meyer and Hall [1992].

Exercises 4.5

1. Verify that $H(x, y)$ in (5.14) is a Hamiltonian function for Eq. (5.13).

2. Prove Theorem 4.5.5.

3. Check if $x_1' = 2x_1 + \sin x_2$, $x_2' = 2x_1 x_2 + \cos x_1 x_2$ is a Hamiltonian system.

4. Verify that $x_1' = x_2$, $x_2' = -x_1 + x_1^2$ is a Newtonian system. Then use the energy function to obtain the phase portrait.

5. Verify (5.26).

6. Complete the proof of Theorem 4.5.10 by analyzing the case when $\frac{\partial^2 H}{\partial x_1^2}(0,0)$ and $\frac{\partial^2 H}{\partial x_2^2}(0,0)$ are negative and the case when $\det A < 0$.

7. Let $H(x_1, x_2) = (x_1 - 3)^2 + x_2^2(x_2 - 2)^2$. Use Theorem 4.5.10 to obtain the phase portrait of the corresponding gradient system.

Chapter 5

Stability. Part I

5.1 Introduction

In Chapter 3, we analyzed structure of solutions of linear differential equations in $\Re^n$ and found that eigenvalues, when the coefficient matrix is constant, can be used to determine long-term properties of solutions. Then in Chapter 4, we studied phase portraits for some autonomous differential equations in $\Re^2$. According to the properties revealed for a center, a node, or a spiral point, we briefly discussed stability properties for autonomous differential equations in $\Re^2$. In this chapter, we extend the study of stability to some differential equations in $\Re^n$. Although one could study stability after Chapter 3 (Linear Differential Equations), we believe analyzing the centers and spiral points in $\Re^2$ will provide some concrete models when we introduce stability notions.

First, let's examine what we mean by a statement that "a solution of a differential equation is stable." Note that a differential equation is typically used to model the movement of a certain physical system or experiment. In running a system or experiment, one needs to deal with some initial measurements, such as putting one gallon of water initially for some experiment, which inevitably involves some errors in measurements or approximations. If the behavior of a system or experiment is stable, then a small change in initial data will result in a small change in the behavior for future time. Thus, by a statement that "a solution ϕ of a differential equation is stable" we mean that other solutions with initial data close to the solution ϕ will remain close to ϕ for future time. For example, for a stable system, if ϕ is the solution corresponding to one gallon of water initially, and x is a solution with its initial value close to one gallon of water, say for example, 1.005 gallons of water, then ϕ and x should be close for the future time, or $|x - \phi|$ should be small for future time.

This is exactly what happened to the phase portraits of autonomous differential equations in $\Re^2$ in Chapter 4. For example, when the origin is a center, the constant vector $\phi = [0, 0]^T$ is a solution, and any periodic orbit x enclosing the origin with an initial point close to the origin $\phi = [0, 0]^T$ will stay close to the origin $\phi = [0, 0]^T$ for future time, or $|x - \phi| = |x|$ will be small for future time if it is small initially. Accordingly, the solution $\phi = [0, 0]^T$ is stable in this case. In some other cases, such as when the origin is a stable spiral point, other solutions will not only stay close to, but also approach the constant solution $\phi = [0, 0]^T$ as $t \to \infty$. In this case, the solution $\phi = [0, 0]^T$ is asymptotically stable.

The above remarks will help us extend the study of stability properties of autonomous differential equations in $\Re^2$ to general differential equations in $\Re^n$. We assume that the time variable t of differential equations is in $[0, \infty)$, so that $t = 0$ can be regarded as the initial time systems start running. In order to consider "long-term" behavior of solutions for "future time," we make the following assumption throughout this chapter.

(A). *Consider the differential equation*

$$x'(t) = f(t,\, x(t)), \tag{1.1}$$

in $D = [0, \infty) \times Q$, where $Q \subset \Re^n$ is a domain containing the zero vector. For any $(t_0,\, x_0) \in D = [0, \infty) \times Q$, Eq. (1.1) has a unique solution $x(t, t_0, x_0)$ existing on $[t_0,\, \infty)$ with $x(t_0) = x_0$.

To begin the study of stability properties, let's recall from Chapter 3 that for linear differential equations with constant coefficients, when the real parts of eigenvalues are negative, then solutions will go to zero. We can follow this to derive stabilities for linear differential equations.

The study of stabilities for nonlinear differential equations is difficult because formulas of solutions and eigenvalues are generally not available or applicable. So far, we have successfully dealt with a nonlinear differential equation in the proof of Theorem 4.3.1(a) in Chapter 4, where we verified that the solutions approach the zero solution without solving the differential equation. Since this idea might provide a clue of how to pursue the stabilities for nonlinear differential equations, we recall the main steps:

First, the existence and uniqueness of solutions is guaranteed. Then define

$$V(t) = \frac{1}{2}[x_1^2(t) + x_2^2(t)]. \tag{1.2}$$

The function V is related to the norm $r(x) = \sqrt{x_1^2 + x_2^2}$ of a solution $(x_1(t), x_2(t))$, that is, the distance from $(x_1(t), x_2(t))$ to the origin $(0,0)$. Then we take a derivative in t by plugging in the differential equation, and obtain

$$V'(t) \leq \alpha V(t) = -[-\alpha V(t)], \quad \alpha < 0,$$

which enables us to verify that

$$V(t) \to 0, \quad t \to \infty.$$

Hence, we conclude that the solutions go to the origin $(0,0)$, or the origin $\phi = (0,0)$ is asymptotically stable.

The most important aspect of this approach is that it is done without solving the differential equation explicitly. The idea of deriving qualitative properties without solving differential equations explicitly was introduced by the independent work of two famous mathematicians, Liapunov [1892] and Poincaré [1892], at the turn of the 20th century when they pioneered the modern qualitative theory of differential equations. Their ideas continue to inspire new research in the area of differential equations and other related areas. In this chapter, we will study some of their most important work concerning stability properties for some linear and nonlinear differential equations in $\Re^n, n \geq 1$.

We point out that in $[0, \infty) \times \Re^2$, a center and a stable spiral point for equations in $\Re^2$ are shown in **Figures 5.1** and **5.2**.

For a center in Figure 5.1, for all $t \geq 0$, the distance from a point on the circle to the origin $(0,0)$ in $\Re^2$ (not the origin in $\Re^3$), or the t-axis, is a constant (radius of the circle). However, for a stable spiral point in Figure 5.2, the distance from a point on the curve to the origin $(0,0)$ in $\Re^2$ (or the t-axis) goes to zero. Accordingly, if we draw these in a plane and use the vertical direction, x, to denote the distance of a solution to the zero solution (t-axis, when the zero is a solution), then a center and a stable spiral point will be shown in **Figures 5.3** and **5.4** in the (t, x) plane.

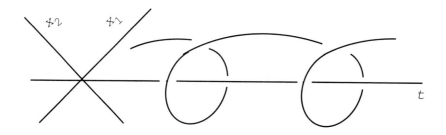

Figure 5.1: A center in $[0, \infty) \times \Re^2$

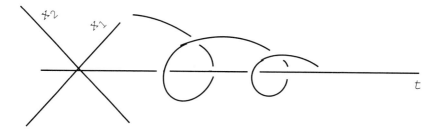

Figure 5.2: A spiral point in $[0, \infty) \times \Re^2$

Now we make the following definitions regarding **stabilities in the sense of Liapunov**. This extends the related studies for autonomous differential equations in $\Re^2$ in Chapter 4. Recall that for a vector $x = [x_1, x_2, \cdots, x_n]^T$ in $\Re^n$, $|x| = \sum_{i=1}^{n} |x_i|$, and for an $n \times n$ matrix $A = [a_{ij}]$, $|A| = \sum_{i,j=1}^{n} |a_{ij}|$.

Definition 5.1.1 *Let $\phi(t) = \phi(t, t_\phi)$ be a solution of Eq. (1.1) on $[t_\phi, \infty)$, $t_\phi \geq 0$.*

*(a). $\phi(t, t_\phi)$ is said to be **stable** if for any $t_0 \geq t_\phi$ and any $\varepsilon > 0$, there exists a $\delta = \delta(\varepsilon, t_0) > 0$, (typically $\delta(\varepsilon, t_0) \leq \varepsilon$), such that $|x_0 - \phi(t_0)| \leq \delta$ implies $|x(t, t_0, x_0) - \phi(t)| \leq \varepsilon$ for $t \geq t_0$.*

*(b). $\phi(t, t_\phi)$ is said to be **uniformly stable** if it is stable and δ in the definition of "stable" can be chosen to be independent of $t_0 \geq t_\phi$. That*

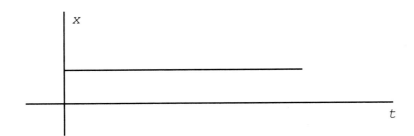

Figure 5.3: A center in the (t, x) plane

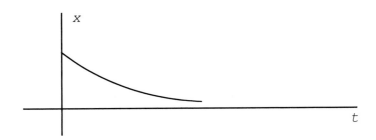

Figure 5.4: A spiral point in the (t, x) plane

is, for any $\varepsilon > 0$, there exists a $\delta = \delta(\varepsilon) > 0$, $(\delta(\varepsilon) \leq \varepsilon,)$ such that $t_0 \geq t_\phi$ and $|x_0 - \phi(t_0)| \leq \delta$ imply $|x(t, t_0, x_0) - \phi(t)| \leq \varepsilon$ for $t \geq t_0$.

*(c). $\phi(t, t_\phi)$ is said to be **asymptotically stable** if it is stable and in addition, for any $t_0 \geq t_\phi$, there exists an $r(t_0) > 0$ such that $|x_0 - \phi(t_0)| \leq r(t_0)$ implies $\lim_{t \to \infty} |x(t, t_0, x_0) - \phi(t)| = 0$.*

*(d). $\phi(t, t_\phi)$ is said to be **uniformly asymptotically stable** if it is uniformly stable and in addition, there exists an $r > 0$ independent of $t_0 \geq t_\phi$, such that $|x_0 - \phi(t_0)| \leq r$ implies $\lim_{t \to \infty} |x(t, t_0, x_0) - \phi(t)| = 0$ uniformly for $t_0 \geq t_\phi$ in the following sense: For any $\varepsilon > 0$, there exists a $T = T(\varepsilon) > 0$ such that $\{t_0 \geq t_\phi, |x_0 - \phi(t_0)| \leq r, t \geq t_0 + T\}$ imply $|x(t, t_0, x_0) - \phi(t)| \leq \varepsilon$.*

*(e). $\phi(t, t_\phi)$ is said to be **unstable** if it is not stable.*

*(f). In particular, if $\phi(t) = 0$, $t \geq 0$, is a solution of Eq. (1.1), or equiv-
alently when $f(t,0) = 0$, $t \geq 0$, then the above give the corresponding
definitions concerning stability properties for the zero solution $\phi = 0$.*

Definitions concerning other stabilities can be found in reference books,
such as Yoshizawa [1966] and Burton [1985]. But here, we concentrate on
the above four types because they are the most important ones. Note the
distinction between the notions of stability and uniform stability. If ϕ is
uniformly stable, then $\delta = \delta(\varepsilon)$ in the definition is such that if a solution
x differs from ϕ initially by at most $\delta(\varepsilon)$, then x will stay in the "ε-tube"
enclosing ϕ for the future time, no matter where the solution x initially
started. However, if ϕ is only stable, then $\delta = \delta(\varepsilon, t_0)$ in the definition is
good only for the solutions started initially at t_0. That is, if a solution x
started at t_0 differs from ϕ initially by at most $\delta(\varepsilon, t_0)$, then x will stay
in the "ε-tube" enclosing ϕ for future time; but, another solution y started
from a different time than t_0 may initially differ from ϕ by at most $\delta(\varepsilon, t_0)$
and then leaves the "ε-tube" enclosing ϕ at some later times. See **Figures
5.5** and **5.6**. Examples can be found in reference books, such as Yoshizawa
[1966] and Burton [1985] that indicate the notions in Definition 5.1.1 are not
equivalent. Examples are also given in this book.

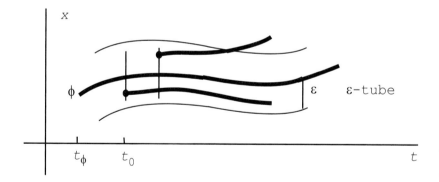

Figure 5.5: Differences among stabilities

However, an inspection of a center or a stable spiral point in Chapter 4 for
planar autonomous differential equations reveals that the constant solution
$\phi = [0,0]^T$ is not only stable or asymptotically stable, but also "uniformly"
stable or "uniformly" asymptotically stable. The following result says that

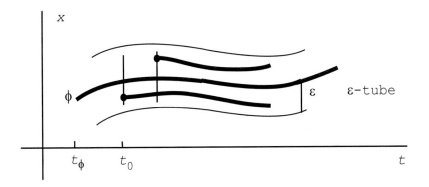

Figure 5.6: Differences among stabilities

---

the same conclusion is also true for general autonomous or periodic differential equations. Its proof will be given in Chapter 9.

Theorem 5.1.2 *Assume that f in Eq. (1.1) is autonomous or periodic in t (that is, there is a constant $T > 0$ such that $f(t, x) = f(t + T, x)$), and satisfies a Lipschitz condition (or weak or local Lipschitz) with respect to x on D. Assume further that $\phi(t) = 0$, $t \geq 0$, is a solution of Eq. (1.1). If $\phi = 0$ is stable, then it is uniformly stable. If $\phi = 0$ is asymptotically stable, then it is uniformly asymptotically stable.* ♠

From Definition 5.1.1, it seems that it is a little easier to work with the zero solution when studying stability properties. We verify next that this can be accomplished: If $\phi(t)$ is a solution of Eq. (1.1), then, to talk about the stability properties of $\phi(t)$ is to talk about the smallness of $|y(t) - \phi(t)|$ with y being another solution of Eq. (1.1). If we define $x(t) = y(t) - \phi(t)$, then it is the same as the smallness of $|x(t)| = |x(t) - 0|$, or how close $x(t)$ is to the zero. Now, let $\phi(t)$ and $y(t)$ be solutions of Eq. (1.1), then we have, for $x(t) = y(t) - \phi(t)$,

$$
\begin{aligned}
x'(t) &= y'(t) - \phi'(t) = f(t, y(t)) - f(t, \phi(t)) \\
&= f(t, x(t) + \phi(t)) - f(t, \phi(t)).
\end{aligned} \tag{1.3}
$$

Thus, for the solution ϕ of Eq. (1.1), if we define

$$
\overline{f}(t, x) = f(t, x + \phi(t)) - f(t, \phi(t)),
$$

then, with y a solution of Eq. (1.1), we find that $x(t) = y(t) - \phi(t)$ is a solution of

$$x'(t) = \overline{f}(t, x(t)). \tag{1.4}$$

Moreover, since $\overline{f}(t, 0) = f(t, \phi(t)) - f(t, \phi(t)) = 0$, the zero is indeed a solution of (1.4), and this zero solution corresponds to the solution $\phi(t)$ of Eq. (1.1), and the stabilities for $\phi(t)$ of Eq. (1.1) is equivalent to the stabilities for the zero solution of Eq. (1.4). Therefore, in most cases, we can simply study stability properties of the zero solution for Eq. (1.1). Consequently, we need to assume $f(t, 0) = 0$ when we do so. This is why in the assumption (A) for Eq. (1.1) we ask the domain $Q \subset \Re^n$ to contain the zero vector, because we will state results for the zero solution. Note that the above transformation in (1.3) may change an autonomous differential equation into a nonautonomous differential equation, unless $\phi(t)$ is a constant.

This chapter is organized as follows: In Section 2, we study stabilities for linear differential equations with constant coefficients and show that eigenvalues of the coefficient matrices determine stability properties. In Section 3, stabilities of linear equations with linear or nonlinear perturbations are studied using the variation of parameters formula and Gronwall's inequality. The results include some planar autonomous nonlinear differential equations studied in Chapter 4 as special cases. Therefore, some unproven results in Chapter 4 can now get a partial proof. In Section 4, linear periodic differential equations are treated. The Floquet theory from Chapter 3 is used to transform linear periodic equations into linear equations with constant coefficients and the results from Section 2 can then be applied. In Section 5, we introduce Liapunov's method for autonomous nonlinear differential equations and prove their stability properties under the assumption that there exist appropriate Liapunov functions. Thus, we can obtain stabilities without explicitly solving differential equations. In Section 6, we provide examples to demonstrate how the Liapunov theory is applied by constructing Liapunov functions in specific applications. Liapunov's method for general (nonautonomous) differential equations will be given in Chapter 9.

Exercises 5.1

1. Determine the stabilities of $\phi = (0, 0)$ for $x'(t) = Ax(t)$, where

(a) $A = \begin{bmatrix} 1 & 3 \\ 3 & 1 \end{bmatrix}$; (b) $A = \begin{bmatrix} 1 & 3 \\ 0 & -2 \end{bmatrix}$; (c) $A = \begin{bmatrix} 1 & -3 \\ 1 & 1 \end{bmatrix}$;

(d) $A = \begin{bmatrix} 1 & -3 \\ 0 & 1 \end{bmatrix}$.

2. Verify directly that if constant $a > 0$, then the zero solution of the scalar equation $x' = -ax$ is uniformly asymptotically stable. Discuss the case when $a = 0$.

3. Discuss the stabilities of the given solutions of the following scalar equations.

   (a). $x' = 0$, $\phi(t) = 2$.   (b). $x' = x$, $\phi(t) = 0$; $\phi(t) = e^t$.

   (c). $x' = -x$, $\phi(t) = 0$; $\phi(t) = e^{-t}$.

4. Denote by $x(t, t_0, x_0)$ the solutions of some system for which $\phi = 0$ is a solution. Let $f(t)$ be a positive function. Show that $|x(t, t_0, x_0)| \leq f(t_0)|x_0|$, $t \geq t_0 \geq 0$, implies that $\phi = 0$ is stable.

5. Denote by $x(t, t_0, x_0)$ the solutions of some system for which $\phi = 0$ is a solution. Let $C > 0$ be a constant. Show that $|x(t, t_0, x_0)| \leq C|x_0|$, $t \geq t_0 \geq 0$, implies that $\phi = 0$ is uniformly stable.

6. Denote by $x(t, t_0, x_0)$ the solutions of some system for which $\phi = 0$ is a solution. Let $f(t, s)$ be a positive function such that for any fixed t_0, $f(t, t_0)$ is bounded for $t \geq t_0$ and $f(t, t_0) \to 0$ as $t \to \infty$. Show that $|x(t, t_0, x_0)| \leq f(t, t_0)|x_0|$, $t \geq t_0 \geq 0$, implies that $\phi = 0$ is asymptotically stable.

7. Denote by $x(t, t_0, x_0)$ the solutions of some system for which $\phi = 0$ is a solution. Let $\alpha > 0$ and $C > 0$ be constants. Show that $|x(t, t_0, x_0)| \leq C|x_0|e^{-\alpha(t-t_0)}$, $t \geq t_0 \geq 0$, implies that $\phi = 0$ is uniformly asymptotically stable.

8. Rewrite $x^{(n)} = f(t, x, x', x'', \cdots, x^{(n-1)})$ as a system of first-order differential equations and then define the corresponding stabilities.

5.2   Linear Equations with Constant Coefficients

In this section, we study stabilities for the linear differential equation with constant coefficients,

$$x'(t) = Ax(t) + f(t), \quad x(t_0) = x_0, \quad t \geq t_0 \geq 0, \quad x \in \Re^n, \tag{2.1}$$

where $f(t)$ is continuous on $\Re^+ = [0, \infty)$.

Using the results of Chapter 3, we know that the unique solution of Eq. (2.1) passing through (t_0, x_0) is given by the variation of parameters formula

$$
\begin{aligned}
x(t) &= e^{(t-t_0)A}\left[x_0 + \int_{t_0}^{t}\{e^{(s-t_0)A}\}^{-1}f(s)ds\right] \\
&= e^{(t-t_0)A}x_0 + \int_{t_0}^{t}e^{(t-s)A}f(s)ds, \quad t \geq t_0,
\end{aligned}
\tag{2.2}
$$

where the fundamental matrix solution of Eq. (2.1) (when $f = 0$) is given by $U(t, t_0) = e^{(t-t_0)A}$.

We first study the linear homogeneous differential equation

$$
x'(t) = Ax(t),
\tag{2.3}
$$

where the unique solution is given by $x(t) = e^{(t-t_0)A}x_0$ for the initial data (t_0, x_0). Note that now $\phi = 0$ is a solution of Eq. (2.3).

Recall from the Jordan canonical form theorem in Chapter 3, there exists a nonsingular constant matrix P (may be complex valued) such that

$$
e^{(t-t_0)A} = Pe^{(t-t_0)J}P^{-1}, e^{(t-t_0)J} = diag(e^{(t-t_0)J_1}, e^{(t-t_0)J_2}, \cdots, e^{(t-t_0)J_s}), \tag{2.4}
$$

with each square matrix $e^{(t-t_0)J_i}$ given by

$$
(e^{\lambda_i(t-t_0)})
\begin{bmatrix}
1 & t-t_0 & \frac{(t-t_0)^2}{2!} & \frac{(t-t_0)^3}{3!} & \cdot & \cdot & \cdot & \frac{(t-t_0)^{m_i-1}}{(m_i-1)!} \\
0 & 1 & t-t_0 & \frac{(t-t_0)^2}{2!} & \cdot & \cdot & \cdot & \cdot \\
0 & 0 & 1 & \cdot & \cdot & \cdot & \cdot & \cdot \\
\cdot & \cdot & \cdot & \cdot & \cdot & \cdot & \cdot & \cdot \\
\cdot & \cdot & \cdot & \cdot & \cdot & \cdot & \cdot & \cdot \\
\cdot & \cdot & \cdot & \cdot & \cdot & \cdot & \frac{(t-t_0)^2}{2!} & \frac{(t-t_0)^3}{3!} \\
\cdot & \cdot & \cdot & \cdot & \cdot & \cdot & t-t_0 & \frac{(t-t_0)^2}{2!} \\
0 & 0 & 0 & \cdot & \cdot & \cdot & 1 & t-t_0 \\
0 & 0 & 0 & \cdot & \cdot & \cdot & 0 & 1
\end{bmatrix}.
\tag{2.5}
$$

Now, Eq. (2.3) is autonomous, so based on Theorem 5.1.2, stability is equivalent to uniform stability, and asymptotic stability is equivalent to uniform asymptotic stability. Also, using Theorem 3.3.11 in Chapter 3, we find that eigenvalues can be used to determine stability properties. For example, for the scalar linear differential equation

$$
x'(t) = ax(t),
$$

the constant a is the eigenvalue and solutions are given by $x(t) = x(0)e^{at}$. Then $\phi = 0$ of the equation is stable if $a = 0$; asymptotically stable if $a < 0$; and unstable if $a > 0$. Similar results for equations in $\Re^n$ are given in the following theorem. We provide some details in its proof in order to assist you understand different types of stabilities and become familiar with the proofs using $\varepsilon - \delta$ format.

Theorem 5.2.1 *Let λ be a complex number and denote by $R(\lambda)$ the real part of λ, and let $\phi = 0$ be the zero solution of Eq. (2.3).*

(A). The following statements are equivalent:

    (1). $\phi = 0$ is stable or uniformly stable;

    (2). For each eigenvalue λ of the matrix A, either $R(\lambda) < 0$, or $R(\lambda) = 0$ but in this case λ appears only in matrices J_i (in the Jordan canonical form for A) such that J_i is a 1×1 matrix;

    (3). There is an (independent or generic) constant $C > 1$ such that

$$|e^{tA}| \leq C, \quad 0 \leq t < \infty. \tag{2.6}$$

(B). The following statements are equivalent:

    (1). $\phi = 0$ is asymptotically stable or uniformly asymptotically stable;

    (2). Each eigenvalue of matrix A has a negative real part;

    (3). There are (independent or generic) constants $C > 1$ and $\alpha > 0$ such that

$$|e^{tA}| \leq Ce^{-\alpha t}, \quad 0 \leq t < \infty. \tag{2.7}$$

(C). The following statements are equivalent:

    (1). $\phi = 0$ is unstable;

    (2). There is an eigenvalue λ of matrix A with $R(\lambda) = 0$ and λ appears in a matrix J_i that is at least 2×2;

    (3). There is an eigenvalue of matrix A with a positive real part.

Proof. (A): The equivalency between (2) and (3) is from Theorem 3.3.11 in Chapter 3.

    (3) $\Rightarrow$ (1): Suppose that (3) is true, then any solution of Eq. (2.3) satisfies $|x(t, t_0, x_0)| = |e^{(t-t_0)A}x_0| \leq C|x_0|, t \geq t_0$. For any $t_0 \geq 0$ and

any $\varepsilon > 0$, we can choose $\delta = \delta(\varepsilon, t_0) = \frac{\varepsilon}{C}$ such that $|x_0| \leq \delta$ implies $|x(t, t_0, x_0)| \leq C|x_0| \leq C\delta = \varepsilon$, proving the stability of $\phi = 0$.

(1) $\Rightarrow$ (3): Suppose that $\phi = 0$ is stable, then for $t_0 = 0$ and $\varepsilon = 1$, there exists a $\delta = \delta(\varepsilon, t_0) = \delta(1, 0) > 0$, $(\delta(1, 0) < 1,)$ such that $|x_0| \leq \delta$ implies $|x(t, 0, x_0)| = |e^{tA}x_0| \leq \varepsilon = 1$ for $t \geq 0$. Now take $x_0 = \delta e_i$, where $e_i, 1, 2, \cdots, n$, form the standard unit basis for $\Re^n$. Then $|x_0| = |\delta e_i| = \delta$ and hence $|e^{tA}\delta e_i| \leq 1$ or $|e^{tA}e_i| \leq \frac{1}{\delta}$. But $e^{tA}e_i$ is the ith column of e^{tA}, therefore

$$|e^{tA}| \leq n(\frac{1}{\delta}) \overset{\text{def}}{=} C, \ t \geq 0,$$

where $C > 1$ is an independent constant.

(B): (1) $\Rightarrow$ (2): Suppose that $\phi = 0$ is asymptotically stable, then for $t_0 = 0$, there exists a $r(0) > 0$ such that $|x_0| \leq r(0)$ implies $\lim_{t \to \infty} |x(t, 0, x_0)| = \lim_{t \to \infty} |e^{tA}x_0| = 0$. Similar to the proof of part (A), we find that every entry in e^{tA} goes to zero, thus $\lim_{t \to \infty} |e^{tA}| = 0$. Then from Theorem 3.3.11, we know that (2) is true.

(2) $\Rightarrow$ (3): Suppose that each eigenvalue of matrix A has a negative real part, then using the structure of e^{tA} given in (2.4) and (2.5), if we let -2α be the largest negative real part, then $|e^{\lambda t}| \leq e^{-2\alpha t}$, $t \geq 0$, for any eigenvalue λ of the matrix A. Hence, $|e^{tA}| \leq P(t)e^{-2\alpha t}$ for a polynomial $P(t)$ in t of degree less than n. Now, l'Hôpital's rule can be applied to verify that $P(t)e^{-\alpha t}$ is bounded for $t \geq 0$. Thus there is an independent constant $C > 1$ such that $|e^{tA}| \leq Ce^{-\alpha t}$, $t \geq 0$.

(3) $\Rightarrow$ (1): Suppose that (3) is true, then $|e^{tA}|$ is bounded and hence from part (A) we have just proved, $\phi = 0$ is stable. Next, $|x(t, t_0, x_0)| = |e^{(t-t_0)A}x_0| \leq Ce^{-\alpha(t-t_0)}|x_0|$. Thus for any $t_0 \geq 0$, we can choose $r(t_0) = 1$ such that $|x_0| \leq r(t_0)$ implies $|x(t, t_0, x_0)| \leq Ce^{-\alpha(t-t_0)} \to 0$ as $t \to \infty$. This proves the asymptotic stability of $\phi = 0$.

(C): See an exercise.

This completes the proof of the theorem.   ♠

Therefore, for linear differential equations with constant coefficients, stabilities are completely determined by eigenvalues of the coefficient matrices. Let's look at some examples.

Example 5.2.2 Consider the following matrices

$$A_1 = \begin{bmatrix} 0 & 0 \\ 0 & 0 \end{bmatrix}, \ A_2 = \begin{bmatrix} 0 & 1 \\ 0 & 0 \end{bmatrix}, \ A_3 = \begin{bmatrix} -1 & 1 \\ 0 & -1 \end{bmatrix}, \ A_4 = \begin{bmatrix} -1 & 0 \\ 0 & -1 \end{bmatrix}.$$

The matrix A_1 has a repeated eigenvalue 0, and the corresponding J_0 in the Jordan canonical form are two 1×1 matrices, or $\begin{bmatrix} 0 & 0 \\ 0 & 0 \end{bmatrix} = \begin{bmatrix} J_0 & 0 \\ 0 & J_0 \end{bmatrix}$, thus the zero solution of the linear equation with the coefficient matrix A_1 is uniformly stable. (In fact, the solutions now are constants, hence the zero solution is uniformly stable.) The matrix A_2 has a repeated eigenvalue 0, and the corresponding J_0 in the Jordan canonical form is A_2 itself, a 2×2 matrix, thus the zero solution is unstable. (Now, the solutions are $\{x(t) = x_0 + y_0 t, \ y(t) = y_0\}$, hence the zero solution is unstable.) For the matrix A_3 or A_4, the zero solution is uniformly asymptotically stable because the eigenvalues are all negative. For example, for the matrix A_3, the solutions are $\{x(t) = x_0 e^{-t} + y_0 t e^{-t}, \ y(t) = y_0 e^{-t}\}$, so that $y(t) \to 0$ as $t \to \infty$, and l'Hôpital's rule implies that $x(t) \to 0$ as $t \to \infty$. Note that for the matrix A_3 or A_4, the origin is a stable node studied in Chapter 4. ♠

Theorem 5.2.1 says that stability properties of Eq. (2.3) can be reduced to the signs of the real parts of eigenvalues of the matrix A. To determine these signs, we present an important and commonly used criterion. Its proof can be found in reference books such as Marden [1966].

Theorem 5.2.3 (Routh-Hurwitz criterion) *Let* $P(\lambda) = \lambda^n + a_1 \lambda^{n-1} + \cdots + a_{n-1}\lambda + a_n$ *be a polynomial with real coefficients, and define*

$$D_1 = a_1, \quad D_2 = \begin{vmatrix} a_1 & a_3 \\ 1 & a_2 \end{vmatrix}, \quad D_k = \begin{vmatrix} a_1 & a_3 & a_5 & a_7 & . & . & . & . & a_{2k-1} \\ 1 & a_2 & a_4 & a_6 & . & . & . & . & a_{2k-2} \\ 0 & a_1 & a_3 & a_5 & . & . & . & . & a_{2k-3} \\ 0 & 1 & a_2 & a_4 & . & . & . & . & a_{2k-4} \\ . & . & . & . & . & . & . & . & . \\ . & . & . & . & . & . & . & . & . \\ . & . & . & . & . & . & . & . & . \\ . & . & . & . & . & . & . & . & a_{k+1} \\ 0 & . & . & . & . & . & . & & a_k \end{vmatrix},$$

for $k = 3, 4, \cdots, n$, *where* $a_j = 0$ *if* $j > n$. *If* $D_k > 0$ *for* $k = 1, 2, \cdots, n$, *then each solution of* $P(\lambda) = 0$ *has a negative real part.* ♠

The following matrices demonstrate how the Routh-Hurwitz criterion is applied.

Example 5.2.4 Consider

$$A = \begin{bmatrix} -2 & 0 & 0 \\ 0 & -1 & 1 \\ 0 & 0 & -1 \end{bmatrix}, \quad \lambda E - A = \begin{bmatrix} \lambda + 2 & 0 & 0 \\ 0 & \lambda + 1 & -1 \\ 0 & 0 & \lambda + 1 \end{bmatrix}. \quad (2.8)$$

The eigenvalues for matrix A are $-2, -1, -1$, which are negative. To check it out with the Routh-Hurwitz criterion, we look at

$$\det(\lambda E - A) = \lambda^3 + 4\lambda^2 + 5\lambda + 2$$

(since the coefficient of λ^n in $P(\lambda)$ is 1, we use $\det(\lambda E - A)$ instead of $\det(A - \lambda E)$). Hence, $D_1 = a_1 = 4$,

$$D_2 = \begin{vmatrix} a_1 & a_3 \\ 1 & a_2 \end{vmatrix} = \begin{vmatrix} 4 & 2 \\ 1 & 5 \end{vmatrix} = 18, \quad D_3 = \begin{vmatrix} a_1 & a_3 & a_5 \\ 1 & a_2 & a_4 \\ 0 & a_1 & a_3 \end{vmatrix} = \begin{vmatrix} 4 & 2 & 0 \\ 1 & 5 & 0 \\ 0 & 4 & 2 \end{vmatrix} = 36,$$

thus, we have $D_i > 0, i = 1, 2, 3$; therefore, the Routh-Hurwitz criterion applies. ♠

Example 5.2.5 Consider

$$A = \begin{bmatrix} 2 & 0 & 0 \\ 0 & -1 & 1 \\ 0 & 0 & -1 \end{bmatrix}, \quad \lambda E - A = \begin{bmatrix} \lambda - 2 & 0 & 0 \\ 0 & \lambda + 1 & -1 \\ 0 & 0 & \lambda + 1 \end{bmatrix}. \quad (2.9)$$

Now, $\det(\lambda E - A) = \lambda^3 - 3\lambda - 2$, then $D_1 = a_1 = 0$, thus the Routh-Hurwitz criterion does not apply. In fact, 2 is a positive eigenvalue. ♠

For the linear nonhomogeneous differential equation (2.1) with $f \neq 0$, the zero is not a solution of Eq. (2.1). Now we look at stabilities of nonzero solutions of Eq. (2.1), which can be reduced to those of the zero solution of Eq. (2.3) because of the affine structure of Eq. (2.1). In other words, the solution set of Eq. (2.1) is some particular solution of Eq. (2.1) plus any solution of Eq. (2.3), or equivalently, the difference of any two solutions of Eq. (2.1) is a solution of Eq. (2.3).

Theorem 5.2.6 *Assume that $f(t)$ is continuous on $\Re^+$. The zero solution of Eq. (2.3) is stable if and only if every solution of Eq. (2.1) is stable. The same statement is true for uniform stability, asymptotic stability, and uniform asymptotic stability.*

Proof. First, assume that the zero solution of Eq. (2.3) is stable. Let $\phi(t, t_\phi)$ be a solution of Eq. (2.1) on $[t_\phi, \infty)$, $t_\phi \geq 0$, and let $x(t, t_0, x_0)$ be another solution of Eq. (2.1) on $[t_0, \infty)$, $t_0 \geq t_\phi$. Then on $[t_0, \infty)$, $y(t) = x(t) - \phi(t)$ is a solution of Eq. (2.3). As the zero solution of Eq. (2.3) is stable, for any $t_0 \geq t_\phi$ and any $\varepsilon > 0$, there exists a $\delta = \delta(\varepsilon, t_0) > 0$, such that $|y(t_0)| = |x_0 - \phi(t_0)| \leq \delta(\varepsilon, t_0)$ implies $|y(t)| = |x(t, t_0, x_0) - \phi(t)| \leq \varepsilon$ for $t \geq t_0$, proving that $\phi(t, t_\phi)$ is stable for Eq. (2.1).

Next, assume that every solution of Eq. (2.1) is stable, and let $\phi(t)$ be a stable solution of Eq. (2.1) on $[0, \infty)$. If $y(t, t_0)$ is a solution of Eq. (2.3) on $[t_0, \infty)$, then $x(t) = y(t) + \phi(t)$ is a solution of Eq. (2.1) on $[t_0, \infty)$, and $y(t) = x(t) - \phi(t)$. Similar to the above, the description concerning the stability of the solution $\phi(t)$ of Eq. (2.1) gives exactly the description concerning the stability of the zero solution of Eq. (2.3). Other cases can also be treated this way. ♠

Exercises 5.2

1. Prove Part (C) of Theorem 5.2.1.

2. Assume that all eigenvalues of the matrices A and B have negative real parts. Show that the integral $\int_0^\infty e^{tA}Ce^{tB}dt$ exists, and is a solution of the matrix equation $AX + XB = -C$.

3. Use the Routh-Hurwitz criterion to determine the asymptotic stability of the zero solution for $X' = \begin{bmatrix} 1 & 2 & 3 \\ 4 & 5 & 6 \\ 7 & 8 & 9 \end{bmatrix} X$.

4. Rewrite the following equations as systems and then use the Routh-Hurwitz criterion to determine the asymptotic stability of the zero solution.

(a) $x'' + 3x' + 4x = 0$.

(b) $x'' - 3x' - 4x = 0$.

(c) $x^{(3)} + 3x'' + 4x = 0$.

(d) $x^{(4)} + 5x'' - 2x = 0$.

5. Verify that the solution $\phi = 0$ of $X' = \begin{bmatrix} -2 & e^{4t} \\ 0 & -2 \end{bmatrix} X$ is unstable.

Note that for any $t \in \Re$, the eigenvalues of the coefficient matrix are -2, which indicates that Theorem 5.2.1 does not apply when $A(t)$ is not a constant matrix.

6. Complete the proof of Theorem 5.2.6.

5.3 Perturbations on Linear Equations

In this section, we use the variation of parameters formula and Gronwall's inequality to derive stability properties for some perturbations on linear differential equations. The idea is to use Gronwall's inequality to estimate solutions of the perturbed equations and then show, under certain conditions, that small perturbations will not change the stability properties of the original linear differential equations.

Theorem 5.3.1 *If the zero solution of Eq. (2.3) is stable, and if the $n \times n$ continuous matrix function $B(t)$ satisfies $\int_0^\infty |B(t)|dt < \infty$, then the zero solution of*

$$x'(t) = Ax(t) + B(t)x(t) = [A + B(t)]x(t) \tag{3.1}$$

is uniformly stable.

Proof. First, since $B(t)$ is continuous, the existence and uniqueness for Eq. (3.1) is guaranteed. Let $x(t)$ be a solution of Eq. (3.1) and treat $B(t)x(t)$ as $f(t)$ in Eq. (2.1), then $x(t)$ is given by the variation of parameters formula

$$x(t) = e^{(t-t_0)A}x(t_0) + \int_{t_0}^t e^{(t-s)A}B(s)x(s)ds, \ \ t \geq t_0 \geq 0. \tag{3.2}$$

From Theorem 5.2.1, the uniform stability of the zero solution of Eq. (2.3) implies that there is an (independent) constant $C > 1$ such that $|e^{(t-t_0)A}| \leq C$, $0 \leq t_0 \leq t < \infty$. Therefore, we have

$$|x(t)| \leq C|x(t_0)| + \int_{t_0}^t C|B(s)||x(s)|ds, \ \ t \geq t_0 \geq 0. \tag{3.3}$$

Now, Gronwall's inequality implies that

$$|x(t)| \leq C|x(t_0)| \exp\left(\int_{t_0}^t C|B(s)|ds\right) \leq \left\{C \exp\left(\int_0^\infty C|B(s)|ds\right)\right\}|x(t_0)|$$

$$\stackrel{\text{def}}{=} C_1|x(t_0)|, \ \ t \geq t_0 \geq 0, \tag{3.4}$$

where C_1 is an independent constant. This implies the uniform stability of the zero solution for Eq. (3.1) because the proof of Theorem 5.2.1(A) can be used. ♠

Theorem 5.3.1.a *If the zero solution of Eq. (2.3) is stable, and if the continuous function $f(t, x)$ satisfies a weak Lipschitz condition with respect to x and $f(t, 0) = 0$, which implies $|f(t, x)| \leq k(t)|x|$, where $k(t)$ is from the weak Lipschitz condition. If $\int_0^\infty k(t)dt < \infty$, then the zero solution of*

$$x'(t) = Ax(t) + f(t, x(t)) \tag{3.5}$$

is uniformly stable. ♠

When the zero solution of Eq. (2.3) is asymptotically stable, the condition on $B(t)$ can be relaxed.

Theorem 5.3.2 *If the zero solution of Eq. (2.3) is asymptotically stable, and if the $n \times n$ continuous matrix function $B(t)$ satisfies*

$$\int_{t_0}^t |B(s)|ds \leq m(t - t_0) + r, \ \ t \geq t_0 \geq 0, \tag{3.6}$$

for some positive constants m and r, then there is an $m_0 > 0$ such that if $m \leq m_0$, then the zero solution of Eq. (3.1) is uniformly asymptotically stable.

Proof. Similar to the beginning part in the proof of Theorem 5.3.1, any solution of Eq. (3.1) is given by

$$x(t) = e^{(t-t_0)A}x(t_0) + \int_{t_0}^t e^{(t-s)A}B(s)x(s)ds, \ \ t \geq t_0 \geq 0. \tag{3.7}$$

From Theorem 5.2.1, the asymptotic stability of the zero solution of Eq. (2.3) implies that there are (independent) constants $C > 1$ and $\alpha > 0$ such that $|e^{(t-t_0)A}| \leq Ce^{-\alpha(t-t_0)}$. Therefore, we have

$$|x(t)| \leq Ce^{-\alpha(t-t_0)}|x(t_0)| + \int_{t_0}^t Ce^{-\alpha(t-s)}|B(s)||x(s)|ds, \ \ t \geq t_0 \geq 0, \tag{3.8}$$

or

$$|x(t)|e^{\alpha t} \leq Ce^{\alpha t_0}|x(t_0)| + \int_{t_0}^t Ce^{\alpha s}|B(s)||x(s)|ds, \ \ t \geq t_0 \geq 0. \tag{3.9}$$

Define $u(t) = |x(t)|e^{\alpha t}$, then $u(t)$ satisfies

$$u(t) \leq Cu(t_0) + \int_{t_0}^{t} C|B(s)|u(s)ds, \ \ t \geq t_0 \geq 0, \qquad (3.10)$$

hence Gronwall's inequality implies that

$$\begin{aligned} u(t) &\leq Cu(t_0)\exp\left(\int_{t_0}^{t} C|B(s)|ds\right) \leq Cu(t_0)\exp\left(C[m(t - t_0) + r]\right) \\ &= Cu(t_0)e^{Cr}\exp\left[Cm(t - t_0)\right]. \qquad (3.11) \end{aligned}$$

This means

$$|x(t)| \leq Ce^{Cr}|x(t_0)|\exp\left[-(\alpha - Cm)(t - t_0)\right]. \qquad (3.12)$$

If we let $m_0 = \frac{\alpha}{2C}$, then $m \leq m_0$ implies that

$$|x(t)| \leq Ce^{Cr}|x(t_0)|\exp\left[-\frac{1}{2}\alpha(t - t_0)\right]. \qquad (3.13)$$

This guarantees the uniform asymptotic stability of the zero solution for Eq. (3.1) because the proof of Theorem 5.2.1(B) can be used. ♠

Theorem 5.3.2.a *If the zero solution of Eq. (2.3) is asymptotically stable, and if the continuous function $f(t,x)$ satisfies a weak Lipschitz condition with respect to x and $f(t,0) = 0$, which implies $|f(t,x)| \leq k(t)|x|$, where $k(t)$ is from the weak Lipschitz condition. If*

$$\int_{t_0}^{t} k(s)ds \leq m(t - t_0) + r, \ \ t \geq t_0 \geq 0, \qquad (3.14)$$

for some positive constants m and r, then there is an $m_0 > 0$ such that if $m \leq m_0$, then the zero solution of Eq. (3.5) is uniformly asymptotically stable. ♠

The following result concerning uniform asymptotic stability with non-linear perturbations can be used to explain certain results of Theorem 4.3.1 in Chapter 4.

Theorem 5.3.3 *If the zero solution of Eq. (2.3) is asymptotically stable, and if the continuous function $f(t,x)$ satisfies a Lipschitz condition (or weak or local Lipschitz) with respect to x on D, and*

$$\lim_{x \to 0} \frac{|f(t,x)|}{|x|} = 0, \quad \text{uniformly for } t \in [0, \infty), \qquad (3.15)$$

then the zero solution of

$$x'(t) = Ax(t) + f(t, x(t)) \tag{3.16}$$

is uniformly asymptotically stable.

Proof. Now, the conditions guarantee the existence and uniqueness for Eq. (3.16), and they also imply that the zero is a solution of Eq. (3.16). By the variation of parameters formula, any solution of Eq. (3.16) is given by

$$x(t) = e^{(t-t_0)A}x(t_0) + \int_{t_0}^{t} e^{(t-s)A} f(s, x(s))ds, \ \ t \geq t_0 \geq 0. \tag{3.17}$$

From Theorem 5.2.1, the asymptotic stability of the zero solution of Eq. (2.3) implies that there are (independent) constants $C > 1$ and $\alpha > 0$ such that $|e^{(t-t_0)A}| \leq Ce^{-\alpha(t-t_0)}$. From (3.15), for any η with $0 < \eta < \frac{\alpha}{C}$, there is a $\Delta = \Delta(\eta) > 0$ such that if $|x| \leq \Delta$, then $|f(t, x)| \leq \eta|x|$ uniformly for $t \geq 0$.

We first verify that if $|x(t_0)| \leq \frac{\Delta}{C}$, then $|x(t)| < \Delta$ for $t \geq t_0$. If this is not true, then as $C > 1$ and $|x(t_0)| \leq \frac{\Delta}{C} < \Delta$, there is a $t_1 > t_0$ such that $|x(t)| < \Delta$ for $t \in [t_0, t_1)$ and $|x(t_1)| = \Delta$. Then $|f(t, x(t))| \leq \eta|x(t)|$ for $t \in [t_0, t_1]$. Thus, for $t \in [t_0, t_1]$, we have, from (3.17),

$$\begin{aligned} |x(t)| &\leq Ce^{-\alpha(t-t_0)}|x(t_0)| + \int_{t_0}^{t} Ce^{-\alpha(t-s)}|f(s, x(s))|ds \\ &\leq Ce^{-\alpha(t-t_0)}|x(t_0)| + \int_{t_0}^{t} Ce^{-\alpha(t-s)}\eta|x(s)|ds. \end{aligned} \tag{3.18}$$

The following is similar to the final part in the proof of Theorem 5.3.2,

$$|x(t)| \leq C|x(t_0)| \exp\Big[-(\alpha - C\eta)(t - t_0)\Big], \ \ t \in [t_0, t_1]. \tag{3.19}$$

However,

$$\begin{aligned} |x(t_1)| &\leq C|x(t_0)| \exp\Big[-(\alpha - C\eta)(t_1 - t_0)\Big] \\ &\leq \Delta \exp\Big[-(\alpha - C\eta)(t_1 - t_0)\Big] < \Delta, \end{aligned} \tag{3.20}$$

contradicting $|x(t_1)| = \Delta$. Thus $|x(t_0)| \leq \frac{\Delta}{C}$ implies $|x(t)| < \Delta$ for $t \geq t_0$. Consequently, if $|x(t_0)| \leq \frac{\Delta}{C}$, then for $t \geq t_0$, one has $|f(t, x(t))| \leq \eta|x(t)|$. Now, similar to (3.18) and (3.19), we have, for $t \geq t_0 \geq 0$,

$$|x(t)| \leq C|x(t_0)| \exp\Big[-(\alpha - C\eta)(t - t_0)\Big], \ \ \text{if} \ |x(t_0)| \leq \frac{\Delta}{C}. \tag{3.21}$$

To prove the uniform asymptotic stability of the zero solution for Eq. (3.16), note that Δ and C are independent constants, thus for any $\varepsilon > 0$, we can choose $\delta = \min\{\frac{\varepsilon}{C}, \frac{\Delta}{C}\} = \delta(\varepsilon)$ such that $|x_0| = |x(t_0)| \le \delta$ and $t \ge t_0$ imply

$$|x(t)| \le C|x(t_0)|\exp\Big[-(\alpha - C\eta)(t - t_0)\Big] \le C|x(t_0)| \le C\delta \le \varepsilon,$$

proving the uniform stability of the zero solution.

Next, to prove the uniform asymptotic stability of the zero solution, let $r = \frac{\Delta}{C}$. (This is needed for (3.21) to be valid; otherwise, the rest of the proof is not required because the proof of Theorem 5.2.1(B) can be used.) Now, for any $0 < \varepsilon < \Delta$, solve $T = T(\varepsilon) > 0$ from $\Delta \exp\Big[-(\alpha - C\eta)T\Big] = \varepsilon$; then from (3.21), we find that $\{|x_0| \le r,\ t_0 \ge 0,\ t \ge t_0 + T\}$ imply

$$
\begin{aligned}
|x(t, t_0, x_0)| &\le C|x(t_0)|\exp\Big[-(\alpha - C\eta)(t - t_0)\Big] \\
&\le Cr\exp\Big[-(\alpha - C\eta)T\Big] = \Delta\exp\Big[-(\alpha - C\eta)T\Big] = \varepsilon,
\end{aligned}
$$

proving the uniform asymptotic stability of the zero solution for Eq. (3.16).
♠

Theorem 5.3.3.a *In Theorem 5.3.3, the condition (3.15) can be relaxed to $|f(t, x)| \le \eta|x|$ with $0 < \eta < \frac{\alpha}{C}$, where α and C are from Theorem 5.2.1.*

Remark 5.3.4 Note that from Chapter 2, $|x| = \sum_{i=1}^{n}|x_i|$ is equivalent to $r(x) = \sqrt{x_1^2 + x_2^2 + \cdots + x_n^2}$. Therefore, when f is autonomous, the condition (3.4) in the Hypothesis (H) in Section 3 of Chapter 4 (concerning perturbations given by $\varepsilon_i(x_1, x_2)$ on linear equations in $\Re^2$) is equivalent to the condition (3.15). Therefore, Theorem 5.3.3 explains why Theorem 4.3.1 in Chapter 4 is true, for example, why stable spiral points and stable nodes for linear differential equations in $\Re^2$ are preserved under small perturbations.
♠

In the next example, we will analyze an equation with details and find the parameters such as ε, δ, r, and T in the definitions of stabilities.

Example 5.3.5 Consider the scalar differential equation

$$x'(t) = -x + x^2,$$

where $f(t, x) = x^2$ satisfies the condition (3.15), thus Theorem 5.3.3 can be applied to obtain the uniform asymptotic stability of the zero solution. To see more details, let's verify this result directly. Note that now $x = 0$ and $x = 1$ are constant solutions. For other solutions, since the equation is autonomous, we only need to consider the solutions starting at 0. They are given by

$$\left|1 - \frac{1}{x(t)}\right| = e^{t+c}, \quad t \geq 0,$$

for some constant c. Now, for $t \geq 0$, either $x(t) > 1$, or $0 < x(t) < 1$, or $x(t) < 0$. Analyzing the three cases, we obtain

$$x(t, 0, x_0) = \left[1 + \frac{1 - x_0}{x_0}e^t\right]^{-1}, \quad t \geq 0, \quad (x_0 \neq 0, \ x_0 \neq 1). \tag{3.22}$$

If $x_0 > 1$, then $x(t)$ exists on a finite interval $[0, \ln\frac{x_0}{x_0 - 1})$, and $x(t) \to \infty$ as $t \nearrow \ln\frac{x_0}{x_0 - 1}$. If $0 < x_0 < 1$, then $x(t)$ exists on $[0, \infty)$ and $x(t) \searrow 0$ as $t \to \infty$. Finally, if $x_0 < 0$, then $x(t, 0, x_0)$ exists on $[0, \infty)$ and $x(t) \nearrow 0$ as $t \to \infty$. See **Figure 5.7**.

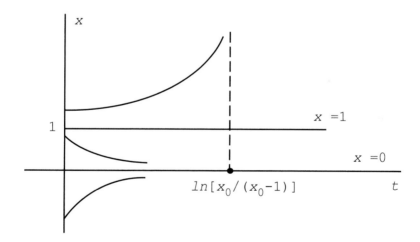

Figure 5.7: Solutions of $x'(t) = -x + x^2$

Next, let's find the parameters such as $\delta(\varepsilon)$, r, and $T(\varepsilon)$ in the definition of uniform asymptotic stability. From Figure 5.7, we see that the solutions are monotone, thus for $0 < \varepsilon < 1$, we can choose $\delta(\varepsilon) = \varepsilon$ to get the uniform stability of the zero solution. For the r in the uniform asymptotic stability of the zero solution, we take it to be a constant in $(\frac{1}{2}, 1)$. Then for $0 < \varepsilon < 1$, we need to find $T = T(\varepsilon) > 0$ such that $\{|x_0| \leq r, \ t \geq T\}$ imply

$$|x(t, 0, x_0)| = \left|\left[1 + \frac{1 - x_0}{x_0}e^t\right]^{-1}\right| \leq \varepsilon, \ \ t \geq T. \tag{3.23}$$

This can be done as follows: For $0 < x_0 \leq r < 1$, (3.23) is true if

$$\left(\frac{1}{\varepsilon} - 1\right)\frac{x_0}{1 - x_0} \leq e^t, \ \ t \geq T. \tag{3.24}$$

Now, since $0 < x_0 \leq r < 1$, we have

$$\left(\frac{1}{\varepsilon} - 1\right)\frac{x_0}{1 - x_0} \leq \left(\frac{1}{\varepsilon} + 1\right)\frac{r}{1 - r}, \tag{3.25}$$

where $\frac{r}{1-r} > 1$ since $r \in (\frac{1}{2}, 1)$. Thus, we can choose $T = T(\varepsilon) > 0$ from $(\frac{1}{\varepsilon} + 1)\frac{r}{1-r} = e^T$, that is, we can define

$$T \overset{\text{def}}{=} \ln\left\{(\frac{1}{\varepsilon} + 1)\frac{r}{1 - r}\right\} = T(\varepsilon) > 0 \tag{3.26}$$

to guarantee (3.24). Hence (3.23) is true if $0 < x_0 \leq r < 1$. When $-r \leq x_0 < 0$, the same can be done, and we leave it as an exercise. Therefore, we have found the parameters $\delta(\varepsilon)$, r, and $T(\varepsilon)$ for the zero solution to be uniformly asymptotically stable.   ♠

Remark 5.3.6 From the analysis of Example 5.3.5, we find that the stability of the zero solution is a local property about the origin. That is, we only look at the solutions with small initial values. The solutions with "large" initial values could behave differently. For example, the solutions in Example 5.3.5 with $x_0 > 1$ actually blow up (approaching ∞) at finite times.   ♠

Exercises 5.3

1. Consider $x'(t) = Ax(t) + f(t)$ where all eigenvalues of A have negative real parts, f is continuous, and $f(t) \to 0$ as $t \to \infty$. Show that for any solution $x(t)$, $x(t) \to 0$ as $t \to \infty$.

2. Consider $x'(t) = Ax(t) + B(t)x(t)$ where all eigenvalues of A have negative real parts, B is continuous and $B(t) \to 0$ as $t \to \infty$. Show that the zero solution is asymptotically stable.

3. Assume that all eigenvalues of A have negative real parts. Prove that there exists a constant $\eta > 0$ such that if $f(t, x)$ is continuous and satisfies a Lipschitz condition and $|f(t, x)| \leq \eta|x|$ for all (t, x), then the zero is a solution of $x'(t) = Ax(t) + f(t, x(t))$ and is uniformly asymptotically stable.

4. Prove Theorem 5.3.1.a.

5. Prove Theorem 5.3.2.a.

6. Prove Theorem 5.3.3.a.

7. Verify (3.22).

8. Verify that solutions of $x'(t) = -x + x^2$ in Example 5.3.5 are given in Figure 5.7.

9. Verify for $0 < x_0 \leq r < 1$ that (3.23) is true if (3.24) is true.

10. Verify (3.25).

11. Complete Example 5.3.5 by finding $T = T(\varepsilon) > 0$ when $-r \leq x_0 < 0$.

5.4 Linear Equations with Periodic Coefficients

In this section, we study stabilities of the linear periodic differential equation

$$x'(t) = A(t)x(t) + f(t), \quad x(t_0) = x_0, \quad t \geq t_0 \geq 0, \quad x \in \Re^n, \qquad (4.1)$$

where $A(t)$, $f(t)$ are continuous on $\Re^+ = [0, \infty)$, and

$$A(t + T) = A(t), \quad t \in \Re^+,$$

for some constant $T > 0$.

First, we study the linear periodic homogeneous differential equation

$$x'(t) = A(t)x(t). \qquad (4.2)$$

From the Floquet theory in Chapter 3, we know that there exists a constant matrix C and a nonsingular continuous T-periodic matrix function $P(t)$ such that $x(t) = P(t)y(t)$ transforms Eq. (4.2) into

$$y'(t) = Cy(t). \qquad (4.3)$$

Since $P(t)$ is periodic and continuous, it is bounded. Thus the transformation $x(t) = P(t)y(t)$ reduces the stability properties of Eq. (4.2) to those of Eq. (4.3), for which the results from Section 5.2 can be applied. Therefore we have the following result based on Theorems 5.1.2 and 5.2.1. The proof is left as an exercise.

Theorem 5.4.1 *Let λ be a complex number and denote by $R(\lambda)$ the real part of λ. Let $\phi = 0$ be the zero solution of Eq. (4.2) where $A(t)$ is T-periodic and let $\psi = 0$ be the zero solution of Eq. (4.3) where C is from the Floquet theory.*

(A). The following statements are equivalent:

> *(1). $\phi = 0$ of Eq. (4.2) is stable or uniformly stable;*

> *(1a). $\psi = 0$ of Eq. (4.3) is stable or uniformly stable;*

> *(2). For each characteristic exponent λ (eigenvalue of matrix C), either $R(\lambda) < 0$, or $R(\lambda) = 0$ but in this case λ appears only in matrices J_i (Jordan canonical form for C) such that J_i is a 1×1 matrix;*

> *(2a). For each characteristic multiplier η (eigenvalue of matrix e^{TC}), either $|\eta| < 1$, or $|\eta| = 1$ but in this case η appears only in matrices J_i (Jordan canonical form for e^{TC}) such that J_i is a 1×1 matrix;*

> *(3). There is an (independent or generic) constant $M > 1$ such that*
> $$|e^{tC}| \le M, \quad 0 \le t < \infty. \tag{4.4}$$

(B). The following statements are equivalent:

> *(1). $\phi = 0$ of Eq. (4.2) is asymptotically stable or uniformly asymptotically stable;*

> *(1a). $\psi = 0$ of Eq. (4.3) is asymptotically stable or uniformly asymptotically stable;*

> *(2). Each characteristic exponent (eigenvalue of matrix C) has a negative real part;*

> *(2a). Each characteristic multiplier η (eigenvalue of matrix e^{TC}) satisfies $|\eta| < 1$;*

(3). There are (independent or generic) constants $M > 1$ and $\alpha > 0$ such that

$$|e^{tC}| \leq Me^{-\alpha t}, \quad 0 \leq t < \infty. \tag{4.5}$$

(C). The following statements are equivalent:

(1). $\phi = 0$ of Eq. (4.2) is unstable;

(1a). $\psi = 0$ of Eq. (4.3) is unstable;

(2). There is a characteristic exponent λ (eigenvalue of matrix C) with $R(\lambda) = 0$ and λ appears in a matrix J_i that is at least 2×2;

(2a). There is a characteristic multiplier η (eigenvalue of matrix e^{TC}) with $|\eta| = 1$ and η appears in a matrix J_i that is at least 2×2;

(3). There is a characteristic exponent with a positive real part;

(3a). There is a characteristic multiplier η with $|\eta| > 1$. ♠

Next, based on Theorem 5.1.2 and the results in the previous section concerning perturbations on linear differential equations, we obtain the following similar results for perturbations on linear periodic differential equations. The proofs are left as exercises.

Theorem 5.4.2 *Assume that $A(t)$ is continuous and periodic on $\Re^+$. If the zero solution of Eq. (4.2) is stable, and if the $n \times n$ continuous matrix function $B(t)$ satisfies $\int_0^\infty |B(t)|dt < \infty$, then the zero solution of*

$$x'(t) = A(t)x(t) + B(t)x(t) = [A(t) + B(t)]x(t) \tag{4.6}$$

is uniformly stable. ♠

Theorem 5.4.3 *Assume that $A(t)$ is continuous and periodic on $\Re^+$. If the zero solution of Eq. (4.2) is asymptotically stable, and if the $n \times n$ continuous matrix function $B(t)$ satisfies*

$$\int_{t_0}^t |B(s)|ds \leq m(t - t_0) + r, \quad t \geq t_0 \geq 0,$$

for some positive constants m and r, then there is an $m_0 > 0$ such that if $m \leq m_0$, then the zero solution of Eq. (4.6) is uniformly asymptotically stable. ♠

Again, a result concerning nonlinear perturbations is also possible under some conditions.

Theorem 5.4.4 *Assume that $A(t)$ is continuous and periodic on $\Re^+$. If the zero solution of Eq. (4.2) is asymptotically stable, and if the continuous function $f(t, x)$ satisfies a Lipschitz condition (or weak or local Lipschitz) with respect to x on D, and*

$$\lim_{x \to 0} \frac{|f(t, x)|}{|x|} = 0, \quad \text{uniformly for } t \in [0, \infty), \tag{4.7}$$

then the zero solution of

$$x'(t) = A(t)x(t) + f(t, x(t)) \tag{4.8}$$

is uniformly asymptotically stable.   ♠

Next, we apply these results to Hill's equation introduced in Section 4 of Chapter 3.

Example 5.4.5 (Hill's equation) Let $\phi(t)$ be real, continuous, and periodic, and consider the scalar linear periodic differential equation, called Hill's equation,

$$y''(t) + \phi(t)y(t) = 0, \quad \phi(t + T) = \phi(t), \quad t \in \Re. \tag{4.9}$$

Following Chapter 3, we define $x = [y, y']^T$, then Eq. (4.9) becomes $x'(t) = A(t)x(t)$, where

$$A(t) = \begin{bmatrix} 0 & 1 \\ -\phi(t) & 0 \end{bmatrix}. \tag{4.10}$$

Let $U(t, 0)$ be the fundamental matrix solution, then $U(T, 0) = e^{TC}$, where C is from the Floquet theory. And we know from Chapter 3 that

$$\eta_1 \eta_2 = \det e^{TC} = \exp\left(\int_0^T tr\, A(s)ds\right) = 1, \tag{4.11}$$

where η_1 and η_2 are the two characteristic multipliers of $A(t)$ (eigenvalues of matrix e^{TC}). Hence

$$\begin{aligned} \det[e^{TC} - \eta E] &= (\eta - \eta_1)(\eta - \eta_2) = \eta^2 - (tr[e^{TC}])\eta + 1 \\ &= \eta^2 - (tr[U(T, 0)])\eta + 1. \end{aligned} \tag{4.12}$$

Due to (4.11), we cannot have $|\eta_1| < 1$ and $|\eta_2| < 1$, thus the conditions in Theorem 5.4.1(B) are not satisfied. Therefore, the zero solution of Hill's equation (4.9) is not asymptotically stable.

To determine other stability properties, let $\eta_1 = e^{\lambda_1}$ and $\eta_2 = e^{\lambda_2}$, where λ_1 and λ_2 are the eigenvalues of the matrix C (see spectral mapping theorem 3.3.14 in Chapter 3). From (4.11), we must have

$$\lambda_1 = a + bi, \quad \lambda_2 = -a + b_1 i, \quad a, b, b_1 \in \Re, \tag{4.13}$$

where $b \in [0, 2\pi]$ and $b + b_1 = 2\pi k$ for some integer k.

If $a \neq 0$, then the real part of λ_1 or λ_2 is positive. Thus from Theorem 5.4.1(C), the zero solution of Hill's equation (4.9) is unstable.

If $a = 0$ and $b \neq 0$, $b \neq \pi$, $b \neq 2\pi$, then $\eta_1 = e^{bi}$ is a complex number. From (4.12), which is a polynomial in η with real coefficients, η_2 must be the complex conjugate of $\eta_1 = e^{bi}$, that is, $\eta_2 = e^{-bi}$ and $\eta_2 \neq \eta_1$. Thus the Jordan canonical form for e^{TC} is

$$\begin{bmatrix} \eta_1 & 0 \\ 0 & \eta_2 \end{bmatrix} = \begin{bmatrix} e^{bi} & 0 \\ 0 & e^{-bi} \end{bmatrix} = \begin{bmatrix} J_1 & 0 \\ 0 & J_2 \end{bmatrix},$$

where J_1 and J_2 are all 1×1 matrices and $|\eta_1| = |\eta_2| = 1$. Therefore, from Theorem 5.4.1(A), the zero solution of Hill's equation (4.9) is stable.

If $a = 0$ and $b = 0$ or $b = 2\pi$, then $\eta_1 = e^{\lambda_1} = 1$, and hence $\eta_2 = 1$. Now, if the Jordan canonical form for e^{TC} is the identity matrix, then e^{TC} itself is the identity matrix. Therefore, if

$$e^{TC} = \begin{bmatrix} 1 & 0 \\ 0 & 1 \end{bmatrix} = \begin{bmatrix} J_1 & 0 \\ 0 & J_2 \end{bmatrix},$$

then the zero solution of Hill's equation (4.9) is stable. Otherwise, if the Jordan canonical form for e^{TC} is

$$\begin{bmatrix} 1 & 1 \\ 0 & 1 \end{bmatrix} = J_1,$$

which is a 2×2 matrix, then the zero solution of Hill's equation (4.9) is unstable.

If $a = 0$ and $b = \pi$, then $\eta_1 = -1$, and hence $\eta_2 = -1$. Now, if the Jordan canonical form for e^{TC} is the negative of the identity matrix, then e^{TC} itself is the negative of the identity matrix. Therefore, if

$$e^{TC} = \begin{bmatrix} -1 & 0 \\ 0 & -1 \end{bmatrix} = \begin{bmatrix} J_1 & 0 \\ 0 & J_2 \end{bmatrix},$$

then the zero solution of Hill's equation (4.9) is stable. Otherwise, if the Jordan canonical form for e^{TC} is

$$\begin{bmatrix} -1 & 1 \\ 0 & -1 \end{bmatrix} = J_1,$$

then the zero solution of Hill's equation (4.9) is unstable.

To summarize, we have, for λ_1 and λ_2 given in (4.13)

(A). The zero solution of Hill's equation (4.9) is not asymptotically stable.

(B). The zero solution of Hill's equation (4.9) is stable if and only if one of the following cases occurs.

(1). $a = 0$ and $b \neq 0$, $b \neq \pi$, $b \neq 2\pi$.

(2). $a = 0$ and $b = 0$ or $b = 2\pi$, (or $\eta_1 = \eta_2 = 1$), and

$$e^{TC} = \begin{bmatrix} 1 & 0 \\ 0 & 1 \end{bmatrix}.$$

(3). $a = 0$ and $b = \pi$, (or $\eta_1 = \eta_2 = -1$), and

$$e^{TC} = \begin{bmatrix} -1 & 0 \\ 0 & -1 \end{bmatrix}. \qquad \spadesuit \qquad (4.14)$$

The above is an analysis of stabilities for Hill's equation (4.9). We point out that descriptions of stabilities for Hill's equation (4.9) using other quantities are also possible. For example, one could use $tr[U(T,0)]$ and obtain, for λ_1 and λ_2 given in (4.13),

$$\begin{aligned} tr[U(T,0)] &= \eta_1 + \eta_2 = e^{\lambda_1} + e^{\lambda_2} \\ &= e^a(\cos b + i \sin b) + e^{-a}(\cos b_1 + i \sin b_1) \\ &= e^a(\cos b + i \sin b) + e^{-a}(\cos b - i \sin b) \\ &= [e^a + e^{-a}] \cos b + i[e^a - e^{-a}] \sin b. \end{aligned} \qquad (4.15)$$

Since $tr[U(T,0)]$ is real, we get

$$\begin{cases} [e^a - e^{-a}] \sin b = 0, \\ tr[U(T,0)] = [e^a + e^{-a}] \cos b. \end{cases} \qquad (4.16)$$

Then, we can verify (see an exercise) that

$$
\begin{cases}
a \neq 0 \text{ if and only if } |tr[U(T,0)]| > 2. \\
a = 0 \text{ and } b \neq 0, \, b \neq \pi, \, b \neq 2\pi \text{ if and only if } |tr[U(T,0)]| < 2. \\
a = 0 \text{ and } b = 0 \text{ or } b = 2\pi \text{ if and only if } tr[U(T,0)] = 2. \\
a = 0 \text{ and } b = \pi \text{ if and only if } tr[U(T,0)] = -2.
\end{cases}
\tag{4.17}
$$

Therefore, the zero solution of Hill's equation (4.9) is stable if and only if one of the following cases occurs.

(1). $|tr[U(T,0)]| < 2.$

(2). $tr[U(T,0)] = 2$, (or $\eta_1 = \eta_2 = 1$), and

$$
e^{TC} = \begin{bmatrix} 1 & 0 \\ 0 & 1 \end{bmatrix}.
$$

(3). $tr[U(T,0)] = -2$, (or $\eta_1 = \eta_2 = -1$), and

$$
e^{TC} = \begin{bmatrix} -1 & 0 \\ 0 & -1 \end{bmatrix}.
$$

Exercises 5.4

1. Prove Theorem 5.4.1.

2. Prove Theorem 5.4.2.

3. Prove Theorem 5.4.3.

4. Prove Theorem 5.4.4.

5. Assume that $A(t+T) = A(t)$ for some constant $T > 0$ and that all solutions of $x'(t) = A(t)x(t)$ go to zero as $t \to \infty$. Prove that there exists a constant $\eta > 0$ such that if $B(t)$ is continuous and satisfies $|B(t)| \leq \eta$ for $t \geq 0$, then all solutions of $x'(t) = A(t)x(t) + B(t)x(t)$ go to zero as $t \to \infty$.

6. Prove (4.17) for Hill's equation in Example 5.4.5.

5.5   Liapunov's Method for Autonomous Equations

In this section, we study stabilities for the autonomous nonlinear differential equation

$$x'(t) = f(x(t)), \quad \text{or} \quad x' = f(x), \tag{5.1}$$

in $D = [0, \infty) \times Q$, where $Q \subset \Re^n$ is a domain containing the zero vector. (Results for the zero solution will be stated again. Therefore we assume that $Q \subset \Re^n$ contains the zero vector.) We also assume that for any $(t_0, x_0) \in D$, Eq. (5.1) has a unique solution $x(t, t_0, x_0)$ existing on $[t_0, \infty)$ with $x(t_0) = x_0$.

Since solutions of Eq. (5.1) cannot generally be found or written out by formulas, and the method using eigenvalues for linear differential equations is not applicable either, we exploit some different approaches than those for linear differential equations.

To begin the study of autonomous nonlinear differential equations, let's start with scalar cases, where the planar geometry can suggest and interpret the results and assist us understand the subjects. The study of scalar differential equations may also provide a direction for us to pursue for differential equations in $\Re^n, n \geq 1$. Therefore, we assume that $f : \Re \to \Re$ for now. Based on the analysis of a logistic equation in Chapter 1, we find that the signs of f can be used to determine the flows of solutions on the x-axis. For example, when f is shown in **Figure 5.8**, we get $x' = f(x) < 0$ for $x > 0$ and $x' = f(x) > 0$ for $x < 0$. Hence a solution x decreases if $x > 0$ and increases if $x < 0$. Therefore, the solutions flow to the zero on the x-axis, as shown in the second picture of Figure 5.8. Accordingly, the zero solution is asymptotically stable.

Another way to look at this situation is given in the following: Treat the numbers x and x' as vectors and let ϕ be the **angle** between them. Then, from Figure 5.8,

$$|x||x'| \cos \phi = xx' = xf(x) < 0, \quad x \neq 0,$$

which indicates that $\phi = \pi$, or the numbers x and x' are on opposite sides of the origin of the x-axis, thus the solutions will flow to the origin from two sides on the x-axis. So far, these arguments are based on intuity. Next, we prove that this intuity is true in this case. Note that in Figure 5.8, $xf(x) < 0$ for $x \neq 0$.

Theorem 5.5.1 *Let Eq. (5.1) be a scalar differential equation and let $\overline{x}$ be a critical point (or a constant solution) of Eq. (5.1), that is, $f(\overline{x}) = 0$.*

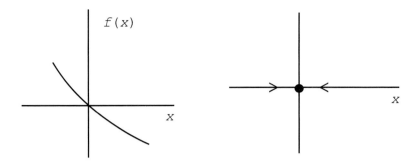

Figure 5.8: A function f with $xf(x) < 0$, $x \neq 0$, and its flows

---

1. $\bar{x}$ *is uniformly asymptotically stable if there is a* $\delta > 0$ *such that* $(x - \bar{x})f(x) < 0$ *for* $0 < |x - \bar{x}| < \delta$.

2. $\bar{x}$ *is uniformly stable if there is a* $\delta > 0$ *such that* $(x - \bar{x})f(x) \leq 0$ *for* $|x - \bar{x}| < \delta$.

3. $\bar{x}$ *is unstable if there is a* $\delta > 0$ *such that* $(x - \bar{x})f(x) > 0$ *for either* $\bar{x} < x < \bar{x} + \delta$ *or* $\bar{x} - \delta < x < \bar{x}$.

Proof. (1): Assume that $(x - \bar{x})f(x) < 0$ for $0 < |x - \bar{x}| < \delta$. First, we let $\bar{x} < x_0 < \bar{x} + \delta$ and prove that for the solution $x(t) = x(t, t_0, x_0)$, one has $x(t) \searrow \bar{x}$ as $t \to \infty$. From uniqueness, we obtain $x(t) > \bar{x}$, $t \geq t_0$. Note that now, as long as $|x(t) - \bar{x}| < \delta$, we have $(x(t) - \bar{x})f(x(t)) < 0$, hence $x'(t) = f(x(t)) < 0$. Thus we conclude that $x(t)$ is decreasing from x_0 for $t \geq t_0$. If $x(t) \nrightarrow \bar{x}$, $t \to \infty$, then $x(t) \to \bar{\bar{x}} > \bar{x}$, $t \to \infty$, or $x(t) \in [\bar{\bar{x}}, x_0]$, $t \geq t_0$. Now, since $[\bar{\bar{x}}, x_0] \subset (\bar{x}, \bar{x} + \delta)$, we obtain that f is negative on $[\bar{\bar{x}}, x_0]$ since $(x - \bar{x})f(x) < 0$ for $0 < |x - \bar{x}| < \delta$. Then, as f is continuous, there is a constant $k > 0$ such that

$$f(\tau) \leq -k, \quad \tau \in [\bar{\bar{x}}, x_0].$$

Now, from

$$x(t) - x(t_0) = \int_{t_0}^{t} f(x(s))ds \leq -k(t - t_0) \to -\infty, \quad t \to \infty,$$

we get a contradiction because as $t \to \infty$, one has $x(t) - x(t_0) \to \bar{\bar{x}} - x(t_0)$, which is a finite number. Therefore, $\bar{x} < x_0 < \bar{x} + \delta$ implies $x(t) \searrow \bar{x}$ as

$t \to \infty$. Similarly, one can show that $\overline{x} - \delta < x_0 < \overline{x}$ implies $x(t) \nearrow \overline{x}$ as $t \to \infty$, thus $\overline{x}$ is uniformly asymptotically stable.

The same idea can be used to prove parts (2) and (3), which are left as an exercise. This completes the proof of the theorem.   ♠

Another way to determine the signs of a function f is to use its derivatives. For example, in Figure 5.8, if $f'(0)$ exists, then $f'(0) < 0$, which indicates in geometry that for x near 0, $f(x) < 0$ for $x > 0$ and $f(x) > 0$ for $x < 0$; or $xf(x) < 0$ for x near 0 but $x \neq 0$. Accordingly, we have

Corollary 5.5.2 *Let Eq. (5.1) be a scalar differential equation and let $\overline{x}$ be a critical point (or a constant solution) of Eq. (5.1), that is, $f(\overline{x}) = 0$. Suppose that f' is continuous at $\overline{x}$.*

1. *$\overline{x}$ is uniformly asymptotically stable if $f'(\overline{x}) < 0$.*

2. *$\overline{x}$ is uniformly stable if there is a $\delta > 0$ such that $f'(x) \leq 0$ for $|x - \overline{x}| < \delta$.*

3. *$\overline{x}$ is unstable if $f'(\overline{x}) > 0$.*

Proof. (1): Since f' is continuous at $\overline{x}$ and $f'(\overline{x}) < 0$, there is a $\delta > 0$ such that $f'(x) < 0$ for $|x - \overline{x}| < \delta$. Using the mean value theorem, for any x with $|x - \overline{x}| < \delta$,

$$f(x) = f(\overline{x}) + f'(c)(x - \overline{x}) = f'(c)(x - \overline{x}), \quad |c - \overline{x}| < \delta,$$

thus,

$$(x - \overline{x})f(x) = f'(c)(x - \overline{x})^2 < 0, \quad 0 < |x - \overline{x}| < \delta,$$

therefore, results in Theorem 5.5.1 can be used. The proofs of (2) and (3) are left as an exercise. This completes the proof of the theorem.   ♠

Theorem 5.5.1 is more useful than Corollary 5.5.2 in applications. For example, Theorem 5.5.1 can be used to verify the uniform asymptotic stability of the zero solution $\overline{x} = 0$ of $x' = -x^3$, which is not obtainable using Corollary 5.5.2 since $f(x) = -x^3$ and $f'(0) = 0$.

Therefore, we find that phase portraits in $\Re$ (on the x-axis) for scalar autonomous differential equations are very simple: Solutions are either critical points (constant solutions), or they flow to critical points or $\pm\infty$ and cannot stop at the middle. A typical function and the corresponding flows on the

x-axis are shown in **Figure 5.9**. Later, we will see that for autonomous differential equations in $\Re^n$, $n \geq 2$, more complicated things can occur. For example, in $\Re^2$, some solutions could flow to finite nonconstant solutions; in $\Re^3$, some solutions may be bounded but never settle down or pile up at some places in a predictable fashion.

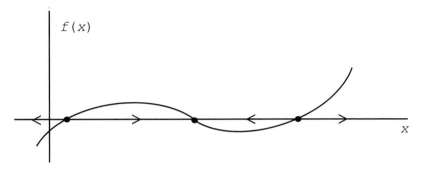

Figure 5.9: A function f and the corresponding flows on the x-axis

The above is a brief account of the situations for scalar autonomous differential equations, where the signs of $x' = f(x)$ can be used to determine stability properties. For autonomous differential equations in $\Re^n$, $n \geq 2$, $x' = f(x)$ is a vector, hence $f(x)$ cannot tell anything directly. One method to study stability now is to extend the idea used in the proof of Theorem 4.3.1(a) in Chapter 4. Let's review the idea in that proof: We first define a function

$$V(t) = V(x_1(t),\, x_2(t)) = \frac{1}{2}[x_1^2(t) + x_2^2(t)], \qquad (5.2)$$

where $(x_1(t), x_2(t))$ is a solution whose existence and uniqueness is guaranteed. $V(t)$ is related to the distance $r(x) = \sqrt{x_1^2 + x_2^2}$ from the solution $(x_1(t),\, x_2(t))$ to the origin $(0,0)$. Then, without solving the differential equation explicitly, we can take a derivative of $V(t)$ in t along the solution, that is, we plug in the differential equation when taking the derivative. The derivative then satisfies

$$V'(t) \leq \alpha V(t) = -[-\alpha V(t)],$$

for a constant $\alpha < 0$. Accordingly, we conclude that

$$V(t) \longrightarrow 0, \quad t \to \infty.$$

Hence,
$$x_1^2(t) + x_2^2(t) \longrightarrow 0, \quad t \to \infty,$$

therefore, the solution will go to the origin, or the zero solution $\phi = (0,0)$ is asymptotically stable.

This demonstrates the key idea of the so-called "Liapunov's method," which allows us to derive qualitative properties, especially stability and boundedness, without solving differential equations explicitly.

Next, let's see why this idea works in geometry. First, let Eq. (5.1) be a scalar equation, and assume that $\phi = 0$ is a solution of Eq. (5.1), or $f(0) = 0$. To study the stability of $\phi = 0$, if we let

$$V(t) = V(x(t)) = \frac{x^2(t)}{2},$$

then, taking a derivative in t and plugging in Eq. (5.1), we get

$$V'(t) = x(t)x'(t) = x(t)f(x(t)) = xf(x).$$

Now, if $V'(t) < 0$, then $xf(x) < 0$ for $x \neq 0$ (or the angle between $x(t)$ and $x'(t)$ is π). This matches exactly with Figure 5.8 which we have just analyzed. Therefore, $V'(t) < 0$ implies that $x(t) \to 0$ as $t \to \infty$. This is why in geometry we use such a V function to derive stabilities.

For the geometry in $\Re^2$, we assume again that the zero is a solution of Eq. (5.1), or $f(0) = 0$, and assume that $(x_1(t), x_2(t))$ is a solution of Eq. (5.1). Consider $V(t) = V(x_1(t), x_2(t)) = \frac{1}{2}[x_1^2(t) + x_2^2(t)]$, or

$$V(x_1, x_2) = \frac{1}{2}[x_1^2 + x_2^2],$$

then, taking a derivative in t, we obtain

$$\begin{aligned}
\frac{d}{dt}V(t) &= x_1 x_1' + x_2 x_2' = [x_1, x_2]^T \cdot [x_1', x_2']^T \\
&= \sqrt{x_1^2 + x_2^2}\sqrt{(x_1')^2 + (x_2')^2}\cos\phi, \qquad (5.3)
\end{aligned}$$

where "$\cdot$" denotes the inner product and ϕ is the angle between the vectors $[x_1, x_2]^T$ and $[x_1', x_2']^T$ (T means the transpose).

For any $c > 0$ fixed,
$$V(x_1, x_2) = c$$

or $\frac{1}{2}[x_1^2 + x_2^2] = c$ is a circle in the (x_1, x_2) plane centered at $(0,0)$ with radius

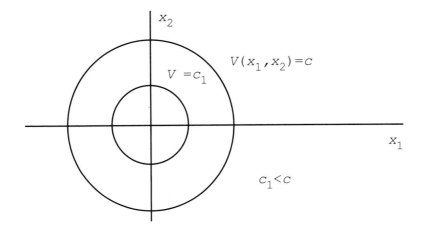

Figure 5.10: Circles of $V(x_1, x_2) = c$

$\sqrt{2c}$, as shown in **Figure 5.10**.

Now, for $V(x_1, x_2) = \frac{1}{2}[x_1^2 + x_2^2]$, the vector $\nabla V = [V_{x_1}, V_{x_2}]^T = [x_1, x_2]^T$ (known as a gradient in multivariable calculus) is outward normal (perpendicular) to the circle at point (x_1, x_2). Also, from multivariable calculus, we know that the vector $[x_1'(t), x_2'(t)]^T$ is tangent to the trajectory $(x_1(t), x_2(t))$ at each point on the trajectory, or vector $[x_1'(t), x_2'(t)]^T$ is the direction of the trajectory. If $V'(t) < 0$, then from (5.3), $\cos \phi < 0$, hence

$$\phi \in (\frac{\pi}{2}, \frac{3\pi}{2}). \tag{5.4}$$

In geometry, (5.4) indicates that if a trajectory intersects the circle $V(x_1, x_2) = c$, then the direction of motion of the trajectory is **inward** with respect to $V(x_1, x_2) = c$, or the trajectory will enter the interior of the circle $V(x_1, x_2) = c$, as shown in **Figure 5.11**.

Consequently, the trajectory will intersect another circle $V(x_1, x_2) = c_1$ with $c_1 < c$, and when it does so, it enters the interior of the circle $V(x_1, x_2) = c_1$. This procedure will continue so that eventually the trajectory approaches the origin and the zero solution is asymptotically stable.

The above gives the geometry in $\Re^2$ why $V'(t) < 0$ can drive the trajectories to the origin. Next, we provide a three-dimensional view of why the idea works. In doing so, we let $V(x_1, x_2) = \frac{1}{2}[x_1^2 + x_2^2]$ be a function defined on the (x_1, x_2) plane, shown as a paraboloid in **Figure 5.12**.

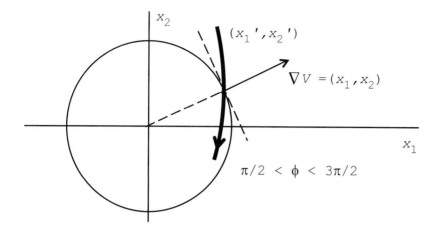

Figure 5.11: A trajectory moves inward of $V = c$ when $V'(t) < 0$

Now, for every $c > 0$, $V(x_1, x_2) = c$ is a "level curve," or a circle in the (x_1, x_2) plane in Figure 5.10. When $V'(t) < 0$, we know in $\Re^2$ that "a trajectory moves inward of a circle," which is now translated in $\Re^3$ to "a trajectory slides down the paraboloid to a lower level." This procedure will continue so that eventually the trajectory slides down the paraboloid and approaches the origin. Therefore, the zero solution is asymptotically stable. See the second picture in Figure 5.12.

The above is a geometric interpretation of the usefulness of $V = \frac{1}{2}[x_1^2 + x_2^2]$ for autonomous differential equations in $\Re^2$, showing why $V'(t) < 0$ implies that solutions approach the origin.

If $V' \leq 0$, then a trajectory will either move inward of a circle $V(x_1, x_2) = c$ or, at worst, tangent to it. Thus, trajectories started inside a circle $V(x_1, x_2) = c$ cannot escape from it. Therefore, the origin is stable. Asymptotic stability is not expected now because trajectories could be circles around the origin, hence the origin is a center and is stable but not asymptotically stable.

If $V' > 0$, then a trajectory inside a circle $V(x_1, x_2) = c$ will move outside of it (no matter how small the value c is). This indicates that the origin is unstable.

Next, we use some examples to demonstrate how this V function is used, this will help you understand the general theory we will present later.

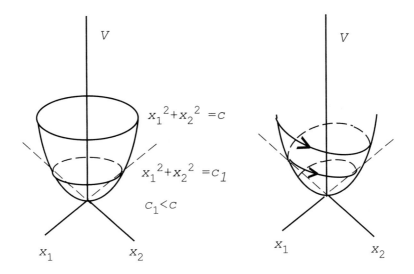

Figure 5.12: A trajectory slides down the paraboloid of $V = \frac{x_1^2 + x_2^2}{2}$

Example 5.5.3 Consider

$$\begin{cases} x_1' = -x_1 + x_2^2 \\ x_2' = -x_2 - x_1 x_2, \end{cases} \tag{5.5}$$

and $V(t) = V(x_1(t), x_2(t)) = \frac{1}{2}[x_1^2(t) + x_2^2(t)]$. Now,

$$\begin{aligned} \frac{dV}{dt} &= x_1(t)x_1'(t) + x_2(t)x_2'(t) \\ &= x_1(-x_1 + x_2^2) + x_2(-x_2 - x_1 x_2) \\ &= -(x_1^2 + x_2^2) = -2V(t). \end{aligned} \tag{5.6}$$

Hence,

$$V(t) = V(t_0)e^{-2(t-t_0)} \longrightarrow 0, \;\; t \to \infty, \tag{5.7}$$

therefore,

$$x_1^2(t) + x_2^2(t) \longrightarrow 0, \;\; t \to \infty.$$

Since this limit applies to every solution, we conclude that every solution approaches the origin as $t \to \infty$. This implies that the zero solution is asymptotically stable. ♠

Example 5.5.4 Consider

$$\begin{cases} x_1' &= -\frac{x_1}{2} + x_2^2 \\ x_2' &= -x_2 - x_1 x_2, \end{cases} \tag{5.8}$$

and $V(t) = V(x_1(t), x_2(t)) = \frac{1}{2}[x_1^2(t) + x_2^2(t)]$. Now,

$$\begin{aligned} \frac{dV}{dt} &= x_1 x_1' + x_2 x_2' \\ &= x_1(-\frac{x_1}{2} + x_2^2) + x_2(-x_2 - x_1 x_2) \\ &= -(\frac{x_1^2}{2} + x_2^2) \\ &\leq -(\frac{x_1^2}{2} + \frac{x_2^2}{2}) = -V(t). \end{aligned} \tag{5.9}$$

Using a differential inequality (see an exercise in Chapter 4), one has

$$V(t) \leq V(t_0) e^{-(t-t_0)} \longrightarrow 0, \ \ t \to \infty, \tag{5.10}$$

therefore, the zero solution is asymptotically stable.   ♠

Example 5.5.5 Consider

$$\begin{cases} x_1' &= x_2 - x_1 x_2^2 \\ x_2' &= -x_1 + x_1^2 x_2, \end{cases} \tag{5.11}$$

and $V(t) = V(x_1(t), x_2(t)) = \frac{1}{2}[x_1^2(t) + x_2^2(t)]$. Now,

$$\begin{aligned} \frac{dV}{dt} &= x_1 x_1' + x_2 x_2' \\ &= x_1(x_2 - x_1 x_2^2) + x_2(-x_1 + x_1^2 x_2) = 0, \end{aligned} \tag{5.12}$$

which means that the distance of a solution and the origin will not increase. Therefore, the zero solution is stable. In this case, asymptotic stability is not obtainable. In fact, this example was analyzed in Section 3 of Chapter 4, where it was found that the origin was a center.   ♠

Example 5.5.6 Consider

$$\begin{cases} x_1' &= x_1 - x_1 x_2^2 \\ x_2' &= x_2 + x_1^2 x_2, \end{cases} \tag{5.13}$$

and $V(t) = V(x_1(t), x_2(t)) = \frac{1}{2}[x_1^2(t) + x_2^2(t)]$. Now,

$$
\begin{aligned}
\frac{dV}{dt} &= x_1 x_1' + x_2 x_2' \\
&= x_1(x_1 - x_1 x_2^2) + x_2(x_2 + x_1^2 x_2) = x_1^2 + x_2^2 = 2V(t), \quad (5.14)
\end{aligned}
$$

hence,

$$
V(t) = V(t_0) e^{2(t-t_0)} \longrightarrow \infty, \ \ t \to \infty, \tag{5.15}
$$

therefore, the zero solution is unstable. In fact, the origin of the linearization is an unstable proper node and using the results of Section 3 in Chapter 4, the origin of Eq. (5.13) is also unstable. ♠

Some applications in physics are discussed in Example 1.3.11 in Chapter 1, where a V function is defined to be the total energy of a system and then used to verify the **law of conservation of energy**. This example will be analyzed again in the next section and the related stability properties will be derived.

Now, we are ready to present a general theory of **Liapunov's method** concerning stability properties for autonomous differential equations in $\Re^n$. The study of nonautonomous differential equations will be given in Chapter 9. Additional results, remarks, and examples can be found in reference books, such as Yoshizawa [1966] and Burton [1985].

Definition 5.5.7 *Let $Q \subset \Re^n$ be a domain containing the zero vector and consider Eq. (5.1). For a continuous function $V : Q \to [0, \infty)$ that is Lipschitz (or weak or local Lipschitz) with respect to x, define*

$$
V_{(5.1)}'(x) \overset{\text{def}}{=} \limsup_{h \to 0^+} \frac{V(x + hf(x)) - V(x)}{h}, \ \ x \in Q. \tag{5.16}
$$

Next, for a solution $x(t)$ of Eq. (5.1), define

$$
V'(x(t)) \overset{\text{def}}{=} \limsup_{h \to 0^+} \frac{V(x(t+h)) - V(x(t))}{h}. \tag{5.17}
$$

For $V_{(5.1)}'(x)$ and $V'(x(t))$ in Definition 5.5.7, we have the following result.

Lemma 5.5.8 *Let $Q \subset \Re^n$ be a domain containing the zero vector. Let $V : Q \to [0, \infty)$ have continuous first partial derivatives and let $x = x(t)$ be a solution of Eq. (5.1), then*

$$
V_{(5.1)}'(x) = V'(x(t)). \tag{5.18}
$$

Moreover, if $V'(x(t)) \leq 0$, then

$$V(x(b)) - V(x(a)) \leq \int_a^b V'(x(t))dt, \quad 0 \leq a \leq b. \tag{5.19}$$

Proof. See the proof of Theorem 9.1.1 in Chapter 9 for general cases, including autonomous or nonautonomous differential equations.   ♠

Lemma 5.5.8 says that the derivative of V with respect to Eq. (5.1) defined by (5.16) is actually the derivative of V along a solution of Eq. (5.1) defined by (5.17). That is, to find $V'_{(5.1)}(x)$, we can take a derivative of $V(x(t))$ in t by plugging in the differential equation (5.1).

The following terminology will be used in Liapunov's method.

Definition 5.5.9 (Positive definite) *Let $Q \subset \Re^n$ be a domain containing the zero vector. A continuous function $V : Q \to [0, \infty)$ is called positive definite if $V(x) > 0$ for $x \neq 0$.*

Definition 5.5.10 (Liapunov function) *Let $Q \subset \Re^n$ be a domain containing the zero vector. A function $V : Q \to [0, \infty)$ is called a Liapunov function if $V(0) = 0$, V is positive definite and has continuous first partial derivatives.*

Example 5.5.11 Based on $V(t) = \frac{1}{2}[x_1^2(t) + x_2^2(t)]$ in the proof of Theorem 4.3.1(a) in Chapter 4, if we define

$$V(x) = \frac{1}{2}[x_1^2 + x_2^2], \quad x = [x_1,\, x_2]^T \in \Re^2,$$

then $V(x)$ is positive definite. Moreover, in that proof,

$$V'(x(t)) \leq -\frac{-\alpha}{2}\Big[x_1^2(t) + x_2^2(t)\Big], \tag{5.20}$$

or,

$$-V'(x(t)) \geq \frac{-\alpha}{2}\Big[x_1^2(t) + x_2^2(t)\Big], \quad -\alpha > 0, \tag{5.21}$$

thus $-V'(x)$ is also positive definite.   ♠

Note, a general Liapunov function has essentially the same properties that $\frac{x_1^2 + x_2^2}{2}$ has, at least locally. Thus a geometric interpretation similar to

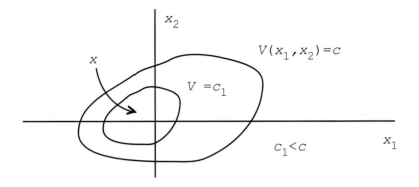

Figure 5.13: Level curves $V = c$ for a general Liapunov function

what we did for $\frac{x_1^2 + x_2^2}{2}$ can be given for a general Liapunov function. For example, for a general Liapunov function $V(x_1, x_2)$ in $\Re^2$, $V(x_1, x_2) = c$ is still a "level curve," or a simple closed curve due to the continuity of V. See **Figure 5.13**.

Now, for $V(t) = V(x_1(t), x_2(t))$, one has

$$
\begin{aligned}
V'(t) &= V_{x_1}x_1' + V_{x_2}x_2' = [V_{x_1}, V_{x_2}]^T \cdot [x_1', x_2']^T \\
&= \sqrt{V_{x_1}^2 + V_{x_2}^2}\sqrt{(x_1')^2 + (x_2')^2}\cos\phi, \qquad (5.22)
\end{aligned}
$$

where ϕ is the angle between the vectors $[V_{x_1}, V_{x_2}]^T$ and $[x_1', x_2']^T$. From multivariable calculus, the gradient vector $\nabla V = [V_{x_1}, V_{x_2}]^T$ (which is $[x_1, x_2]^T$ when $V = \frac{x_1^2 + x_2^2}{2}$) is outward normal (perpendicular) to a level curve $V(x_1, x_2) = c$. Thus the angle between ∇V and the direction $[x_1', x_2']^T$ of a trajectory can still be used to examine the direction a trajectory will follow, hence the stability properties can be determined accordingly.

The following theorem utilizes Liapunov's method to derive stability properties for autonomous differential equations.

Theorem 5.5.12 *Let $Q \subset \Re^n$ be a domain containing the zero vector. Consider Eq. (5.1) on $[0, \infty) \times Q$ with $f(0) = 0$ so that $\phi = 0$ is a solution of Eq. (5.1). Assume that V is a Liapunov function.*

(A). If $V'_{(5.1)}(x) \leq 0$, then $\phi = 0$ is uniformly stable.

(B). If $V'_{(5.1)}(x) < 0$, $x \neq 0$, (or $-V'_{(5.1)}(x)$ is positive definite), then $\phi = 0$ is uniformly asymptotically stable.

(C). If $V'_{(5.1)}(x) > 0$, $x \neq 0$, then $\phi = 0$ is unstable.

Proof. Because the stability of $\phi = 0$ is a local property about the origin, so without loss of generality, we assume that $\{x \in \Re^n : |x| \leq 1\} \subset Q$. Also, the parameters used in the proof, such as ε and δ, are assumed to be less than or equal to 1.

(A): For any $t_0 \geq 0$ and any $\varepsilon > 0$, let $x(t) = x(t, t_0, x_0)$ be a solution of Eq. (5.1), and we need to find a $\delta = \delta(\varepsilon) > 0$ such that $\{|x_0| \leq \delta, t \geq t_0\}$ imply $|x(t, t_0, x_0)| \leq \varepsilon$. Now, we have $V'(x(t)) \leq 0$, hence $V(x(t)) \leq V(x_0)$, $t \geq t_0$. Since $V(x)$ is continuous and positive definite, it is continuous and positive on the closed and bounded set $|x| = \varepsilon$. Then,

$$V_0 = \min_{|x|=\varepsilon} V(x) > 0,$$

(or in geometry, $V(x)$ has a positive minimum on the surface of a ball centered at the origin with radius ε). Next, since $V(x)$ is continuous and $V(0) = 0$, there is a $\delta = \delta(\varepsilon) > 0$ and $\delta < \varepsilon$, such that $|x| \leq \delta$ implies $V(x) < V_0$. Therefore, if $\{|x_0| \leq \delta, t \geq t_0\}$, then

$$V(x(t)) \leq V(x_0) < V_0 = \min_{|x|=\varepsilon} V(x), \tag{5.23}$$

which implies that $|x(t)| < \varepsilon$ for $t \geq t_0$ (otherwise, since $x(t)$ started with $|x(t_0)| = |x_0| \leq \delta < \varepsilon$, there is a $t_1 > t_0$ with $|x(t_1)| = \varepsilon$; but then $V(x(t_1)) \geq \min_{|x|=\varepsilon} V(x) = V_0$, a contradiction to (5.23)). This proves the uniform stability of $\phi = 0$.

(B): From part (A) we just proved, $\phi = 0$ is uniformly stable. The uniform asymptotic stability of $\phi = 0$ can be proved directly (see an exercise); however, we apply Theorem 5.1.2 here and only prove the asymptotic stability of $\phi = 0$. That is, we will prove that there is an $r > 0$ such that $|x_0| \leq r$ implies $\lim_{t \to \infty} |x(t, t_0, x_0)| = 0$. Let $\varepsilon = 1$, then from the uniform stability of $\phi = 0$, there is a $\delta > 0$ such that $|x_0| \leq \delta$ implies $|x(t, t_0, x_0)| \leq 1$ for $t_0 \geq 0$ and $t \geq t_0$. Let $r = \delta$ and let $|x_0| \leq r$, then $|x(t, t_0, x_0)| \leq 1$ for $t_0 \geq 0$ and $t \geq t_0$.

Now, for $x(t) = x(t, t_0, x_0)$ with $x_0 \neq 0$, we have $x(t) \neq 0$ by uniqueness. Thus, $\frac{d}{dt} V(x(t)) < 0$ and hence $V(x(t))$ is monotone decreasing. We first prove that $\lim_{t \to \infty} V(x(t)) = 0$. If this is not true, then $V(x(t)) \geq L$, $t \geq t_0$, for some constant $L > 0$. Since $V(x)$ is continuous and $V(0) = 0$, one concludes that no sequence of $x(t)$ could go to zero. Therefore, there is a $k \in (0, 1)$ such that $k \leq |x(t)| \leq 1$, $t \geq t_0$. Since $V'_{(5.1)}(x) < 0$, $x \neq 0$, and

$V'_{(5.1)}(x)$ is continuous on the bounded and closed set $k \leq |x| \leq 1$, one has $p = \max_{k \leq |x| \leq 1} V'_{(5.1)}(x) < 0$. Hence $\frac{d}{dt}V(x(t)) \leq p < 0$, $t \geq t_0$. Then

$$V(x(t)) - V(x(t_0)) = \int_{t_0}^{t} \frac{d}{ds}V(x(s))ds \leq \int_{t_0}^{t} pds = p(t - t_0) \longrightarrow -\infty,\ t \to \infty,$$

which contradicts the fact that $V(x(t)) \geq 0$. Thus, $\lim_{t \to \infty} V(x(t)) = 0$.

Next, we prove that $\lim_{t \to \infty} x(t) = 0$. If this is not true, then there is a constant $d > 0$ and a sequence $t_m \to \infty$ such that $d \leq |x(t_m)| \leq 1$. Then $q = \min_{d \leq |x| \leq 1} V(x) > 0$ since V is positive definite, hence $V(x(t_m)) \geq q > 0$, which contradicts $\lim_{t \to \infty} V(x(t)) = 0$. This proves the asymptotic stability of $\phi = 0$, and hence the uniform asymptotic stability of $\phi = 0$ using Theorem 5.1.2.

(C): If $\phi = 0$ is stable, then for $\varepsilon = 1$ and $t_0 = 0$, there is a $\delta > 0$ such that $|x_0| \leq \delta$ implies $|x(t, 0, x_0)| \leq 1$, $t \geq 0$. Hence for $x(t) = x(t, 0, x_0)$, $V(x(t))$ is bounded for $t \geq 0$ when $|x_0| \leq \delta$. Fix an x_0 with $|x_0| = \frac{\delta}{2}$, then $V(x_0) > 0$. Now, $V(x(t))$ is increasing, thus $V(x(t)) \geq V(x_0) > 0$, $t \geq 0$. As $V(x)$ is continuous and $V(0) = 0$, there is a $k \in (0, 1)$ such that $k \leq |x(t)| \leq 1$, $t \geq 0$. Next, $V'_{(5.1)}(x)$ is continuous and positive on the closed and bounded set $k \leq |x| \leq 1$, one has $h = \min_{k \leq |x| \leq 1} V'_{(5.1)}(x) > 0$, thus

$$V(x(t)) - V(x_0) = \int_{0}^{t} \frac{d}{ds}V(x(s))ds \geq \int_{0}^{t} hds = h(t - t_0) \longrightarrow \infty,\ t \to \infty,$$

which contradicts to the fact that $V(x(t))$ is bounded for $t \geq 0$ when $|x_0| \leq \delta$. Therefore, $\phi = 0$ is unstable.

This completes the proof of the theorem. ♠

Some conditions in Theorem 5.5.12 are also necessary conditions. For example, if $\phi = 0$ is uniformly asymptotically stable, then there exists a Liapunov function such that its derivative along the solutions is negative definite. Refer to "converse of stabilities" in Chapter 9.

The above presents a brief coverage of the Liapunov theory concerning stability properties for autonomous differential equations. Roughly speaking, it reduces the study of stability properties to the problem of constructing appropriate Liapunov functions. However, constructing these functions is not an easy task; it requires a great deal of experience and skill. In the next section, we demonstrate basic techniques used in the construction of Liapunov functions for some applications.

Exercises 5.5

1. In the proof of Theorem 5.5.1(1), complete the case when $\bar{x} - \delta < x_0 < \bar{x}$. Then prove parts (2) and (3) of Theorem 5.5.1.

2. Prove parts (2) and (3) of Corollary 5.5.2.

3. Determine if $V : Q \to [0, \infty)$ is positive definite on some domain Q.

  (a) $V(x) = x^2 + \cos x$.

  (b) $V(x) = x^2 + \cos^2 x$.

  (c) $V(x) = x^2 + \sin x$.

  (d) $V(x) = x^2 + \sin^2 x$.

  (e) $V(x) = -x^2 + x^4$.

  (f) $V(x) = -x^2 + x^3$.

  (g) $V(x_1, x_2) = x_1^2 + x_2^4$.

  (h) $V(x_1, x_2, x_3) = x_1^2 + x_2^2 + x_3^4$.

  (i) $V(x_1, x_2, x_3) = x_1^2 + x_2^2 - x_3^4$.

  (j) $V(x_1, x_2, x_3) = x_1^2 + x_2^2 + x_3^2 + x_1 x_2$.

4. Examine the stabilities of the zero solution for the logistic equation $x' = ax[C - x]$.

5. Give a geometric interpretation for a general Liapunov function in $\Re^2$. That is, discuss the relationship between V' and a level curve $V(x_1, x_2) = c$.

6. Let $Q \subset \Re^n$ be a domain containing the zero vector and let P be a function defined on Q. Prove that $P(x) < 0$, $x \neq 0$, if and only if $-P$ is positive definite.

7. Prove the uniform asymptotic stability of the zero solution in Theorem 5.5.12(B) directly.

8. Use Liapunov's method to study the stabilities of the zero solution for

  (a) $\begin{cases} x_1' &= -x_1 + x_2^2 - x_1^3, \\ x_2' &= -x_2 - x_1 x_2 - x_2^3. \end{cases}$ (b) $\begin{cases} x_1' &= -\frac{x_1}{2} + x_2^2 - x_1^3, \\ x_2' &= -x_2 - x_1 x_2 - x_2^3. \end{cases}$

  (c) $\begin{cases} x_1' &= x_2 - x_1 x_2^2 + x_1^4 x_2, \\ x_2' &= -x_1 + x_1^2 x_2 - x_1^5. \end{cases}$ (d) $\begin{cases} x_1' &= x_1 - x_1 x_2^2 + x_1^4 x_2, \\ x_2' &= x_2 + x_1^2 x_2 - x_1^5. \end{cases}$

5.6 Some Applications

In this section, we apply Liapunov's method by constructing Liapunov functions for the following differential equations. Note that $\frac{1}{2}[x_1^2 + x_2^2 + \cdots + x_n^2]$ works for most cases, such as in Examples 5.5.3–5.5.6. At times, some variations of this are needed. Also, when considering applications in physics, certain laws are useful in the construction of Liapunov functions.

Example 5.6.1 Revisit Example 5.3.5 and consider the scalar differential equation $x'(t) = -x(t) + x^2(t)$. Theorem 5.3.3 can be applied to obtain the uniform asymptotic stability of $\phi = 0$. Here, we will apply Liapunov's method. Try $V(x) = \frac{1}{2}x^2$. Then V is positive definite. Next,

$$\begin{aligned} V'(x(t)) &= x(t)x'(t) = x(t)[-x(t) + x^2(t)] \\ &= -x^2(t)[1 - x(t)]. \end{aligned} \tag{6.1}$$

Accordingly, we can take the domain $Q = \{x \in \Re : |x| < \frac{1}{2}\}$, and on Q, one has

$$V'(x(t)) = -x^2(t)[1 - x(t)] \leq -\frac{1}{2}x^2(t), \tag{6.2}$$

hence $-V'(x)$ is positive definite. From Theorem 5.5.12(B), $\phi = 0$ is uniformly asymptotically stable. ♠

Example 5.6.2 Consider the scalar differential equation

$$x'(t) = -x(t) + x^k(t).$$

For x^k to satisfy a local Lipschitz condition in an open interval containing 0, we must have $k \geq 1$. Try $V(x) = \frac{1}{2}x^2$, then

$$\begin{aligned} V'(x(t)) &= x(t)x'(t) = x(t)[-x(t) + x^k(t)] \\ &= -x^2(t)[1 - x^{k-1}(t)]. \end{aligned} \tag{6.3}$$

If $k = 1$, then Theorem 5.5.12(A) implies that $\phi = 0$ is uniformly stable. (In fact, in this case we have $x'(t) = 0$, then the solutions are constants, hence $\phi = 0$ is uniformly stable.) If $k > 1$, then let's assume that k is an integer (to guarantee that x^k is a real number). Now, define $Q = \{x \in \Re : |x| < (\frac{1}{2})^{1/(k-1)}\}$, and on Q,

$$V'(x(t)) \leq -x^2(t)[1 - x^{k-1}(t)] \leq -\frac{1}{2}x^2(t). \tag{6.4}$$

Thus, similar to Example 5.6.1, $\phi = 0$ is uniformly asymptotically stable. ♠

Example 5.6.3 Consider

$$
\begin{bmatrix} x_1(t) \\ x_2(t) \end{bmatrix}' = \begin{bmatrix} -2 & 0 \\ 0 & -1 \end{bmatrix} \begin{bmatrix} x_1(t) \\ x_2(t) \end{bmatrix} + \begin{bmatrix} 2x_1^2(t) - x_1(t)x_2(t) \\ 4x_1(t)x_2(t) + x_2^2(t) \end{bmatrix}. \qquad (6.5)
$$

Try $V(x) = \frac{1}{2}[x_1^2 + x_2^2]$ where $x = [x_1, x_2]^T \in \Re^2$. Then V is positive definite. Next,

$$
\begin{aligned}
V'(x(t)) &= x_1(t)x_1'(t) + x_2(t)x_2'(t) \\
&= x_1[-2x_1 + 2x_1^2 - x_1x_2] + x_2[-x_2 + 4x_1x_2 + x_2^2] \\
&= -x_1^2[2 - 2x_1 + x_2] - x_2^2[1 - 4x_1 - x_2]. \qquad (6.6)
\end{aligned}
$$

Now, when x_1 and x_2 are small, say for example $2|x_1| + |x_2| < \frac{3}{2}$ and $4|x_1| + |x_2| < \frac{1}{2}$, then

$$
V'(x(t)) \le -\frac{1}{2}[x_1^2 + x_2^2]. \qquad (6.7)
$$

Thus, we can let $Q = \{x \in \Re^2 : |x_1| + |x_2| < \frac{1}{8}\}$ and apply Theorem 5.5.12(B) to conclude that $\phi = 0$ is uniformly asymptotically stable. ♠

Example 5.6.4 Consider

$$
\begin{bmatrix} x_1(t) \\ x_2(t) \end{bmatrix}' = \begin{bmatrix} 0 & -1 \\ 1 & 0 \end{bmatrix} \begin{bmatrix} x_1(t) \\ x_2(t) \end{bmatrix} + \begin{bmatrix} -x_1^3(t) \\ -x_2^3(t) \end{bmatrix}. \qquad (6.8)
$$

Try $V(x) = \frac{1}{2}[x_1^2 + x_2^2]$. Then V is positive definite. Next,

$$
\begin{aligned}
V'(x(t)) &= x_1(t)x_1'(t) + x_2(t)x_2'(t) \\
&= x_1(t)[-x_2(t) - x_1^3(t)] + x_2(t)[x_1(t) - x_2^3(t)] \\
&= -[x_1^4(t) + x_2^4(t)], \qquad (6.9)
\end{aligned}
$$

thus, from Theorem 5.5.12(B), $\phi = 0$ is uniformly asymptotically stable. ♠

Note that in Example 5.6.4, the origin for the linearization is a center and hence is uniformly stable but not uniformly asymptotically stable; however, the nonlinear perturbation makes it uniformly asymptotically stable.

Example 5.6.5 Consider

$$
\begin{bmatrix} x_1(t) \\ x_2(t) \\ x_3(t) \end{bmatrix}' = \begin{bmatrix} 0 & -1 & 0 \\ 2 & 0 & 0 \\ 0 & 0 & 0 \end{bmatrix} \begin{bmatrix} x_1(t) \\ x_2(t) \\ x_3(t) \end{bmatrix} + \begin{bmatrix} x_2(t)x_3(t) - x_1^3(t) \\ -x_2^3(t) \\ -x_1(t)x_2(t) - x_3^3(t) \end{bmatrix}. \qquad (6.10)
$$

Try $V(x) = \frac{1}{2}[c_1 x_1^2 + c_2 x_2^2 + c_3 x_3^2]$ where $x = [x_1, x_2, x_3]^T \in \Re^3$, with positive constants $c_i, i = 1, 2, 3$. Then V is positive definite. Next,

$$
\begin{aligned}
V'(x(t)) &= c_1 x_1(t) x_1'(t) + c_2 x_2(t) x_2'(t) + c_3 x_3(t) x_3'(t) \\
&= c_1 x_1(t)[-x_2(t) + x_2(t) x_3(t) - x_1^3(t)] + c_2 x_2(t)[2x_1(t) - x_2^3(t)] \\
&\quad + c_3 x_3(t)[-x_1(t) x_2(t) - x_3^3(t)] \\
&= (2c_2 - c_1) x_1 x_2 + (c_1 - c_3) x_1 x_2 x_3 \\
&\quad - c_1 x_1^4 - c_2 x_2^4 - c_3 x_3^4.
\end{aligned}
\tag{6.11}
$$

Now, we choose $2c_2 = c_1 = c_3$, then

$$
V'(x(t)) = -c_1[x_1^4 + \frac{1}{2} x_2^4 + x_3^4], \quad c_1 > 0.
\tag{6.12}
$$

From Theorem 5.5.12(B), $\phi = 0$ is uniformly asymptotically stable. Again, in this case, the origin for the linearization is uniformly stable but not uniformly asymptotically stable, and the nonlinear perturbation makes it uniformly asymptotically stable. ♠

In other applications, Liapunov functions may take different forms, such as $V(x) = x_1^2 + x_2^4$, $V(x) = x_1^4 + x_2^2$, $V(x) = x_1^4 + x_2^4 + \cdots + x_n^4$, and so on.

Next, we look at how to apply the Liapunov theory to some second-order Lienard-type differential equations with applications in physics (see the introduction of the Lienard-type differential equations in Chapter 1). Refer to additional related results in Brauer and Nohel [1969] and Burton [1985].

Example 5.6.6 Consider the second-order scalar differential equation

$$
u'' + g(u) = 0, \quad u = u(t) \in \Re.
\tag{6.13}
$$

Here, g is continuously differentiable for $|u| < k$ where $k > 0$ is a constant, and $ug(u) > 0$ if $u \neq 0$, (this means $g(u) > 0$ for $u > 0$ and $g(u) < 0$ for $u < 0$). These conditions are satisfied, for example, if $g(u) = \sin u$, $|u| < \pi$, in the case of an undamped simple pendulum discussed in Chapter 4. We let $x_1 = u$, $x_2 = u'$, and change the second-order differential equation (6.13) into the following differential equation in $\Re^2$,

$$
\begin{cases}
x_1' = x_2, \\
x_2' = -g(x_1).
\end{cases}
\tag{6.14}
$$

Now, the conditions on g guarantee the existence and uniqueness of solutions for Eq. (6.13) or Eq. (6.14), and imply $g(0) = 0$. Therefore, $u = 0$ is a solution of Eq. (6.13), or $(x_1, x_2) = (0, 0)$ is a solution of Eq. (6.14). In physics applications, we can think of $g(u)$ as the restoring force acting on a particle at a displacement u from the equilibrium $(u = 0)$, and of u' as the velocity of the particle. Then the potential energy at a displacement u from the equilibrium $(u = 0)$ is $\int_0^u g(s)ds$, and the kinetic energy is $\frac{1}{2}(u')^2$. Thus, the **total energy** is

$$V(t) = \frac{1}{2}[u'(t)]^2 + \int_0^{u(t)} g(s)ds. \tag{6.15}$$

Now, the **law of conservation of energy** indicates that $\frac{d}{dt}V(t) = 0$. Indeed,

$$\frac{d}{dt}V(t) = u'u'' + g(u)u' = u'[u'' + g(u)] = 0. \tag{6.16}$$

Therefore, for Eq. (6.14), if we try a Liapunov function

$$V(x) = V(x_1, x_2) = \frac{1}{2}x_2^2 + \int_0^{x_1} g(s)ds, \quad |x_1| < k, \ |x_2| < \infty \tag{6.17}$$

for $x = [x_1, x_2]^T$, then we obtain

$$V'_{(6.14)}(x) = x_2 x_2' + g(x_1)x_1' = x_2[-g(x_1)] + g(x_1)x_2 = 0.$$

Next, from the assumption that $ug(u) > 0$ for $u \neq 0$, we find that

$$\int_0^{x_1} g(s)ds > 0 \ \text{ for } \ x_1 \neq 0,$$

thus, $V(x_1, x_2)$ is positive definite. Hence $\phi = 0$ is uniformly stable from Theorem 5.5.12(A). ♠

Compare Eq. (6.13) in Example 5.6.6 with the general second-order differential equation (2.8) in Chapter 1 derived using Newton's second law of motion. Then we find for Eq. (6.13) that the externally applied force is zero. Thus, the interpretation in physics for the result in Example 5.6.6 is that when the externally applied force is zero, certain conditions can be given to the restoring force, in terms of the function g, so as to make the zero solution become uniformly stable.

Note that in Example 5.6.6, the asymptotic stability of $\phi = 0$ is not expected because in the case of an undamped simple pendulum, for example, the pendulum will "keep oscillating indefinitely" about the equilibrium $\phi = 0$. Therefore the origin is a center and hence is not asymptotically stable. However, adding a damping term may change the situation, see the following example.

Example 5.6.7 Add u' to the equation in Example 5.6.6 and consider

$$u'' + u' + g(u) = 0, \quad u = u(t) \in \Re. \tag{6.18}$$

In the case of a simple pendulum, Eq. (6.18) means that a damping term is added, thus the pendulum will encounter air resistance or friction, so we expect $\phi = 0$ to be uniformly asymptotically stable. However, if we try the same Liapunov function $V(x_1, x_2) = \frac{1}{2}x_2^2 + \int_0^{x_1} g(s)ds$ from Example 5.6.6 for

$$\begin{cases} x_1' = x_2, \\ x_2' = -x_2 - g(x_1), \end{cases} \tag{6.19}$$

then we get

$$V'_{(6.19)}(x) = x_2 x_2' + g(x_1)x_1' = x_2[-x_2 - g(x_1)] + g(x_1)x_2 = -x_2^2, \tag{6.20}$$

which does not imply that $-V'_{(6.19)}(x) = x_2^2$ is positive definite because it doesn't have x_1 terms (for example, it is zero at $(x_1, x_2) = (1, 0)$). Therefore, Theorem 5.5.12(B) cannot be applied to this V function to derive the uniform asymptotic stability of $\phi = 0$.

This example indicates that some skills are needed to construct "good" Liapunov functions. Next, we modify the above V by considering

$$V(x_1, x_2) = \frac{1}{2}x_2^2 + \frac{1}{2}[x_1 + x_2]^2 + q \int_0^{x_1} g(s)ds, \tag{6.21}$$

where the constant $q > 0$ will be determined in order to cancel some terms, and $[x_1 + x_2]^2$ is used so that V is still positive definite and that x_1 terms may appear in the derivative. Now,

$$\begin{aligned} V'_{(6.19)}(x) &= x_2 x_2' + [x_1 + x_2][x_1' + x_2'] + qg(x_1)x_1' \\ &= x_2[-x_2 - g(x_1)] + [x_1 + x_2][x_2 - x_2 - g(x_1)] + qg(x_1)x_2 \\ &= -x_2^2 + (q - 2)g(x_1)x_2 - x_1 g(x_1). \end{aligned} \tag{6.22}$$

Accordingly, we choose $q = 2$, so that $-V'_{(6.19)}(x) = x_2^2 + x_1 g(x_1)$ is positive definite. Therefore, Theorem 5.5.12(B) can be applied to derive the uniform asymptotic stability of $\phi = 0$. ♠

From Examples 5.6.6 and 5.6.7, we see that for a simple pendulum without a damping term, a Liapunov function can be constructed to obtain stability but not asymptotic stability; and when a damping exists, another Liapunov function can be constructed to obtain asymptotic stability. This demonstrates that Liapunov's method is a very useful tool in deriving qualitative properties in physics applications.

From the above examples, we see that $\frac{1}{2}[x_1^2 + x_2^2 + \cdots + x_n^2]$ can be used as a Liapunov function in most cases, however, modifications are needed sometimes. In physics applications, knowledge of the subjects is very useful in constructing the corresponding Liapunov functions, such as the total energy for a simple pendulum.

Finally, note that Theorem 5.3.3 can be applied to Examples 5.6.1, 5.6.2, and 5.6.3, but not Examples 5.6.4–5.6.7. Therefore, we find that if a stability result is established by some other methods, then probably it also can be established by Liapunov's method. However, for a stability result established by Liapunov's method, it may not be establishable by other methods, such as Examples 5.6.4–5.6.7. This makes Liapunov's method very important since it derives certain properties that are not obtainable using other methods.

Exercises 5.6

1. Discuss the stabilities of the zero solution for the Volterra equation

$$\begin{cases} x'(t) = ax(t) - bx^2(t) - cx(t)y(t), \\ y'(t) = -dy(t) + ex(t)y(t), \end{cases}$$

where a, b, c, d, e are positive constants.

2. Discuss the stabilities of the zero solution for the Lotka-Volterra equation

$$\begin{cases} x_1' = \beta_1 x_1 (K_1 - x_1 - \mu_1 x_2), \\ x_2' = \beta_2 x_2 (K_2 - x_2 - \mu_2 x_1), \end{cases}$$

where β_i, K_i, μ_i, $i = 1, 2$, are positive constants.

3. Discuss the stabilities of the zero solution for

$$\begin{cases} x' = -2y^3, \\ y' = 5x^3, \end{cases}$$

with $V = \frac{1}{4}[c_1 x^4 + c_2 y^4]$.

4. Discuss the stabilities of the zero solution for

$$\begin{cases} x' = -x - xy^2 - x^3, \\ y' = -7y + 3x^2 y - 2yz^2 - y^3, \\ z' = -5z + y^2 z - z^3, \end{cases}$$

with $V = \frac{1}{2}[c_1 x^2 + c_2 y^2 + c_3 z^2]$.

5. Discuss the stabilities of the zero solution for

(a) $\begin{cases} x' = y, \\ y' = -x - y. \end{cases}$ (b) $\begin{cases} x' = -x^2 + y + 5, \\ y' = 2y^2 - 2xy. \end{cases}$

(c) $\begin{cases} x' = -y - xy^2 + z^2 - x^3, \\ y' = x - y^3 + z^3, \\ z' = -xz - x^2 z - yx^2 - z^5. \end{cases}$

6. In Example 5.6.6, use l'Hôpital's rule to show that

$$\lim_{x_1 \to 0} \frac{g^2(x_1)}{\int_0^{x_1} g(s)ds} = \lim_{x_1 \to 0} \frac{2g(x_1)g'(x_1)}{g(x_1)} = 2g'(0),$$

where the limit exists since g is continuously differentiable. Then show that there exist constants $C > 0$ and $0 < k_1 \le k$ such that

$$\int_0^{x_1} g(s)ds \ge Cg^2(x_1), \quad |x_1| \le k_1.$$

7. In $u'' + u' + g(u) = 0$, $u \in \Re$, if $g(u) = au + f(u)$ with the constant $a > 0$ and $\frac{f(u)}{u} \to 0$ as $u \to 0$, then show that the zero solution is uniformly asymptotically stable.

8. Discuss the stabilities of the zero solution for $u'' + u' + au + bu^2 - cu^3 = 0$, $u \in \Re$, where a, b, c are positive constants.

9. Discuss the stabilities of the zero solution for $u'' + f(u, u')u' + g(u) = 0$, $u \in \Re$, where $f > 0$ and g satisfies the conditions in Example 5.6.6.

Chapter 6

Bifurcation

6.1 Introduction

In this chapter, we look at the autonomous differential equations with parameters,

$$x'(t) = f(x(t), \mu), \quad \text{or} \quad x' = f(x, \mu), \tag{1.1}$$

where $(t, x) \in \Re \times \Re^n$ and the parameter μ is in a domain $D_0 \subset \Re^k$, $k \geq 1$. We have already seen this type of differential equations. For example, in Chapter 1, the logistic equation $x' = rx[1 - x]$ (or $x' = ax[C - x]$) is a differential equation with parameter r (or a and C); the Lotka-Volterra competition equation can be regarded as a differential equation with parameters $\beta_1, \mu_1, \beta_2, \mu_2$; and the motion of a simple pendulum can be regarded as a differential equation with parameters k and q. In general, differential equations are derived in physical problems using scientific principles, and a parameter usually represents something that is fixed in each application but that varies between applications. For example, the carrying capacity of a population, or the length of a pendulum may be different in different situations. One of the fundamental issues in using differential equations for modeling the qualitative properties of a system is how the system changes with parameters.

In Chapter 2 we have also seen that under some appropriate conditions, for each fixed $\mu \in D_0$, Eq. (1.1) has a unique solution, and the dependence of the solution with respect to μ is smooth, that is, the solution is continuous and differentiable with respect to the parameter $\mu \in D_0$.

Although solutions enjoy the continuous dependence with respect to parameters, when some parameters are varied, a system may experience **abrupt qualitative changes** after some **smooth quantitative changes**. For example, if you squeeze a small piece of straight thin wood, after a while the wood is nearly straight (this corresponds to the quantitative changes). Then there will be a moment the wood will buckle. Here, "buckle" corresponds to the abrupt qualitative changes. A similar example of Euler's buckling beam is given in Chapter 1.

Although it is initially puzzling, we now expect that certain qualitative properties of Eq. (1.1) may be changed abruptly when some parameters are varied and pass some critical values. In fact, from the following examples, we will see that the qualitative properties subject to change include the number of critical points and their stability properties, and, in the case of planar equations, the nature of the origin (in terms of a center, a spiral point, a saddle point, or a node). Generally, the qualitative changes in a system are called **bifurcations**, and the parameter values at which bifurcations occur are called **bifurcation points** or **bifurcation values**.

The following examples indicate the different types of bifurcations that could occur.

Example 6.1.1 Consider the linear differential equation with constant coefficients,

$$
\begin{bmatrix} x \\ y \end{bmatrix}' = \begin{bmatrix} a & b \\ c & d \end{bmatrix} \begin{bmatrix} x \\ y \end{bmatrix}. \tag{1.2}
$$

We could regard Eq. (1.2) as a differential equation with parameters a, b, c, and d, but the analysis is complicated with that many parameters. Recall that in Chapter 4 we used $p = tr A = a + d$ and $q = \det A = ad - bc$ to characterize Eq. (1.2) by using the Distribution Diagram 4.16 in the (p, q) plane, where the nature of the origin (in terms of a center, a spiral point, a saddle point, or a node) can be determined easily. Therefore, we will use p and q as the parameters of Eq. (1.2), and we copy Figure 4.16 here as **Figure 6.1** for easy reference.

Now, if we start with $q > 0, p < 0$ and move across the positive q-axis (where $p = 0$) to become $q > 0, p > 0$, then the zero solution of the system will change from asymptotically stable to unstable, and when the parameters (p, q) cross the positive q-axis at $p = 0$, a stable center will be created. In this case, the stability property of the zero solution of Eq. (1.2) is a qualitative property. Thus Eq. (1.2) undergoes a qualitative property change, or a bifurcation, at the bifurcation value $p = 0$.

Next, if we start with $q < 0$ and move up and cross the p-axis (where $q = 0$), then the origin will change from a saddle point to a node. Now, the nature of the origin (in terms of a center, a spiral point, a saddle point, or a node) is also a qualitative property of Eq. (1.2). Thus Eq. (1.2) also undergoes a qualitative property change, or a bifurcation, at the bifurcation value $q = 0$.   ♠

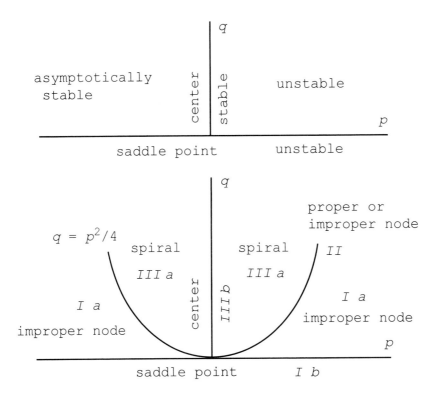

Figure 6.1: Distribution Diagram for the origin of Eq. (1.2)

From Example 6.1.1, we find that in the cases of having complex eigenvalues (such as when the origin is a spiral point or a center), when a pair of conjugate complex eigenvalues cross the imaginary axis (where the real part is zero, or the origin is a center), there is an exchange of stability for the origin (from asymptotically stable to stable and then to unstable, or vice versa). Periodic orbits also will be created when the origin is a center. In the cases of having real eigenvalues (for nodes and saddle points), when real eigenvalues change signs or cross 0 on the real line, the origin will change from a saddle point to a node, or vice versa.

These results are related to the findings of Poincaré [1885]. His pioneer work in modern qualitative theory of differential equations has generated a great deal of interest in the research of dynamical systems and other related areas. Some of these results will be discussed in Section 5.

The following example shows a "branching effect" that explains why the word "bifurcation" is used.

Example 6.1.2 Consider the scalar differential equation

$$x' = f(x, \mu) = \mu + x^2, \tag{1.3}$$

where $\mu \in \Re$ is a parameter. If $\mu > 0$, then Eq. (1.3) has no critical point (that is, $x' = \mu + x^2 = 0$ has no solution), or the curve $y = \mu + x^2$ will not intersect the x-axis, see **Figure 6.2**.

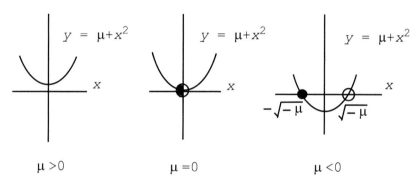

Figure 6.2: Graph of $y = \mu + x^2$ and the critical points of $x' = \mu + x^2$

When μ decreases to 0 from above, the graph of $y = \mu + x^2$ moves down and then intersects the x-axis tangentially when $\mu = 0$, in which case one critical point appears at $x = 0$. If μ continues to decrease, then $\mu < 0$ and hence the graph of $y = \mu + x^2$ crosses the x-axis and two critical points appear at

$$x = \pm\sqrt{-\mu}.$$

From the stability analysis in Chapter 5, using

$$f'_x = \frac{\partial}{\partial x} f(x, \mu) = 2x,$$

we find that $x = -\sqrt{-\mu}$ is asymptotically stable and $x = \sqrt{-\mu}$ is unstable. Thus we have the arrows in Figure 6.2 that indicate the flows of the solutions on the x-axis. We use a solid point at $(-\sqrt{-\mu}, 0)$ to denote that $-\sqrt{-\mu}$ is a stable critical point, and use an open point at $(\sqrt{-\mu}, 0)$ to denote that $\sqrt{-\mu}$ is an unstable critical point. Note that at $x = 0$ (or $\mu = 0$) we have a "half-stable" critical point, which is really delicate because it disappears as soon as $\mu \neq 0$. The number of critical points and their stabilities in this case are qualitative properties of Eq. (1.3), thus Eq. (1.3) undergoes some qualitative property changes, or a bifurcation takes place at the bifurcation value $\mu = 0$.

Because we let the parameter μ to vary first, then μ is treated as an independent variable, and the corresponding critical point x (if any) is treated as a function of μ. In Eq. (1.3), for $\mu > 0$, there is no critical point; for $\mu = 0$, $x = 0$ is the only corresponding critical point; for $\mu < 0$, the critical points are given by two functions $x = \sqrt{-\mu}$ and $x = -\sqrt{-\mu}$. It is customary to draw all these functions in one (μ, x) plane. Now from **Figure 6.3** we find that for $\mu > 0$, there is no critical point; however, when μ decreases and passes 0, then suddenly, "two branches" of critical points appear according to $x = \sqrt{-\mu}$ and $x = -\sqrt{-\mu}$. This means that a "**bi**"-furcation takes place, explaining why the word "bifurcation" is used to describe the situation. Similar to Figure 6.2, we use a solid curve in Figure 6.3 to denote the stable critical points and a broken curve to denote the unstable critical points. Figures with this kind of information are called **bifurcation diagrams**.   ♠

In summary, when some parameters of a system are varied, certain qualitative properties of the system may go through a period of smooth quantitative changes, and then suddenly, at some critical bifurcation values, abrupt qualitative changes, or bifurcations, take place. These qualitative properties include the number of critical points and their stability properties, and, in the case of a planar equation, the nature of the origin (in terms of a center, a spiral point, a saddle point, or a node). We will discuss these subjects in this chapter.

This chapter is organized as follows: In Section 2, we study saddle-node bifurcations and use some examples to explain why saddle and node appear

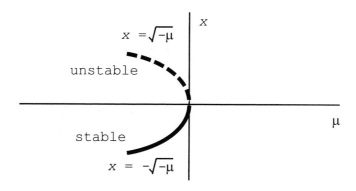

Figure 6.3: A bifurcation diagram determined by $x' = \mu + x^2$

for this type of bifurcations. We analyze the geometric aspects of some scalar differential equations that undergo saddle-node bifurcations and use them to formulate and prove a result concerning saddle-node bifurcations for scalar differential equations. In Section 3, we study transcritical bifurcations and apply them to a solid-state laser in physics. Again, the geometric aspects of some examples are analyzed and used to formulate and prove a result concerning transcritical bifurcations for scalar differential equations. In Section 4, we study pitchfork bifurcations and apply them to Euler's buckling beam and calculate Euler's first buckling load, which is the value the buckling takes place. The hysteresis effect with applications in physics is also discussed. A result concerning pitchfork bifurcations for scalar differential equations is formulated using the geometric interpretation. In Section 5, we analyze the situations where a pair of two conjugate complex eigenvalues cross the pure imaginary axis when some parameters are varied. We introduce the Poincaré-Andronov-Hopf bifurcation theorem and apply it to van der Pol's oscillator in physics.

Exercises 6.1

1. Find the critical points for the following equations, where $\mu \in \Re$ is a parameter.

(a) $\begin{cases} x' = 2 + \mu + x^2, \\ y' = 3 - y. \end{cases}$ (b) $\begin{cases} x' = y + 3, \\ y' = -2\mu + 4y + x^2. \end{cases}$

(c) $\begin{cases} x' & = & (\mu - 4)x + 2x^3, \\ y' & = & -4y + 5. \end{cases}$ (d) $x' = (\mu + 3)x + x^2.$

6.2   Saddle-Node Bifurcation

In order to understand why some bifurcations are called "saddle-node bifurcations," let's begin with some examples.

Example 6.2.1 Consider

$$\begin{cases} x' & = & \mu + x^2, \\ y' & = & -y, \end{cases} \tag{2.1}$$

where $\mu \in \Re$ is a parameter. The first equation in x is the same as in Example 6.1.2, the second equation in y means that $y = 0$ is asymptotically stable. For $\mu > 0$, Eq. (2.1) has no critical point. For $\mu < 0$, two critical points of Eq. (2.1) appear at

$$(-\sqrt{-\mu}, 0), \quad (\sqrt{-\mu}, 0).$$

At $(-\sqrt{-\mu}, 0)$, it is asymptotically stable for both x and y, hence $(-\sqrt{-\mu}, 0)$ is a stable node, (a linearization can be used to verify this, see an exercise). At $(\sqrt{-\mu}, 0)$, y is asymptotically stable but x is unstable, thus $(\sqrt{-\mu}, 0)$ is a saddle.

When $\mu = 0$, the stable node and saddle will collide and create a half-stable point, see **Figure 6.4**.   ♠

The following example needs a little more analysis than Example 6.2.1.

Example 6.2.2 Consider

$$\begin{cases} x' & = & y, \\ y' & = & -\mu - y + x^2, \end{cases} \tag{2.2}$$

where $\mu \in \Re$ is a parameter. For $\mu < 0$, Eq. (2.2) has no critical point. For $\mu > 0$, Eq. (2.2) has two critical points $(\sqrt{\mu}, 0)$ and $(-\sqrt{\mu}, 0)$. For $(\sqrt{\mu}, 0)$, let's use the change of variables $x_1 = x - \sqrt{\mu}$ and $x_2 = y$ to transform $(\sqrt{\mu}, 0)$ to the origin. Then we obtain

$$\begin{cases} x_1' & = & x' = x_2, \\ x_2' & = & y' = -\mu - x_2 + (x_1 + \sqrt{\mu})^2 = 2\sqrt{\mu}x_1 - x_2 + x_1^2. \end{cases} \tag{2.3}$$

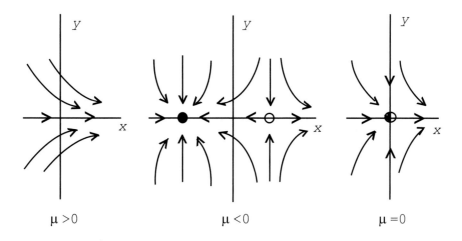

$\mu > 0$ $\hspace{4em}$ $\mu < 0$ $\hspace{4em}$ $\mu = 0$

Figure 6.4: A planar saddle-node bifurcation

From the linearization of Eq. (2.3), we get $p = trA = -1 < 0$ and $q = \det A = -2\sqrt{\mu} < 0$. Thus the origin of Eq. (2.3), or the critical point $(\sqrt{\mu}, 0)$ of Eq. (2.2), is a saddle point.

For $(-\sqrt{\mu}, 0)$, let's change the variables $x_1 = x + \sqrt{\mu}$ and $x_2 = y$, and obtain

$$\begin{cases} x_1' = x' = x_2, \\ x_2' = y' = -\mu - x_2 + (x_1 - \sqrt{\mu})^2 = -2\sqrt{\mu}x_1 - x_2 + x_1^2. \end{cases} \quad (2.4)$$

From the linearization of Eq. (2.4), we get $p = trA = -1 < 0$ and $q = \det A = 2\sqrt{\mu} > 0$. Assume μ is near 0 such that $q = 2\sqrt{\mu} < \frac{1}{4} = \frac{p^2}{4}$, then the origin of Eq. (2.4), or the critical point $(-\sqrt{\mu}, 0)$ of Eq. (2.2), is a stable node.

When $\mu = 0$, the stable node and saddle will collide and create a half-stable point, see **Figure 6.5**. ♠

In Examples 6.2.1 and 6.2.2, when the parameter μ is varied, a **saddle** and a **node** appear at the same time, and then they collide and finally disappear altogether. Accordingly, this type of bifurcations are called **saddle-node bifurcations**.

Next, let's look at saddle-node bifurcations from a different point of view. In Eq. (2.2) of Example 6.2.2, the critical points are from $x' = 0$ and $y' = 0$,

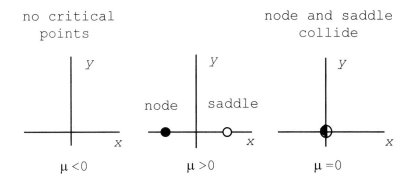

Figure 6.5: A planar saddle-node bifurcation

which imply $y = 0$ and $-\mu - y + x^2 = 0$. They are two curves in the (x, y) plane: the x-axis and the curve determined by the function $y = -\mu + x^2$. See **Figure 6.6**.

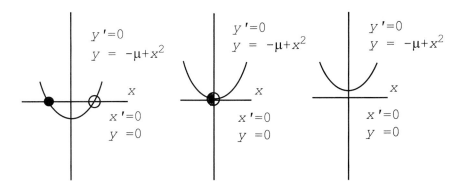

Figure 6.6: Curves determined by $y = 0$ and $y = -\mu + x^2$

In Figure 6.6, the two intersections of the curves $y = 0$ and $-\mu - y + x^2 = 0$ correspond to the two critical points of the equation because they are from $x' = 0$ and $y' = 0$. When the parameter μ is varied, the two curves pull away from each other and the two critical points move toward each other and then collide at $\mu = 0$ when the two curves become tangent to each other. After the two curves pull apart, critical points disappear since there are no intersections. This is typically what will happen locally for **all** saddle-node bifurcations. See **Figure 6.7**, where a planar differential equation is written as $x' = P(x, y, \mu)$, $y' = Q(x, y, \mu)$, and the two curves determined by $P(x, y, \mu) = 0$ and $Q(x, y, \mu) = 0$ in the (x, y) plane intersect at two points and then pull apart when the parameter μ is varied.

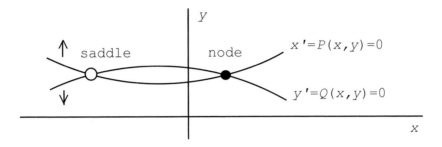

Figure 6.7: Curves determined by $x' = P(x, y, \mu) = 0$ and $y' = Q(x, y, \mu) = 0$

In Example 6.2.1, the situation is similar, but the difference is that for $\mu > 0$, $\mu + x^2 = 0$ will not determine a curve in the (x, y) plane; for $\mu < 0$, $\mu + x^2 = 0$ determines two vertical straight lines $x = \pm\sqrt{-\mu}$ in the (x, y) plane. These lines intersect the curve $-y = 0$ (the x-axis) to create two critical points; and for $\mu = 0$, $\mu + x^2 = 0$ is the y-axis and $-y = 0$ is the x-axis, thus the intersection is the origin. See **Figure 6.8**.

Additional examples, remarks, and some general results concerning the saddle-node bifurcations for equations in $\Re^n$, $n \geq 2$, can be found in Hale and Kocak [1991], Hubbard and West [1995], and Chicone [1999].

In order to incorporate the geometric aspects of the qualitative changes of a system, let's look at saddle-node bifurcations for scalar differential equations. Recall that in Example 6.1.2 (and a similar equation $x' = \mu - x^2$ in Example 1.3.2 in Chapter 1), for some parameter values, there are two critical points for those scalar equations, one is stable and the other unstable. When the parameter is varied, the two critical points move toward each other and collide and then disappear altogether. That is, for scalar differential equations, if we think of a stable critical point as a "stable node" and an unstable critical point as a "saddle," then this situation is the same as that described for the saddle-node bifurcations of differential equations in $\Re^2$. In this sense, we say that the bifurcations in Examples 6.1.2 and 1.3.2

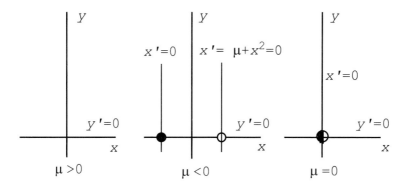

Figure 6.8: Curves determined by $\mu + x^2 = 0$ and $-y = 0$ in the (x, y) plane

are "saddle-node bifurcations for scalar differential equations," and Figure 6.3 is the saddle-node bifurcation diagram of $x' = \mu + x^2$ in Example 6.1.2. We also provide the saddle-node bifurcation diagram for $x' = \mu - x^2$ in **Figure 6.9**. It illustrates also the stability properties that were not done in Example 1.3.2 in Chapter 1.

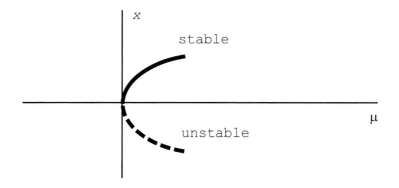

Figure 6.9: A bifurcation diagram determined by $x' = \mu - x^2$

The next example may indicate when a bifurcation will not occur.

Example 6.2.3 Consider the scalar differential equation

$$x' = f(x, \mu) = x - \mu, \tag{2.5}$$

where $\mu \in \Re$ is a parameter. Now, for any $\mu \in \Re$ fixed, $f(x, \mu) = x - \mu = 0$ has a unique solution $x = \mu$, thus the number of critical points is one and will never change. Next, $\frac{\partial}{\partial x} f(x, \mu) = 1 > 0$, thus

$$\frac{\partial f}{\partial x}(\mu, \mu) = \frac{\partial}{\partial x} f(x, \mu)|_{x=\mu} = 1 > 0,$$

which indicates, using the stability analysis in Chapter 5, that every critical point $x = \mu$ is unstable, or the stability property of the critical point will never change. Therefore, no bifurcation can occur. (Notice the different notations for the partial derivatives, that is, $\frac{\partial}{\partial x} f(x, \mu)$ means taking a partial derivative with respect to x, and $\frac{\partial f}{\partial x}(\mu, \mu)$ means evaluating the partial derivative with respect to x at (μ, μ).) ♠

Example 6.2.3 states that there is no bifurcation at $(x, \mu) = (0, 0)$. One reason is that

$$\frac{\partial f}{\partial x}(0, 0) = \frac{\partial}{\partial x} f(x, \mu)|_{x=0, \mu=0} = \frac{\partial}{\partial x} x|_{x=0} = 1 \neq 0,$$

thus for (x, μ) near $(0, 0)$, the relationship between x and μ from $f(x, \mu) = 0$ (for solving the critical points) is one-to-one, or the critical point $x = x(\mu)$ can be solved uniquely in terms of μ. (For Example 6.2.3, $x(\mu) = \mu$.) Hence the number of critical points will not change. Moreover,

$$\frac{\partial f}{\partial x}(x(\mu), \mu) \approx \frac{\partial f}{\partial x}(0, 0) = 1 > 0 \quad \text{for} \quad (x(\mu), \mu) \approx (0, 0),$$

thus for $\mu \approx 0$, the stability property of the critical point $x = x(\mu)$ will not change either.

Next, review Example 6.1.2 and then compare it with Example 6.2.3. Now, for Example 6.1.2, $f(x, \mu) = \mu + x^2$, hence,

$$\frac{\partial f}{\partial x}(0, 0) = \frac{\partial}{\partial x} f(x, \mu)|_{x=0, \mu=0} = \frac{\partial}{\partial x} x^2|_{x=0} = 2x|_{x=0} = 0,$$

which means that for (x, μ) near $(0, 0)$, the relationship between x and μ from $f(x, \mu) = 0$ is not one-to-one. (In fact, for each $\mu < 0$, $x = \pm\sqrt{-\mu}$ are two solutions of $f(x, \mu) = 0$.)

If we observe the phenomena carefully, we realize the situations are exactly the ones governed by the "implicit function theorem," where $\frac{\partial f}{\partial x}(0,0)$ can be used to check if $x = x(\mu)$ can be solved uniquely near the origin. Therefore, we present the following implicit function theorem which can be applied to scalar differential equations. See **Figure 6.10** for a local picture explaining this. A proof of the implicit function theorem can be found in Chow and Hale [1982].

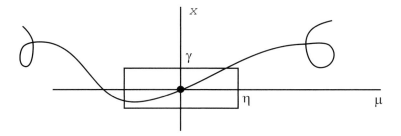

Figure 6.10: A picture explaining the implicit function theorem

---

Theorem 6.2.4 (Implicit function theorem) *Assume that $f : (x, \mu) \in \Re^2 \to \Re$ have continuous first partial derivatives such that*

$$f(0,0) = 0, \quad \frac{\partial f}{\partial x}(0,0) \neq 0. \tag{2.6}$$

Then there exist constants $\gamma > 0$, $\eta > 0$, and $f(x, \mu) = 0$ has a unique solution $x = x(\mu)$ such that $x(0) = 0$, $x(\mu)$ is continuously differentiable in μ, and $|x(\mu)| \leq \gamma$ when $|\mu| \leq \eta$. ♠

For the scalar differential equation $x' = f(x, \mu)$, the implicit function theorem implies, under the condition (2.6), that $x = x(\mu)$ can be uniquely solved near the origin from $f(x, \mu) = 0$. Thus the number of critical points will not change. Next, $\frac{\partial f}{\partial x}(x(\mu), \mu) \approx \frac{\partial f}{\partial x}(0,0) \neq 0$ when $\mu \approx 0$, hence (for μ small) the stability property of the critical point $x(\mu)$ will not change either. Thus, we have the same situation as in Example 6.2.3 and no bifurcation can occur.

Therefore, we conclude that bifurcations may occur only when the implicit function theorem fails to apply to solve x uniquely in terms of μ, which is the case for $f(x,\mu) = \mu + x^2$ in Example 6.1.2, where $\frac{\partial f}{\partial x}(0,0) = 0$. Next, a further analysis of $f(x,\mu) = \mu + x^2$ in Example 6.1.2 shows that

$$\frac{\partial^2 f}{\partial x^2}(0,0) = 2 \neq 0.$$

Now, for a general function $f(x)$ that satisfies

$$\frac{\partial f}{\partial x}(0) = 0, \quad \frac{\partial^2 f}{\partial x^2}(0) \neq 0,$$

it follows from the second derivative test in calculus that the graph of the function $f(x)$ for x near 0 is concave down or concave up with a horizontal tangent line at $x = 0$. Thus the graph of the function $f(x)$ looks like a **parabola** near $x = 0$. Therefore, the graph of the function $f(x,\mu) = \mu + x^2$ in Example 6.1.2 looks like a parabola near the origin. (This is obvious when you look at the formula of $f(x,\mu) = \mu + x^2$, but we can derive the same conclusion using derivatives.)

Next, for $f(x,\mu) = \mu + x^2$,

$$\frac{\partial f}{\partial \mu}(0,0) = \frac{\partial}{\partial \mu} f(x,\mu)|_{x=0, \mu=0} = 1 \neq 0,$$

thus $f(x,\mu)$ changes sign as μ passes zero, and the positive or negative values of μ will move the parabola $f(x,\mu) = \mu + x^2$ above or below the x-axis. Thus, as the parameter μ is varied and passes $\mu = 0$, the critical points are created or destroyed. See Figure 6.2.

This is the main reason that caused the saddle-node bifurcation in Example 6.1.2. Using the above remarks and geometric interpretation, we can now prove the following result concerning saddle-node bifurcations for scalar differential equations.

Theorem 6.2.5 (Saddle-node bifurcation) *Let $f(x,\mu)$ in Eq. (1.1) be a scalar function defined on $\Re^2$. If f has continuous second partial derivatives such that*

$$f(0,0) = 0, \quad \frac{\partial f}{\partial x}(0,0) = 0, \quad \frac{\partial^2 f}{\partial x^2}(0,0) \neq 0, \quad \frac{\partial f}{\partial \mu}(0,0) \neq 0. \qquad (2.7)$$

Then Eq. (1.1) undergoes a saddle-node bifurcation at $(x,\mu) = (0,0)$.

Proof. We verify that for small μ, the graph of the function $f(x,\mu)$ looks like a parabola in x near $x = 0$, which can be achieved by showing the

existence of a unique local maximum or minimum value for the function
$f(x, \mu)$ in x near $x = 0$. The local extreme points (where the maximum or
minimum values may occur) of $f(x, \mu)$ are the solutions x of

$$\frac{\partial}{\partial x} f(x, \mu) = 0. \tag{2.8}$$

To Solve Eq. (2.8), let's define $F(x, \mu) = \frac{\partial}{\partial x} f(x, \mu)$. Then condition
(2.7) implies

$$F(0, 0) = 0, \quad \frac{\partial F}{\partial x}(0, 0) = \frac{\partial^2 f}{\partial x^2}(0, 0) \neq 0. \tag{2.9}$$

Thus the implicit function theorem 6.2.4 can be applied to F (not to f)
so that there exist constants $\gamma > 0$, $\eta > 0$, and $F(x, \mu) = 0$ has a unique
solution $x = x(\mu)$, that is,

$$\frac{\partial f}{\partial x}(x(\mu), \mu) = 0, \tag{2.10}$$

such that $x(0) = 0$, $x(\mu)$ is continuously differentiable in μ, and $|x(\mu)| \leq \gamma$
when $|\mu| \leq \eta$. Next, the concavity at $(x(\mu), \mu)$ for the function $f(x, \mu)$
in x is determined by $\frac{\partial^2 f}{\partial x^2}(x(\mu), \mu)$, which is, by the continuity, close to
$\frac{\partial^2 f}{\partial x^2}(0, 0) \neq 0$. Therefore, for μ small,

$$\frac{\partial^2 f}{\partial x^2}(x(\mu), \mu) \neq 0. \tag{2.11}$$

Thus, based on (2.10) and (2.11), the second derivative test in calculus
can be used to argue that for (x, μ) near $(0, 0)$, the local extreme for the
function $f(x, \mu)$ in x is uniquely determined at $x = x(\mu)$ and the graph of
the function $f(x, \mu)$ in x looks like a parabola that opens up or down. More
specifically, $f(x, \mu)$ has a local minimum (concave up) if $\frac{\partial^2 f}{\partial x^2}(0, 0) > 0$; or a
local maximum (concave down) if $\frac{\partial^2 f}{\partial x^2}(0, 0) < 0$.

Next, let's examine the function values of $f(x, \mu)$ at the local extreme
points $x(\mu)$, given by $h(\mu) = f(x(\mu), \mu)$. From (2.7) and (2.10), we have

$$h(0) = f(0, 0) = 0,$$

and

$$\begin{aligned}
\frac{dh}{d\mu}(0) &= \left(\frac{\partial f}{\partial x}(x(\mu), \mu) \frac{d}{d\mu} x(\mu) + \frac{\partial f}{\partial \mu}(x(\mu), \mu) \right)\Big|_{\mu=0} \\
&= \frac{\partial f}{\partial \mu}(x(\mu), \mu)\Big|_{\mu=0} = \frac{\partial f}{\partial \mu}(0, 0) \neq 0, \tag{2.12}
\end{aligned}$$

therefore, $h(\mu) = f(x(\mu), \mu)$ changes sign as μ passes zero. Next, we look at the different cases for $h(\mu)\frac{\partial^2 f}{\partial x^2}(0,0)$. When $h(\mu)\frac{\partial^2 f}{\partial x^2}(0,0) = 0$, we have $\mu = 0$ and $h(0) = 0$. Thus the local extreme function value of $f(x, \mu)$ is zero, or the parabola is tangent to the x-axis, therefore there is only one critical point for $\mu = 0$. When $h(\mu)\frac{\partial^2 f}{\partial x^2}(0,0) < 0$, or when $h(\mu)$ and $\frac{\partial^2 f}{\partial x^2}(0,0)$ have the opposite signs, the function $f(x, \mu)$ in x is either concave up with the local minimum function value below the x-axis, or concave down with the local maximum function value above the x-axis. Hence there are two nonzero critical points such that one is stable and the other unstable. Note that the condition $\frac{\partial f}{\partial \mu}(0,0) \neq 0$ implies that $x = 0$ is not a critical point for $\mu \neq 0$ and small. Finally, when $h(r)\frac{\partial^2 f}{\partial x^2}(0,0) > 0$, the function $f(x, \mu)$ in x is either concave up with the local minimum function value above the x-axis, or concave down with the local maximum function value below the x-axis, in which cases there are no critical points. See **Figure 6.11** for the local pictures for the positions of the parabolas. Therefore, Eq. (1.1) undergoes a saddle-node bifurcation at $(x, \mu) = (0,0)$. This completes the proof. ♠

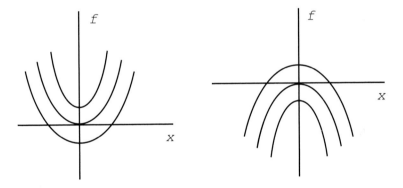

Figure 6.11: Local pictures showing the positions of the parabolas

Example 6.2.6 Let's show that the scalar differential equation

$$x' = f(x, \mu) = \mu + x - e^x + 1, \tag{2.13}$$

undergoes a saddle-node bifurcation at $(x, \mu) = (0,0)$. Now, $f(0,0) = 0$, $\frac{\partial f}{\partial x}(0,0) = (1 - e^x)|_{x=0} = 0$, $\frac{\partial^2 f}{\partial x^2}(0,0) = -e^x|_{x=0} = -1 \neq 0$, $\frac{\partial f}{\partial \mu}(0,0) = 1 \neq 0$, therefore Theorem 6.2.5 can be applied. In this case, the geometry concerning the locations of the functions $y = x + \mu$ and $y = e^x - 1$ in the (x, y)

plane can be used to obtain the saddle-node bifurcation at $(x, \mu) = (0, 0)$, see an exercise.   ♠

Finally, for scalar differential equations, we point out that if a saddle-node bifurcation occurs at $(x, \mu) = (0, 0)$, then $f(0, 0) = 0$ since $x = 0$ is a critical point, and $\frac{\partial f}{\partial x}(0, 0) = 0$ because of the tangency condition of a saddle-node bifurcation. Thus, if in the Taylor expansion of $f(x, \mu)$ we neglect quadratic terms in μ and cubic terms in x, then near $(x, \mu) = (0, 0)$, we obtain

$$
\begin{aligned}
f(x, \mu) &= \mu \frac{\partial f}{\partial \mu}(0, 0) + x \frac{\partial f}{\partial x}(0, 0) + \frac{x^2}{2} \frac{\partial^2 f}{\partial x^2}(0, 0) + \cdots \\
&= a\mu + bx^2 + \cdots \tag{2.14}
\end{aligned}
$$

where $a = \frac{\partial f}{\partial \mu}(0, 0)$ and $b = \frac{1}{2} \frac{\partial^2 f}{\partial x^2}(0, 0)$. The form $a\mu + bx^2$ in (2.14) is called a **normal form** or **representative** for scalar saddle-node bifurcations. Therefore, we know now that $f(x, \mu) = \mu + x^2$ in Example 6.1.2 and $f(x, \mu) = \mu - x^2$ in Example 1.3.2 in Chapter 1 are actually representatives of **all** saddle-node bifurcations for scalar equations. In other words, all scalar saddle-node bifurcations behave in essentially the same fashion as those for $f(x, \mu) = \mu \pm x^2$. More results on the normal forms and the related subjects can be found in Birkhoff [1927], Chow and Hale [1982], Guckenheimer and Holmes [1986], and Wiggins [1990].

Exercises 6.2

1. Use linearizations to examine each critical point for Example 6.2.1 when $\mu < 0$.

2. Obtain the saddle-node bifurcation at $(x, \mu) = (0, 0)$ for Example 6.2.6 using the geometry concerning the locations of the functions $y = x + \mu$ and $y = e^x - 1$ in the (x, y) plane.

3. Use a Taylor expansion to find the normal form for the equation in Example 6.2.6.

6.3   Transcritical Bifurcation

We have discussed saddle-node bifurcations in the previous section, where a key feature is that critical points can be created and destroyed when some parameters are varied. However, for some systems in applications, a critical point may exist for all values of parameters, or will never disappear. For example, for the logistic equation $x' = rx(1-x)$ studied in Chapter 1, $x = 0$ is a critical point for all values of r.

Example 6.3.1 Consider the scalar differential equation

$$x' = f(x, \mu) = \mu x - x^2, \tag{3.1}$$

which looks like the logistic equation $x' = rx(1-x)$, but here x and μ are allowed to be positive or negative, (recall that in applications using the logistic equations in biology, $x \geq 0$ is assumed to denote a population). Now, $x = 0$ is a critical point for all μ values and $x = \mu$ gives a different critical point if $\mu \neq 0$. See **Figure 6.12** which shows the positions of the function $y = \mu x - x^2$ for different μ values, and hence the stability properties.

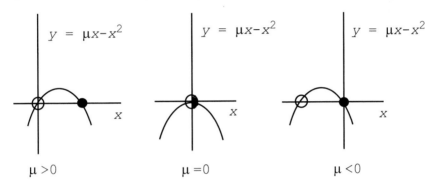

Figure 6.12: Pictures of the function $y = \mu x - x^2$ showing the stability properties

Using the stability analysis from Chapter 5, we find that when $\mu > 0$, $x = \mu$ is a stable critical point; when $\mu < 0$, $x = \mu$ is an unstable critical point; and finally, when $\mu = 0$, the two critical points collide such that the critical point $x = 0$ is half-stable.

Some of the above descriptions are similar to those of saddle-node bifurcations. However, the difference is that here, the two critical points do not disappear when $\mu > 0$ (or $\mu < 0$), instead they switch their stabilities. That is, $x = 0$ changes its stability by **transferring** its stability property to another critical point. Accordingly, these bifurcations are called **transcritical**

bifurcations. See **Figure 6.13** for a transcritical bifurcation diagram of $x' = \mu x - x^2$, where the bifurcation value is $\mu = 0$.   ♠

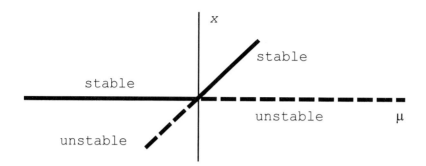

Figure 6.13: A transcritical bifurcation diagram of $x' = \mu x - x^2$

Next, we look at an application of transcritical bifurcations in physics.

Example 6.3.2 (A solid-state laser) The following is a brief account of the situation following Haken [1983] and Strogatz [1994]. A collection of laser-active atoms are embedded in a solid-state matrix, and an external energy source is applied to pump or excite the atoms out of their ground states to produce a laser.

When the strength of the pumping is increased, the system goes through a period of quantitative changes during which the excited atoms oscillate independently of each other and emit randomly phased light waves. This effect is just like an ordinary lamp. Then, suddenly, when the strength of the pumping exceeds a critical level, called a **laser threshold**, a beam of radiation that is much more coherent and intense will appear, or a **laser** is produced.

Let $n(t)$ be the number of photons and $N(t)$ the number of excited atoms of the system. The time rate of $n(t)$ is given as

$$n' = \text{Gain} - \text{Loss}.$$

The "Gain" comes from the stimulated emission, in which photons stimulate excited atoms to emit more photons. This activity happens when photons encounter excited atoms whose process is random, thus "Gain"

$= Gn(t)N(t)$, where $G > 0$ is a "gain coefficient." Similarly, "Loss" $= Ln(t)$, giving the escape of photons with a "loss coefficient" $L > 0$. Therefore,

$$n' = \text{Gain} - \text{Loss} = GnN - Ln.$$

Next, let's find the relationship between n and N. After an excited atom emits a photon, it drops back to its ground state and is no longer excited. Thus, if we assume that the pump keeps the number of excited atoms fixed at N_0 before a laser is produced, then during the laser process, the actual number of excited atoms will be determined according to

$$N(t) = N_0 - \alpha n(t),$$

where $\alpha > 0$ is the rate at which atoms drop back to their ground states. Therefore, we obtain

$$n' = \text{Gain} - \text{Loss} = GnN - Ln = (GN_0 - L)n - (\alpha G)n^2,$$

which is similar to Eq. (3.1), and hence indicates that a transcritical bifurcation takes place when $GN_0 - L = 0$, or at the **laser threshold** $N_0 = \frac{L}{G}$, which is the loss coefficient divided by the gain coefficient.

In physics, this means that when the pump strength is weak, that is, $N_0 < \frac{L}{G}$, the critical point $n = 0$ is stable and no laser is produced. When the pump strength is increased and passes the laser threshold $\frac{L}{G}$, or when $N_0 > \frac{L}{G}$, the origin $n = 0$ loses its stability and a stable critical point appears at $n = \frac{GN_0 - L}{\alpha G} > 0$: spontaneous laser action is taking place, or a laser is produced. See **Figure 6.14** for the function $y = (GN_0 - L)n - (\alpha G)n^2$, showing the stability properties (where $n \geq 0$ is assumed to denote the number of photons); see also **Figure 6.15** for a transcritical bifurcation diagram of the laser. ♠

Transcritical bifurcations, where an origin transfers its stability to another critical point, can also occur for differential equations in higher dimensions.

Example 6.3.3 Consider

$$\begin{cases} x' &= \mu x + x^2, \\ y' &= -y. \end{cases} \tag{3.2}$$

Similar to Example 6.2.1, all actions are taken in the first equation, because $y = 0$ in the second equation is asymptotically stable. Now, $(0, 0)$

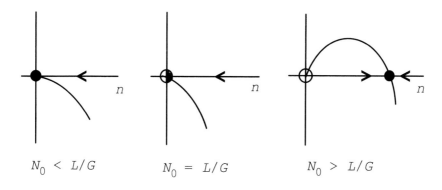

$$N_0 < L/G \qquad\qquad N_0 = L/G \qquad\qquad N_0 > L/G$$

Figure 6.14: Pictures of the function $y = (GN_0 - L)n - (\alpha G)n^2$, showing the stability properties

is a critical point and will never disappear. Next, for $\mu < 0$, the origin $(0,0)$ is asymptotically stable (a stable node) and the other critical point $(-\mu, 0)$ is unstable (a saddle). For $\mu > 0$, the origin $(0,0)$ becomes unstable and the other critical point $(-\mu, 0)$ becomes asymptotically stable. Therefore, the origin transfers its stability to another critical point $(-\mu, 0)$, and the system undergoes a transcritical bifurcation at the bifurcation value $\mu = 0$, see **Figure 6.16**.   ♠

Note in Example 6.3.3 that even though saddles and nodes appear, however, since the critical point $(0,0)$ will never disappear, the bifurcation is transcritical.

Next, let's look at the similarity and difference between the saddle-node and transcritical bifurcations given in Examples 6.1.2 and 6.3.1. Locally, both of the graphs of $f(x, \mu)$ in $x' = f(x, \mu)$ look like parabolas. But the difference is that for the saddle-node bifurcation, the change of the parameter μ near the bifurcation value causes the corresponding parabola to move above or below the x-axis, while the change of μ near the bifurcation value for the transcritical bifurcation causes the corresponding parabola to move to the left or the right of the y-axis and keeps $x = 0$ as a critical point. See Figures 6.2 and 6.12. Consequently, we find that the maximum value of the function $f(x, \mu) = \mu x - x^2$ for the transcritical bifurcation, when treated as a function of μ, looks also like a parabola that is tangent to the μ-axis at $\mu = 0$. In fact, for $f(x, \mu) = \mu x - x^2$, the maximum value occurs when $\frac{\partial f}{\partial x} = \mu - 2x = 0$, or

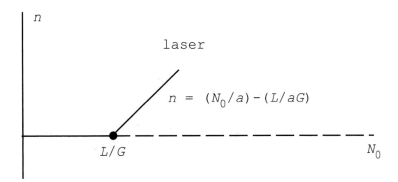

Figure 6.15: A transcritical bifurcation diagram for a laser

when $x = \frac{\mu}{2}$, and the maximum function value at $x = \frac{\mu}{2}$ is

$$f(x,\mu) = f(\frac{\mu}{2}, \mu) = \mu\frac{\mu}{2} - (\frac{\mu}{2})^2 = \frac{\mu^2}{4},$$

which is a parabola in μ. The general case is shown in **Figure 6.17**.

According to this geometric interpretation, we can now prove the following result concerning transcritical bifurcations for scalar differential equations.

Theorem 6.3.4 (Transcritical bifurcation) *Let $f(x,\mu)$ in Eq. (1.1) be a scalar function defined on $\Re^2$. If f has continuous second partial derivatives such that*

$$f(0,\mu) = 0, \ \mu \in \Re, \ \ \frac{\partial f}{\partial x}(0,0) = 0, \ \ \frac{\partial^2 f}{\partial x^2}(0,0) \neq 0,$$

$$\frac{\partial f}{\partial \mu}(0,0) = 0, \ \ \frac{\partial^2 f}{\partial \mu^2}(0,0) = 0, \ \ \frac{\partial^2 f}{\partial x \partial \mu}(0,0) \neq 0. \tag{3.3}$$

Then Eq. (1.1) undergoes a transcritical bifurcation at $(x,\mu) = (0,0)$.

Proof. Because the conditions in (3.3) imply the first three conditions in (2.7), then the first part of the proof of Theorem 6.2.5 shows that for μ small, the local maximum or minimum value for the function $f(x,\mu)$ in x near $x = 0$ exists uniquely, and the graph of $f(x,\mu)$ looks like a parabola in x that opens up or down.

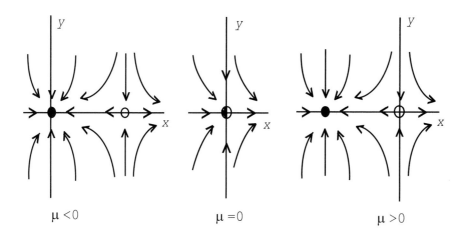

$$\mu < 0 \qquad\qquad \mu = 0 \qquad\qquad \mu > 0$$

Figure 6.16: A planar transcritical bifurcation

Assume that the parabola opens down, or f has a unique local maximum value. (The case for a unique local minimum value is similar and is left as an exercise.) Now, $f(0, \mu) = 0$ for $\mu \in \Re$ in (3.3) means that $x = 0$ is a critical point for any μ, or, as a function in x, $f(x, \mu)$ will cross $(x, y) = (0, 0)$ for any μ. Thus the graph of $f(x, \mu)$ in x crosses $(x, y) = (0, 0)$ and goes above the x-axis and reaches the maximum function value and then goes down and crosses the x-axis, creating another critical point. See **Figure 6.18**.

Next, we verify that as the parameter μ is varied and passes $\mu = 0$, the maximum value of $f(x, \mu)$ will appear on both sides of the y-axis. From the proof of Theorem 6.2.5, the maximum value of $f(x, \mu)$ occurs at $x = x(\mu)$ (where $x(0) = 0$). We will examine the function value at $x(\mu)$, given by $h(\mu) = f(x(\mu), \mu)$, and verify that $h(\mu)$ is a parabola as a function in μ. From (2.10) in the proof of Theorem 6.2.5, we take a derivative in μ and obtain

$$\frac{\partial f^2}{\partial x^2}(x(\mu), \mu)\frac{d}{d\mu}x(\mu) + \frac{\partial f^2}{\partial x \partial \mu}(x(\mu), \mu) = 0, \tag{3.4}$$

which implies, from (3.3),

$$\frac{d}{d\mu}x(\mu)|_{\mu=0} \neq 0. \tag{3.5}$$

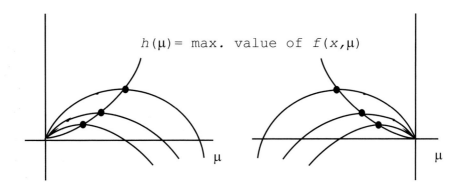

$h(\mu) = $ max. value of $f(x,\mu)$

Figure 6.17: The maximum value of $f(x, \mu)$ as a function in μ looks like a parabola

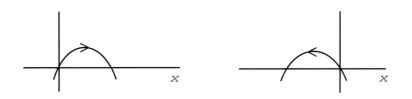

Figure 6.18: The function $f(x, \mu)$ goes up and down, creating another critical point

Now, for $h(\mu) = f(x(\mu), \mu)$, we have $h(0) = f(0, 0) = 0$ and, from (2.10),

$$\frac{d}{d\mu} h(\mu) = \frac{\partial f}{\partial x}(x(\mu), \mu)\frac{d}{d\mu}x(\mu) + \frac{\partial f}{\partial \mu}(x(\mu), \mu)$$

$$= \frac{\partial f}{\partial \mu}(x(\mu), \mu), \qquad (3.6)$$

then from (3.3) and (3.6),

$$\frac{dh}{d\mu}(0) = \frac{\partial f}{\partial \mu}(x(\mu), \mu)|_{\mu=0} = \frac{\partial f}{\partial \mu}(0, 0) = 0. \qquad (3.7)$$

Next, from (3.3), (3.5), and (3.6),

$$\frac{d^2h}{d\mu^2}(0) = \left(\frac{\partial}{\partial \mu}\left[\frac{\partial f}{\partial \mu}(x(\mu), \mu)\right]\right)|_{\mu=0}$$

$$= \frac{\partial f^2}{\partial \mu \partial x}(x(\mu), \mu) \frac{d}{d\mu} x(\mu)|_{\mu=0} + \frac{\partial f^2}{\partial \mu^2}(x(\mu), \mu)|_{\mu=0}$$

$$= \frac{\partial f^2}{\partial \mu \partial x}(x(\mu), \mu) \frac{d}{d\mu} x(\mu)|_{\mu=0} \neq 0. \tag{3.8}$$

Thus, using the second derivative test, we find that as a function in μ, the graph of the maximum value of $f(x, \mu)$ at $x(\mu)$, which is given by $h(\mu) = f(x(\mu), \mu)$, looks locally like a parabola that is tangent to the μ-axis when $\mu = 0$. See Figure 6.17. Therefore, as the parameter μ is varied and passes $\mu = 0$, the maximum value of f will occur on both sides of the y-axis creating stable and unstable critical points, see Figure 6.18. Therefore, a transcritical bifurcation takes place. This completes the proof.   ♠

Exercises 6.3

1. Show that $x' = \mu x + 3x^2$ undergoes a transcritical bifurcation and find the bifurcation value.

2. Show that $x' = (2 + \mu)x - 5x^2$ undergoes a transcritical bifurcation and find the bifurcation value.

3. Make up some examples that undergo transcritical bifurcations in higher dimensions and verify your claims.

4. Complete the case when $f(x, \mu)$ has a unique local minimum value in the proof of Theorem 6.3.4.

6.4   Pitchfork Bifurcation

In some applications, an object may move in the directions that are **symmetric** with respect to a certain position. For example, the buckling beam introduced in Chapter 1 could buckle to any direction from its vertical position. In those cases, the critical points tend to appear and disappear symmetrically, creating a type of bifurcations that are different from saddle-node and transcritical bifurcations. Let's begin with some examples.

Example 6.4.1 Consider the scalar differential equation

$$x' = f(x, \mu) = \mu x - x^3, \tag{4.1}$$

where $\mu \in \Re$ is a parameter. Note that for any fixed μ, function $y = \mu x - x^3$ is an odd function in x, which indicates that critical point will appear and disappear symmetrically with respect to the origin $x = 0$, or the system has the left-right symmetry. See **Figure 6.19** for the function $y = \mu x - x^3$, showing the stability properties.

Now, $x = 0$ is a critical point that will never disappear. For $\mu \leq 0$, $x = 0$ is the only critical point and is asymptotically stable; for $\mu > 0$, two new asymptotically stable critical points appear at $x = \pm\sqrt{\mu}$, symmetrically about the origin, and $x = 0$ loses its stability. Therefore, a bifurcation takes place at the bifurcation value $\mu = 0$. ♠

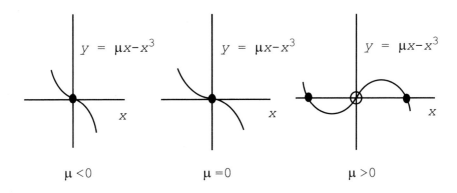

Figure 6.19: Pictures of the function $y = \mu x - x^3$, showing the stability properties

If we draw the bifurcation diagram for $f(x, \mu) = \mu x - x^3$ in Example 6.4.1, we get **Figure 6.20**, which looks like a "pitchfork." Therefore, due to their appearance, these bifurcations are called **pitchfork bifurcations**.

In an exercise, you are asked to verify that

$$x' = f(x, \mu) = \mu x + x^3, \tag{4.2}$$

also undergoes a pitchfork bifurcation at $\mu = 0$, and that the bifurcation diagram is given in **Figure 6.21**.

Next, let's look at the difference between equations (4.1) and (4.2). In Eq. (4.1), $x = 0$ of $x' = \mu x$ is unstable for $\mu > 0$, and $-x^3$ is used as a **stabilizing term** that changes the direction field and pulls the solutions back to the origin. In Eq. (4.2), however, things are so different, because

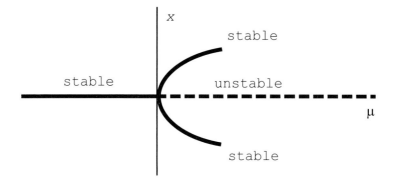

Figure 6.20: A bifurcation diagram determined by $x' = \mu x - x^3$

$x = 0$ of $x' = \mu x$ is already unstable for $\mu > 0$, in addition, x^3 is added as a **destabilizing term** that makes it even "more unstable," which in fact may cause "blow-up." That is, the solutions may approach ∞ at finite times, similar to what happened for $x' = x^2$ discussed in Chapter 2.

The above discussions have applications in engineering, where the pitchfork bifurcations caused by "stabilizing terms," such as $-x^3$, are referred to as "safe" or "soft," or the bifurcating critical points $x = \pm\sqrt{\mu}$ are stable, and are called **supercritical pitchfork bifurcations**. On the other hand, the pitchfork bifurcations caused by "destabilizing terms," such as x^3, are referred to as "dangerous" or "hard," or the bifurcating critical points $x = \pm\sqrt{-\mu}$ are unstable, and are called **subcritical pitchfork bifurcations**. "Supercritical" and "subcritical" are used in different texts with different reasons. For example, in Golubitsky and Schaeffer [1979] they denote the directions in which bifurcations take place. That is, a pitchfork bifurcation at $\mu = \mu_0$ is said to be supercritical if there is locally only one critical point for $r < r_0$, or, "supercritical" bifurcates to the right. Accordingly, "subcritical" bifurcates to the left. (Here, the μ-axis points to the right.)

Sometimes, notations and terminology in Bifurcation Theory are quite confusing and different texts use different words for the same phenomena. For example, "saddle-node" bifurcations here are called "turning point" bifurcations or "fold" bifurcations in other texts. They are called "blue sky" bifurcations in Abraham and Shaw [1988] to mean that two branches of critical points appear "out of the clear blue sky" as a parameter is varied.

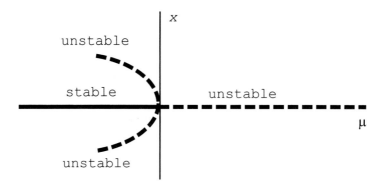

Figure 6.21: A bifurcation diagram determined by $x' = \mu x + x^3$

The following is an equation in $\Re^2$ that undergoes a pitchfork bifurcation. The analysis is left as an exercise.

Example 6.4.2 Consider

$$\begin{cases} x' &= \mu x - x^3, \\ y' &= -y. \end{cases} \tag{4.3}$$

Similar to Example 6.2.1, one only needs to analyze the first equation. It is left as an exercise to verify that Eq. (4.3) undergoes a pitchfork bifurcation, see **Figure 6.22**. ♠

Next, let's analyze the Euler's buckling beam introduced in Chapter 1.

Example 6.4.3 (Euler's buckling beam) Revisit Euler's buckling beam introduced in Chapter 1. To analyze the beam better, we place the beam on the interval $[0, 1]$ on the horizontal w-axis, shown in **Figure 6.23**, where a compressive force $\Gamma \geq 0$ is applied along its axis.

The viscous damping is denoted by $\delta > 0$ and the membrane stiffness by $K > 0$. Then, from Huang and Nachbar [1968], Holmes [1979], and Seydel [1988], the small deflections $v(w, t)$ of the beam satisfy the following partial differential equation

$$v_{wwww} + \left[\Gamma - K \int_0^1 v_w^2(h, t)dh \right] v_{ww} + \delta v_t + v_{tt} = 0. \tag{4.4}$$

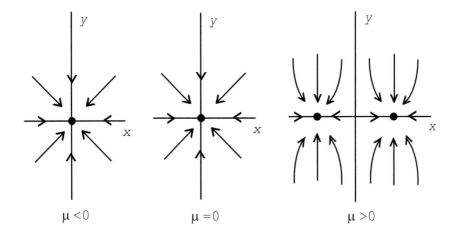

Figure 6.22: A planar pitchfork bifurcation

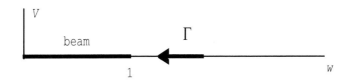

Figure 6.23: Euler's buckling beam

To simplify Eq. (4.4), we assume that $v(w,t)$ is perfectly symmetric in the spatial variable w about $w = 0.5$ and $v(w,t)$ has only one extreme (maximum or minimum) in w at $w = 0.5$. Thus we may take $v(w,t)$ to be

$$v(w,t) = u(t)\sin\pi w, \quad w \in [0,1], \ t \geq 0,$$

where $u(t)$ denotes the amplitude in time t. (This is a very common practice in physics where the spatial variable w and the time variable t are separated in this way.) Plugging this into Eq. (4.4), one obtains an ordinary differential equation of the amplitude u in the time variable t, given by

$$u''(t) + \delta u'(t) - \pi^2(\Gamma - \pi^2)u(t) + \frac{1}{2}K\pi^4 u^3(t) = 0. \tag{4.5}$$

To analyze Eq. (4.5), we change the variables $x = u$ and $y = u'$, then we get

$$\begin{cases} x' &= y, \\ y' &= \pi^2(\Gamma - \pi^2)x - \delta y - \frac{1}{2}K\pi^4 x^3, \end{cases} \tag{4.6}$$

where $\delta > 0$ and $K > 0$ are regarded as constants and $\Gamma \geq 0$ as the only parameter. Now, $(0,0)$ is a critical point of Eq. (4.6) for all δ, K, and Γ, and other candidates for critical points are given by

$$(\pm\sqrt{\frac{2(\Gamma - \pi^2)}{K\pi^2}}, \ 0).$$

Therefore, for $\Gamma \leq \pi^2$, Eq. (4.6) has only one critical point $(0,0)$ and for $\Gamma > \pi^2$, Eq. (4.6) has three critical points. Accordingly, a bifurcation takes place at $\Gamma = \pi^2$, which is called **Euler's first buckling load**.

Next, let's determine the stabilities of the critical points. For $(0,0)$, the matrix of the linearization is given by

$$A = \begin{bmatrix} 0 & 1, \\ \pi^2(\Gamma - \pi^2) & -\delta \end{bmatrix}, \tag{4.7}$$

thus, $p = tr A = -\delta < 0$ and $q = \det A = -\pi^2(\Gamma - \pi^2) = \pi^2(\pi^2 - \Gamma)$. Therefore, from the Distribution Diagram 6.1 in Section 1 for planar equations, we find that $(0,0)$ is stable for $\Gamma < \pi^2$ and unstable (saddle) for $\Gamma > \pi^2$. In practice, this means that the beam will buckle if the compressive force $\Gamma > \pi^2$.

For the critical point $(\sqrt{\frac{2(\Gamma-\pi^2)}{K\pi^2}}, \ 0)$ under the condition $\Gamma > \pi^2$, we change the variables $x_1 = x - \sqrt{\frac{2(\Gamma-\pi^2)}{K\pi^2}}$ and $y_1 = y$, then (left as an exercise) the matrix of the linearization is given by

$$A = \begin{bmatrix} 0 & 1, \\ -2\pi^2(\Gamma - \pi^2) & -\delta \end{bmatrix}, \tag{4.8}$$

thus, $p = tr A = -\delta < 0$ and $q = \det A = 2\pi^2(\Gamma - \pi^2) > 0$. Therefore, from the Distribution Diagram 6.1, we find that $(\sqrt{\frac{2(\Gamma-\pi^2)}{K\pi^2}}, \ 0)$ is stable. It is also left as an exercise to verify that $(-\sqrt{\frac{2(\Gamma-\pi^2)}{K\pi^2}}, \ 0)$ is also stable. Accordingly, the bifurcation is a pitchfork bifurcation, (in fact, it is supercritical).

In Eq. (4.6), which is an equation in $\Re^2$, we are only interested in the quantity $x = u$ (amplitude). Therefore, we have the bifurcation diagram in the (Γ, u) plane in **Figure 6.24**. The diagram in Figure 6.24 shows the **symmetry** of the two branches around the Γ-axis, which reveals our basic assumption of perfect symmetry. This means that the beam could buckle in any direction in space. When we treat the beam only in the (w, v) plane, then the beam may buckle "up" (when u is positive) or "down" (when u is negative), as shown in **Figure 6.25**.   ♠

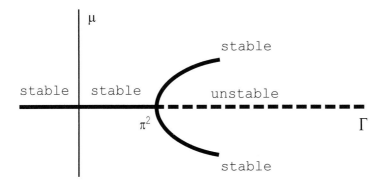

Figure 6.24: A bifurcation diagram of Euler's buckling beam in the (Γ, u) plane using $u = \pm\sqrt{\frac{2(\Gamma - \pi^2)}{K\pi^2}}$

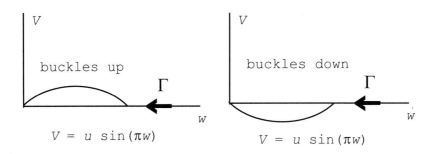

Figure 6.25: The beam may buckle "up" or "down"

Next, let's examine a phenomenon that frequently occurs in physics applications.

Example 6.4.4 (Hysteresis) We have indicated earlier that the term x^3 in the subcritical pitchfork bifurcation of $x' = \mu x + x^3$ will destabilize the system and cause some danger. In physics applications, a higher order term is then added to stabilize the system. For simplicity, we assume $-x^5$ is added such that $f(x, \mu) = \mu x + x^3 - x^5$ is still an odd function in x. Now, $x = 0$ is a critical point for any μ. To get other nonzero (real) critical points, we solve

$$\mu + x^2 - x^4 = 0, \quad \text{or} \quad x^2 = \frac{1 \pm \sqrt{1 + 4\mu}}{2},$$

from which we find none such critical points for $\mu < -\frac{1}{4}$, two such critical points if $\mu = -\frac{1}{4}$, four such critical points if $\mu \in (-\frac{1}{4}, 0)$, and two such critical points if $\mu \geq 0$. Next, since $\lim_{x \to \infty} f(x, \mu) = -\infty$ and f is odd in x, we must have **Figure 6.26**, which also shows the stabilities.

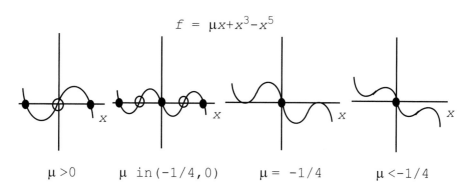

$f = \mu x + x^3 - x^5$

$\mu > 0 \qquad \mu \text{ in } (-1/4, 0) \qquad \mu = -1/4 \qquad \mu < -1/4$

Figure 6.26: Pictures of $f(x, \mu) = \mu x + x^3 - x^5$, showing the critical points and stabilities

Accordingly, we derive the bifurcation diagram in **Figure 6.27**.

Now, let's start the system in the state $x = 0$ and increase the parameter μ; see the arrows in Figure 6.27. The state remains at the origin until $\mu = 0$. Then, as μ is increased, the origin loses its stability and the state will **jump** to one of the two large-amplitude branches. If μ is decreased now, the state will not jump back to $x = 0$ at $\mu = 0$, instead it will remain on the large-amplitude branch until $\mu = -\frac{1}{4}$, at which the state jumps back to the

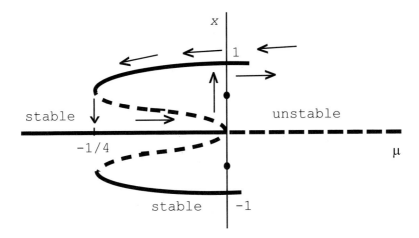

Figure 6.27: The bifurcation diagram of $x' = \mu x + x^3 - x^5$, showing the hysteresis effect

origin. This phenomenon, involving the jumps and nonreversible behavior, is referred to as the **hysteresis** in physics applications. It is also one expression stating that the choice of the stable critical point by a system depends on the **past history** of the system. The arrows in Figure 6.27 are said to form a **hysteresis loop**. Applications in physics, such as magnetic hysteresis of a magnetic field, can be found in Erber, Guralnik, and Latal [1972]. ♠

Finally, if we analyze the forms of equations in (4.1) and (4.2), we find that

$$\mu x \pm x^3 = x(\mu \pm x^2), \tag{4.9}$$

and $\mu \pm x^2$ are the typical cases, or normal forms, for saddle-node bifurcations. Thus, we consider

$$f(x,\mu) = xg(x,\mu) \tag{4.10}$$

for pitchfork bifurcations. Now, if we can argue that $g(x,\mu)$ undergoes a saddle-node bifurcation near $(0,0)$, or the graph of $g(x,\mu)$ is a parabola around the origin $x = 0$ for x and μ small, and the parabola moves above or below the x-axis as μ is varied and passes $\mu = 0$, then $xg(x,\mu)$ will cause the two critical points to have the same stability property. That is, a pitchfork bifurcation will take place, see **Figure 6.28**.

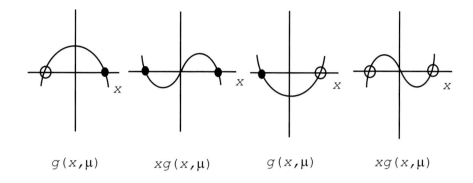

$$g(x,\mu) \qquad xg(x,\mu) \qquad g(x,\mu) \qquad xg(x,\mu)$$

Figure 6.28: Local pictures for $g(x,\mu)$ and $xg(x,\mu)$

According to this remark and geometric interpretation, we have the following result concerning pitchfork bifurcations for scalar differential equations. The proof is left as an exercise.

Theorem 6.4.5 (Pitchfork bifurcation) *Let $f(x,\mu)$ in Eq. (1.1) be a scalar function defined on $\Re^2$. If $f(x,\mu) = xg(x,\mu)$ and g has continuous second partial derivatives such that*

$$g(0,0) = 0, \quad \frac{\partial g}{\partial x}(0,0) = 0, \quad \frac{\partial^2 g}{\partial x^2}(0,0) \neq 0, \quad \frac{\partial g}{\partial \mu}(0,0) \neq 0. \qquad (4.11)$$

Then Eq. (1.1) undergoes a pitchfork bifurcation at $(x,\mu) = (0,0)$.   ♠

Exercises 6.4

1. Verify that

$$x' = f(x,\mu) = \mu x + x^3 \qquad (4.12)$$

undergoes a pitchfork bifurcation at $\mu = 0$, and that the bifurcation diagram is given in Figure 6.21.

2. Verify that the equation

$$\begin{cases} x' &= \mu x - x^3, \\ y' &= -y, \end{cases} \qquad (4.13)$$

undergoes a pitchfork bifurcation, and that the phase portraits are given in Figure 6.22.

3. In Example 6.4.3 of Euler's buckling beam, for the critical point $(\sqrt{\frac{2(\Gamma-\pi^2)}{K\pi^2}}, 0)$ when $\Gamma > \pi^2$ and the change of variables $x_1 = x - \sqrt{\frac{2(\Gamma-\pi^2)}{K\pi^2}}$ and $y_1 = y$, verify that the matrix of linearization is given by

$$A = \begin{bmatrix} 0 & 1 \\ -2\pi^2(\Gamma - \pi^2) & -\delta \end{bmatrix}. \tag{4.14}$$

4. In Example 6.4.3 of Euler's buckling beam, verify that $(-\sqrt{\frac{2(\Gamma-\pi^2)}{K\pi^2}}, 0)$ is also stable.

5. In Example 6.4.4 concerning hysteresis, verify for $f(x,\mu) = \mu x + x^3 - x^5$ that we must have Figure 6.26.

6. In Example 6.4.4 concerning hysteresis, verify that the bifurcation diagram is given in Figure 6.27.

7. Prove Theorem 6.4.5.

6.5   Poincaré-Andronov-Hopf Bifurcation

So far, we have seen saddle-node, transcritical, and pitchfork bifurcations, where critical points can be created and destroyed, and their stability properties can be changed when some parameters are varied. If we analyze the examples given above for those bifurcations, we find that the eigenvalues of the linearizations for those equations are all real numbers. Next, we use the following example to introduce some other possibilities.

Example 6.5.1 Consider the differential equation in $\Re^2$,

$$\begin{cases} x' & = & \mu x + y - x(x^2 + y^2), \\ y' & = & -x + \mu y - y(x^2 + y^2), \end{cases} \tag{5.1}$$

where $\mu \in \Re$ is a parameter. When $\mu = 0$, the matrix of the linearization has a pair of pure imaginary eigenvalues. Following Chapter 4, we change

Eq. (5.1) to polar coordinates in order to better analyze the system. That is, let $x(t) = r(t) \cos \theta(t)$, $y(t) = r(t) \sin \theta(t)$, and obtain

$$\begin{cases} r' &= r(\mu - r^2) = \mu r - r^3, \\ \theta' &= -1. \end{cases} \tag{5.2}$$

Now, we find that for $\mu \leq 0$, the origin is an asymptotically stable spiral point since $r \searrow 0$ and $\theta \to -\infty$ as $t \to \infty$. However, as soon as μ becomes positive, $r' = r(\mu - r^2) = 0$ has a positive constant solution $r = \sqrt{\mu}$, hence a **circle** appears with radius $\sqrt{\mu}$. Moreover, when we regard the equation $r' = r(\mu - r^2)$ as a scalar differential equation on the r-axis, we find that for a solution $r(t)$, $0 < r(0) < \sqrt{\mu}$ implies $r'(t) > 0$, and $r(0) > \sqrt{\mu}$ implies $r'(t) < 0$. Thus, $r = \sqrt{\mu}$ is asymptotically stable for the scalar differential equation $r' = r(\mu - r^2)$. Translating this result back to Eq. (5.2) or Eq. (5.1), we conclude that the circle $r = \sqrt{\mu}$ (when $\mu > 0$) for Eq. (5.2), or Eq. (5.1) is an asymptotically stable periodic solution and attracts all nonzero solutions, and the origin (which is still a solution now) becomes unstable. See **Figure 6.29**.

That is, as the parameter μ increases through $\mu = 0$, the critical point $r = 0$ (origin) of the system splits into the critical point $r = 0$ and the circle $r = \sqrt{\mu}$, and the stability of $r = 0$ is "transferred" to the circle $r = \sqrt{\mu}$. These changes are qualitative property changes, thus we say that Eq. (5.1) undergoes a bifurcation at the bifurcation value $\mu = 0$. ♠

Remark 6.5.2 The asymptotic stability of periodic solutions (orbits) requires further explanation. For a general autonomous differential equation $x' = f(x)$ in $\Re^n$, if $x(t) = x(t, 0, x_0)$, $t \geq 0$, is a periodic orbit, then $y(t) = x(t + t_1, 0, x_0)$, $t \geq 0$, is also a periodic orbit for a constant $t_1 \in \Re$. Now, t_1 can be chosen so that $y(0) = x(t_1, 0, x_0)$ is different from, but close to x_0, hence $|x(t) - y(t)|$ will not approach zero as $t \to \infty$ (because the initial difference will be carried after each period). Therefore, $x(t)$ is not asymptotically stable using the definition given in Chapter 5. However, the circle $r = \sqrt{\mu}$ in Example 6.5.1 *should be asymptotically stable in some sense!* In fact, $r = \sqrt{\mu}$ *is* asymptotically stable using the definition given in Chapter 5 *if* other solutions used to compare the differences do not have the same solution curve of $r = \sqrt{\mu}$ in $\Re^2$. Therefore, when we speak of "asymptotic stability of a periodic orbit $x(t)$" in this book, we slightly modify the definition given in Chapter 5 by requiring that other solutions used to compare the differences do not have the same solution curve of $x(t)$ in $\Re^n$, and we call it *L-asymptotically stable*. ♠

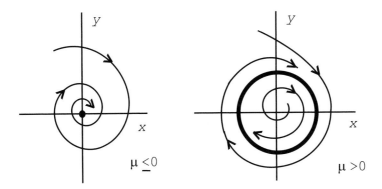

Figure 6.29: The stabilities for the circle $r = \sqrt{\mu}$ (when $\mu > 0$) and the origin

The bifurcation in Example 6.5.1 is different from all other bifurcations we have seen so far, and it can only happen for equations in $\Re^n$ with $n \geq 2$, because a pair of conjugate complex eigenvalues for the linearization are needed to cross the pure imaginary axis as the parameter is varied.

The geometric interpretation of the bifurcation in Example 6.5.1 is that for $\mu \leq 0$, the origin is a stable spiral point that attracts local solutions; as soon as $\mu > 0$, then, all of a sudden, **a balloon pops out**, and $x = 0$ transfers its stability to the circles on the balloon and repels solutions to those circles. See **Figure 6.30**.

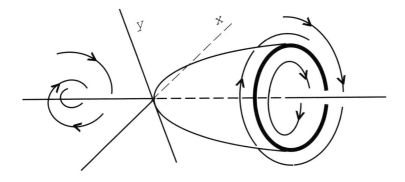

Figure 6.30: A geometric interpretation of bifurcation in Eq. (5.1)

The following **Figure 6.31** is the bifurcation diagram for Eq. (5.1) according to the radius $r = \sqrt{\mu}$, where $r = 0$ corresponds to the origin $(x, y) = (0, 0)$.

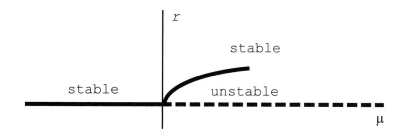

Figure 6.31: A bifurcation diagram of Eq. (5.1) according to $r = \sqrt{\mu}$

Another explanation of the bifurcation in Example 6.5.1 is that when μ increases through $\mu = 0$, the two conjugate complex eigenvalues cross the pure imaginary axis (from the left to the right). Then, from the Distribution Diagram 6.1, the origin will change from "spiral in" to "spiral out," which indicates that a circle or a periodic orbit surrounding the origin should be created to "compromise" or "balance" the two different flows.

The above phenomenon was discovered by Poincaré [1892] and proved by Andronov [1929] for equations in $\Re^2$ and then proved by Hopf [1942] for equations in $\Re^n$ for any $n \geq 2$ (where there is only one pair of pure imaginary eigenvalues and no other eigenvalues with zero real parts). To give all of them credit, we call this type of bifurcations the **Poincaré-Andronov-Hopf bifurcations**. Since these bifurcations address the creation of periodic orbits, it can be applied to physical problems where oscillations can be turned on or off.

Next, we state the celebrated Poincaré-Andronov-Hopf bifurcation theorem for equations in $\Re^2$ without proof, because the proof is beyond the scope of this book since the complicated analysis involves higher order nonlinear terms. See Marsden and McCracken [1976], Hassard, Kazarinoff, and Wan [1980], Chow and Hale [1982], Guckenheimer and Holmes [1986], Wiggins [1990], Arrowsmith and Place [1990], Hale and Kocak [1991], and Glendinning [1994] for additional references.

Theorem 6.5.3 (Poincaré-Andronov-Hopf bifurcation) *Let $f(x, \mu)$ in $x' = f(x, \mu)$ be defined on $\Re^2 \times \Re$ with $f(0, \mu) = 0$. Assume that f has continuous third partial derivatives such that the linearization of f in x at $x = 0$, given by $\frac{\partial f}{\partial x}(0, \mu)$, has the eigenvalues*

$$\alpha(\mu) \pm i\beta(\mu),$$

with $\alpha(0) = 0$, $\beta(0) \neq 0$. Moreover, if the eigenvalues cross the pure imaginary axis with nonzero speed, that is,

$$\frac{d}{d\mu}\alpha(\mu)|_{\mu=0} \neq 0,$$

then, for any domain D in $\Re^2$ containing the origin and any given $\mu_0 > 0$, there is a μ^* with $|\mu^*| < \mu_0$ such that the differential equation $x' = f(x, \mu^*)$ has a (nontrivial) periodic orbit in domain D. ♠

Example 6.5.4 (van der Pol's oscillator) Let's apply Theorem 6.5.3 to the equation of van der Pol's oscillator, given by

$$y'' - (2\mu - y^2)y' + y = 0,$$

for a small scalar parameter μ, which is equivalent to

$$\begin{cases} x_1' &= x_2, \\ x_2' &= -x_1 + 2\mu x_2 - x_1^2 x_2. \end{cases} \tag{5.3}$$

The eigenvalues of the linearization at the origin are given by $\mu \pm i\sqrt{1 - \mu^2}$ for μ small. Now, the conditions of Theorem 6.5.3 are satisfied, thus there exist periodic orbits of Eq. (5.3) near the origin for small values of μ. ♠

Although we are not going to prove the Poincaré-Andronov-Hopf bifurcation theorem, we will briefly outline the major steps of how this theorem is derived. This will help determine the stability properties of the bifurcating periodic orbits. According to the reference books Wiggins [1990], Hale and Kocak [1991], and Glendinning [1994], we assume that a series of reductions have been performed, such that $x' = f(x, \mu)$ is transformed to its **normal form**, given by

$$\begin{cases} x_1' = \alpha(\mu)x_1 - \beta(\mu)x_2 + \left[a(\mu)x_1 - b(\mu)x_2\right](x_1^2 + x_2^2) + O(|x_1|^5, |x_2|^5), \\ x_2' = \beta(\mu)x_1 + \alpha(\mu)x_2 + \left[b(\mu)x_1 + a(\mu)x_2\right](x_1^2 + x_2^2) + O(|x_1|^5, |x_2|^5), \end{cases} \tag{5.4}$$

where O (called "big O") denotes the terms of the same or higher orders than the arguments. Similar to Example 6.5.1, Eq. (5.4) is changed to polar coordinates, given by

$$\begin{cases} r' &= \alpha(\mu)r + a(\mu)r^3 + O(r^5), \\ \theta' &= \beta(\mu) + b(\mu)r^2 + O(r^4). \end{cases} \tag{5.5}$$

Next, since we are concerned with the system near $\mu = 0$, we use Taylor expansions for functions in μ about $\mu = 0$ and obtain

$$\begin{cases} r' &= k\mu r + ar^3 + O(\mu^2 r, \ \mu r^3, \ r^5), \\ \theta' &= \beta + c\mu + br^2 + O(\mu^2, \ \mu r^2, \ r^4), \end{cases} \tag{5.6}$$

where $\alpha(0) = 0$ is used in deriving Eq. (5.6) (since we assume that the eigenvalues of the linearization cross the pure imaginary axis when $\mu = 0$), and

$$k = \alpha'(0) \neq 0, \ a = a(0), \ \beta = \beta(0) \neq 0, \ c = \beta'(0), \ b = b(0),$$

(here α' and β' denote the differentiation with respect to μ). The **truncated normal form** of Eq. (5.6) up to order three is given by

$$\begin{cases} r' &= k\mu r + ar^3 = r(\mu k + ar^2), \\ \theta' &= \beta + c\mu + br^2. \end{cases} \tag{5.7}$$

Eq. (5.7) is similar to Eq. (5.2) of Example 6.5.1, thus the results about Eq. (5.2) indicate that for Eq. (5.7), if the first equation in r undergoes a bifurcation and the θ in the second equation goes to $\pm\infty$, then circles will appear, or a Poincaré-Andronov-Hopf bifurcation will take place for Eq. (5.7). Finally, for Eq. (5.6) which contains higher order terms, the results in the aforementioned reference books show that if a Poincaré-Andronov-Hopf bifurcation happens to Eq. (5.7), then it also happens to Eq. (5.6). (This part of dealing with higher order terms is the most difficult step in the proof of the Poincaré-Andronov-Hopf bifurcation theorem.)

The above is an outline of how the Poincaré-Andronov-Hopf bifurcation theorem is derived. Next, we analyze Eq. (5.7) in order to find more details about Poincaré-Andronov-Hopf bifurcations and the stabilities of the bifurcating periodic orbits. First, we have the following result. The proof is left as an exercise.

Lemma 6.5.5 *Consider Eq. (5.7) for sufficiently small μ.*

1. *Assume $a = 0$. Then the origin is a stable spiral point when $\mu k < 0$; a center when $\mu = 0$; and an unstable spiral point when $\mu k > 0$. And a Poincaré-Andronov-Hopf bifurcation will not occur (that is, no periodic orbits when $\mu \neq 0$).*

2. *If $a \neq 0$ and $\frac{\mu k}{a} < 0$, then*

$$\Big(r(t),\, \theta(t)\Big) = \Big(\sqrt{\frac{-\mu k}{a}},\ \ [\beta + c\mu + \frac{-b\mu k}{a}]t + \theta_0\Big) \qquad (5.8)$$

   is a periodic orbit of Eq. (5.7).

3. *The periodic orbit given in (5.8) (when $\frac{\mu k}{a} < 0$) is L-asymptotically stable if $a < 0$; unstable if $a > 0$.*

4. *The origin is a critical point of Eq. (5.7) for all μ. At $\mu = 0$, the origin is asymptotically stable if $a < 0$; unstable if $a > 0$.* ♠

Accordingly, in order for a Poincaré-Andronov-Hopf bifurcation to take place, we must have $a \neq 0$. Also note that if a Poincaré-Andronov-Hopf bifurcation does happen to Eq. (5.7), then, since we need the radius $r > 0$, Eq. (5.7) has a unique periodic orbit having the amplitude $O(\sqrt{|\mu|})$ for each fixed $\mu \neq 0$ and small. Now, if the bifurcating periodic orbit is stable, then the origin is unstable since it repels the solutions to the circle, and vice versa.

Based on these observations, and knowing that the properties concerning bifurcations for Eq. (5.6) and Eq. (5.7) are the same, we can state the Poincaré-Andronov-Hopf bifurcation theorem for Eq. (5.6) in the following format, depending on the signs of the constants $a \neq 0$ and $k \neq 0$.

Theorem 6.5.6 (Poincaré-Andronov-Hopf bifurcation) *Consider the full normal form Eq. (5.6). For μ sufficiently small, one of the following four cases holds.*

1. *$a < 0$ and $k > 0$. Now, for $\mu \leq 0$, the origin is an asymptotically stable critical point; for $\mu > 0$, the origin is an unstable critical point and there exists a L-asymptotically stable periodic orbit. See* **Figure 6.32**.

2. *$a < 0$ and $k < 0$. Now, for $\mu \geq 0$, the origin is an asymptotically stable critical point; for $\mu < 0$, the origin is an unstable critical point and there exists a L-asymptotically stable periodic orbit. See* **Figure 6.33**.

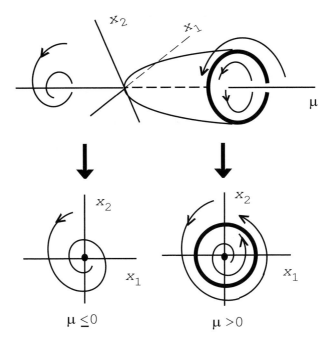

Figure 6.32: Case 1: $a < 0$ and $k > 0$. Supercritical

---

3. *$a > 0$ and $k > 0$. Now, for $\mu \geq 0$, the origin is an unstable critical point; for $\mu < 0$, the origin is an asymptotically stable critical point and there exists an unstable periodic orbit. See* **Figure 6.34**.

4. *$a > 0$ and $k < 0$. Now, for $\mu \leq 0$, the origin is an unstable critical point; for $\mu > 0$, the origin is an asymptotically stable critical point and there exists an unstable periodic orbit. See* **Figure 6.35**. ♠

The first two cases are called "supercritical" because the bifurcating periodic orbits are stable; the last two cases are called "subcritical" because the bifurcating periodic orbits are unstable. The results for the above four cases can be verified for Eq. (5.7), see an exercise.

Next, we point out that hysteresis could also happen to Poincaré-Andronov-Hopf bifurcations, similar to what happened to subcritical pitchfork bifurcations.

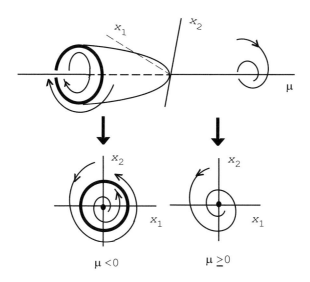

Figure 6.33: Case 2: $a < 0$ and $k < 0$. Supercritical

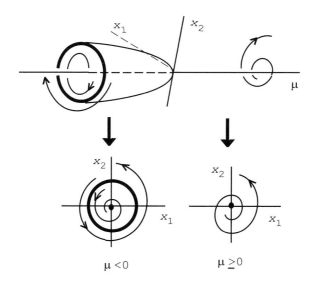

Figure 6.34: Case 3: $a > 0$ and $k > 0$. Subcritical

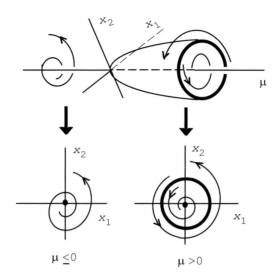

Figure 6.35: Case 4: $a > 0$ and $k < 0$. Subcritical

Example 6.5.7 (Hysteresis) Consider the differential equation in polar coordinates,

$$\begin{cases} r' = \mu r + r^3 - r^5, \\ \theta' = 1, \end{cases} \tag{5.9}$$

where the first equation for r is the same as the equation in Example 6.4.4 for x. Hence the analysis and Figure 6.27 of Example 6.4.4 can be used here, with the difference that now we must have $r \geq 0$. Accordingly, the origin is a stable critical point for $\mu < 0$, and an unstable critical point for $\mu \geq 0$. For $\mu \in (-\frac{1}{4}, 0)$, two circles appear with one stable and the other unstable. See **Figure 6.36**.

As μ increases toward zero, the unstable periodic orbit shrinks to become the origin when $\mu = 0$ and thus makes the origin unstable afterward for $\mu > 0$. At this point the stable periodic orbit is the only periodic orbit left that now attracts all solutions (including even those that were attracted to the origin when $\mu < 0$).

In physics applications, hysteresis effect says that once large-amplitude oscillations have turned on for $\mu \geq 0$, they cannot be turned off by bringing μ back to zero. In fact, when μ is decreasing from positive to negative, the large-amplitude oscillations will be kept on until $\mu = -\frac{1}{4}$, where the stable and unstable periodic orbits collide and then, when $\mu < -\frac{1}{4}$, the large-amplitude oscillations are turned off and the system jumps to the origin, see Figure 6.27. Applications of Poincaré-Andronov-Hopf bifurcations with hysteresis effects can be found in Drazin and Reid [1981], Dowell and Ilgamova [1988], and Strogatz [1994].   ♠

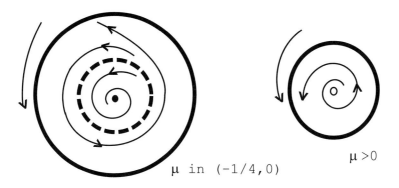

μ in (-1/4,0)   μ >0

Figure 6.36: A Poincaré-Andronov-Hopf bifurcation with hysteresis

---

Exercises 6.5

1. In Example 6.5.1, verify that for $\mu \le 0$, the origin is an asymptotically stable spiral point.

2. In Remark 6.5.2, prove that a periodic orbit $x(t)$ is not asymptotically stable using the definition given in Chapter 5. Then prove that it is asymptotically stable using the definition given in Chapter 5 if other solutions used to compare the differences do not have the same solution curve of $x(t)$. Give some details for the circle $r = \sqrt{\mu}$ in Example 6.5.1.

3. Prove Lemma 6.5.5.

4. Use the analysis in Example 6.5.1 to verify the four cases in Theorem 6.5.6 for Eq. (5.7).

Chapter 7

Chaos

7.1 Introduction

Consider the autonomous differential equation with parameter,

$$x'(t) = f(x(t), \mu), \quad \text{or} \quad x' = f(x, \mu), \tag{1.1}$$

where $(t, x) \in \Re \times \Re^n$ and the parameter μ is in a domain $D_0 \subset \Re^k$, $k \geq 1$. In Chapter 6, we obtained some results concerning bifurcations for one-dimensional and two-dimensional differential equations. For example, we know that when some parameters are varied, critical points and periodic orbits can be created or destroyed, and their stabilities can be changed. In this chapter, we will see that things that are even more "severe" or "worse" than bifurcations can also occur.

First, based on our understanding of differential equations thus far, let's ask what do we "expect" of solutions of differential equations. We know from Chapter 2 that, under some appropriate conditions, the dependence of solutions with respect to initial data and parameters is smooth, so we expect that for the initial values that are nearby, the corresponding solutions on any finite interval should be "reasonably close," or display "similar" qualitative properties. Next, differential equations are used to model the movements of certain systems in applications, thus we expect that the long term properties of these systems can be predicted using the solutions. Briefly, we expect that solutions of differential equations behave in an "orderly fashion."

However, we will see that even for some innocuous looking differential equations, the solutions could behave very "badly," or behave in some very strange and complex fashion. In other words, the solutions that started with the initial values nearby display radically different qualitative properties, and for all practical purposes, prediction of the state of the system is limited to relatively short time intervals since long term prediction is impossible. This situation is nowadays generally described as **chaos**, a word that to the ancient Greeks denoted the infinite formless space that existed before the creation of the universe.

Let's begin with some examples.

Example 7.1.1 Consider the scalar differential equation

$$x' = 2x,$$

whose solutions are given by

$$x(t, 0, x_0) = x_0 e^{2t}, \quad t \geq 0,$$

and $\phi(t) = 0$, $t \geq 0$, is also a solution. Now, for $x_0 \neq 0$ but small, solutions $x(t, 0, x_0)$ and ϕ start with their corresponding initial values x_0 and 0 that are nearby, but their difference,

$$|x(t, 0, x_0) - \phi(t)| = |x_0 e^{2t} - 0| = |x_0| e^{2t},$$

grows to ∞ as $t \to \infty$. Is there anything strange about this system? The answer is no, because even though the small initial discrepancy grows large, we understand the behavior of the solution $x_0 e^{2t}$. That is, the long term behavior of the solution $x_0 e^{2t}$ is **predictable**. (It goes to $\pm\infty$ depends on the sign of x_0.) Therefore, there is nothing strange about this system.   ♠

Next, let's look at some cases that are even "worse" than the bifurcations in Chapter 6. For example, if you continue to squeeze a small piece of straight thin wood, the wood will not only buckle, but will eventually break. Or if you continue to inflate a tire, after a while the tire will burst. Here, "break" or "burst" indicates that the system undergoes a drastic and irreversible qualitative property change, which is more severe than bifurcations. If there is a solution describing the wood or tire, then after it breaks or bursts, the solution even ceases to exist. Another way to think about this is that when the parameters are varied, a system goes from "having orders" to "without orders." The next example will pave a way toward some strange and complex phenomena.

Example 7.1.2 Let x_0 be any fixed number in $[0, 1]$ and consider a recursion relation

$$x_1 = \mu \sin \pi x_0, \; x_2 = \mu \sin \pi x_1, \cdots, \; x_{m+1} = \mu \sin \pi x_m, \quad m = 0, 1, 2, \cdots,$$

where $\mu \in [0, 1]$ is regarded as a parameter. We will demonstrate later that recursion relations are related to differential equations. For $x_0 = 0.5$ and $\mu = 0.7$, a calculation using the software called Maple with the code

```
x[0]:=0.5;
for i from 1 by 1 to 100 do
x[i]:=evalf(0.7*sin(Pi*x[i-1]));
od;
```

gives the following result (note that we use $x[m]$ for x_m because it is what you will see in Maple),

x[1] = .7, x[2] = .566311, x[3] = .684865, x[4] = .585227, x[5] = .675057, x[6] = .596781, x[7] = .667892, x[8] = .604865, x[9] = .662355, x[10] = .610902, x[11] = .657941, x[12] = .615582, x[13] = .654356, x[14] = .619296, x[15] = .651411, x[16] = .622289, x[17] = .648973, x[18] = .624726, x[19] = .646945, x[20] = .626724, ······
x[81] = .636603, x[82] = .636523, x[83] = .636596, x[84] = .636529, x[85] = .636590, x[86] = .636534, x[87] = .636586, x[88] = .636539, x[89] = .636582, x[90] = .636542, x[91] = .636578, x[92] = .636545, x[93] = .636575, x[94] = .636548, x[95] = .636573, x[96] = .636550, x[97] = .636571, x[98] = .636552, x[99] = .636569, x[100] = .636553.

Now, the sequence $\{x_m\}$ will approach $x^* = .6365$, thus we call $x^* = .6365$ an **attractor**, or a **period-1 cycle**. See **Figure 7.1**, where a "corner point" denotes a point (m, x_m), and the straight line segments linking the points are used to guide you see the trend; that is, only the corner points in the figure are meaningful.

For $x_0 = 0.5$ and $\mu = 0.8$, we replace 0.7 in the above code by 0.8 and get

x[1] = .8, x[2] = .470228, x[3] = .796503, x[4] = .477309, x[5] = .797968, x[6] = .474349, x[7] = .797403, x[8] = .475490, x[9] = .797629, x[10] = .475034, x[11] = .797540, x[12] = .475214, x[13] = .797576, x[14] = .475143, x[15] = .797562, x[16] = .475171, x[17] = .797567, x[18] = .475160, x[19] = .797565, x[20] = .475164, ······

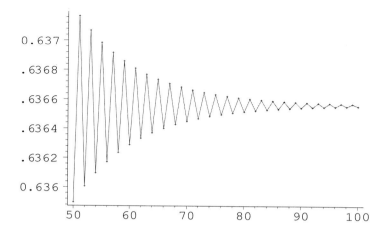

Figure 7.1: An attractor $x^* = .6365$ of the recursion relation $x_{m+1} = 0.7 \sin \pi x_m$

---

x[81] = .797565, x[82] = .475163, x[83] = .797565, x[84] = .475163, x[85] = .797565, x[86] = .475163, x[87] = .797565, x[88] = .475163, x[89] = .797565, x[90] = .475163, x[91] = .797565, x[92] = .475163, x[93] = .797565, x[94] = .475163, x[95] = .797565, x[96] = .475163, x[97] = .797565, x[98] = .475163, x[99] = .797565, x[100] = .475163.

Now,

$$\{x_1^* = .475163, \ x_2^* = .797565\}$$

consists of two attractors, or it forms a **period-2 cycle** since $\{x_m\}$ oscillate and approach each of the two numbers after every two iterations, see **Figure 7.2**.

So far, the sequences of iterations are still behaving in an orderly fashion and the properties of the iterations can be predicted. Next, we will see that for some other parameter μ values, there are no orders at all. We still take $x_0 = 0.5$ but use $\mu = 0.95$, then we get

x[1] = .95, x[2] = .148612, x[3] = .427597, x[4] = .925530, x[5] = .220232, x[6] = .606088, x[7] = .897723, x[8] = .300019, x[9] = .768599, x[10] = .631374, x[11] = .870229, x[12] = .376661, x[13] = .879571, x[14] = .350907, x[15] = .847681, $\cdots\cdots$

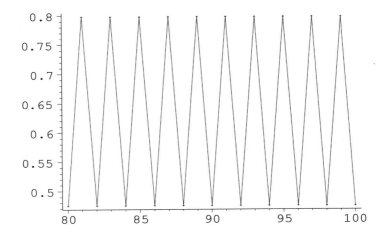

Figure 7.2: Attractors $x_1^* = .475163$ and $x_2^* = .797565$ of the recursion relation $x_{m+1} = 0.8 \sin \pi x_m$

---

x[81] = .497496, x[82] = .949970, x[83] = .148699, x[84] = .427828, x[85] = .925685, x[86] = .219782, x[87] = .605052, x[88] = .898730, x[89] = .297167, x[90] = .763567, x[91] = .642518, x[92] = .856359, x[93] = .414295, x[94] = .915772, x[95] = .248453, x[96] = .668480, x[97] = .820004, x[98] = .509025, x[99] = .949618, x[100] = .149738.

Now, unlike the first two cases, the iterations $\{x_m\}$ never settle down to a single attractor or a periodic cycle. Or, there are **no orders** and the behavior of the iterations is **unpredictable**. In fact, if we plot more points of (m, x_m), then it seems that the x_m values will nearly cover all numbers of some x-interval in $[0, 1]$, see **Figure 7.3**. In an exercise, you are asked to start with some initial values that are nearby for $x_{m+1} = 0.95 \sin \pi x_m$ and then compare how different the sequences become after some iterations. A good question to ponder is how could these strange and complex things happen to simple and innocuous looking systems, such as $x_{m+1} = \mu \sin \pi x_m$. ♠

To summarize, when some parameters are varied, such as μ in $x_{m+1} = \mu \sin \pi x_m$, certain qualitative properties of a system may undergo some drastic (sometimes irreversible) changes that are more severe than bifurcations. The changes will cause the system to go from "having orders" to "without

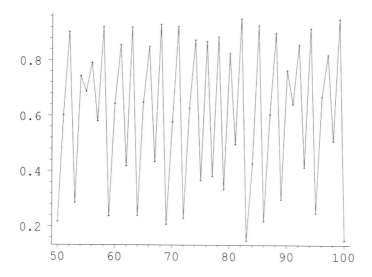

Figure 7.3: Placements of $x_{m+1} = 0.95 \sin \pi x_m$ showing no orders and unpredictability

orders," or to a situation that is unpredictable and does not fit any conventional types. This situation is generally called **chaos**.

The above reveals some chaotic behavior for discrete systems (iterations). For continuous systems (that is, differential equations), if they are in $\Re$ or $\Re^2$, then based on what we have seen for scalar or planar equations, we do expect that the trajectories in $\Re$ or $\Re^2$ behave orderly (this will be discussed further in Chapter 8). However, for continuous systems in $\Re^n$, $n \geq 3$, the trajectories will be in spaces of at least three dimensions. Therefore they could get twisted and twisted and become complex beyond recognition.

A famous example in $\Re^3$, along with its important contributions, is given by the **Lorenz system**,

$$\begin{cases} \frac{dx}{dt} &= 10(y - x), \\ \frac{dy}{dt} &= 28x - y - xz, \\ \frac{dz}{dt} &= xy - (8/3)z, \end{cases} \qquad (1.2)$$

which is used to model the weather forecast (additional details will be given later). Again, despite of its innocuous looks, the numerical experiments in a milestone paper of Lorenz [1963] show that the (x, z) plane projection of a three-dimensional trajectory of Eq. (1.2) is given in **Figure 7.4**.

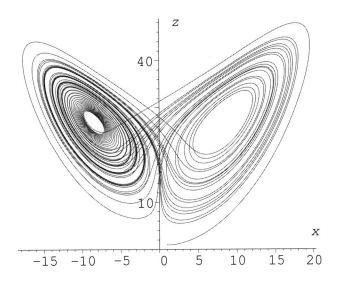

Figure 7.4: The (x, z) plane projection of a three-dimensional trajectory of the Lorenz system (1.2)

The trajectory in Figure 7.4 does not intersect itself in $\Re^3$, so the crossings in Figure 7.4 are the result of projection in $\Re^2$. Moreover, the trajectory will cruise a few circuits on one side, suddenly move to the other side, cruise a few circuits, and then suddenly move back – this process will continue forever, in a way that the trajectory will wind around the two sides infinitely many times without ever settling down. In addition, the way the trajectory moves around the two sides is unpredictable. Nowadays, the Lorenz system has become an icon of nonlinear dynamical systems due to its rich source of examples of various types of bifurcations and chaos that could occur in these systems. Lorenz showed that the system (1.2) has an attractor whose properties are so strange and complex, hence it is called a **strange attractor**. This is a very important subject in the study of chaos.

This chapter is organized as follows: In Section 2, we study recursion relations, also called maps, and their bifurcation properties by finding the similarities to the bifurcations of critical points of differential equations, hence the results in Chapter 6 can be carried over. In Section 3, we look at a phenomenon called period-doubling bifurcations cascade, which provides a route to chaos. In Section 4, we introduce some universality results concerning one-dimensional maps. In Section 5, we study some properties of the Lorenz system and introduce the notion of strange attractors. In Section 6, we study the Smale horseshoe which provides an example of a strange invariant set possessing chaotic dynamics.

Exercises 7.1

1. For $x_{m+1} = 0.95 \sin \pi x_m$ in Example 7.1.2, start with some initial values that are nearby and then compare how different the sequences become after some iterations.

2. Use the Maple software to run the iterations for $x_{m+1} = \mu \sin \pi x_m$ with $x_0 = 0.5$ and $\mu = 0.5, 0.65, 0.75, 0.85, 0.93, 0.935, 0.94, 0.945$.

3. Consider a recursion relation $x_{m+1} = \mu x_m (1 - x_m)$. Run the iterations with $x_0 = 0.5$ and $\mu = 2.9, 3, 3.05, 3.45, 3.5, 3.56, 3.7, 3.83, 4$.

7.2   Maps and Their Bifurcations

Recall in Chapter 2 we utilized Euler's method to prove the existence of solutions of

$$x' = f(x), \quad x(t_0) = x_0, \quad t \in [t_0, b), \quad b \le \infty, \tag{2.1}$$

(where we did not assume a Lipschitz condition, thus uniqueness was not guaranteed). To do so, we divide the interval $[t_0, b)$ into smaller intervals $[t_0, t_1], [t_1, t_2], \cdots, [t_m, t_{m+1}], \cdots$, with $h = t_{m+1} - t_m$ being a constant for $m \ge 0$, called the **step size**. Then the key idea of Euler's method is that on every small interval $[t_m, t_{m+1}]$, $m = 0, 1, 2, \cdots$, the derivative

$$x'(t) \text{ is approximated by } \frac{x(t_{m+1}) - x(t_m)}{h}. \tag{2.2}$$

Hence, Eq. (2.1) is approximated by

$$\frac{x(t_{m+1}) - x(t_m)}{h} = f(x(t_m)), \quad m = 0, 1, 2, \cdots. \tag{2.3}$$

Therefore, based on Euler's method, if we define the approximate value of x at t_m to be

$$x_m = x(t_m), \tag{2.4}$$

then from (2.3), we obtain a **recursion relation**

$$x_{m+1} = x_m + hf(x_m), \tag{2.5}$$

also called a **difference equation**, which is a **discretized version** of the differential equation (2.1). Now, for a given value x_0, the sequence $\{x_0, x_1, x_2, \cdots, x_m, \cdots\}$ determined by Eq. (2.5) is called a **solution of the difference equation (2.5) with the initial value** x_0.

With this background, we can generalize the ideas mentioned above and, in general, call a recursion relation (or a difference equation)

$$x_{m+1} = M(x_m), \quad m = 0, 1, 2, \cdots \tag{2.6}$$

as a **map**. If a map involves a parameter μ, then it is written as

$$x_{m+1} = M(x_m, \mu), \quad m = 0, 1, 2, \cdots \tag{2.7}$$

and for a given value x_0, the sequence $\{x_0, x_1(\mu), x_2(\mu), \cdots, x_m(\mu), \cdots\}$ determined by Eq. (2.7) is a **solution of the map (2.7) with the initial value** x_0. Since maps are originated from differential equations, the concepts of critical points and their stabilities and bifurcations for differential equations can be, and should be, extended to maps.

In this regard, we analyze (2.2) and find that the condition for a critical point of a differential equation, that is,

$$x'(t) = 0, \quad t \geq t_0, \tag{2.8}$$

is now translated to

$$x(t_{m+1}) = x(t_m), \quad m \geq 0. \tag{2.9}$$

Using the map in (2.6), we see that (2.4) and (2.9) imply

$$M(x_m) = x_{m+1} = x(t_{m+1}) = x(t_m) = x_m, \quad m \geq 0,$$

which is (see an exercise) equivalent to

$$M(x_0) = x_0. \tag{2.10}$$

Accordingly, we make the following definition.

Definition 7.2.1 *A point x^* in the domain of a map M is said to be a* **fixed point** *of M if $M(x^*) = x^*$; or x^* is a solution of $M(x) - x = 0$.*

Now, based on the relationship between (2.8) and (2.10), we find that **a
critical point for a differential equation is translated to become a
fixed point of a map**. Therefore, we are going to carry the study of critical
points of differential equations to the study of fixed points of maps and their
stabilities and bifurcations. Again, due to the close relationship between
differential equations and maps, the definitions of stabilities for fixed points
of maps are similar to those for critical points of differential equations. For
example, Definition 5.1.1 in Chapter 5 concerning stabilities for differential
equations now reads as follows.

Definition 7.2.2 *Let x^* be a fixed point of the map M in (2.6).*

(a). *x^* is said to be* **stable** *(or* **uniformly stable** *because (2.6) is autonomous) if for any $\varepsilon > 0$, there exists a $\delta = \delta(\varepsilon) > 0$, such that
$|x_0 - x^*| \leq \delta$ implies $|x_m - x^*| \leq \varepsilon$ for $m \geq 0$, where $x_{m+1} =
M(x_m)$, $m \geq 0$.*

(b). *x^* is said to be* **asymptotically stable** *(or* **uniformly asymptotically stable***) if it is stable and in addition, there exists an (indepen-
dent constant) $r > 0$ such that $|x_0 - x^*| \leq r$ implies $\lim_{m \to \infty} |x_m - x^*| =
0$, where $x_{m+1} = M(x_m)$, $m \geq 0$.*

(c). *x^* is said to be* **unstable** *if it is not stable.*

For bifurcations of fixed points of maps, we should understand them in
the same way as we did for those of critical points of differential equations.
That is, we let the parameter μ in the map (2.7) to vary, and then look
at how fixed points are created and destroyed, and how their stabilities are
changed. Despite the similarities between critical points and fixed points,
we point out that solutions of scalar differential equations flow monotonically on the x-axis, showing simple phase portraits. However, a solution
of a scalar map, $\{x_0, x_1, x_2, \cdots\}$, is a set of **discrete values** on the x-axis
that can **jump back and forth**. So we anticipate that phase portraits, or
placements of solution sequences of maps, will be much more complex than
phase portraits of differential equations. In fact, maps, because of their discrete structures, even in one-dimensional form, can exhibit some fascinating
phenomena leading to chaos.

Next, let's introduce the so-called **cobweb** method, which is very effective in visualizing scalar maps in geometry. Let $M(x)$ be a scalar map
and start with x_0 on the x-axis. We draw the graph of $y = M(x)$ and the

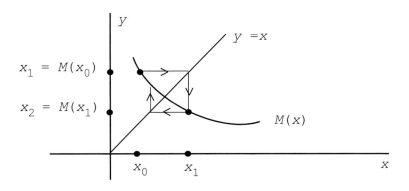

Figure 7.5: Cobwebs for a scalar map $x_{m+1} = M(x_m)$

straight line $y = x$ in the (x, y) plane. Then $x_1 = M(x_0)$ is on the y-axis and $(x_0, M(x_0))$ is on the graph of $y = M(x)$, see **Figure 7.5**.

Now, to perform the second iteration $x_2 = M(x_1)$, we should regard $x_1 = M(x_0)$ as a value on the x-axis. To visualize this, we draw a horizontal straight line from $(x_0, x_1) = (x_0, M(x_0))$ such that the x-component of the intersection of this horizontal straight line and $y = x$ is x_1. Now, use x_1 as the x value to start the second iteration $x_2 = M(x_1)$ such that $(x_1, M(x_1))$ is on the graph of $y = M(x)$. This process can be continued to create something looks like "cobwebs" after we use vertical lines to link points on the graphs of $y = M(x)$ and $y = x$. See Figure 7.5.

Example 7.2.3 Consider the scalar map

$$M(x) = 0.5x, \quad \text{or} \quad x_{m+1} = M(x_m) = 0.5x_m, \quad m = 0, 1, 2, \cdots. \quad (2.11)$$

The cobwebs for the map (2.11) are given in **Figure 7.6**.

In this case, the fixed points are from

$$M(x) - x = 0, \quad \text{or} \quad 0.5x - x = 0,$$

thus $x = 0$ is the only fixed point. Moreover, from Figure 7.6, any solution $\{x_0, x_1, x_2, \cdots\}$ of the map approaches the fixed point $x = 0$, or the fixed point $x = 0$ is asymptotically stable.

Consider the scalar map

$$M(x) = 3x, \quad \text{or} \quad x_{m+1} = M(x_m) = 3x_m, \quad m = 0, 1, 2, \cdots. \quad (2.12)$$

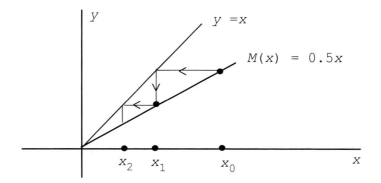

Figure 7.6: Cobwebs for the map $M(x) = 0.5x$

The cobwebs for the map (2.12) are given in **Figure 7.7**.

In this case, $x = 0$ is also the only fixed point. And from Figure 7.7, any solution of the map gets repelled by the fixed point $x = 0$, or the fixed point $x = 0$ is unstable.

Consider the scalar map

$$M(x) = -x, \quad \text{or} \quad x_{m+1} = M(x_m) = -x_m, \quad m = 0, 1, 2, \cdots. \tag{2.13}$$

The cobwebs for the map (2.13) are given in **Figure 7.8**.

In this case, $x = 0$ is also the only fixed point. And from Figure 7.8, for any solution, the distance of each iteration of the map and $x = 0$ is a constant, thus the fixed point $x = 0$ is stable.

Consider the scalar map

$$M(x) = x + x^3, \quad \text{or} \quad x_{m+1} = M(x_m) = x_m + x_m^3, \quad m = 0, 1, 2, \cdots. \tag{2.14}$$

The cobwebs for the map (2.14) are given in **Figure 7.9**.

Again, $x = 0$ is also the only fixed point. And from Figure 7.9 (see an exercise), any solution of the map gets repelled by the fixed point $x = 0$, or the fixed point $x = 0$ is unstable.   ♠

From Example 7.2.3, we find the following interesting things:

1. The fixed point $x = 0$ is asymptotically stable for the map (2.11), where $M'(0) = 0.5 < 1$.

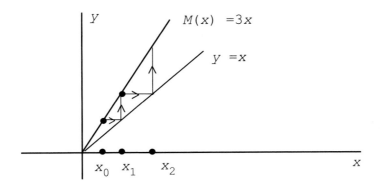

Figure 7.7: Cobwebs for the map $M(x) = 3x$

---

2. The fixed point $x = 0$ is unstable for the map (2.12), where $M'(0) = 3 > 1$.

3. The solutions of the maps (2.11) and (2.12) are **monotone**, that is, the iterations are increasing or decreasing monotonically on one side of the fixed point $x = 0$, where $M'(0) > 0$.

4. The solutions of the map (2.13) jump alternately to both sides of the fixed point $x = 0$, where $M'(0) < 0$. (This can never happen to solutions of scalar differential equations because they flow monotonically on the x-axis to critical points or $\pm\infty$.)

5. According to the results of the maps (2.13) and (2.14), stability properties of the fixed point $x = 0$ cannot be determined when $|M'(0)| = 1$.

Next, let's try to confirm these discoveries in Example 7.2.3 for general cases. We will restrict our analysis to scalar continuously differentiable maps $M(x)$ defined on some closed interval $[a, b]$. That is, we assume $M : [a, b] \to [a, b]$ and $M'(x)$ is continuous, even though some results are also true for other cases or higher order dimensions. Note that the reason we require $M : [a, b] \to [a, b]$ is for the iterations $x_{m+1} = M(x_m)$ to be well defined.

First, let's analyze the stabilities of a fixed point $x^* = M(x^*)$, that is, the properties of $|x_m - x^*|$ for a solution sequence $\{x_0, x_1, x_2, \cdots, x_m, \cdots\}$. Assume that $|M'(x^*)| < 1$, then there is a constant $d > 0$ such that

$$p = \max_{|x-x^*| \le d} |M'(x)| < 1. \tag{2.15}$$

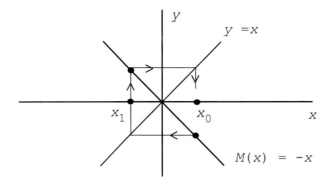

Figure 7.8: Cobwebs for the map $M(x) = -x$

Now, assume $|x_0 - x^*| \leq d$. Then from the mean value theorem, we have

$$x_1 - x^* = M(x_0) - x^* = M(x_0) - M(x^*) = M'(c)[x_0 - x^*], \qquad (2.16)$$

where $|c - x^*| \leq d$. Thus, (2.15) and (2.16) imply

$$|x_1 - x^*| \leq |M'(c)||x_0 - x^*| \leq p|x_0 - x^*| \leq |x_0 - x^*| \leq d, \qquad (2.17)$$

which means x_1 is still in the range of $|x - x^*| \leq d$. Thus, this procedure can be repeated to conclude that $|x_k - x^*| \leq d$, $k = 1, 2, \cdots$. Hence, from (2.15),

$$
\begin{aligned}
|x_m - x^*| &= |M(x_{m-1}) - M(x^*)| \leq p|x_{m-1} - x^*| \leq p^2|x_{m-2} - x^*| \\
&\leq \cdots \\
&\leq p^m|x_0 - x^*| \qquad (2.18)
\end{aligned}
$$

(which can be verified by an induction). As $0 \leq p < 1$, one has $p^m \to 0$, $m \to \infty$. Thus

$$|x_m - x^*| \leq p^m|x_0 - x^*| \to 0, \ m \to \infty.$$

Therefore, the fixed point x^* is asymptotically stable.

To determine if a solution of a map is monotone, we note from the mean value theorem that

$$x_m - x_{m-1} = M(x_{m-1}) - M(x_{m-2}) = M'(c)[x_{m-1} - x_{m-2}], \ c \in [a, b].$$

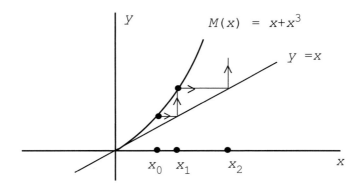

Figure 7.9: Cobwebs for the map $M(x) = x + x^3$

Thus, if $M'(x) > 0$ for $x \in [a, b]$, then $x_m - x_{m-1}$ will have the same sign as that of $x_1 - x_0$. Also, we find from (2.16) that if $M'(x) < 0$ for $x \in [a, b]$, then $x_m - x^*$ will have the opposite sign as that of $x_{m-1} - x^*$. If we combine all these analyses, we obtain the following result. The proof is left as an exercise.

Lemma 7.2.4 *Let $M : [a, b] \to [a, b]$ be a scalar continuously differentiable map with a fixed point x^*.*

1. *x^* is asymptotically stable if $|M'(x^*)| < 1$.*

2. *x^* is unstable if $|M'(x^*)| > 1$.*

3. *The derivative test for the stabilities of the fixed point x^* is inconclusive when $|M'(x^*)| = 1$.*

4. *If $M'(x) > 0$ for $x \in [a, b]$, then any solution of the map (2.6) is a monotone sequence. In this case, the map M is called a* **monotone map***.*

5. *If $M'(x) < 0$ for $x \in [a, b]$, then any solution of the map (2.6) with $x_0 \neq x^*$ will jump alternately to both sides of the fixed point x^*.* ♠

Now, let's look at bifurcations of the fixed points of the map (2.7), which are the solutions of

$$M(x, \mu) = x, \quad \text{or} \quad M(x, \mu) - x = 0. \tag{2.19}$$

That is, we analyze how the fixed points of the map (2.7) are created or destroyed, and how their stabilities are changed when the parameter μ is varied. Note in geometry, (2.19) means that fixed points are the x components of intersections of the graphs $y = M(x, \mu)$ and $y = x$ in the (x, y) plane, see **Figure 7.10**.

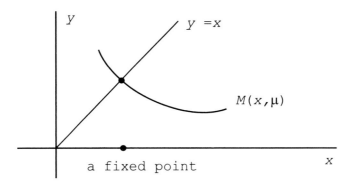

Figure 7.10: Fixed points are the solutions of $M(x, \mu) = x$

Observe that if we define

$$f(x, \mu) = M(x, \mu) - x,$$

then, finding fixed points of the map (2.7) is equivalent to finding solutions of

$$f(x, \mu) = 0,$$

which is the same as finding critical points of the differential equation

$$x' = f(x, \mu).$$

Now, we look at some examples that reveal the connection between bifurcations for critical points and fixed points.

Example 7.2.5 Consider the map

$$M(x, \mu) = x + \mu - x^2, \tag{2.20}$$

from which we find that $M(x, \mu) - x = 0$ is equivalent to $\mu - x^2 = 0$. Thus, the corresponding results from Chapter 6 show that the critical points of

$x' = \mu - x^2$, or the fixed points of M, undergo a saddle-node bifurcation at $(x, \mu) = (0, 0)$. See **Figure 7.11**, where the stabilities are determined from $\frac{\partial M}{\partial x} = (1 - 2x)|_{x=\pm\sqrt{\mu}} = 1 \mp 2\sqrt{\mu}$ when $\mu > 0$. ♠

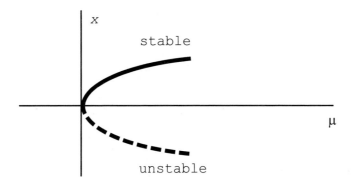

Figure 7.11: A saddle-node bifurcation for $M(x, \mu) = x + \mu - x^2$

Example 7.2.6 Consider the map

$$M(x, \mu) = x + \mu x - x^2. \tag{2.21}$$

Then the fixed points of M are the solutions of $M(x, \mu) - x = \mu x - x^2 = 0$, which, using the corresponding results in Chapter 6, undergo a transcritical bifurcation at $(x, \mu) = (0, 0)$. See **Figure 7.12**, where the stabilities are determined from $\frac{\partial M}{\partial x} = 1 + \mu - 2x$ at $x = 0$ and $x = \mu$. ♠

Example 7.2.7 Consider the map

$$M(x, \mu) = x + \mu x - x^3. \tag{2.22}$$

Then the fixed points of M are the solutions of $M(x, \mu) - x = \mu x - x^3 = x(\mu - x^2) = 0$, which, from Chapter 6 again, undergo a pitchfork bifurcation at $(x, \mu) = (0, 0)$. See **Figure 7.13**, where the stabilities are determined from $\frac{\partial M}{\partial x} = (1 + \mu - 3x^2)|_{x=\pm\sqrt{\mu}} = 1 - 2\mu$ when $\mu > 0$. ♠

From the above three examples, we find that bifurcations for maps can be handled in a similar way as those for differential equations. Therefore,

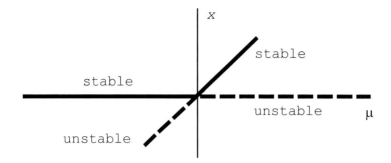

Figure 7.12: A transcritical bifurcation for $M(x, \mu) = x + \mu x - x^2$

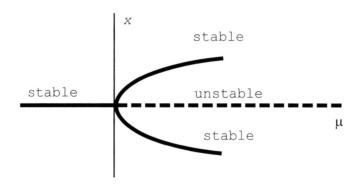

Figure 7.13: A pitchfork bifurcation for $M(x, \mu) = x + \mu x - x^3$

the results in Chapter 6 concerning bifurcations for differential equations can be applied here to derive the results concerning bifurcations for maps. Consequently, we have the following results. The proofs are left as exercises.

Theorem 7.2.8 (Saddle-node bifurcation) *Let $M(x, \mu)$ in the map (2.7) be a scalar function defined on $\Re^2$ with continuous second partial derivatives such that*

$$M(0,0) = 0, \quad \frac{\partial M}{\partial x}(0,0) = 1, \quad \frac{\partial^2 M}{\partial x^2}(0,0) \neq 0, \quad \frac{\partial M}{\partial \mu}(0,0) \neq 0. \quad (2.23)$$

Then the map (2.7) undergoes a saddle-node bifurcation at $(x, \mu) = (0, 0)$. ♠

Theorem 7.2.9 (Transcritical bifurcation) *Let $M(x, \mu)$ in the map (2.7) be a scalar function defined on $\Re^2$ with continuous second partial derivatives such that*

$$M(0, \mu) = 0, \ \mu \in \Re, \quad \frac{\partial M}{\partial x}(0, 0) = 1, \quad \frac{\partial^2 M}{\partial x^2}(0, 0) \neq 0,$$

$$\frac{\partial M}{\partial \mu}(0, 0) = 0, \quad \frac{\partial^2 M}{\partial \mu^2}(0, 0) = 0, \quad \frac{\partial^2 M}{\partial x \partial \mu}(0, 0) \neq 0. \qquad (2.24)$$

Then the map (2.7) undergoes a transcritical bifurcation at $(x, \mu) = (0, 0)$. ♠

Theorem 7.2.10 (Pitchfork bifurcation) *Let $M(x, \mu)$ in the map (2.7) be a scalar function defined on $\Re^2$ with $M(x, \mu) - x = xg(x, \mu)$ and g has continuous second partial derivatives such that*

$$g(0, 0) = 0, \quad \frac{\partial g}{\partial x}(0, 0) = 0, \quad \frac{\partial^2 g}{\partial x^2}(0, 0) \neq 0, \quad \frac{\partial g}{\partial \mu}(0, 0) \neq 0. \qquad (2.25)$$

Then the map (2.7) undergoes a pitchfork bifurcation at $(x, \mu) = (0, 0)$. ♠

Exercises 7.2

1. Verify that if $M(x) = x$, then $M(M(x)) = x$.

2. For $x_{m+1} = M(x_m)$, show that $M(x_m) = x_m$, $m \geq 0$, if and only if $M(x_0) = x_0$.

3. Examine the map defined by $M(x) = 1$, $x \in \Re$.

4. Consider the map $M(x) = x$. Find the fixed points and determine their stabilities.

5. Verify that Figure 7.9 is the cobwebs for the map (2.14).

6. Verify (2.18) using an induction.

7. Prove Lemma 7.2.4.

8. Determine the stabilities for the fixed points in Example 7.2.5.

9. Determine the stabilities for the fixed points in Example 7.2.6.

10. Determine the stabilities for the fixed points in Example 7.2.7.

11. Prove Theorem 7.2.8.

12. Prove Theorem 7.2.9.

13. Prove Theorem 7.2.10.

7.3   Period-Doubling Bifurcations: Route to Chaos

From Lemma 7.2.4(5) of the previous section, we know that if x^* is a fixed point of a map M and $\frac{dM}{dx} < 0$, then a solution sequence (different from x^*) of the map will jump alternately to both sides of the fixed point x^*. Now, if x^* is unstable, then the solution cannot approach x^*. Next, if the solution is a bounded sequence, then the odd iterates, which are on one side of x^*, will converge to a point $x^+ \neq x^*$. Therefore the even iterates, which are on the other side of x^*, will converge to the point $M(x^+)$. In this case, we have $x^+ = \lim_{m\to\infty} x_{2m+1}$ and $M(x^+) \neq x^+$ (since $M(x^+)$ and x^+ are on different sides of x^*), and

$$
\begin{aligned}
M(M(x^+)) &= M(M(\lim_{m\to\infty} x_{2m+1})) = \lim_{m\to\infty} M(M(x_{2m+1})) \\
&= \lim_{m\to\infty} x_{2(m+1)+1} = x^+.
\end{aligned}
\tag{3.1}
$$

Thus,

$$
\begin{aligned}
&\{x^+, \ M(x^+), \ M(M(x^+)), \ M(M(M(x^+))), \cdots\} \\
&= \{x^+, \ M(x^+), \ x^+, \ M(x^+), \cdots\}
\end{aligned}
\tag{3.2}
$$

is a solution that takes the two values x^+ and $M(x^+)$ alternately. Accordingly, (3.2) is called a **solution of period 2**, or a **period-2 cycle** or a **period-2 orbit**.

Based on the above notations, we define

$$
M^{[2]}(x) = M(M(x)), \ \cdots, \ M^{[k]}(x) = M(M^{[k-1]}(x)), \ \cdots
$$

and we have the following definition.

Definition 7.3.1 *A point $x^\star$ is said to be a **period-k point** of a map M if $M^{[j]}(x^\star) \neq x^\star$, $j = 1, 2, \cdots, k-1$, $M^{[k]}(x^\star) = x^\star$. The set $\{x^\star, M(x^\star), M^{[2]}(x^\star), \cdots, M^{[k-1]}(x^\star)\}$ of a period-k point $x^\star$ is said to be a **period-k cycle**, or a **period-k orbit** of the map M.*

For example, x^+ (or $M(x^+)$) in (3.2) is a period-2 point, and $\{x^+, M(x^+)\}$ is a period-2 cycle. Since a period-k point of a map M is a fixed point of the map $M^{[k]}$, its stabilities are defined in the same way as those for fixed points of the map $M^{[k]}$.

Definition 7.3.2 *A period-k point of a map M is said to be stable, asymptotically stable, or unstable if it is, respectively, a stable, an asymptotically stable, or an unstable fixed point of the map $M^{[k]}$.*

Accordingly, the derivative $\frac{d}{dx}M^{[k]}(x)$ and Lemma 7.2.4 can be used to test the stabilities of period-k points of a map M. If a bifurcation happens to a map M in such a way that period-2 points appear, then the bifurcation is called a **period-2 bifurcation**, or a **period-doubling bifurcation**. Some texts also call a period-2 bifurcation as a **flip bifurcation** to mean that the map M will "flip" alternately around the two sides of some value. A typical period-doubling bifurcation diagram is shown in **Figure 7.14**, which looks like a pitchfork bifurcation, but they mean different things. That is, in Figure 7.14, one branch is x and the other is $M(x)$.

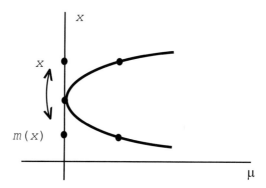

Figure 7.14: A typical period-doubling bifurcation diagram

Next, we study period-doubling bifurcations. In order for a map M to jump alternately, we use Lemma 7.2.4 and assume $\frac{dM}{dx}(x_0) < 0$. Observe that if $\frac{dM}{dx}(x_0) \in (-\infty, -1) \cup (-1, 0)$, then the stabilities of x_0 will be preserved under small perturbations to the map M and to the point x_0. Thus interesting things may occur near x_0 when $\frac{dM}{dx}(x_0) = -1$.

Because period-doubling bifurcations of a map M is related to bifurcations of fixed points of the map $M^{[2]}$, thus, when $M^{[2]}$ is regarded as a general map, Theorems 7.2.8, 7.2.9, and 7.2.10 can be applied to $M^{[2]}$ and obtain results concerning period-doubling bifurcations. For example, we have the following result. The proof is left as an exercise.

Theorem 7.3.3 (Period-Doubling Bifurcation) *Let $M(x,\mu)$ in the map (2.7) be a scalar function defined on $\Re^2$ with continuous partial derivatives and $M(0,\mu) = 0$, $\mu \in \Re$, $\frac{\partial M}{\partial x}(0,0) = -1$, $M^{[2]}(x,\mu) - x = xg(x,\mu)$ and g has continuous second partial derivatives such that*

$$g(0,0) = 0, \quad \frac{\partial g}{\partial x}(0,0) = 0, \quad \frac{\partial^2 g}{\partial x^2}(0,0) \neq 0, \quad \frac{\partial g}{\partial \mu}(0,0) \neq 0. \qquad (3.3)$$

Then the map $M^{[2]}(x,\mu)$ undergoes a pitchfork bifurcation and the map $M(x,\mu)$ undergoes a period-doubling bifurcation at $(x,\mu) = (0,0)$. ♠

Observe that in general, a fixed point of $M^{[2]}$ may also be a fixed point of M. However, we point out that under the conditions in Theorem 7.3.3, the bifurcating nonzero fixed points of $M^{[2]}$ from Theorem 7.3.3 cannot be fixed points of M. That is, there is a $\delta > 0$ such that there is no nonzero fixed point of M for $|x| \leq \delta$ and $|\mu| \leq \delta$. Otherwise, there are $x_m \neq 0$ and $\mu_m \neq 0$ with $x_m \to 0$ and $\mu_m \to 0$, $m \to \infty$, such that $M(x_m, \mu_m) = x_m$. From $\frac{\partial M}{\partial x}(0,0) = -1$ and the continuity of $\frac{\partial M}{\partial x}$, we get

$$\frac{M(x_m, \mu_m) - M(0, \mu_m)}{x_m} \longrightarrow -1, \quad m \to \infty.$$

But

$$\frac{M(x_m, \mu_m) - M(0, \mu_m)}{x_m} = \frac{M(x_m, \mu_m) - 0}{x_m} = \frac{x_m}{x_m} = 1,$$

a contradiction.

After these remarks, we are ready to analyze a very important example, the **logistic map**, given by

$$x \to M(x,\mu) = \mu x(1 - x), \qquad (3.4)$$

which is nowadays regarded as an icon of bifurcations and chaos. The discrete map (3.4) is related to the logistic equation for population growth introduced in Chapter 1. Due to the map's innocuous appearance (3.4), especially after seeing how easy the logistic differential equation is handled in Chapter 1, you probably think that the analysis of the logistic map is also easy. Then just wait: A whole new world is going to unfold.

In an influential article by May [1976], where the map (3.4) is analyzed in detail, he plead "for the introduction of these difference equations into elementary mathematics courses, so that students' intuition may be enriched by seeing the wild things that simple nonlinear equations can do."

The logistic map (3.4), which has become a subject of great interest recently, has benefited from the advancement in modern computer technology. This is evidenced by a surge of research articles in this area in the 1970s. Since "analysis" becomes extremely difficult and unmanageable at certain stages, numerical solutions can lend a helping hand.

In the map (3.4), we assume $x \in [0, 1]$. The maximum value of $x(1-x)$ is $\frac{1}{4}$ at $x = \frac{1}{2}$, hence, to be able to continue the iterations, we assume $0 \leq \mu \leq 4$. Next, the fixed points of the logistic map are the solutions of

$$\mu x(1-x) - x = x(\mu - \mu x - 1),$$

which are given by

$$x^* = 0, \quad x_\mu^* = 1 - \frac{1}{\mu}. \tag{3.5}$$

Now, in order to use $x_\mu^* = 1 - \frac{1}{\mu}$ to denote a fixed point in $[0, 1]$ that is different from $x^* = 0$, we further assume $\mu \in (1, 4]$, see **Figure 7.15**.

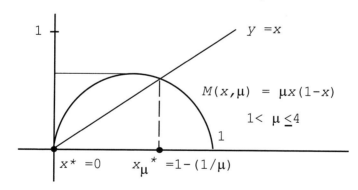

Figure 7.15: Fixed points $x^* = 0$ and $x_\mu^* = 1 - \frac{1}{\mu}$ of the logistic map $\mu x(1-x)$ with $1 < \mu \leq 4$

Since

$$\frac{\partial}{\partial x} M(x, \mu) = \mu(1 - 2x),$$

we have

$$\frac{\partial M}{\partial x}(x^*, \mu) = \frac{\partial M}{\partial x}(0, \mu) = \mu, \quad \frac{\partial M}{\partial x}(x_\mu^*, \mu) = 2 - \mu.$$

Therefore, the fixed point $x^* = 0$ is unstable since we assume $\mu > 1$; the fixed point $x_\mu^* = 1 - \frac{1}{\mu}$ is stable if $1 < \mu < 3$ (where $-1 < 2 - \mu < 1$) and unstable if $3 < \mu \le 4$ (where $2 - \mu < -1$). Note that $2 - \mu > 1$ cannot happen because we assume $\mu > 1$.

Because the stabilities of x_μ^* are different on the two sides of $\mu = 3$, we ask: **What is going to happen when μ is increased through** 3? At $\mu = 3$, $\frac{\partial M}{\partial x}(x_\mu^*, \mu) = 2 - 3 = -1$, thus from the conditions in Theorem 7.3.3, it is a hint that a period-doubling bifurcation might occur. Another reason why a period-doubling bifurcation might occur is that, now, the fixed point $x_\mu^* = 1 - \frac{1}{\mu}$ will lose its stability as soon as $\mu > 3$, so that any solution sequence cannot approach either of the two fixed points $x^* = 0$ and $x_\mu^* = 1 - \frac{1}{\mu}$. But the solution sequence is in $[0, 1]$ and hence is bounded, and, for $\mu \approx 3$, the solution sequence flips on the two sides of x_μ^* since $\frac{\partial M}{\partial x}(x_\mu^*, \mu) = 2 - \mu < 0$. Thus, it is plausible that the fixed point x_μ^* undergoes a period-doubling bifurcation, creating a stable period-2 orbit with one branch on each side of x_μ^*.

Next, we first verify that a period-doubling bifurcation occurs at $\mu = 3$ by using geometry and numerics. Then we will prove this by using Theorem 7.3.3.

The cobwebs for $\mu x(1 - x)$ with different μ values are given in **Figure 7.16**, where $x_0 = 0.705$. From Figure 7.16, we find that the fixed point $x_\mu^* = 1 - \frac{1}{\mu}$ is stable for $\mu \le 3$, but unstable for $\mu = 3.4$. Moreover, part (c) in Figure 7.16 indicates that a period-2 orbit is likely to occur.

Now, some numerics. For $\mu = 3.4$, $x_\mu^* = 1 - \frac{1}{3.4} \approx 0.7059$. We start with $x_0 = 0.705$ such that x_0 and $x_\mu^* \approx 0.7059$ are very close. Similar to the calculations for $x_{m+1} = r \sin \pi x_m$, we use Maple to calculate $x_{m+1} = 3.4 x_m (1 - x_m)$ and obtain

x[1] = .707115, x[2] = .704151, x[3] = .708295, x[4] = .7024843071, x[5] = .7106003584, x[6] = .699201, x[7] = .715083, x[8] = .692712, x[9] = .723730, x[10] = .679811, x[11] = .740070, x[12] = .654045, x[13] = .769318, x[14] = .603390, x[15] = .813655, x[16] = .515508, x[17] = .849182, x[18] = .435443, x[19] = .835830, x[20] = .466540, $\cdots\cdots$
x[81] = .842155, x[82] = .451960, x[83] = .842153, x[84] = .451965, x[85] = .842155, x[86] = .451961, x[87] = .842153, x[88] = .451964, x[89] = .842154,

x[90] = .451962, x[91] = .842154, x[92] = .451963, x[93] = .842154, x[94] = .451962, x[95] = .842154, x[96] = .451963, x[97] = .842154, x[98] = .451962, x[99] = .842154, x[100] = .451963.

From this numerical solution, we find that even though $x_0 = 0.705$ is so close to $x_\mu^* \approx 0.7059$, the iterations $\{x_m\}$ move away from $x_\mu^* \approx 0.7059$ quickly. They appear to settle down, in an alternating fashion, to the two values

$$\{.451963, \quad .842154\}, \tag{3.6}$$

which indicates that a period-2 orbit $\{\overline{x}, M(\overline{x}, 3.4)\} = \{.451963, .842154\}$ comes into being. See **Figure 7.17**.

The above can also be seen from the graph of the second iteration $M^{[2]}(x, \mu)$ of the logistic map M in (3.4), which is given by

$$\begin{aligned} M^{[2]}(x,\mu) &= M(M(x,\mu),\mu) = \mu M(x,\mu)[1 - M(x,\mu)] \\ &= \mu^2[x - (\mu+1)x^2 + 2\mu x^3 - \mu x^4]. \end{aligned} \tag{3.7}$$

By using the sketching techniques in calculus or using Maple software, one obtains the graphs of $M^{[2]}(x,\mu)$ for different μ values in **Figure 7.18**.

Note that the two fixed points $x^* = 0$ and $x_\mu^* = 1 - \frac{1}{\mu}$ of the map $M(x,\mu)$ will be carried over to become two fixed points of the map $M^{[2]}(x,\mu)$. Figure 7.18 shows that $\mu = 3$ is the last time the graph of $y = M^{[2]}(x,\mu)$ intersects the graph $y = x$ twice at $x^* = 0$ and $x_\mu^* = 1 - \frac{1}{\mu}$. When $\mu = 3.4$, two new intersections are created, and cobwebs can be drawn to verify that the two new intersections are stable. This demonstrates in geometry that a stable period-2 orbit is created.

Next, let's do some theoretical analysis. The fixed points of $M^{[2]}(x,\mu)$ are the solutions of $M^{[2]}(x,\mu) - x = 0$, or

$$\mu^2[x - (\mu+1)x^2 + 2\mu x^3 - \mu x^4] - x = 0, \tag{3.8}$$

which is a polynomial of degree 4 in x. Using the fact that the two fixed points $x^* = 0$ and $x_\mu^* = 1 - \frac{1}{\mu}$ of the map $M(x,\mu)$ are also fixed points of the map $M^{[2]}(x,\mu)$, we can perform a long division (dividing (3.8) by $x(1 - \mu + \mu x)$), and obtain

$$x(1 - \mu + \mu x)[\mu^2 x^2 - (\mu + \mu^2)x + 1 + \mu] = 0. \tag{3.9}$$

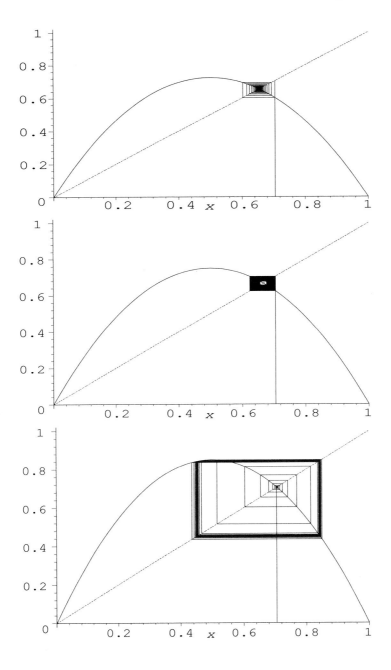

Figure 7.16: Cobwebs for $\mu x(1-x)$: (a) $\mu = 2.9$; (b) $\mu = 3$; (c) $\mu = 3.4$, where $x_0 = 0.705$ and the first 500 iterates are plotted

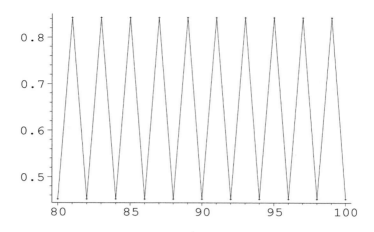

Figure 7.17: The numerical solution of the logistic map with $\mu = 3.4$ and $x_0 = 0.705$

The new fixed points of $M^{[2]}(x, \mu)$ are the solutions of $\mu^2 x^2 - (\mu + \mu^2)x + 1 + \mu = 0$, which are given by (note that we assume $\mu > 1$)

$$x_{\pm}^* = \frac{\mu + 1 \pm \sqrt{(\mu + 1)(\mu - 3)}}{2\mu}. \tag{3.10}$$

Now, for the discriminant to be nonnegative, we must have $\mu \geq 3$. Next, observe that when $\mu = 3$, $x_{\pm}^* = \frac{1+3}{6} = \frac{2}{3} = 1 - \frac{1}{3}$ is the same as the fixed point $x_{\mu}^* = 1 - \frac{1}{\mu}$ of the map M at $\mu = 3$. Thus, new fixed points of $M^{[2]}$ (different from the two of M) appear when $\mu > 3$. It can be verified (see an exercise) that for $\mu > 3$,

$$M(x_+^*, \mu) = x_-^* \quad \text{and} \quad M(x_-^*, \mu) = x_+^*,$$

thus $\{x_+^*, x_-^*\}$ is really a period-2 orbit of period-doubling bifurcating points of the map M.

For the stabilities of $x_{\pm}^*$, we take a derivative of $M^{[2]}(x, \mu)$ with respect to x and evaluate at $x = x_{\pm}^*$, and obtain (see an exercise)

$$\frac{\partial M^{[2]}}{\partial x}(x_{\pm}^*, \mu) = 4 + 2\mu - \mu^2 = -(\mu - 1)^2 + 5. \tag{3.11}$$

Since we now consider $\mu \geq 3$, we see that $\frac{\partial M^{[2]}}{\partial x}(x_{\pm}^*, \mu) = -(\mu - 1)^2 + 5$ is decreasing in μ. To determine the stabilities using $|\frac{\partial M^{[2]}}{\partial x}(x_{\pm}^*, \mu)| < 1$, we

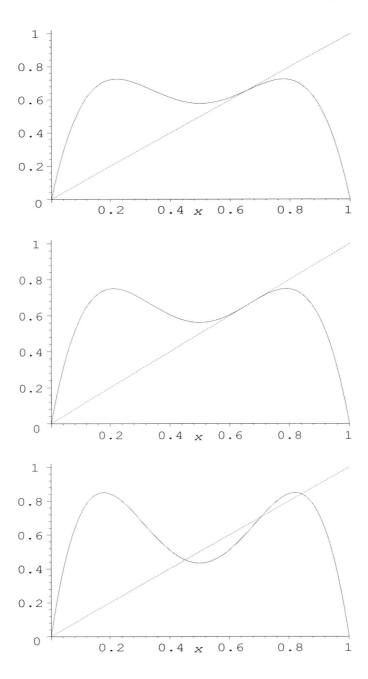

Figure 7.18: Graphs of $M^{[2]}(x, \mu)$: (a) $\mu = 2.9$; (b) $\mu = 3$; (c) $\mu = 3.4$

solve $-(\mu-1)^2 + 5 = \pm 1$ and find that $-(\mu-1)^2 + 5 = 1$ at $\mu = 3$; $|-(\mu-1)^2 + 5| < 1$ for $3 < \mu < 1 + \sqrt{6}$; $-(\mu-1)^2 + 5 = -1$ at $\mu = 1 + \sqrt{6}$; and $-(\mu-1)^2 + 5 < -1$ for $\mu > 1 + \sqrt{6}$. The following **Figure 7.19** is the period-doubling bifurcation diagram near $\mu = 3$. Notice that when $\mu = 3$, $x_\mu^* = 1 - \frac{1}{\mu} = \frac{2}{3}$.

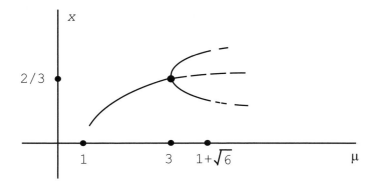

Figure 7.19: Period-doubling bifurcation diagram near $\mu = 3$ for the logistic map

In Figure 7.19, the solid curves represent **attractors**, or stable period-1 or period-2 points of the map M for the corresponding μ values, and the broken curves are for unstable fixed points. For example, for $\mu = 3.4$, the stable period-2 points $x_\pm^*$ in (3.10) are given by

$$x_+^* = \frac{\mu + 1 + \sqrt{(\mu+1)(\mu-3)}}{2\mu} \approx 0.842154,$$

$$x_-^* = \frac{\mu + 1 - \sqrt{(\mu+1)(\mu-3)}}{2\mu} \approx 0.451963,$$

which means that for the map M with $\mu = 3.4$, we can start with any $x_0 \in (0, 1)$, and the solution sequence $\{x_0, x_1, x_2, \cdots\}$ will eventually settle down to the two values 0.842154 and 0.451963, which matches well with a numerical solution we did above using $x_0 = 0.705$ in Figure 7.17.

We summarize these as follows.

Proposition 7.3.4 *Consider the logistic map $M(x, \mu) = \mu x(1-x)$ in (3.4) with $\mu \in (1, 4]$. The fixed point $x_\mu^* = 1 - \frac{1}{\mu}$ of the map $M(x, \mu)$ is stable* *for $1 < \mu < 3$, unstable for $3 < \mu \leq 4$, and undergoes a period-doubling (period-2) bifurcation as μ is increased through 3, (or the map $M^{[2]}(x, \mu)$ undergoes a bifurcation at $\mu = 3$), creating a period-2 orbit that is stable for $3 < \mu < 1 + \sqrt{6} \approx 3.4494897$ and unstable for $1 + \sqrt{6} < \mu \leq 4$.*

The above analysis can be used to verify this proposition. However, we will look at how to apply Theorem 7.3.3 to the map $M^{[2]}(x, \mu)$ to obtain the results concerning the period-doubling bifurcation.

Proof of Proposition 7.3.4. To be able to apply Theorem 7.3.3, we need to transform the fixed point $x_\mu^* = 1 - \frac{1}{\mu}$ to the origin and transform the bifurcation value $\mu = 3$ to zero. Thus, we change (x, μ) to (y, λ) according to

$$y = x - (1 - \frac{1}{\mu}), \quad \lambda = \mu - 3.$$

Then the logistic map $M(x, \mu) = \mu x(1 - x)$ becomes

$$(1 - \frac{1}{3 + \lambda}) - (1 + \lambda)y - (3 + \lambda)y^2 = x_\mu^* - (1 + \lambda)y - (3 + \lambda)y^2. \quad (3.12)$$

Next, in order to analyze the bifurcation from x_μ^*, see Figure 7.19, we should regard x_μ^* as the "origin" in the vertical direction. That is, we should analyze the bifurcation of $-(1 + \lambda)y - (3 + \lambda)y^2$ at $(y, \lambda) = (0, 0)$. Define

$$\overline{M}(y, \lambda) = -(1 + \lambda)y - (3 + \lambda)y^2. \quad (3.13)$$

Then we obtain

$$\overline{M}(0, \lambda) = 0, \quad \lambda \in \Re, \quad \frac{\partial \overline{M}}{\partial y}(0, 0) = -1,$$

and,

$$\begin{aligned}
\overline{M}^{[2]}(y, \lambda) &= \overline{M}(\overline{M}(y, \lambda), \lambda) \\
&= -(1 + \lambda)[\overline{M}(y, \lambda)] - (3 + \lambda)[\overline{M}(y, \lambda)]^2 \\
&= (1 + \lambda)^2 y - \lambda(1 + \lambda)(3 + \lambda)y^2 \\
&\quad -2(3 + \lambda)^2(1 + \lambda)y^3 - (3 + \lambda)^3 y^4, \quad (3.14)
\end{aligned}$$

thus,

$$\begin{aligned}
\overline{M}^{[2]}(y, \lambda) - y &= y[(2\lambda + \lambda^2) - \lambda(1 + \lambda)(3 + \lambda)y \\
&\quad -2(3 + \lambda)^2(1 + \lambda)y^2 - (3 + \lambda)^3 y^3] \\
&\stackrel{\text{def}}{=} yg(y, \lambda). \quad (3.15)
\end{aligned}$$

Next, for the function g defined in (3.15), we have

$$g(0,0) = 0, \quad \frac{\partial g}{\partial y}(0,0) = 0, \quad \frac{\partial^2 g}{\partial y^2}(0,0) = -36 \neq 0, \quad \frac{\partial g}{\partial \lambda}(0,0) = 2 \neq 0, \quad (3.16)$$

therefore, according to Theorem 7.3.3, the map $\overline{M}^{[2]}(y, \lambda)$ undergoes a pitchfork bifurcation at $(y, \lambda) = (0, 0)$. Thus the map $M^{[2]}(x, \mu)$ undergoes a pitchfork bifurcation at $(x, \mu) = (1 - \frac{1}{\mu}, 3) = (\frac{2}{3}, 3)$, or the map $M(x, \mu)$ undergoes a period-doubling bifurcation at $(x, \mu) = (\frac{2}{3}, 3)$. This completes the proof. ♠

This concludes the detailed treatment of the logistic map $M(x, \mu) = \mu x(1-x)$ near $\mu = 3$ using geometry, numerics, and theoretical analysis; and the period-doubling bifurcation diagram near $\mu = 3$ is given in Figure 7.19. In short, we find from the above analysis that the logistic map undergoes a period-doubling bifurcation at $\mu = 3$ because

$$\begin{cases} \frac{\partial M}{\partial x}(x_\mu^*, \mu)|_{\mu=3} = -1, \text{ and} \\ \\ \text{the fixed point } x_\mu^* \text{ of } M(x, \mu) \text{ loses its stability when} \\ \mu \text{ is increased through 3.} \end{cases} \quad (3.17)$$

Based on the above analysis and Proposition 7.3.4, it is natural to ask: **What is going to happen when μ is increased through $1 + \sqrt{6} \approx 3.4494897$?** Now, for the period-2 orbit $x_\pm^* = \frac{\mu+1\pm\sqrt{(\mu+1)(\mu-3)}}{2\mu}$ given in (3.10) (which are fixed points of the map $M^{[2]}(x, \mu)$), we know from (3.11) and Proposition 7.3.4 that

$$\begin{cases} \frac{\partial M^{[2]}}{\partial x}(x_\pm^*, \mu)|_{\mu=1+\sqrt{6}} = -1, \text{ and} \\ \\ \text{the fixed points } x_\pm^* \text{ of } M^{[2]}(x, \mu) \text{ lose their stability when} \\ \mu \text{ is increased through } 1 + \sqrt{6}. \end{cases} \quad (3.18)$$

That is, (3.18) is a **duplication** of (3.17), or the situation for the map $M^{[2]}(x, \mu)$ and its fixed points $x_\pm^*$ at $1 + \sqrt{6}$ is a duplication of the situation for the map $M(x, \mu)$ and its fixed point x_μ^* at $\mu = 3$. Therefore, if we regard $M^{[2]}(x, \mu)$ as $M(x, \mu)$ and $x_\pm^*$ as x_μ^*, then we expect $M^{[2]}(x, \mu)$ to undergo a period-doubling bifurcation at $1 + \sqrt{6}$, such that fixed points of the second iteration of $M^{[2]}$, $M^{[2]}(M^{[2]}) = M^{[4]}$, will appear, or a period-4 orbit will come into being.

How difficult it is to "analyze" $M^{[4]}$ **near** $\mu = 1 + \sqrt{6}$**?**   In the above, we provided a detailed analysis for the map $M^{[2]}$ near $\mu = 3$, which you can use to compare and find what kind of analysis we should perform for the map $M^{[4]}$. If you do, you will find that the analysis for $M^{[4]}$ is extremely difficult and unmanageable, because $M^{[4]}$ is now a **polynomial of degree 16**. Even though the 4 fixed points of $M^{[2]}$ will be carried over to become 4 fixed points of $M^{[4]}$, we are still left with a **polynomial of degree 12 to be factored!** In this sense, we give up "analysis" and leave the stage and the show to "numerical experiments" and "graphical demonstrations."

For example, for $\mu = 3.4495 > 1 + \sqrt{6}$ (≈ 3.4494897), a calculation with $x_0 = 0.5$ using Maple gives

x[1] = .71740, x[2] = .69932, x[3] = .72532, x[4] = .68724, x[5] = .74143, ...

x[981] = .84831, x[982] = .44386, x[983] = .85150, x[984] = .43617, x[985] = .84832, x[986] = .44385, x[987] = .85150, x[988] = .43617, x[989] = .84832, x[990] = .44384, x[991] = .85149, x[992] = .43618, x[993] = .84832, x[994] = .44384, x[995] = .85149, x[996] = .43619, x[997] = .84833, x[998] = .44383, x[999] = .85149, x[1000] = .43619,

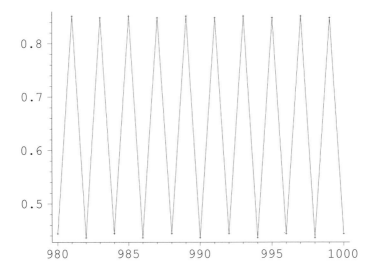

Figure 7.20: A period-4 orbit for the logistic map with $\mu = 3.4495$

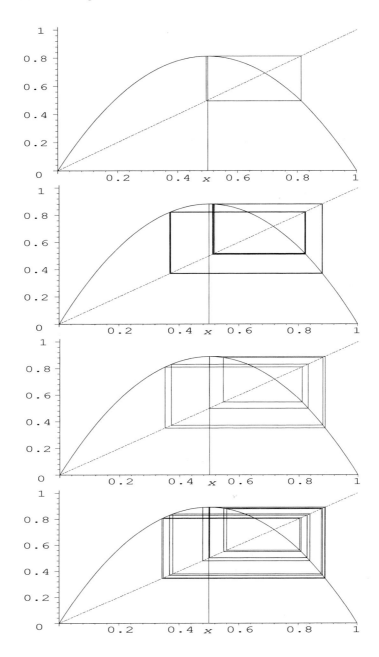

Figure 7.21: Cobwebs for the logistic map with $\mu = 3.25, 3.525, 3.555$, and 3.568 showing the stable period 2, 4, 8, and 16 orbits, where $x_0 = 0.5$ and the first 500 iterates are plotted

which is also plotted in **Figure 7.20**. They show the existence of a period-4
orbit. (Note, compared to previous calculations, it now takes more iterations
for the sequence to settle down.) In an exercise, you are asked to use a
computer software to plot the graph of $M^{[4]}(x, \mu)$ with $\mu = 3.4495$, and then
verify that a period-4 orbit exists.

In fact, for the logistic map $\mu x(1 - x)$, the numerical experiments from
May [1976], Collet and Eckmann [1980], and Cvitanovic [1984] show that
there is a sequence $\mu_m \in [3, 4]$ such that μ_m is the first period-doubling
bifurcation value at which a stable period-2^m orbit appears, and the stable
orbit loses its stability at the next period-doubling bifurcation value μ_{m+1}.
The first few μ_m values are approximately given by the following,

$$\mu_1 = 3, \ \mu_2 = 1 + \sqrt{6} \approx 3.44949, \ \mu_3 = 3.54409, \ \mu_4 = 3.56441, \ \mu_5 = 3.56876, \cdots.$$

See the cobwebs in **Figure 7.21** for the logistic map with $\mu = 3.25, 3.525,$
3.555, and 3.568, showing the corresponding stable period-2^m orbits for $m =$
$1, 2, 3, 4$.

If $\mu_m \to 4$ as $m \to \infty$, then things are not that bad, because then we
can simply say that on the μ-interval $(1, 4]$, there is a sequence of period-
doubling bifurcations that is carried to the end of the interval at $\mu = 4$, such
that for each fixed $\mu \in (1, 4)$, the map has only a **finite** number of attractors
(which are stable period-2^m fixed points of $M^{[2^m]}$) and a **finite** number of
unstable orbits (since a stable orbit loses its stability at the next μ_m value).

However, the "bad news" is that $\mu_m \to \mu_\infty \approx 3.56994$ as $m \to \infty$. See
Figure 7.22 for a period-doubling bifurcations cascade.

Therefore, for μ_m near μ_∞ as $m \to \infty$, there will be $2^m \approx \infty$ number
of attractors distributed inside the x-interval $[0, 1]$, that is why the situation is so "bad." We now see that the period-doubling bifurcations as μ is
increased through 3 and $1 + \sqrt{6}$ are only the beginning of a fascinating sequence of bifurcations that will eventually lead to some very complicated
and unexpected behavior.

The "worst" is yet to come. That is, we now should ask: **What is going to happen when μ is increased through $\mu_\infty \approx 3.56994$?** It turns
out that for some μ values in $(\mu_\infty, 4]$, the corresponding solution sequences
of the map behave in very erratic ways and never settle down to periodic
orbits, or the paths of solution sequences are nonperiodic and completely
unpredictable for all practical purposes. For example, for $\mu = 3.8$, the numerical solution with $x_0 = 0.5$ using Maple gives

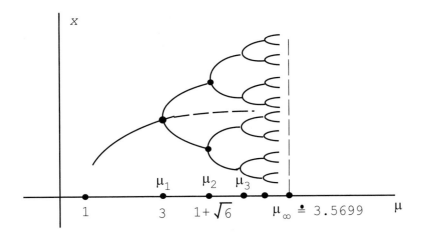

Figure 7.22: $\mu_m \to \mu_\infty \approx 3.56994$ and the period-doubling bifurcations cascade

---

$x[1] = .790305$, $x[2] = .629747$, $x[3] = .886029$, $x[4] = .383729$, $x[5] = .898628$, $x[6] = .346163$, $x[7] = .860069$, $x[8] = .457328$, $\cdots\cdots$
$x[981] = .567515$, $x[982] = .932678$, $x[983] = .238600$, $x[984] = .690347$, $x[985] = .812318$, $x[986] = .579337$, $x[987] = .926081$, $x[988] = .260128$, $x[989] = .731353$, $x[990] = .746606$, $x[991] = .718903$, $x[992] = .767909$, $x[993] = .677253$, $x[994] = .830609$, $x[995] = .534651$, $x[996] = .945437$, $x[997] = .196025$, $x[998] = .598877$, $x[999] = .912848$, $x[1000] = .302313$,

and the cobwebs are given in **Figure 7.23**, showing that the solution sequence is unpredictable or chaotic.

Moreover, these solutions are "sensitively dependent" on initial values: Solutions started with initial values that are very close can diverge widely from each other after some iterations. In an exercise, you are asked to carry out a numerical solution with $\mu = 3.8$ and $x_0 = 0.505$ for the logistic map and compare it with the numerical solution given above for $\mu = 3.8$ and $x_0 = 0.5$. This situation, showing unpredictability and sensitive dependence on initial values, is in general referred to as **chaos** for discrete systems, and it is often said that the **period-doubling bifurcations cascade of the logistic map is a route to chaos**.

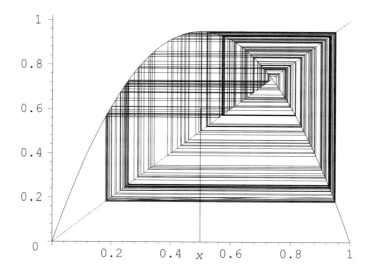

Figure 7.23: Cobwebs for the logistic map with $\mu = 3.8$ showing chaotic behavior

Now, probably you would expect that the behavior of the logistic map become more and more chaotic when μ is increased from $\mu_\infty \approx 3.56994$ to $\mu = 4$. However, the numerical experiments show a truly amazing phenomenon: After seeing the nonperiodic and chaotic iterations that nearly cover some x-intervals in $[0, 1]$ for some μ values, such as $\mu = 3.8$ in Figure 7.23, then, all of a sudden, a stable period-3 orbit emerges for (approximately) $3.8284 \leq \mu \leq 3.8415$. For example, a numerical solution with $x_0 = 0.5$ and $\mu = 3.84$ (which is close to $\mu = 3.8$ in Figure 7.23) using Maple gives

x[1] = .79862, x[2] = .61756, x[3] = .90692, x[4] = .32413, x[5] = .84123, x[6] = .51286, x[7] = .95936, x[8] = .14969, x[9] = .48879, x[10] = .95951, x[11] = .14915, x[12] = .48733, x[13] = .95938, x[14] = .14962, x[15] = .48860, x[16] = .95950, x[17] = .14921, x[18] = .48749, x[19] = .95939, x[20] = .14957, $\cdots\cdots$ x[981] = .48800, x[982] = .95944, x[983] = .14940, x[984] = .48800, x[985] = .95944, x[986] = .14940, x[987] = .48800, x[988] = .95944, x[989] = .14940, x[990] = .48800, x[991] = .95944, x[992] = .14940, x[993] = .48800, x[994] = .95944, x[995] = .14940, x[996] = .48800, x[997] = .95944, x[998] = .14940, x[999] = .48800, x[1000] = .95944,

which shows that
$$\{.14940, \ .48800, \ .95944\}$$
is a stable period-3 orbit for $\mu = 3.84$. The set of these stable period-3 orbits for $3.8284 \le \mu \le 3.8415$ is called a **period-3 window**. To understand why this happens, we need to look at the map $M^{[3]}(x, \mu)$ of the logistic map M because period-3 orbits are from fixed points of $M^{[3]}(x, \mu)$. In an exercise, you are asked to plot the graphs of $M^{[3]}(x, \mu)$ for $\mu = 3.8$ and 3.84, and use them to argue that when μ is increased from $\mu = 3.8$ to 3.84, the graph of $y = M^{[3]}(x, \mu)$ intersects the graph of $y = x$ and creates a stable period-3 orbit. Refer to the graphs in Arrowsmith and Place [1990].

After this period-3 window, a period-doubling bifurcation will occur to this stable period-3 orbit, creating a stable period-6 orbit. Similar to the case for the period-2^m cascade, there follows a cascade of stable period-(3×2^m) orbits. Then these stable period-(3×2^m) orbits accumulate at $\mu \approx 3.8495$, and then a new window comes into being, and period-doubling bifurcations come into play again $\cdots$.

Now, with the above analysis and description, it is time to introduce the so-called **Final State Diagrams** (a name taken from Peitgen, Jurgens, and Saupe [1992]) in **Figure 7.24**, which plots "all" attractors of the logistic map $M(x, \mu) = \mu x(1 - x)$ for "all" parameter μ values. That is, depending on the printing device, the μ-interval $[2.8, 4]$ and the x-interval $[0, 1]$ are divided according to the horizontal and the vertical pixel resolutions. (As a reference, a computer monitor's resolution is about 1000×800 pixels.) For each parameter value μ that is determined by the resolution, iterations $\{x_m\}$ are generated with an arbitrary initial x_0 value. Typically the first few hundreds of iterations are discarded to allow the solution sequence to settle down to its eventual behavior, and the next few hundreds of iterations are plotted. The parameter range is traversed one pixel at a time and eventually all parameter values set by the resolution are covered. See Figure 7.24, which is generated using Maple with the code

```
u[0]:=2.8; N:=300; pix:=(4-2.8)/N; Max:=1000; Min:=500;
for i from 1 by 1 to N do
x[i,0]:=0.5; u[i]:= u[i-1]+pix;
for m from 0 by 1 to Max do
x[i,m+1]:= u[i]*x[i,m]*(1-x[i,m]);
od;
od:
```

```
plot([seq([seq([u[i],x[i,k]],k=Min..Max)],i=1..N)],
    color=black,style=point,symbol=point);
```

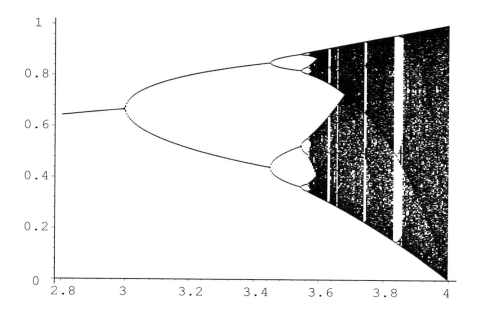

Figure 7.24: The Final State Diagrams for the logistic map

Note that the Final State Diagrams in Figure 7.24 is different from a bifurcation diagram for fixed points (such as Figure 7.19), because unstable fixed points are not shown in Figure 7.24. All of the results we have derived for the logistic map $M(x,\mu) = \mu x(1 - x)$ can be found in Figure 7.24. For example, when $\mu = 3.4$, the period-2 orbit is given by $\{0.451963, 0.842154\}$, which can be determined from Figure 7.24. Also note that the white band near $\mu = 3.835$ is the period-3 window mentioned above. (Now, imagine the Final State Diagrams in Figure 7.24 as a dark house, then you will understand why the word "window" is used.)

Compare the logistic differential equation and the logistic map. It is now clear that the logistic map is much more complex and difficult to handle because unlike scalar differential equations where solutions flow monotonically on the x-axis, the solution sequences of the logistic map can jump back and forth, leading to some strange and unexpected behavior, or chaos.

Exercises 7.3

1. Verify that if $M(x) = x$, then $M^{[m]}(x) = x$, $m \geq 1$.

2. Prove Theorem 7.3.3.

3. Prove that points $\{x, M(x), M^{[2]}(x), \cdots, M^{[k-1]}(x)\}$ of a period-k orbit are distinct.

4. Verify the pictures in Figure 7.15 for different μ values.

5. Divide the μ-interval $[2.8, 4]$ into N=100 subintervals. For each subinterval, arbitrarily select a μ value and calculate (using a computer software such as Maple) the corresponding logistic map with an arbitrary initial value x_0. Plot your calculations in the (μ, x) plane and compare it with Figure 7.24. Do the same with N=50, 200, 300,$\cdots$.

6. Use the sketching techniques in calculus (that is, use derivatives) or use Maple software to sketch the three functions in Figure 7.18.

7. Derive (3.9).

8. For the $x_{\pm}^*$ given by (3.10) for $\mu > 3$, verify that $M(x_+^*, \mu) = x_-^*$ and $M(x_-^*, \mu) = x_+^*$.

9. Verify $\frac{\partial M^{[2]}}{\partial x}(x_{\pm}^*, \mu) = 4 + 2\mu - \mu^2$ in (3.11).

10. Derive (3.12).

11. Derive (3.16).

12. Use a computer software to plot the graph of $M^{[4]}(x, \mu)$ of the logistic map with $\mu = 3.4495$, and then verify that a period-4 orbit exists.

13. Carry out a numerical solution with $\mu = 3.8$ and $x_0 = 0.505$ for the logistic map and compare it with the numerical solution given in this section for $\mu = 3.8$ and $x_0 = 0.5$.

14. Plot the graphs of $M^{[3]}(x, \mu)$ of the logistic map for $\mu = 3.8$ and 3.84, and use them to argue that when μ is increased from $\mu = 3.8$ to 3.84, the graph of $y = M^{[3]}(x, \mu)$ intersects the graph of $y = x$ and creates a stable period-3 orbit.

7.4 Universality

In the previous section, we have seen how strange and complex the Final State Diagrams are for the logistic map in Figure 7.24. Now, the **sine** map

$$M(x) = \mu \sin \pi x, \quad x \in [0, 1], \quad \mu \in [0, 1], \tag{4.1}$$

in Example 7.1.2 is completely different from the logistic map, therefore we may expect the Final State Diagrams for the sine map (4.1) will be a completely different "monster." However, the amazing thing is that the Final State Diagrams for the sine map (4.1) given in **Figure 7.25** resembles Figure 7.24 of the logistic map almost perfectly. The calculations in Example 7.1.2 can be used to support the Final State Diagrams for the sine map (4.1) in Figure 7.25. For example, for $\mu = 0.7$, Example 7.1.2 indicates that $x^* = .6365$ is an attractor, which can be seen in Figure 7.25; for $\mu = 0.8$, Example 7.1.2 indicates that $x_1^* = .475163$ and $x_2^* = .797565$ are two attractors (or a period-2 cycle), which can also be seen in Figure 7.25; finally, for $\mu = 0.95$, Example 7.1.2 indicates that the iterations $\{x_m\}$ never settle down to a single attractor or a periodic cycle, and cover nearly all values from 0.2 to 0.9, which is evident in Figure 7.25.

The difference between the Final State Diagrams for the two maps (the logistic map and the sine map) is in the horizontal axis: $[2.8, 4]$ for the logistic map and $[0.7, 1]$ for the sine map. Otherwise, the qualitative properties of the two diagrams for the two maps are the same: they both start with one single attractor, then period-doubling bifurcations lead them to chaos, then periodic windows occur, then chaos, $\cdots$.

This observation in fact illustrates a remarkable **universality result** of Metropolis, et al. [1973] for any **unimodal map** $\mu f(x)$, where the function $f(x)$ is concave down and has a unique maximum. See **Figure 7.26** for a unimodal map $\mu f(x)$, which includes the logistic map and the sine map as special cases.

Metropolis, et al. [1973] proved that for **all unimodal maps**, as the parameter μ is varied, the numbers of periodic attractors appear in the same order, now called the **universal sequence** or **U-sequence**:

$$1, \; 2, \; 2 \times 2, \; 4 \times 2, \; \cdots, \; 6, \; \cdots, \; 5, \; \cdots, \; 3, \; 3 \times 2, \; \cdots$$

where 1 is for the first attractor; 2 is for the first period-doubling bifurcation; $2 \times 2 = 2^2$ is for the second period-doubling bifurcation, and $4 \times 2 = 2^3$ is for the third period-doubling bifurcation; $\cdots$, 6, 5, and 3 are for the three

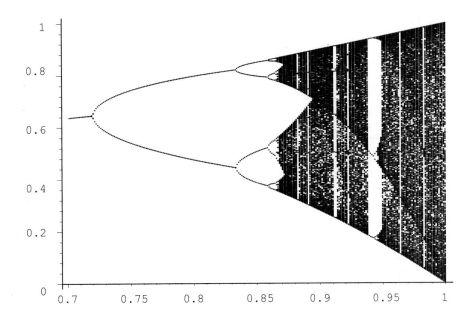

Figure 7.25: The Final State Diagrams for the sine map (4.1)

large visible windows; and 3×2 is for the first period-doubling bifurcation of the visible period-3 window, $\cdots$.

The U-sequence has been found in experiments on the Belousov-Zhabotinsky chemical reaction. That is, within the experimental resolution, the periodic states occurred in the exact order predicted by the U-sequence, see Strogatz [1994] for further details.

Next, let's look at another universality result for unimodal maps. For a given unimodal map, define a sequence μ_m such that μ_m is the first period-doubling bifurcation value at which a stable period-2^m orbit appears. For example, for the logistic map, we know from the previous section that

$$\bar{\mu}_1 = 3, \bar{\mu}_2 = 1+\sqrt{6} \approx 3.44949, \bar{\mu}_3 = 3.54409, \bar{\mu}_4 = 3.56441, \bar{\mu}_5 = 3.56876, \cdots$$

(we use $\bar{\mu}$ here for the logistic map). Now define

$$\Delta_m = \mu_{m+1} - \mu_m,$$

which in geometry measures the distance between consecutive bifurcation values, see **Figure 7.27**.

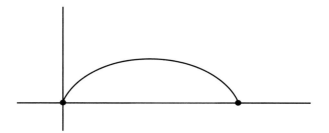

Figure 7.26: A unimodal map $\mu f(x)$

Consider the ratio

$$\delta_m = \frac{\Delta_m}{\Delta_{m+1}}.$$

For example, for the logistic map,

$$\overline{\delta}_1 = \frac{\overline{\Delta}_1}{\overline{\Delta}_2} = \frac{3.44949 - 3}{3.54409 - 3.44949} = 4.7514799;$$

$$\overline{\delta}_2 = \frac{\overline{\Delta}_2}{\overline{\Delta}_3} = \frac{3.54409 - 3.44949}{3.56441 - 3.54409} = 4.6555118;$$

$$\overline{\delta}_3 = \frac{\overline{\Delta}_3}{\overline{\Delta}_4} = \frac{3.56441 - 3.54409}{3.56876 - 3.56441} = 4.6712643; \;\cdots. \qquad (4.2)$$

Feigenbaum [1979] verified (see de Melo and van Strien [1993] for references of a formal proof) that the series $\{\delta_m\}$ converges and

$$\delta = \lim_{m \to \infty} \delta_m = 4.669 \cdots$$

which is an important mathematical constant, such as π and e. This useful result can be applied to any unimodal map and estimate when the next bifurcation will occur.

Finally, we introduce one more astonishing universality result of Sarkovskii [1964] concerning general one-dimensional maps (not necessarily unimodal maps). Let's order the positive integers in the following way, called the **Sarkovskii ordering**:

$$3 \prec 5 \prec 7 \prec 9 \prec \cdots$$
$$2 \cdot 3 \prec 2 \cdot 5 \prec 2 \cdot 7 \prec 2 \cdot 9 \prec \cdots$$

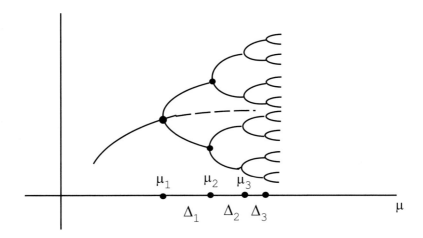

Figure 7.27: The distance between consecutive bifurcation values given by Δ_m

$$2^2 \cdot 3 \prec 2^2 \cdot 5 \prec 2^2 \cdot 7 \prec 2^2 \cdot 9 \prec \cdots$$
$$2^3 \cdot 3 \prec 2^3 \cdot 5 \prec 2^3 \cdot 7 \prec 2^3 \cdot 9 \prec \cdots$$
$$\cdots \cdots$$
$$\cdots \prec 2^4 \prec 2^3 \prec 2^2 \prec 2 \prec 1. \tag{4.3}$$

That is, write all the odd numbers except 1, then 2 times these odd numbers, 2^2 times these odd numbers, 2^3 times these odd numbers, etc, finally, write the powers of 2 in decreasing order. They look strange, but they include all positive integers.

Theorem 7.4.1 (Sarkovskii) *Let* $f : \Re \to \Re$ *be a continuous map. If* f *has a period-p orbit, then* f *has a period-q orbit if* $p \prec q$ *in the Sarkovskii ordering.* ♠

Theorem 7.4.1 is such a powerful result, and indicates, in particular, that if f has a period-3 orbit, then f has periodic orbits of **every** integer period. Applying this to the logistic map, for which we have calculated in the previous section that $\{.14940, .48800, .95944\}$ is a period-3 orbit for $\hat{\mu} = 3.84$, we conclude that for $\hat{\mu} = 3.84$ the logistic map has periodic orbits of every integer period. Note that we don't see them (except period-3) in the Final State Diagrams in Figure 7.24 because they are unstable. Next, for any $\mu \in [\hat{\mu}, 4] = [3.84, 4]$, the corresponding map also has a period-3 orbit (which may be unstable for some μ values); thus the map also has periodic orbits of every integer period. This idea was utilized by Li and Yorke [1975] in their famous paper "Period Three Implies Chaos."

Some related topics concerning universality results can be found in Glendinning [1994], Strogatz [1994], and Smith [1998].

Exercises 7.4

1. For the sine map (4.1), divide the μ-interval $[0, 1]$ into 100 subintervals. For each subinterval, arbitrarily select a μ value and calculate, using a computer software such as Maple, the corresponding sine map with an arbitrary initial value x_0. Plot your calculations in the (μ, x) plane and compare it with Figure 7.25.

2. Use a computer software to plot the graph of $M^{[4]}(x, \mu)$ of the sine map with $\mu = 0.85$, and then verify that a period-4 orbit exists.

3. Plot the graphs of $M^{[3]}(x, \mu)$ of the sine map for $\mu = 0.925$ and 0.94, and use them to argue that when μ is increased from $\mu = 0.925$ to 0.94, the graph of $y = M^{[3]}(x, \mu)$ intersects the graph of $y = x$ and creates a stable period-3 orbit.

4. Use the analysis for the logistic map in the previous section as a guide and carry out similar analysis for the sine map.

5. Consider the map

$$M(x, \mu) = \begin{cases} 2\mu x, & 0 \le x \le \frac{1}{2}, \\ 2\mu(1 - x), & \frac{1}{2} < x \le 1, \end{cases} \quad \mu \in [0, 1],$$

where for each fixed μ the graph of $M(x, \mu)$ in x looks like a "$\wedge$," thus it is called a **tent** map.

   (a) Divide the μ-interval $[0, 1]$ into 100 subintervals. For each subinterval, arbitrarily select a μ value and calculate, using a computer software such as Maple, the corresponding tent map with an arbitrary initial value x_0. Plot your calculations in the (μ, x) plane.

   (b) Use the analysis for the logistic map in the previous section as a guide and carry out similar analysis for the tent map.

7.5 The Lorenz System and Strange Attractors

We have seen that chaotic behavior can occur to discrete systems (maps or difference equations). For continuous systems (differential equations), solutions of scalar differential equations behave orderly because they are monotone flows on the x-axis. For differential equations in $\Re^2$, solutions move in a plane and also behave orderly due to the famous Poincaré-Bendixson theorem to be derived in Chapter 8. For differential equations in $\Re^n, n \geq 3$, solutions are in space, thus chaotic behavior can occur.

A major contribution in this area was given by Lorenz in 1963, at a time computer technology made heavy numerical calculations possible. The so-called Lorenz system has generated a great deal of interest in analysis and numerical solutions of nonlinear dynamical systems and chaos, and has become another icon of nonlinear dynamics. It is also a rich source of examples of various types of bifurcations and chaotic behavior in nonlinear dynamics.

The **Lorenz system** is a simplified version of a set of complicated equations modeling the motion of convective fluid (or weather for short) warmed from below, but still retains the interesting and representative behavior of the original equations. It is a differential equation in $\Re^3$ given by

$$
\begin{cases}
\frac{dx}{dt} & = & \sigma(y - x), \\
\frac{dy}{dt} & = & rx - y - xz, \\
\frac{dz}{dt} & = & xy - bz,
\end{cases}
\tag{5.1}
$$

where x measures the rate of convective overturning, y measures the horizontal temperature variation, and z measures the vertical temperature variation, and σ, r, and b are positive parameters that are proportional to some numbers derived from some physics experiments, such as the Prandtl number and the Rayleigh number.

Similar to the logistic map, the Lorenz system (5.1) looks so innocuous, and were it not for the two nonlinear terms xz and xy, the system can be solved completely using the results of Chapter 3. However, as we will see, the solutions of the Lorenz system can behave very "strangely" or "badly" as the parameters are varied.

First, we outline some results about the Lorenz system that can still be handled by using "analysis," see Lorenz [1963], Sparrow [1982], and Guckenheimer and Holmes [1986] for more details. The origin $(0, 0, 0)$ is a critical point for all parameter values, thus let's look at the linearized equation at

$(0,0,0)$, whose matrix is given by

$$\begin{bmatrix} -\sigma & \sigma & 0 \\ r & -1 & 0 \\ 0 & 0 & -b \end{bmatrix}. \tag{5.2}$$

Now, the eigenvalues are given by

$$-b, \quad \frac{-(1+\sigma) \pm \sqrt{(1+\sigma)^2 - 4\sigma(1-r)}}{2}. \tag{5.3}$$

If $0 < r < 1$, then $-4\sigma(1-r) < 0$, and

$$(1+\sigma)^2 - 4\sigma(1-r) > (1+\sigma)^2 - 4\sigma = (1-\sigma)^2 \geq 0.$$

Thus, all three eigenvalues are real negative numbers. Therefore, the origin $(0,0,0)$ of Eq. (5.1) is asymptotically stable when $0 < r < 1$, using the stability analysis from Chapter 5. This result can also be established by examining the Liapunov function

$$V = \frac{1}{2\sigma}[x^2 + \sigma y^2 + \sigma z^2], \tag{5.4}$$

and its derivative using Eq. (5.1),

$$\begin{aligned} \frac{dV}{dt} &= x(y-x) + y(rx - y - xz) + z(xy - bz) \\ &= -x^2 + (1+r)xy - y^2 - bz^2 \\ &\leq -x^2 + \frac{1+r}{2}(x^2 + y^2) - y^2 - bz^2 \\ &= -(1 - \frac{1+r}{2})x^2 - (1 - \frac{1+r}{2})y^2 - bz^2. \end{aligned} \tag{5.5}$$

Now, for $0 < r < 1$, we have

$$\frac{dV}{dt} \leq -(1 - \frac{1+r}{2})x^2 - (1 - \frac{1+r}{2})y^2 - bz^2 \leq 0, \tag{5.6}$$

and

$$\frac{dV}{dt} < 0 \quad \text{if} \quad (x,y,z) \neq (0,0,0), \tag{5.7}$$

which implies that V and $-V'$ are positive definite. Thus, using the stability analysis from Chapter 5, the origin $(0,0,0)$ of Eq. (5.1) is asymptotically stable when $0 < r < 1$. It is also true that the origin $(0,0,0)$ is the only critical point when $0 < r < 1$, see an exercise.

When $r > 1$, the origin loses its stability because one (and only one) eigenvalue given in (5.3) becomes positive. Now, two new critical points appear at

$$C_1 = (\sqrt{b(r-1)}, \sqrt{b(r-1)}, r-1), \quad C_2 = (-\sqrt{b(r-1)}, -\sqrt{b(r-1)}, r-1),$$

therefore, the system (5.1) undergoes a bifurcation at the bifurcation value $r = 1$. For the parameter values $\sigma = 10$ and $b = \frac{8}{3}$ selected in Lorenz [1963], it is shown that using the linearizations near C_1 and C_2 and then the eigenvalues of the corresponding matrices, the critical points C_1 and C_2 are stable for $1 < r < 24.74$ (approximately), and unstable for $r > 24.74$ (approximately). Moreover, Poincaré-Andronov-Hopf bifurcations take place at C_1 and C_2 when $r \approx 24.74$.

That is, for $r > 24.74$, all critical points of the Lorenz system are unstable, thus the trajectories cannot approach the three critical points. Accordingly, we ask: **Where do the trajectories go when $r > 24.74$?** Based on our current understanding of the trajectories on a line or on a plane of scalar or planar equations, we may guess that the trajectories of the Lorenz system would go to infinity or "pile up" at some periodic orbits. If that is the case, then there is nothing strange about the Lorenz system. However, things are so complex and strange about the Lorenz system, since now the trajectories are in $\Re^3$ and can behave in erratic ways. In fact, the following result says that the trajectories of Eq. (5.1) cannot go to infinity.

Proposition 7.5.1 *Consider the Lorenz system (5.1). For any fixed parameter values σ, b, and r, there is a sphere with radius R, S_R, in $\Re^3$ such that every trajectory of Eq. (5.1) eventually enters S_R and never thereafter leaves it.*

Proof. Consider the Liapunov function

$$V = \frac{1}{2\sigma b}[rx^2 + \sigma y^2 + \sigma(z - 2r)^2], \tag{5.8}$$

which is related to the distance from (x, y, z) to $(0, 0, 2r)$. Its derivative using Eq. (5.1) gives

$$\frac{dV}{dt} = \frac{r}{b}x(y - x) + \frac{1}{b}y(rx - y - xz) + \frac{1}{b}(z - 2r)(xy - bz)$$

$$= -\frac{r}{b}x^2 - \frac{1}{b}y^2 - (z^2 - 2rz)$$

$$= -\frac{r}{b}x^2 - \frac{1}{b}y^2 - (z - r)^2 + r^2. \qquad (5.9)$$

Then $\frac{dV}{dt} \geq 0$ (or $\frac{r}{b}x^2 + \frac{1}{b}y^2 + (z-r)^2 \leq r^2$) defines a bounded and closed set D in $\Re^3$ including the point $(0,0,2r)$. Therefore, there exists a sphere with radius R, S_R, in $\Re^3$ such that D is inside the interior of S_R.

Consequently, $\frac{dV}{dt} < 0$ on the boundary or outside of S_R. Now, a trajectory inside S_R cannot leave S_R because $\frac{dV}{dt} < 0$ on the boundary of S_R. Next, if a trajectory starts from a point Q that is outside of S_R, then Q is inside a sphere with radius $R^* > R$. For the closed and bounded set in $\Re^3$ defined by $R \leq \sqrt{x^2 + y^2 + z^2} \leq R^*$, there is a constant $\delta > 0$ such that $\frac{dV}{dt} \leq -\delta$. Thus for the trajectory from Q, the corresponding V value decreases and (see an exercise) the trajectory enters S_R at some finite t value. This completes the proof.   ♠

Now we know that the trajectories of the Lorenz system are confined in a bounded set in $\Re^3$, so the next question is: **Will the trajectories pile up at some periodic orbits?** The answer is that for the parameter values $\sigma = 10$, $b = \frac{8}{3}$, and r near 28, there are no stable periodic orbits. Thus for those parameter values, the trajectories are bounded but cannot pile up at some periodic orbits. Then the question is: **Where do the trajectories go and how do they behave?** Now, it is time to introduce **Figure 7.28**, showing some pictures of a trajectory of Eq. (5.1) derived using numerical experiments with $\sigma = 10$, $b = \frac{8}{3}$, and $r = 28$. See Lorenz [1963], Sparrow [1982], and Guckenheimer and Holmes [1986] for additional references.

The trajectory in Figure 7.28 does not intersect itself in $\Re^3$, so the crossings in Figure 7.28 are the result of projection in $\Re^2$. Moreover, the numerical experiments indicate that all trajectories are **attracted** to a set that resembles a **butterfly** in motion. That is, the set looks like a "thick" surface with an infinite number of sheets or wings, or looks like a "book" with an infinite number of pages. Next, the most bizarre thing is that the set has zero volume and the dimension of the set is about 2.05, **not even an integer!** Because things are so strange about this set, it is nowadays called a **strange attractor**.

The numerical experiments also indicate that the solutions of Eq. (5.1) are "sensitively dependent" on initial values. That is, solutions started with initial values that are nearby diverge, or display radically different dynamical behavior, after a short time, which makes the long time prediction impos-

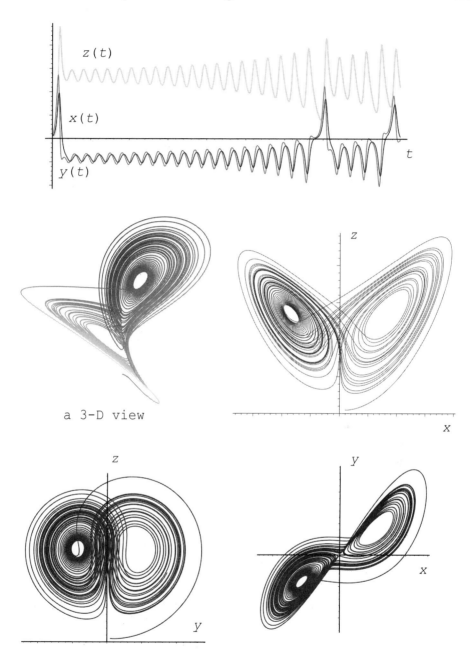

Figure 7.28: A trajectory of the Lorenz system with $\sigma = 10$, $b = \frac{8}{3}$, and $r = 28$

sible. In fact, Lorenz found that every time he tried to recompute a given solution using the same computer and program, he got a different solution, and those solutions looked as if they were not related. The reason was that his recorded values of x, y, and z were less accurate than the internal representation of the computer, so that actually those solutions started with different but nearby initial values.

Moreover, in the projection onto the (x, z) plane in Figure 7.28 that looks like a butterfly, a trajectory will cruise a few circuits on one side, then suddenly move to the other side and cruise a few circuits, and then suddenly move back $\cdots$, and this process will continue forever, such that the trajectory will wind around the two sides infinitely many times without ever settling down. And the fashion in which the trajectory moves around the two sides is unpredictable. Similar to discrete systems, the nonperiodic behavior or unpredictability of solutions and the sensitive dependence on initial values is generally called **chaos** for continuous systems (differential equations).

Since we mentioned the nonperiodic behavior of solutions for the parameter values $\sigma = 10$, $b = \frac{8}{3}$, and r near 28, we also must mention that for other r values, bifurcations can happen and periodic orbits will appear for the Lorenz system, in ways similar to those of the logistic map. In fact, for $145 < r < 167$, the bifurcation diagram of the period-doubling cascade of the Lorenz system is very similar to those of the logistic map given in Section 7.3. See Sparrow [1982] for additional details.

Now, a periodic orbit of the Lorenz system, Γ, is in $\Re^3$ and is not easy to deal with. A main tool here is to find a two-dimensional hypersurface, $\sum$, that is perpendicular to Γ at some point p_0 of Γ, and then look at how trajectories leave and return to $\sum$. More precisely, if $p \in \sum$ is sufficiently near p_0, then using continuous dependence on initial conditions, the trajectory leaving p will return to $\sum$. Thus we can denote $P(p) \in \sum$ the point of first return of the point p to $\sum$. See **Figure 7.29**. We call $\sum$ a **Poincaré section** and the map P a **Poincaré map** (or a **first return map**), because this idea was introduced by Poincaré [1892] in his study of the three body problem in celestial mechanics, where he reduced the study of a continuous time system (differential equation) to the study of an associated discrete time system (map).

With this Poincaré section and Poincaré map, the analysis of trajectories near a periodic orbit Γ of the Lorenz system in dimension three is reduced to a two-dimensional map. Now, p_0 is a fixed point of the Poincaré map, and the collection of points on $\sum$ can provide us with very good information about the trajectories near Γ. For example, if $p_1 \in \sum$ is another fixed point

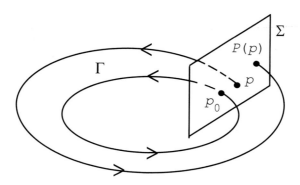

Figure 7.29: A Poincaré section and a Poincaré map

of P, then the corresponding trajectory is a periodic orbit of the Lorenz system with its period close to that of Γ. Otherwise, if $p_2 \in \sum$ gives rise to a chaotic trajectory, then the corresponding points on $\sum$ will not show any pattern. This way, the trajectories near periodic orbits can be analyzed. See Hirsch and Smale [1974], Guckenheimer and Holmes [1986], Wiggins [1990], Perko [1991], Tsonis [1992], and Alligood, Sauer, and Yorke [1997] for additional details.

Finally, we point out that most results about the Lorenz system or Lorenz-like chaotic differential equations are based on numerical experiments, which suggests caution in the interpretation of them before some "mathematical proofs" can be found. Nowadays, people have started to rigorously define "strange attractors" and "chaos" and then prove their existence. For example, the following definitions are given in Wiggins [1990].

Definition 7.5.2 *In $\Re^n$, consider an autonomous system for $t \geq 0$ and denote $\phi(t, x)$ the solution started from x (respectively denote $M(x)$ the map started from x). A set $\Lambda \subset \Re^n$ is said to be **invariant** if $x \in \Lambda$ implies $\phi(t, x) \in \Lambda$ for $t \geq 0$, (respectively $M^{[k]}(x) \in \Lambda$ for $k \geq 0$). A closed invariant set $\Lambda \subset \Re^n$ is called an **attracting set** if there is some neighborhood U of Λ such that for any $x \in U$ and any $t \geq 0$, one has $\phi(t, x) \in U$ and $\phi(t, x) \to \Lambda$ as $t \to \infty$ (respectively for any $x \in U$ and any $k \geq 0$, one has $M^{[k]}(x) \in U$ and $M^{[k]}(x) \to \Lambda$ as $k \to \infty$).*

Definition 7.5.3 *A solution $\phi(t, x)$ (respectively a map $M(x)$) is said to be **topologically transitive** on a closed invariant set $\Lambda \subset \Re^n$ if for any* *two open sets $U, V \subset \Lambda$, there exists a $t \geq 0$ (respectively $k \geq 0$) such that $\phi(t, U) \cap V \neq \emptyset$ (respectively $M^{[k]}(U) \cap V \neq \emptyset$).*

Definition 7.5.4 *A closed invariant set $\Lambda \subset \Re^n$ is said to be an **attractor** if it is a topologically transitive attracting set.*

Definition 7.5.5 *Let $\Lambda \subset \Re^n$ be a compact invariant set. A solution $\phi(t, x)$ (respectively a map $M(x)$) is said to have **sensitive dependence on initial conditions on** Λ if there exists an $\varepsilon > 0$ such that for any $x \in \Lambda$ and any neighborhood U of x, there exists a $y \in U$ and $t > 0$ (respectively $k > 0$) such that $|\phi(t, x) - \phi(t, y)| > \varepsilon$ (respectively $|M^{[k]}(x) - M^{[k]}(y)| > \varepsilon$).*

Definition 7.5.6 *A compact invariant set $\Lambda \subset \Re^n$ is said to be **chaotic** if a solution $\phi(t, x)$ (respectively a map $M(x)$) has sensitive dependence on initial conditions on Λ, and $\phi(t, x)$ (respectively a map $M(x)$) is topologically transitive on Λ, and the periodic orbits are dense in Λ.*

Definition 7.5.7 *Assume that the compact invariant set $\Lambda \subset \Re^n$ is an attractor. Then Λ is called a **strange attractor** if it is chaotic.*

An immediate consequence of the above definitions is that there is nothing strange about $x' = 2x$ in Example 7.1.1, or there is no strange attractor for $x' = 2x$ (see an exercise).

For one-dimensional maps, such as the logistic map, the existence results for the strange attractors as defined in Definition 7.5.7 are quite complete, see Jakobsen [1981], Misiurewicz [1981], and Johnson [1987]. For Lorenz or Lorenz-like chaotic differential equations, some progress has been made in this direction. See Sinai and Vul [1981], Afraimovich, Bykov, and Silnikov [1983] for additional references.

Exercises 7.5

1. Verify that the origin $(0, 0, 0)$ is the only critical point for the Lorenz system (5.1) when $0 < r < 1$.

2. Assume $0 < r < 1$ and consider the function of two variables x and y given by
$$f(x, y) = -x^2 + (1 + r)xy - y^2.$$

Use

$$\begin{cases} \frac{\partial f}{\partial x} = -2x + (1+r)y, \\ \frac{\partial f}{\partial y} = -2y + (1+r)x, \end{cases} \tag{5.10}$$

and

$$\frac{\partial^2 f}{\partial x^2} = -2, \quad \frac{\partial^2 f}{\partial y^2} = -2, \quad \frac{\partial^2 f}{\partial x \partial y} = 1+r,$$

and the second derivative test to verify for $f(x,y)$ that $(x,y) = (0,0)$ is the only maximum point. This is another way to analyze the Liapunov function in (5.4) and confirm that the origin $(0,0,0)$ is asymptotically stable.

3. Verify that if $r > 1$, two new critical points appear at

$$\begin{aligned} C_1 &= (\sqrt{b(r-1)}, \sqrt{b(r-1)}, r-1), \\ C_2 &= (-\sqrt{b(r-1)}, -\sqrt{b(r-1)}, r-1), \end{aligned}$$

for the Lorenz system. Hence, the system undergoes a bifurcation at the bifurcation value $r = 1$.

4. Find all critical points of the Lorenz system.

5. Linearize the Lorenz system near C_1 and C_2, and determine the linear matrices. Then use the matrices to determine the characteristic equations for the eigenvalues, and then analyze the eigenvalues.

6. Verify in the proof of Proposition 7.5.1 that for the closed and bounded set in $\Re^3$ defined by $R \leq \sqrt{x^2 + y^2 + z^2} \leq R^*$, there is a constant $\delta > 0$ such that $\frac{dV}{dt} \leq -\delta$.

7. Provide more details for the last statement in the proof of Proposition 7.5.1, that is, the corresponding V value decreases and the trajectory enters S_R at some finite t value.

8. Analyze $x' = 2x$ in Example 7.1.1 and verify that there is no strange attractor as defined in Definition 7.5.7.

7.6   The Smale Horseshoe

In this section, we provide an example of a strange invariant set possessing chaotic dynamics that resulted from a map first studied by Smale [1963]. Due to the image of the map of its domain, the map is called a **Smale horseshoe**, and will be briefly described here. See Guckenheimer and Holmes [1986], Wiggins [1990], and Arrowsmith and Place [1990] for additional details.
   Let
$$D = \{(x,y) \in \Re^2 : 0 \le x \le 1,\, 0 \le y \le 1\}$$
be the unit square in $\Re^2$ enclosing the two horizontal rectangles

$$
\begin{aligned}
H_0 &= \{(x,y) \in \Re^2 : 0 \le x \le 1,\, 0 \le y \le \eta\},\\
H_1 &= \{(x,y) \in \Re^2 : 0 \le x \le 1,\, 1 - \eta \le y \le 1\}.
\end{aligned}
$$

Define a map $f : D \to \Re^2$ using the following geometry in **Figure 7.30**.

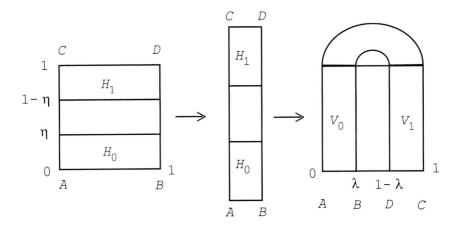

Figure 7.30: The Smale horseshoe map

That is, the map f contracts the unit square D in the x-direction to a width of λ, and expands in the y-direction, and then folds H_1, or the **up end**, by **negative** $180°$ so that $V_0 = f(H_0)$ and $V_1 = f(H_1)$ are inside the unit square D as shown in Figure 7.30, which looks like a *horseshoe* and explains why the map f is called a horseshoe map.
   Since f is one-to-one and onto, f has an inverse f^{-1}, and f^{-1} restricted to $V_0 \cup V_1$ has the following geometry in **Figure 7.31**.

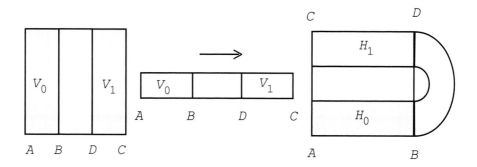

Figure 7.31: The map f^{-1} restricted to $V_0 \cup V_1$

That is, the map f^{-1} contracts in the y-direction to a height of η, and expands in the x-direction, and then folds V_1, or the **right end**, by **positive** 180° so that $f^{-1}(V_0) = H_0$ and $f^{-1}(V_1) = H_1$. Note that the part between V_0 and V_1 is not in the domain of f^{-1}, but it is attached for convenience in geometry, and we write $f^{-1}(D)$ to mean $f^{-1}(V_0 \cup V_1)$.

Accordingly, we have the following important result, which is explained in **Figure 7.32**.

Lemma 7.6.1 *(a). If V is a vertical rectangle inside D with height 1, then $f(V) \cap D$ consists of precisely two vertical rectangles with height 1, one in V_0 and one in V_1. (b). If H is a horizontal rectangle inside D with width 1, then $f^{-1}(H) \cap D$ consists of precisely two horizontal rectangles with width 1, one in H_0 and one in H_1.* ♠

To understand the iterations $f^i, i = \pm 1, \pm 2, \cdots$, we begin with f^2. According to Lemma 7.6.1, $f^2(D) \cap D$ maps $H_0 \cup H_1$ to four vertical rectangles shown in **Figure 7.33**, where V_0 is mapped to V_{00} in V_0 and V_{01} in V_1; V_1 is mapped to V_{10} in V_0 and V_{11} in V_1. That is, the first digit from the left in V_{**} tells where it came from and the second digit from the left tells which $V_i, i = 0, 1$, it belongs now.

Similarly, $f^3(D) \cap D$ maps $H_0 \cup H_1$ to eight vertical rectangles shown in **Figure 7.33**, where, for example, V_{101} resulted from

$$V_1 \to V_{10} = f(V_1) \cap V_0 \to V_{101} = f(V_{10}) \cap V_1,$$

which is completely determined by the sequence (101) of 0's and 1's when we consider the doubling of vertical rectangles. Therefore, we can identify

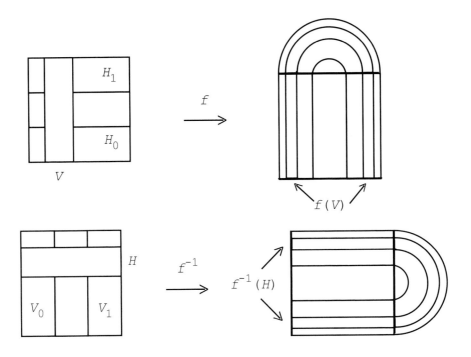

Figure 7.32: $f(V)$ and $f^{-1}(H)$

V_{101} with a sequence $(101)_V$, where the subindex "V" means that we are dealing with vertical rectangles in $V_0 \cup V_1$.

If we continue in this fashion, we find that $f^j(D) \cap D, j \geq 1$, consists of 2^j vertical rectangles in $V_0 \cup V_1$, and can be identified with sequences of 0's and 1's of length j. For example, V_{1010} will result from

$$V_1 \rightarrow V_{10} = f(V_1) \cap V_0 \rightarrow V_{101} = f(V_{10}) \cap V_1 \rightarrow V_{1010} = f(V_{101}) \cap V_0,$$

and can be identified with $(1010)_V$. Accordingly,

$$\bigcap_{j=0}^{\infty} f^j(D) = D \cap f(D) \cap f^2(D) \cap f^3(D) \cap f^4(D) \cdots$$

gives

$$\{V_{s_0 s_1 s_2 \cdots} : \quad s_j \in \{0, 1\}, \quad j \geq 0\},$$

which can be identified with infinite sequences $(s_0 s_1 s_2 \cdots)_V$ of 0's and 1's. In geometry, $\bigcap_{j=0}^{\infty} f^j(D)$ consists of an infinite number of vertical lines (rectangles are sliced into lines after a *limit* process), and has a structure that

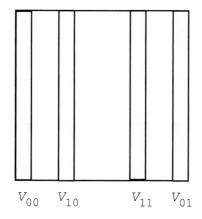 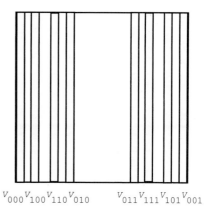

$$V_{00} \quad V_{10} \qquad V_{11} \quad V_{01} \qquad V_{000} V_{100} V_{110} V_{010} \qquad V_{011} V_{111} V_{101} V_{001}$$

Figure 7.33: $f^2(D) \cap D$ and $f^3(D) \cap D$

looks like a **Cantor set**. (A one-dimensional Cantor set is constructed by deleting the middle third open interval $(\frac{1}{3}, \frac{2}{3})$ from the interval $[0, 1]$, then deleting the middle third open interval of each remaining piece, and continue in this fashion.)

Next, let's look at f^{-2}. From Lemma 7.6.1, $f^{-2}(D) \cap D$ maps $V_0 \cup V_1$ to four horizontal rectangles shown in **Figure 7.34**, where H_0 is mapped to H_{00} in H_0 and H_{01} in H_1; H_1 is mapped to H_{10} in H_0 and H_{11} in H_1.

Similar to V_{**} of f^2, H_{**} can be identified with sequences $(t_1 t_2)_H$ of 0's and 1's, where the subindex "H" means that we are dealing with horizontal rectangles in $H_0 \cup H_1$. Accordingly, $f^{-j}, j > 2$, can be defined, and

$$\bigcap_{j=0}^{\infty} f^{-j}(D) = D \cap f^{-1}(D) \cap f^{-2}(D) \cap f^{-3}(D) \cap f^{-4}(D) \cdots$$

gives

$$\{H_{t_1 t_2 t_3 \cdots} : \quad t_j \in \{0, 1\}, \quad j \geq 1\},$$

which can be identified with infinite sequences $(t_1 t_2 t_3 \cdots)_H$ of 0's and 1's, and $\bigcap_{j=0}^{\infty} f^{-j}(D)$ consists of an infinite number of horizontal lines.

Next, let's look at the intersections of these vertical and horizontal rectangles and lines. For example,

$$f^{-2}(D) \cap f^{-1}(D) \cap D \cap f(D) \cap f^2(D)$$

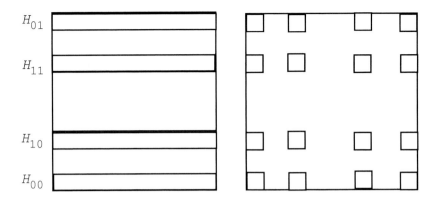

Figure 7.34: $f^{-2}(D) \cap D$ and its intersection with $f^2(D) \cap D$

is given, in geometry, as the intersection of $f^2(D) \cap D$ in Figure 7.33 and $f^{-2}(D) \cap D$ in Figure 7.34, and is shown as sixteen small rectangles in the second picture of Figure 7.34.

Now, if we define

$$\Lambda = \bigcap_{j=-\infty}^{\infty} f^j(D) = \cdots f^{-2}(D) \cap f^{-1}(D) \cap D \cap f(D) \cap f^2(D) \cdots, \quad (6.1)$$

then, since a decreasing intersection of compact sets is nonempty, Λ is well defined and nonempty. And, due to its construction, Λ is (see an exercise) invariant under any map $f^j, j \in \{\pm 1, \pm 2, \cdots\}$. In geometry, Λ consists of an infinite number of points which are the intersections of the vertical lines in $\cap_{j=0}^{\infty} f^j(D)$ and the horizontal lines in $\cap_{j=0}^{\infty} f^{-j}(D)$. Therefore, each point $p \in \Lambda$ can be identified with a pair of perpendicular lines with the vertical line from $\cap_{j=0}^{\infty} f^j(D)$ and the horizontal line from $\cap_{j=0}^{\infty} f^{-j}(D)$. Thus each point $p \in \Lambda$ can be identified with an infinite sequence $(s_0 s_1 s_2 \cdots)_V$ and an infinite sequence $(t_1 t_2 t_3 \cdots)_H$ of 0's and 1's, and hence we can define a map, ϕ, from Λ to $\sum$, the collection of all bi-infinite sequences of 0's and 1's, by

$$\phi : p \in \Lambda \longrightarrow \{\cdots s_{-3} s_{-2} s_{-1} \, ; \, s_0 s_1 s_2 \cdots\} \in \sum, \quad (6.2)$$

where we write $(t_1 t_2 t_3 \cdots)_H$ as $\{\cdots s_{-3} s_{-2} s_{-1}\}$ to indicate that it is related to "negative" maps $f^{-j}, j \geq 1$. (Here, $s_{-1} = t_1, s_{-2} = t_2$ and so on.) It can be shown (see an exercise) that the map ϕ is one-to-one, onto, continuous, and ϕ^{-1} is also continuous. Therefore, the study of the dynamics of Λ is now the same as the study of the dynamics of the bi-infinite sequences $\{\cdots s_{-3}s_{-2}s_{-1}\,;\,s_0s_1s_2\cdots\}$ of 0's and 1's, which is called **symbolic dynamics**.

Consider $f(\Lambda)$, which maps the position of $f^{-1}(D)$ to the position of D, the position of D to the position of $f(D)$ and so on, we find that the effect of f on Λ is to shift the position of D to the left by one position. Now, if we regard the position of D in (6.1) as ";" of the bi-infinite sequences $\{\cdots s_{-3}s_{-2}s_{-1}\,;\,s_0s_1s_2\cdots\}$, then the counterpart of f on Λ will be a **shift map** σ on the bi-infinite sequences $\{\cdots s_{-3}s_{-2}s_{-1}\,;\,s_0s_1s_2\cdots\}$ of $\sum$ such that σ shifts the ";" of a sequence one place to the left. For example, for a bi-infinite sequence $s = \{\cdots 1010\,;\,1010\cdots\}$ with 0 and 1 alternating, we have

$$\sigma(s) = \{\cdots 0101\,;\,0101\cdots\},$$
$$\sigma^2(s) = \sigma(\{\cdots 0101\,;\,0101\cdots\}) = \{\cdots 1010\,;\,1010\cdots\} = s,$$

which implies that s is a period-2 point of the map σ, or $\{s,\,\sigma(s)\}$ is a period-2 orbit of the map σ. Similarly, $\{\cdots 100100\,;\,100100\cdots\}$ (constructed using "100") is a period-3 point of the map σ. In fact, in this fashion, we can find period-j points of the map σ for any $j \geq 1$. Moreover, if a bi-infinite sequence periodically repeats after some fixed length, k, then the sequence is a period-k point of the map σ. And each period-j point of the map σ is given by a bi-infinite sequence periodically repeating a block of length j.

Next, let's look at whether the map σ has nonperiodic orbits. To this end, we rewrite bi-infinite sequences of $\sum$ as follows

$$\{\cdots s_{-3}s_{-2}s_{-1}\,;\,s_0s_1s_2\cdots\} \longrightarrow 0.s_0s_1s_{-1}s_2s_{-2}\cdots, \tag{6.3}$$

which are the base-2 expansions of numbers in the interval $[0,1]$. Now, it is known that $[0,1]$ has an uncountable number of irrational numbers given as nonrepeating base-2 expansions in (6.3), which correspond to nonrepeating bi-infinite sequences of $\sum$. Since these nonrepeating bi-infinite sequences of $\sum$ are now nonperiodic points of the map σ, we conclude that σ has an uncountable number of nonperiodic orbits.

Moreover, for

$$s = \{\cdots s_{-3}s_{-2}s_{-1}\,;\,s_0s_1s_2\cdots\}, \quad \bar{s} = \{\cdots \bar{s}_{-3}\bar{s}_{-2}\bar{s}_{-1}\,;\,\bar{s}_0\bar{s}_1\bar{s}_2\cdots\} \in \sum,$$

the distance between s and $\bar{s}$ can be defined to be

$$d(s,\bar{s}) = \sum_{i=-\infty}^{\infty} \frac{\delta_i}{2^{|i|}}, \quad \delta_i = \begin{cases} 0 & \text{if } s_i = \bar{s}_i, \\ 1 & \text{if } s_i \neq \bar{s}_i, \end{cases} \tag{6.4}$$

which can be used to show that the map σ has **sensitive dependence on initial conditions on** $\sum$, and that σ on $\sum$ has a dense orbit from some $s \in \sum$ such that for any $s' \in \sum$ and any $\varepsilon > 0$, there exists some integer j with $d(\sigma^j(s),\ s') < \varepsilon$.

Now, with the above discussions, we can state the following result of Smale [1963].

Theorem 7.6.2 *Consider the Smale horseshoe map f on* Λ *(or the shift map* σ *on* $\sum$*), it has*

1. *a countable infinity of periodic orbits of arbitrarily large period,*

2. *an uncountable infinity of nonperiodic orbits,*

3. *a dense orbit,*

4. *a chaotic invariant set* Λ *(or* $\sum$*).*   ♠

Therefore, in addition to the logistic map and the Lorenz system, the Smale horseshoe map provides an example of an invariant chaotic set having the properties stated in Theorem 7.6.2. In particular, property 1 of Theorem 7.6.2 is the same as a property of the logistic map $x \to \mu x(1-x)$ with μ near 3.84, where the logistic map has a period-3 orbit, and then the Sarkovskii theorem 7.4.1 and Li and Yorke's [1975] paper "Period Three Implies Chaos" imply that the logistic map has period-j orbits for any $j \geq 1$.

Exercises 7.6

1. Verify that f on H_0 and H_1 are given by

$$H_0 \quad : \quad \begin{bmatrix} x \\ y \end{bmatrix} \longrightarrow \begin{bmatrix} \lambda & 0 \\ 0 & \frac{1}{\eta} \end{bmatrix} \begin{bmatrix} x \\ y \end{bmatrix},$$

$$H_1 \quad : \quad \begin{bmatrix} x \\ y \end{bmatrix} \longrightarrow \begin{bmatrix} -\lambda & 0 \\ 0 & -\frac{1}{\eta} \end{bmatrix} \begin{bmatrix} x \\ y \end{bmatrix} + \begin{bmatrix} 1 \\ \frac{1}{\eta} \end{bmatrix}.$$

2. Prove Lemma 7.6.1.

3. Verify that f^{-1} restricted to $V_0 \cup V_1$ has the geometry in Figure 7.31.

4. Prove that Λ in (6.1) is well defined, nonempty, and is invariant under any map $f^j, j \in \{\pm 1, \pm 2, \cdots\}$.

5. Prove that ϕ in (6.2) is one-to-one, onto, continuous, and ϕ^{-1} is also continuous.

6. Find all period-j points of the map σ for $j = 1, 2, 3$.

7. Verify that we can find period-j points of the map σ for any $j \geq 1$.

8. Verify that if a bi-infinite sequence periodically repeats after some fixed length, k, then the sequence is a period-k point of the map σ; and each period-j point of the map σ is given by a bi-infinite sequence periodically repeating a block of length j.

9. Prove that $d(s, \bar{s})$ in (6.4) defines a metric on Σ.

10. Prove that the map σ on Σ has sensitive dependence on initial conditions.

11. Prove that the map σ on Σ has a dense orbit.

Chapter 8

Dynamical Systems

8.1 Introduction

From Section 1 of Chapter 4, we know that the solutions of

$$x' = f(x), \quad x \in \Re^2, \tag{1.1}$$

define a dynamical system, which describes how a point $x_0 \in \Re^2$ moves with respect to the time t according to the solution $x(t, x_0)$ of Eq. (1.1). We also know from Theorem 4.1.9 in Chapter 4 that all possible trajectories in $\Re^2$ are either critical points, periodic orbits (simple closed curves), or nonintersecting curves.

In sections 2 and 3 of Chapter 4, we also discussed the phase portraits of Eq. (1.1) near critical points. The study there can be characterized as the **local properties**, because they only describe the trajectories that are sufficiently close to the critical points.

In this chapter, we will do two things. First, we continue the study of Eq. (1.1) and look at the movements of the trajectories "globally" over the entire $\Re^2$ and try to derive some **global properties**. These are properties concerning the geometrical relationship between critical points, periodic orbits, and nonintersecting curves. Next, we will extend certain "local" properties in $\Re^2$ to differential equations in $\Re^n, n \geq 1$.

In the study of the local properties, critical points play an important role. For the global properties, however, we will see that periodic orbits are the main subject of study. We have seen that solutions of differential equations in $\Re^n, n \geq 3$, can exhibit some chaotic behavior, such as the solutions of the Lorenz system in $\Re^3$. However, we will see for Eq. (1.1) in $\Re^2$ that we have a relatively simple geometry for the relationship between critical points, periodic orbits, and nonintersecting curves, due to the famous Poincaré-Bendixson theorem that we will derive in this chapter.

Next, we state the Jordan curve theorem, because it is used either explicitly or implicitly when we deal with the planar geometry. See **Figure 8.1**.

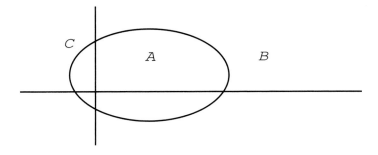

Figure 8.1: The Jordan curve theorem

Theorem 8.1.1 (Jordan curve theorem) *Let C be a simple closed curve in $\Re^2$. Then*

$$\Re^2 - C = A \cup B,$$

where A and B are disjoint nonempty connected open sets such that

1. *The curve C is the boundary of A and of B.*

2. *One of the open set, say for example, A, is bounded (called the interior of C) and the other, B, is unbounded (called the exterior of C).* ♠

The Jordan curve theorem looks "obvious." However, a rigorous proof is lengthy and difficult, because we are dealing with the planar geometry, so sometimes distinguishing between intuitive and rigorous arguments is not easy. See Cronin [1994], for example, for a proof.

This chapter is organized as follows: In Section 2, we study the dynamics in $\Re^2$ and prove the Poincaré-Bendixson theorem. In Section 3, we use the Poincaré-Bendixson theorem, together with other results, to obtain existence and nonexistence of limit cycles, which in turn help us determine the global properties of planar systems. In Section 4, we apply the results to a Lotka-Volterra competition equation. In Section 5, we study invariant manifolds and the Hartman-Grobman theorem, which generalize certain results for planar equations in Chapter 4 to differential equations in $\Re^n$.

Exercises 8.1

1. Find the solutions of

$$\begin{cases} x'(t) = & y - x(x^2 + y^2 - 1), \\ y'(t) = & -x - y(x^2 + y^2 - 1), \end{cases}$$

in a polar system. Show that the circle $x^2 + y^2 = 1$ is a trajectory of the equation. Next, if $(x(t), y(t))$ is a solution of the equation, find $\lim_{t \to \infty} [x^2(t) + y^2(t)]$.

8.2   Poincaré-Bendixson Theorem in $\Re^2$

To begin, we introduce the first, and maybe the most important definition here: "limit points." The definition is given here for planar equations and is also valid for equations in $\Re^n, n \geq 1$.

Definition 8.2.1 *Let $x(t, x_0)$ be the unique solution of the planar equation (1.1) that exists on $\Re$ with $x(0) = x_0$. A point $p \in \Re^2$ is called an ω-**limit point** of the trajectory $x(t, x_0)$ if there exists a sequence $t_m \to \infty$ as $m \to \infty$ such that*

$$\lim_{m \to \infty} x(t_m, x_0) = p.$$

*The set of all ω-limit points of the trajectory $x(t, x_0)$ is called the ω-**limit set** of $x(t, x_0)$, and is denoted by $\omega(x_0)$ or $\omega(\Gamma)$ where Γ denotes the trajectory $x(t, x_0)$. If we want to specify that Γ is the trajectory starting at x_0 when $t = 0$, then we write Γ as Γ_{x_0}.*

*Similarly, a point $q \in \Re^2$ is called an α-**limit point** of the trajectory $x(t, x_0)$ if there exists a sequence $t_m \to -\infty$ as $m \to \infty$ such that*

$$\lim_{m \to \infty} x(t_m, x_0) = q.$$

*The set of all α-limit points of the trajectory $x(t, x_0)$ is called the α-**limit set** of $x(t, x_0)$, and is denoted by $\alpha(x_0)$ or $\alpha(\Gamma)$ where Γ denotes the trajectory $x(t, x_0)$.*

If $x(t)$ is a solution of Eq. (1.1), then $y(t) = x(-t)$ satisfies $y'(t) = -x'(-t) = -f(x(-t)) = -f(y(t))$. Thus an α-limit point of $x(t)$ can be regarded as an ω-limit point of a solution of another system, therefore, any result about an ω-limit set is also true for an α-limit set.

Example 8.2.2 If a trajectory of a differential equation is shown in **Figure 8.2**, then the ω-limit set is given by the straight lines $x_2 = \pm 1$. ♠

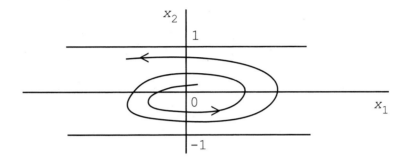

Figure 8.2: An ω-limit set given by the straight lines $x_2 = \pm 1$

Example 8.2.3 If the trajectories of a differential equation are shown in **Figure 8.3**, then the ω-limit set is given by the unit circle. ♠

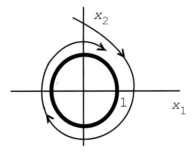

Figure 8.3: An ω-limit set given by the unit circle

Example 8.2.4 If the origin $(0,0)$ is a saddle point shown in **Figure 8.4**, then for a point p that is on the x_1-axis, the ω-limit set $\omega(p)$ is $(0,0)$; otherwise, the ω-limit set $\omega(p)$ is empty if p is not on the x_1-axis. ♠

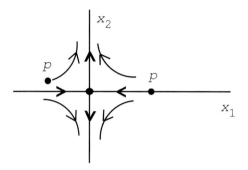

Figure 8.4: $\omega(p) = (0,0)$ for p on the x_1-axis; otherwise $\omega(p)$ is empty

---

Other related definitions concerning the trajectories are given below.

Definition 8.2.5 *Let $x(t, x_0)$ be the unique solution of the planar equation (1.1) that exists on $\Re$ with $x(0) = x_0$, and let*

$$\Gamma_{x_0} = \{x(t, x_0) \in \Re^2 : t \in \Re\},$$

be the corresponding trajectory. We use

$$\Gamma_{x_0}^+ = \{x(t, x_0) \in \Re^2 : t \geq 0\},$$

and

$$\Gamma_{x_0}^- = \{x(t, x_0) \in \Re^2 : t \leq 0\},$$

*to denote the **positive half-trajectory** and the **negative half-trajectory** through x_0 respectively.*

Definition 8.2.6 *A set $M \subset \Re^2$ is called a **positively (negatively) in-variant set** of Eq. (1.1) if for each $x_0 \in M$, $\Gamma_{x_0}^+ \subset M$ ($\Gamma_{x_0}^- \subset M$). A set $M \subset \Re^2$ is called **an invariant set** of Eq. (1.1) if for each $x_0 \in M$, $\Gamma_{x_0} \subset M$.*

Accordingly, any trajectory is itself an invariant set. The following are some general results concerning limit sets.

Theorem 8.2.7 *The ω-limit set $\omega(\Gamma)$ and the α-limit set $\alpha(\Gamma)$ of a trajectory Γ are closed and invariant. Furthermore, if Γ^+ (Γ^-) is a bounded set of $\Re^2$, then $\omega(\Gamma)$ ($\alpha(\Gamma)$) is a nonempty, compact (that is, bounded and closed), and connected set in $\Re^2$.*

Proof. We only prove the results for $\omega(\Gamma)$ since the case for $\alpha(\Gamma)$ is similar. To show that $\omega(\Gamma)$ is closed, we let $q_m \in \omega(\Gamma)$ with $q_m \to q$, $m \to \infty$, and verify that $q \in \omega(\Gamma)$. Let $\Gamma = \Gamma_{x_0}$. Since each q_m, $m = 1, 2, \cdots$, is an ω-limit point of Γ_{x_0}, there exists, for each m, a sequence $t_m \geq m$ such that

$$|x(t_m, x_0) - q_m| \leq \frac{1}{m}.$$

Now,

$$|x(t_m, x_0) - q| \leq |x(t_m, x_0) - q_m| + |q_m - q| \leq \frac{1}{m} + |q_m - q| \to 0, \quad m \to \infty,$$

therefore, $q \in \omega(\Gamma)$, hence $\omega(\Gamma)$ is closed.

To show that $\omega(\Gamma)$ is invariant, let $\Gamma = \Gamma_{x_0}$ be the trajectory of $x(t, x_0)$ and let $q \in \omega(\Gamma)$. Then there is a sequence $t_m \to \infty$ such that $x(t_m, x_0) \to q$ as $m \to \infty$. Now, consider $x(\bar{t}, q)$ for any fixed $\bar{t} \in \Re$. We have, according to the properties of a dynamical system,

$$x(\bar{t} + t_m, x_0) = x(\bar{t}, x(t_m, x_0)) \to x(\bar{t}, q), \quad m \to \infty,$$

thus, $x(\bar{t}, q) \in \omega(\Gamma)$, and then $\Gamma_q \subset \omega(\Gamma)$ since $\bar{t} \in \Re$ is arbitrary. Therefore, $\omega(\Gamma)$ is invariant.

Next, if Γ^+ is bounded in $\Re^2$, then any sequence $x(t_m, x_0)$, $t_m \geq 0$, is bounded and hence has a convergent subsequence. This implies that $\omega(\Gamma)$ is nonempty. Now, $\omega(\Gamma)$ is bounded since Γ^+ is bounded, hence $\omega(\Gamma)$ is compact since we have just proved that $\omega(\Gamma)$ is closed.

If $\omega(\Gamma)$ is not connected when Γ^+ is bounded in $\Re^2$, then, as $\omega(\Gamma)$ is compact, there exist nonempty disjoint compact sets L and R such that $\omega(\Gamma) = L \cup R$. Now, the distance between L and R,

$$d_0 = d(L, R) = \inf_{a \in L, b \in R} |a - b|,$$

is positive. Let Γ be the trajectory of $x(t, x_0)$. Accordingly, since $\omega(\Gamma) = L \cup R$, there exist $t'_m \to \infty$ and $t''_m \to \infty$ with $t'_m < t''_m < t'_{m+1}$ such that

$$d\Big(x(t'_m, x_0), L\Big) < \frac{d_0}{3}, \quad \text{and} \quad d\Big(x(t''_m, x_0), R\Big) < \frac{d_0}{3}, \tag{2.1}$$

see **Figure 8.5**. Now, as the curve of $x(t, x_0)$ is connected and the distance function is continuous, there exists a sequence t_m^* with $t_m' < t_m^* < t_m''$ such that

$$d\Big(x(t_m^*,\ x_0),\ L\Big) \geq \frac{d_0}{3}, \quad \text{and} \quad d\Big(x(t_m^*,\ x_0),\ R\Big) \geq \frac{d_0}{3}. \qquad (2.2)$$

Note that the sequence $x(t_m^*,\ x_0)$, $m = 1, 2, \cdots$, is bounded, hence it has a convergent subsequence, whose limit is in $\omega(\Gamma)$, but not in $L \cup R$ according to (2.2), a contradiction since $L \cup R = \omega(\Gamma)$. This completes the proof. ♠

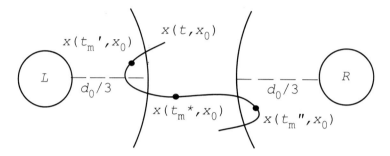

Figure 8.5: $\omega(\Gamma) = L \cup R$ with L and R disconnected, and the solution $x(t, x_0)$

Corollary 8.2.8 *If Γ^+ (Γ^-) is bounded in $\Re^2$ and $\omega(\Gamma)$ $(\alpha(\Gamma))$ contains a periodic orbit, then $\omega(\Gamma)$ $(\alpha(\Gamma))$ coincides with this periodic orbit.* ♠

In Figure 8.2 in Example 8.2.2 where the trajectory is unbounded, the ω-limit set, shown as the straight lines $x_2 = \pm 1$, is nonempty, invariant, closed, but not bounded, and not connected. In Example 8.2.3, the ω-limit set, shown as the unit circle in Figure 8.3, is invariant, compact, and connected.

We have seen in Chapter 4 that a critical point of a planar differential equation means "directionless" in geometry, thus the trajectories near a critical point could go in all different directions; while for a regular point, the trajectories nearby will go in essentially the same direction. The following two results clarify this even further. Recall that $B_d(p) = \{x \in \Re^n : |x - p| \leq d\}$ is used to denote the ball centered at p with radius d. However, if we look at $|x| = |x_1| + |x_2|$ carefully for $x = (x_1, x_2) \in \Re^2$, we find that the "ball" $B_d(p)$ is actually a square. To clear this up, we note that $|x_1| + |x_2|$ is equivalent to the commonly used norm $\sqrt{x_1^2 + x_2^2}$, thus all results stated with $|x_1| + |x_2|$ will also be true if $|x_1| + |x_2|$ is replaced by $\sqrt{x_1^2 + x_2^2}$. Therefore, in the rest of this chapter, especially when we draw graphs in $\Re^2$, we will use $|x|$ to denote $\sqrt{x_1^2 + x_2^2}$. Hence, the "ball" $B_d(p)$ is still a round ball.

Theorem 8.2.9 *If p is a regular point of Eq. (1.1), then there exists a $\delta > 0$ such that for any $q \in B_\delta(p)$, the trajectory of $x(t, q)$ will leave $B_\delta(p)$ as $|t| \to \infty$. That is, there exists a $T > 0$ such that $x(\pm T, q) \notin B_\delta(p)$.*

Proof. Since p is a regular point, we know from Corollary 4.1.10 in Chapter 4 that there exists a $T > 0$ such that $p_1 = x(-T, p) \neq p$. Now we must have $p_2 = x(T, p) \neq p$, because otherwise, $y(t) = x(T + t, p)$ is also a solution of Eq. (1.1) with $y(0) = x(T, p) = p$. Then uniqueness implies that $x(t, p) = y(t) = x(T + t, p)$, or $x(t, p)$ is T-periodic on $\Re$, and hence $x(-T, p) = p$, a contradiction.

Let $d = \min\{|p - p_1|, |p - p_2|\} > 0$. From the continuity of a dynamical system (or continuous dependence on initial data), we know that when w is in a compact set of $\Re^2$, $x(t, w)$ is continuous in w uniformly for $t \in [-T, T]$. Accordingly, there exists a $\delta = \delta(d) > 0$ and $\delta \leq \frac{d}{3}$ such that if $|p - q| \leq \delta$ (or $q \in B_\delta(p)$) then $|x(t, p) - x(t, q)| \leq \frac{d}{3}$ for $t \in [-T, T]$, see **Figure 8.6**. Hence, $|p_2 - x(T, q)| = |x(T, p) - x(T, q)| \leq \frac{d}{3}$. Since

$$d \leq |p - p_2| \leq |p - x(T, q)| + |x(T, q) - p_2| \leq |p - x(T, q)| + \frac{d}{3},$$

we have

$$|p - x(T, q)| \geq \frac{2d}{3} > \frac{d}{3} \geq \delta.$$

That is, $x(t, q)$ leaves $B_\delta(p)$ at T. The same is true for $-T$. This completes the proof. ♠

For a regular point p of Eq. (1.1), $f(p) \neq 0$. Thus in geometry, the direction vector $f(p)$ at p is a nonzero vector. We can now draw a normal (perpendicular) line N_p with respect to $f(p)$, and call the side in the direction of $f(p)$ as the **positive side** of N_p and the other side as the **negative side** of N_p, see **Figure 8.7**.

Theorem 8.2.10 *If p is a regular point of Eq. (1.1), then there exists an $\varepsilon > 0$ such that any trajectory of Eq. (1.1) cannot enter the ball $B_\varepsilon(p)$ from*

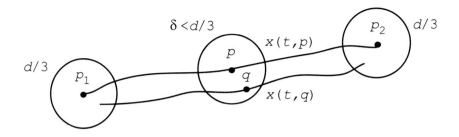

Figure 8.6: $|x(t,p) - x(t,q)| \leq \frac{d}{3}$ for $t \in [-T, T]$

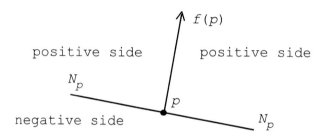

Figure 8.7: The positive and negative sides of N_p of a regular point

the positive side of the normal line N_p. If a trajectory of Eq. (1.1) enters $B_{\varepsilon/2}(p)$ from the negative side, then it intersects the normal line N_p and then leaves $B_\varepsilon(p)$. See figures 8.8 and 8.9.

Proof. Since $f(p) \neq 0$ and $f(x)$ is continuous in x, we have, as $q \to p$,

$$\cos\theta = \frac{f(p) \cdot f(q)}{|f(p)||f(q)|} = \frac{f(p) \cdot f(p)}{|f(p)||f(q)|} + \frac{f(p) \cdot [f(q) - f(p)]}{|f(p)||f(q)|} \to 1,$$

where θ is the angle between $f(p)$ and $f(q)$. Hence, there exists an $\varepsilon > 0$ such that if $q \in B_{3\varepsilon}(p)$ then $f(q) \neq 0$ and the angle θ between $f(p)$ and $f(q)$ is at most $30°$. Now draw **Figure 8.8**, where p is the center, the small circle is of radius ε, the middle half circle is of radius 2ε, and the larger half circle is of radius 3ε.

Draw the straight lines $\overline{AB}$ and $\overline{BC}$ as shown in Figure 8.8. Then from the planar geometry, the two straight lines will not intersect the small circle

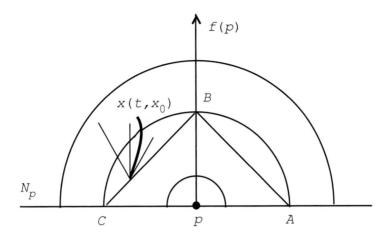

Figure 8.8: A trajectory cannot enter $B_\varepsilon(p)$ from positive side of N_p

with radius ε (see an exercise). If there is an $x_0 \in \Re^2$ and $t_0 \in \Re$ such that $x(t_0, x_0)$ is on one of the two straight lines, then when t increases a little, that is, when $t > t_0$ and near t_0, the corresponding solution curve $x(t, x_0)$ will remain inside a sector whose angle is $60°$ and whose center line is parallel to $f(p)$, as shown in 8.8. In other words, the straight lines $\overline{AB}$ and $\overline{BC}$ serve as a "shield" or a "wall" that bounces back any trajectory trying to get close to p from the positive side of the normal line N_p. Now, the circle with center p and radius ε is protected, therefore, any trajectory of Eq. (1.1) cannot enter the ball $B_\varepsilon(p)$ from the positive side of the normal line N_p.

If a trajectory $x(t, q)$ enters $B_{\varepsilon/2}(p)$ from the negative side of N_p at $t = t_1$, then at the point $x(t_1, q)$ we draw a sector whose angle is $60°$ and whose center line is parallel to $f(p)$, as shown in **Figure 8.9**. According to the planar geometry, the side lines of this sector intersect N_p at the points a and b that are at most away from p by a distance of ε.

Similar to the above, when t increases from t_1, $x(t, q)$ will remain inside of this sector (as long as $x(t, q)$ is inside $B_{3\varepsilon}(p)$). Finally, from Theorem 8.2.9, when ε is small (say for example, $\varepsilon \leq \delta$ where δ is from Theorem 8.2.9), $x(t, q)$ will leave $B_\varepsilon(p)$ as $t \to \infty$. Therefore, $x(t, q)$ has to intersect the normal line N_p and then leave $B_\varepsilon(p)$. This completes the proof. ♠

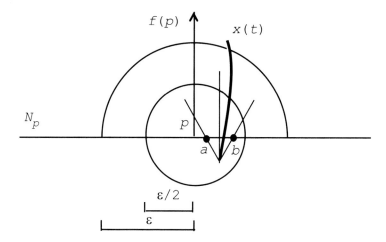

Figure 8.9: A trajectory from $B_{\varepsilon/2}(p)$ intersects N_p and then leaves $B_\varepsilon(p)$

We will see that Theorem 8.2.10 is a very useful tool in analyzing the trajectories near a regular point.

Note that if $x(t, x_0)$ is a critical point or a periodic orbit, then any point on $x(t, x_0)$ is an ω-limit point and an α-limit point of $x(t, x_0)$. The next result says that they are the only possible cases in $\Re^2$.

Theorem 8.2.11 *If a trajectory $x(t, p)$ of Eq. (1.1) contains an ω-limit point or an α-limit point of $x(t, p)$, then $x(t, p)$ is either a critical point or a periodic orbit.*

Proof. Assume that the trajectory $x(t, p)$ contains an ω-limit point Q of $x(t, p)$, (the case for an α-limit point is similar). If $x(t, p)$ is neither a critical point nor a periodic orbit, then from Theorem 4.1.9 in Chapter 4, $x(t, p)$ is a nonintersecting curve. Now, Q is a regular point (otherwise $x(t, p)$ is a critical point), so from Theorems 8.2.9 and 8.2.10, there exists an $\varepsilon > 0$ such that $x(t, Q)$ leaves $B_\varepsilon(Q)$ from the positive side of the normal line N_Q as t increases. Since $x(t, p)$ and $x(t, Q)$ describe the same trajectory, and Q is an ω-limit point of $x(t, p)$, $x(t, Q)$ will get close to Q for some large t. Using Theorem 8.2.10, $x(t, Q)$ has to enter $B_{\varepsilon/2}(Q)$ from the negative side, then $x(t, Q)$ intersects the normal line N_Q at some point denoted by Z before leaving $B_\varepsilon(Q)$, and $|Q - Z| \leq \varepsilon$. Since $x(t, p)$ is a nonintersecting curve,

$Q \neq Z$. Then, since $x(t,Q)$ cannot enter $B_\varepsilon(Q)$ from the positive side of N_Q, we only have the two possibilities shown in **Figure 8.10**.

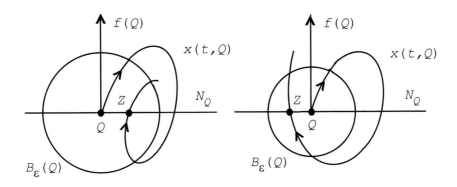

Figure 8.10: Two possibilities when $Q \neq Z$

According to Figure 8.10, where Jordan curves can be formed, $x(t,Q)$ cannot get close to Q when t is large because $x(t,Q)$ cannot intersect itself and cannot enter $B_\varepsilon(Q)$ from the positive side of the normal line N_Q. Therefore, Q cannot be an ω-limit point of $x(t,Q)$ (or $x(t,p)$). This is a contradiction, and hence we complete the proof. ♠

Now, we are ready to prove the most important result for planar autonomous differential equations: the Poincaré-Bendixson theorem, which describes the relationship between critical points, periodic orbits, and non-intersecting curves of Eq. (1.1), and indicates that trajectories for planar autonomous differential equations behave in an **orderly** fashion as compared to what may happen for differential equations in $\Re^n$, $n \geq 3$.

Theorem 8.2.12 (Poincaré-Bendixson theorem) *If a trajectory $x(t,p)$ of Eq. (1.1) is bounded in $\Re^2$ for $t \geq 0$ (the corresponding results for $t \leq 0$ are also true), then one of the following three things must happen:*

  1. the ω-limit set of $x(t,p)$, $\omega(p)$, contains a critical point,

  2. $x(t,p)$ is a periodic orbit,

  3. $\omega(p)$ is a periodic orbit, and $x(t,p)$ approaches $\omega(p)$ spirally as $t \to \infty$.

Proof.   Assume that $\omega(p)$ contains no critical points and that $x(t, p)$ is not a periodic orbit, then we show that $\omega(p)$ is a periodic orbit and $x(t, p)$ approaches this periodic orbit in a spiral way. From Theorem 8.2.7, $\omega(p)$ is nonempty. Let $A \in \omega(p)$, then A is a regular point. Since $\omega(p)$ is invariant from Theorem 8.2.7, $x(t, A) \in \omega(p)$, $t \in \Re$. As Γ_p^+ is bounded, $\omega(p)$ is also bounded, hence the trajectory $x(t, A) \in \omega(p)$ is bounded. Therefore, $\omega(A)$ is nonempty. Let $B \in \omega(A)$. As $x(t, A) \in \omega(p)$ and $\omega(p)$ is closed from Theorem 8.2.7, $B \in \omega(p)$, and hence B is a regular point.

We first verify that the trajectory $x(t, p)$ does not intersect the trajectory $x(t, A)$. Otherwise the two trajectories coincide using uniqueness, and hence $x(t, p)$ contains its ω-limit point A. Then, from Theorem 8.2.11, $x(t, p)$ must be a critical point or a periodic orbit, a contradiction.

Next, we prove that $x(t, A)$ is a periodic orbit. To do this, we show that the point B is on the trajectory $x(t, A)$, which implies that $x(t, A)$ contains its ω-limit point B, hence, from Theorem 8.2.11, $x(t, A)$ must be a periodic orbit. Since $B \in \omega(A)$, $x(t, A)$ will get close to B for some large t. Using Theorem 8.2.10, $x(t, A)$ has to enter $B_{\varepsilon/2}(B)$ from the negative side of the normal line N_B and then intersect the normal line N_B at some point denoted by Z before leaving $B_\varepsilon(B)$. If B is not on the trajectory $x(t, A)$, then $B \neq Z$, so we only have the two possibilities shown in **Figure 8.11**, where Jordan curves can be formed.

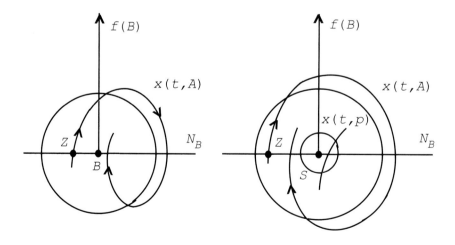

Figure 8.11: Two possibilities when $B \neq Z$

In the first case, similar to the last part of the proof of Theorem 8.2.11, we find that $B \notin \omega(A)$, a contradiction (unless $x(t, A)$ intersects itself, in which case $x(t, A)$ must be a periodic orbit). In the second case, for a sufficiently small circle S around B as shown in Figure 8.11, $x(t_*, p) \in S$ for some $t_* > 0$ since $B \in \omega(p)$. Now, $x(t, p)$ cannot get close to Z by entering the ball $B_\varepsilon(B)$ from the positive side of the normal line N_B; and $x(t, p)$ cannot intersect $x(t, A)$, thus $x(t, p)$ cannot get close to Z by entering $B_\varepsilon(B)$ from the negative side of N_B either. Therefore, $Z \notin \omega(p)$, a contradiction to the fact that $Z \in x(t, A) \subset \omega(p)$. Thus, B must be on the trajectory $x(t, A)$, therefore $x(t, A)$ is a periodic orbit. Since $x(t, A) \in \omega(p)$, $\omega(p)$ must be the same as this periodic orbit $x(t, A)$, using Corollary 8.2.8.

Next, we prove that $x(t, p)$ approaches the periodic orbit $\omega(p)$ in a spiral way. Assume that $x(t, p)$ is inside the periodic orbit $\omega(p)$. (The case when $x(t, p)$ is outside is the same.) Let H be any point on $\omega(p)$, then H is a regular point. There exists an $\varepsilon > 0$ such that $x(t, p)$ gets close to H as shown in **Figure 8.12**.

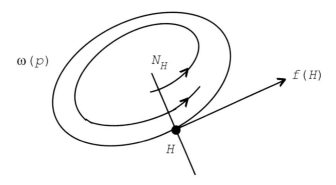

Figure 8.12: $x(t, p)$ approaches the periodic orbit $\omega(p)$ spirally

---

Accordingly, there exist $t_m \to \infty$, $t_m < t_{m+1}$, such that $x(t_m, p)$ intersects the normal line N_H and $x(t_{m+1}, p)$ is between $x(t_m, p)$ and H. Therefore, as H is an arbitrary point on $\omega(p)$, $x(t, p)$ approaches the periodic orbit $\omega(p)$ spirally. The proof is complete. ♠

From the study of phase portraits in sections 2 and 3 in Chapter 4, we find that when the origin $(0, 0)$ is an isolated critical point and also a stable spiral point, then $(0, 0)$ is the ω-limit set of trajectories. That is, trajectories "pile up" at the critical point $(0,0)$, and periodic orbits do not exist. However, if $x(t, p)$, $t \geq 0$, is bounded and has no critical points as its ω-limit points, then $x(t, p)$ has to "pile up" at somewhere other than a single point. According to the Poincaré-Bendixson theorem, $x(t, p)$ piles up at a periodic orbit. Therefore, solutions of the planar autonomous differential equations behave in an orderly fashion, compare to what may happen to solutions of differential equations in $\Re^n$, $n \geq 3$, such as the Lorenz system.

An important consequence of the Poincaré-Bendixson theorem is the following result describing the geometrical relationship between periodic orbits and critical points.

Theorem 8.2.13 *Every periodic orbit of Eq. (1.1) has a critical point inside its interior.*

Proof. Let C be the periodic orbit and denote R the interior of C together with C. Then periodic orbits exist in R, since C is at least one such periodic orbit. If C_* is a periodic orbit in R, then we use R_* to denote the interior of C_* together with C_*. Consider a nonempty set $\{C_\mu\}$ of periodic orbits in R and a nonempty set $\{R_\mu\}$ of compact regions that is partially ordered by the inclusion relation. By Hausdorff's maximality principle, there exists a maximal chain $\{R_\tau\}$ of this partially ordered set $\{R_\mu\}$. Let $\Delta = \cap R_\tau$, then Δ is nonempty since each R_τ is compact. Let $p \in \Delta$, then the trajectory $x(t, p)$, $t \in \Re$, is contained in Δ since it is in each R_τ.

Suppose that Δ does not contain critical points. If $x(t, p)$ is not a periodic orbit, then from the Poincaré-Bendixson theorem, its ω-limit set $\omega(p)$ and α-limit set $\alpha(p)$ are both periodic orbits in Δ. Now, $\omega(p)$ and $\alpha(p)$ must be different and one is entirely inside the interior of the other one (see an exercise). Assume $\alpha(p)$ is inside $\omega(p)$. Then p cannot be on $\alpha(p)$ because $x(t, p)$ is not periodic; and p cannot be inside $\alpha(p)$ because otherwise $x(t, p)$ cannot get out of $\alpha(p)$ to approach $\omega(p)$ for large t. Therefore, $\alpha(p)$ is not in $\{C_\tau\}$ since $p \in \Delta = \cap R_\tau$. Now, $\alpha(p)$ can be added to $\{C_\tau\}$ using the inclusion relation since $\alpha(p)$ is inside each R_τ. This contradicts the fact that $\{R_\tau\}$ is a maximal chain.

If $x(t, p)$ is a periodic orbit, then, as Δ contains no critical points, there is a point q in the interior of $x(t, p)$ which will result in at least one periodic orbit C_* that is in the interior of $x(t, p)$. Now, $p \in \Delta = \cap R_\tau$ and p is not on or inside C_*, therefore, similar to the above case, C_* is not in $\{C_\tau\}$ and hence can be added to $\{C_\tau\}$. Thus, $\{R_\tau\}$ is not maximal, a contradiction. Therefore, $\Delta \subset R$ must contain critical points. This completes the proof. ♠

The above proof uses Hausdorff's maximality principle. Next, let's give a concrete proof when the region in $\Re^2$ enclosed by the periodic orbit is convex or can be made to become convex in a continuous one-to-one fashion.

Proof of Theorem 8.2.13 when the periodic orbit encloses a convex region. Let R be the compact region in $\Re^2$ enclosed by the periodic orbit such that R is convex. For any $T_1 > 0$ fixed, consider the mapping $p \to x(T_1, p)$. This mapping is from R to R because a trajectory from inside of R cannot get outside of R using uniqueness. From Brouwer's first fixed point theorem (see the Appendix), there is a $p_1 \in R$ such that $x(T_1, p_1) = p_1$. Thus, there are $T_m > 0$, $T_m \to 0$, and $p_m \in R$ such that $x(T_m, p_m) = p_m$. Hence $x(t, p_m)$ is T_m-periodic. Using a subsequence if necessary, we may assume that p_m converges to a $p_0 \in R$. Let $t > 0$ be fixed. For any integer m, there is a $\tau = \tau(t, m)$ and an integer $k = k(t, m)$ such that $t = kT_m + \tau$, $0 \le \tau < T_m$. Now, as $m \to \infty$,

$$
\begin{aligned}
|x(t, p_0) - p_0| &\le |x(t, p_0) - x(t, p_m)| + |x(t, p_m) - p_m| + |p_m - p_0| \\
&\le |x(t, p_0) - x(t, p_m)| + |x(kT_m + \tau, p_m) - p_m| + |p_m - p_0| \\
&\le |x(t, p_0) - x(t, p_m)| + |x(\tau, p_m) - p_m| + |p_m - p_0| \to 0,
\end{aligned}
$$

since $0 \le \tau < T_m \to 0$ as $m \to \infty$. Therefore, $x(t, p_0) = p_0$ for any $t \ge 0$, hence p_0 is a critical point, using Corollary 4.1.10 in Chapter 4. ♠

Example 8.2.14 Consider

$$
x' = y + x[1 - x^2 - y^2], \quad y' = -x + y[1 - x^2 - y^2]. \tag{2.3}
$$

Taking a derivative in t on both sides of $r^2(t) = x^2(t) + y^2(t)$ and $\tan \theta(t) = \frac{y(t)}{x(t)}$, we obtain

$$
\begin{aligned}
r'(t) &= \frac{x(t)x'(t) + y(t)y'(t)}{r(t)} = r(t)[1 - r^2(t)], \\
\theta'(t) &= \frac{x(t)y'(t) - y(t)x'(t)}{r^2(t)} = -1.
\end{aligned}
$$

Then, using partial fractions, the solutions are given by

$$
r(t) = \frac{1}{\sqrt{1 + ce^{-2t}}}, \quad c > -1, \quad \theta(t) = -t + d, \quad t \ge 0, \tag{2.4}
$$

where c and d are constants. Now, when $c = 0$, $(x(t), y(t)) = (\cos t, -\sin t)$ is a periodic solution with its trajectory given by the unit circle. If $c > 0$, then, since $r(t) \nearrow 1$ as $t \to \infty$, the corresponding solution approaches the unit circle from inside of the circle in a spiral way as $t \to \infty$; if $-1 < c < 0$, then, since $r(t) \searrow 1$ as $t \to \infty$, the corresponding solution approaches the unit circle from outside of the circle in a spiral way as $t \to \infty$. Therefore, the unit circle is the ω-limit set. Note that the only critical point of Eq. (2.3) is the origin $(0, 0)$, which is inside the unit circle.

Another way to derive a periodic orbit in this case is to consider the annular ring $r_1 \le r \le r_2$ with $0 < r_1 < 1$ and $1 < r_2 < \infty$. Since the ring is free of critical points and the solutions entering the ring will stay in the ring according to the increasing or decreasing of $r(t) = \frac{1}{\sqrt{1 + ce^{-2t}}}$, the Poincaré-Bendixson theorem is applicable to obtain a periodic orbit inside the ring. See **Figure 8.13**.   ♠

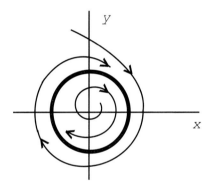

Figure 8.13: The solutions approach the ω-limit set (the unit circle) spirally

Next, we briefly discuss the stability of periodic orbits. Let's first get some ideas from Example 8.2.14, where for any point near the periodic orbit $\Gamma : r = 1$, the corresponding trajectory will approach Γ as $t \to \infty$, thus Γ (regarded as a periodic solution) is L-asymptotically stable (see Remark 6.5.2).

To study the L-asymptotic stability of a periodic orbit, Γ, of a general planar system, a main tool is to use a Poincaré section $\sum$ and a Poincaré map P, where $\sum$ is now a line segment perpendicular to Γ at some point $p_0 \in \Gamma$, and P is a one-dimensional map. We first use an example to demonstrate some of the details.

Example 8.2.15 Consider the equation in Example 8.2.14 again and write the solution (r, θ) as

$$r = r(t, r_0) = \left[1 + \left(\frac{1}{r_0^2} - 1\right)e^{-2t}\right]^{-1/2}, \quad t \geq 0,$$

$$\theta = \theta(t, \theta_0) = -t + \theta_0, \quad t \geq 0.$$

For any angle θ_0, we may let $\sum$ be the ray $\theta = \theta_0$. See **Figure 8.14**, where the trajectory from (r_0, θ_0) to (r, θ) implies that $\theta = \theta_0 - 2\pi$, or $t = 2\pi$. Therefore, the one-dimensional Poincaré map P is given by

$$r_0 \rightarrow P(r_0) = r(2\pi, r_0) = \left[1 + \left(\frac{1}{r_0^2} - 1\right)e^{-4\pi}\right]^{-1/2}$$

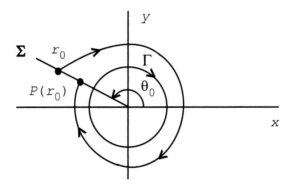

Figure 8.14: A Poincaré section and a Poincaré map near $r = 1$

Since $\Gamma : r = 1$ is a periodic orbit, we have $P(1) = 1$, that is, 1 is a fixed point of the map P. Next, we have

$$P'(r_0) = e^{-4\pi}r_0^{-3}\left[1 + \left(\frac{1}{r_0^2} - 1\right)e^{-4\pi}\right]^{-3/2},$$

hence

$$P'(1) = e^{-4\pi} < 1, \quad P(1) = 1.$$

Now, recall from Chapter 7 that $|P'(1)| < 1$ implies that the fixed point 1 of the map P is asymptotically stable. Therefore, for points (r, θ_0) with $r \approx 1$, we have $P^j(r) \rightarrow 1$, $j \rightarrow \infty$. Since the choice of the ray $\theta = \theta_0$

is arbitrary, similar things happen near every point on Γ, thus, trajectories started with points near Γ will approach Γ, which verifies that the periodic orbit Γ is L-asymptotically stable.   ♠

In the previous example, a Poincaré section can be constructed everywhere on the periodic orbit. For a general planar system, if *one* Poincaré section can be constructed at some point p_0 of a periodic orbit Γ such that p_0 is an asymptotically stable fixed point of the corresponding Poincaré map P, then the periodic orbit Γ is L-asymptotically stable, due to the following result of Hubbard and West [1995].

Theorem 8.2.16 *If $|P'(p_0)| < 1$, there is a neighborhood of Γ such that every solution starting in that neighborhood converges to Γ as $t \to \infty$.*   ♠

Therefore, we have the following result.

Theorem 8.2.17 *Let Γ be a periodic orbit of a planar system and let $\sum$ be a Poincaré section (a line segment) that is perpendicular to Γ at some point $p_0 \in \Gamma$. If the corresponding Poincaré map P satisfies $|P'(p_0)| < 1$, then the periodic orbit Γ is L-asymptotically stable. If $|P'(p_0)| > 1$, then Γ is unstable.*   ♠

For planar systems, $P'(p_0)$ can be calculated, based on the following result of Andronov, Leontovich, Gordon, and Maier [1973](2).

Theorem 8.2.18 *Let $r(t)$ be a T-periodic orbit of a planar system $x' = f(x)$ where f is differentiable. Let $\sum$ be a Poincaré section (a line segment) that is perpendicular to $r(t)$ at the point $p_0 = r(0)$. Then the corresponding Poincaré map P satisfies*

$$P'(p_0) = \exp\left(\int_0^T \left[\frac{\partial f}{\partial x_1}(r(t)) + \frac{\partial f}{\partial x_2}(r(t))\right]dt\right).$$

Accordingly, $r(t)$ is L-asymptotically stable if $\int_0^T \left[\frac{\partial f}{\partial x_1}(r(t)) + \frac{\partial f}{\partial x_2}(r(t))\right]dt < 0$, and $r(t)$ is unstable if $\int_0^T \left[\frac{\partial f}{\partial x_1}(r(t)) + \frac{\partial f}{\partial x_2}(r(t))\right]dt > 0$.   ♠

We can check this result with Example 8.2.15, where $r(t) = (\cos t, -\sin t)$, $T = 2\pi$, and (let $x_1 = x, x_2 = y$)

$$\frac{\partial f}{\partial x_1} = 1 - 3x_1^2 - x_2^2, \quad \frac{\partial f}{\partial x_2} = 1 - x_1^2 - 3x_2^2,$$

then

$$P'(p_0) = \exp\left(\int_0^{2\pi}\left[1 - 3\cos^2 t - \sin^2 t + 1 - \cos^2 t - 3\sin^2 t\right]dt\right)$$

$$= \exp\left(\int_0^{2\pi}[2 - 3 - 1]dt\right) = e^{-4\pi},$$

which is the same as $P'(1)$ found in Example 8.2.15.

The same ideas can be extended to systems in $\Re^n$, where a Poincaré section at a point p_0 of a periodic orbit Γ is an $(n-1)$-dimensional hypersurface and the corresponding Poincaré map is an $(n - 1)$-dimensional map, whose derivative $P'(p_0)$ at p_0, now an $(n - 1) \times (n - 1)$ matrix, can still be used to determine the stability of the periodic orbit Γ. We only outline some of the results here, see Hartman [1964], Hirsch and Smale [1974], Guckenheimer and Holmes [1986], Wiggins [1990], Perko [1991], Tsonis [1992], and Alligood, Sauer, and Yorke [1997] for additional details.

Let $v_1 \in \sum$ and define $v_{i+1} = P(v_i)$, $i = 1, 2, \cdots$, and write $u_i = v_i - p_0, i = 1, 2, \cdots$. Then, since $P(p_0) = p_0$, we have

$$
\begin{aligned}
u_2 &= v_2 - p_0 = P(v_1) - p_0 = P(p_0 + u_1) - p_0 \\
&= P(p_0) + P'(p_0)u_1 + O(|u_1|^2) - p_0 = P'(p_0)u_1 + O(|u_1|^2) \\
&\approx P'(p_0)u_1.
\end{aligned}
$$

Now, if $u_k \approx [P'(p_0)]^{k-1}u_1$, then

$$
\begin{aligned}
u_{k+1} &= v_{k+1} - p_0 = P(v_k) - p_0 = P(p_0 + u_k) - p_0 \\
&= P(p_0) + P'(p_0)u_k + O(|u_k|^2) - p_0 = P'(p_0)u_k + O(|u_k|^2) \\
&\approx P'(p_0)u_k \approx [P'(p_0)][P'(p_0)]^{k-1}u_1 \approx [P'(p_0)]^k u_1,
\end{aligned}
$$

therefore, we obtain $u_j \approx [P'(p_0)]^{j-1}u_1$, $j \geq 2$, using an induction.

Let $\lambda_1, \lambda_2, \cdots, \lambda_{n-1}$ and $\eta_1, \eta_2, \cdots, \eta_{n-1}$ be the $n - 1$ eigenvalues and their corresponding eigenvectors of the $(n - 1) \times (n - 1)$ matrix $P'(p_0)$, and assume we can write $u_1 = \sum_{i=1}^{n-1} c_i\eta_i$, a linear combination of η_i, then

$$
\begin{aligned}
u_{k+1} &\approx [P'(p_0)]^k u_1 = [P'(p_0)]^{k-1}P'(p_0)\sum_{i=1}^{n-1} c_i\eta_i \\
&= [P'(p_0)]^{k-1}\sum_{i=1}^{n-1} c_i P'(p_0)\eta_i = [P'(p_0)]^{k-1}\sum_{i=1}^{n-1} c_i\lambda_i\eta_i \\
&= [P'(p_0)]^{k-2}\sum_{i=1}^{n-1} c_i\lambda_i^2\eta_i = \sum_{i=1}^{n-1} c_i\lambda_i^k\eta_i.
\end{aligned}
$$

Now, if $|\lambda_i| < 1$, $i = 1, 2, \cdots, n - 1$, then $|u_k| \to 0$, $k \to \infty$. Thus we obtain $v_k \to p_0$, $k \to \infty$, therefore p_0 is an asymptotically stable fixed point of the map P, and consequently the periodic orbit is L-asymptotically stable. On the other hand, if $|\lambda_j| > 1$ for some j, then perturbations along η_j grow large, thus p_0 is unstable. This way, the stability of periodic orbits can be analyzed. Finally, we point out that for a planar system, the conditions on eigenvalues reduce to the condition on the scalar $P'(p_0)$ itself.

Exercises 8.2

1. Prove Theorem 8.2.7 for the α-limit set.

2. Find the ω and α limit sets for
   (a) $x' = 2x$; (b) $x' = -2x$; (c) $x' = 0$.

3. Find the ω-limit sets and α-limit sets for the solutions in Example 8.2.4.

4. Prove that the ω-limit set and α-limit set of a critical point is itself. Then prove the same result for a periodic orbit.

5. Prove Corollary 8.2.8.

6. Draw $B_d(p) = \{x \in \Re^2 : |x - p| \leq d\}$ where $p = (0,0)$ and $|x| = |x_1| + |x_2|$.

7. Verify in the proof of Theorem 8.2.9 that $x(t, q)$ leaves $B_\delta(p)$ at $-T$.

8. Verify in Figure 8.8 that the two straight lines connecting points A, B and B, C will not intersect the small circle with radius ε.

9. Verify in Figure 8.9 that the side lines of the sector intersect N_p at the points a and b that are at most away from p by a distance of ε.

10. Prove Theorem 8.2.11 for an α-limit point.

11. In the last part of the proof of Theorem 8.2.12, complete the case when $x(t, p)$ is outside the periodic orbit $\omega(p)$.

12. In the proof of Theorem 8.2.13, verify that $\Delta = \cap R_\tau$ is nonempty.

13. Prove that the ω-limit set of a bounded solution contains a critical point if the system has no periodic orbits.

14. Prove that if $x(t,p)$ is not a periodic orbit and is bounded in a set free of critical points, then its ω-limit set $\omega(p)$ and α-limit set $\alpha(p)$ are different periodic orbits and one is entirely inside the interior of the other one.

15. In the proof of Theorem 8.2.13 when the periodic orbit encloses a convex region, show that $x(t, p_m)$ is T_m-periodic.

16. Find the critical points of Eq. (2.3).

17. Prove that if the ω-limit set of a bounded solution contains an asymptotically stable critical point, then it contains nothing else.

18. Derive a periodic orbit for

$$ x' = -y + x[1 - x^2 - y^2], \quad y' = x + y[1 - x^2 - y^2]. \tag{2.5} $$

19. Prove Theorem 8.2.16.

20. Prove Theorem 8.2.17.

8.3   Limit Cycles

An interesting feature of Example 8.2.14 is that the ω-limit set, $r = 1$, is a "cycle" and is also the "limit" of some other trajectories, thus we make the following definition.

Definition 8.3.1 *A **limit cycle** L of Eq. (1.1) in $\Re^2$ is a periodic orbit of Eq. (1.1) which is also the ω-limit set or α-limit set of some trajectory of Eq. (1.1) other than L.*

Accordingly, $r = 1$ in Example 8.2.14 is a limit cycle. To determine the local properties, critical points are important subjects. But in order to determine the global behavior of trajectories of a planar dynamical system, besides the critical points, it is also crucial to know how many limit cycles are around each critical point of the system. Therefore, limit cycles are important in applications and also in theoretical analysis. However, determining the exact number of limit cycles is extremely difficult. In 1900, the world-famous mathematician Hilbert presented to the Second International Congress of Mathematicians a list of twenty three frontline mathematical research problems. Part of Hilbert's sixteenth problem asks for the maximum number of limit cycles, H_m, of the mth degree polynomial system in $\Re^2$ given by,

$$x' = \sum_{i+j=0}^{m} a_{ij}x^i y^j, \quad y' = \sum_{i+j=0}^{m} b_{ij}x^i y^j.$$

This problem is still open. So far, the best results about the Hilbert number H_m are given below:

$$\begin{cases} H_0 = 0; \; H_1 = 0; \; H_2 \geq 4; \; H_3 \geq 11; \\ H_m \geq \frac{m-1}{2} \text{ if } m \text{ is odd}; \\ \text{For any } m, \; H_m < \infty. \end{cases}$$

The example showing the existence of 4 limit cycles when $m = 2$ (that is, $H_2 \geq 4$) is given by Shi [1980] as

$$x' \;\; = \;\; -\frac{1}{10^{200}}x - y - 10x^2 + (5 - \frac{1}{10^{13}})xy + y^2, \qquad (3.1)$$

$$y' \;\; = \;\; x + x^2 + (\frac{9}{10^{13}} - \frac{8}{10^{52}} - 25)xy, \qquad (3.2)$$

which may give you a sense of how delicate the analysis there would be.

Next, we first give some results that can be used to eliminate the existence of periodic orbits, and hence limit cycles; then we present some results concerning the existence of limit cycles. Recall that a domain D is said to be "simply connected" if the interior of any simple closed curve in D is in D, or in other words, there are no holes in D. Let's write a planar autonomous differential equation as

$$\begin{cases} x'(t) \;\; = \;\; P(x(t), y(t)), \\ y'(t) \;\; = \;\; Q(x(t), y(t)), \;\; x, \, y, \, t \in \Re. \end{cases} \qquad (3.3)$$

Theorem 8.3.2 (Bendixson's criterion)   *Let $P(x, y)$ and $Q(x, y)$ have continuous first partial derivatives in a simply connected domain $D \subset \Re^2$ and assume that $\frac{\partial P}{\partial x} + \frac{\partial Q}{\partial y}$ is not identically zero and does not change sign in any open set of D. Then Eq. (3.3) has no periodic orbit in D.*

Proof. Suppose Eq. (3.3) has a periodic orbit Γ with period T in D and denote R the interior of Γ. Then along Γ, we have, from Green's theorem,

$$\iint_R \left[\frac{\partial P}{\partial x} + \frac{\partial Q}{\partial y}\right] dx dy \;\; = \;\; \int_\Gamma (P dy - Q dx) = \int_0^T (P y' - Q x') dt$$

$$= \;\; \int_0^T (PQ - QP) dt = 0. \qquad (3.4)$$

But this is impossible since $\frac{\partial P}{\partial x} + \frac{\partial Q}{\partial y}$ is not identically zero and does not change sign in R. The proof is complete. ♠

Example 8.3.3 Consider

$$\begin{cases} x' &= g(y), \\ y' &= (1 + x^2)y + h(x), \end{cases} \tag{3.5}$$

where g and h are any differentiable functions. Now,

$$\frac{\partial P}{\partial x} + \frac{\partial Q}{\partial y} = 1 + x^2 > 0, \quad x \in \Re,$$

thus the equation has no periodic orbit. ♠

Example 8.3.4 Consider $x'' + g(x)x' + h(x) = 0$, which is equivalent to

$$\begin{cases} x' &= y, \\ y' &= -g(x)y - h(x). \end{cases} \tag{3.6}$$

Now,

$$\frac{\partial P}{\partial x} + \frac{\partial Q}{\partial y} = -g(x).$$

Thus, if $g(x) > 0$ or $g(x) < 0$ for all x, then the equation has no periodic orbit. Later we will see that when g satisfies other conditions, the equation may have periodic orbits. ♠

Example 8.3.5 The motion of a simple pendulum, $\theta''(t) + k\theta'(t) + q\sin\theta(t) = 0$, is a special case of Example 8.3.4, where $g(x) = k$. Accordingly, if $k \neq 0$, then the equation has no periodic orbit, which is already obtained in Chapter 4. If $k = 0$, then Bendixson's criterion does not apply. In fact, it is verified in Chapter 4 that the equation has periodic orbits in this case. ♠

The following is an extension of Bendixson's criterion. The proof is left an an exercise.

Theorem 8.3.6 (Bendixson-Dulac's criterion) *Let $P(x,y)$, $Q(x,y)$, and $B(x,y)$ have continuous first partial derivatives in a simply connected domain $D \subset \Re^2$ and assume that $\frac{\partial(PB)}{\partial x} + \frac{\partial(QB)}{\partial y}$ is not identically zero and does not change sign in any open set of D. Then Eq. (3.3) has no periodic orbit in D.* ♠

Example 8.3.7 Consider

$$\begin{cases} x' &=& (y-a)x, \\ y' &=& bx + cy + ky^2, \end{cases} \qquad (3.7)$$

where a, b, c, k are constants. Note that

$$\frac{\partial P}{\partial x} + \frac{\partial Q}{\partial y} = y - a + c + 2ky,$$

thus Bendixson's criterion does not apply. Let $B(x,y) = x^{-(2k+1)}$, then

$$\frac{\partial (PB)}{\partial x} + \frac{\partial (QB)}{\partial y} = (c + 2ka)x^{-(2k+1)}.$$

Thus, if $c + 2ka \neq 0$, then Bendixson-Dulac's criterion implies that the equation has no periodic orbit in the half plane $x > 0$ or $x < 0$. Next, if $x = 0$, then starting from any point on the y-axis, the equation has a solution that lies on the y-axis. Thus we conclude that the equation has no periodic orbit in $\Re^2$ (otherwise uniqueness will be violated).   ♠

You probably have noticed that finding $B(x,y)$, just like finding a Liapunov function, requires certain skills and experience. Typically one such $B(x,y)$ emerges after many trials and errors.

Example 8.3.8 Consider

$$\begin{cases} x' &=& y, \\ y' &=& -x - y + x^2 + y^2. \end{cases} \qquad (3.8)$$

Again, Bendixson's criterion does not apply. Choose $B(x,y) = e^{-2x}$, then

$$\frac{\partial (PB)}{\partial x} + \frac{\partial (QB)}{\partial y} = -2ye^{-2x} - e^{-2x} + 2ye^{-2x} = -e^{-2x} < 0,$$

hence the equation has no periodic orbit in $\Re^2$.   ♠

The above are some results about the nonexistence of periodic orbits or limit cycles. Next, we look at the existence of limit cycles. First, using the idea of Example 8.2.14, we find that if a solution enters and then stays in a bounded annular region that is free of critical points, then this solution will pile up at a limit cycle. In fact, we have the following result. The proof is left as an exercise.

Theorem 8.3.9 (Poincaré-Bendixson annular region theorem) *Let R be a bounded region in $\Re^2$ enclosed by two simple closed curves C_{in} and C_{out}, as shown in* **Figure 8.15**. *If R is free of critical points and if there is a solution of Eq. (3.3) that enters the interior of R from the boundary and stays in the interior of R as t increases (or decreases), then R has a limit cycle of Eq. (3.3).* ♠

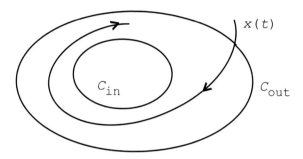

Figure 8.15: An annular region for the existence of limit cycles

Accordingly, to apply Theorem 8.3.9, all we need is to construct such curves C_{in} and C_{out}. To do so, two things are commonly used. One is the direction field, or "slopes" $\frac{dy}{dx}$, that can "guide" solutions to enter and stay in the region R. Another one is the Liapunov function $V(t) = \frac{x^2(t)+y^2(t)}{2}$ that indicates the "tendency" of a solution $(x(t), y(t))$.

Example 8.3.10 Revisit Example 8.2.14,

$$x' = y + x[1 - x^2 - y^2], \quad y' = -x + y[1 - x^2 - y^2]. \tag{3.9}$$

Define $V(t) = \frac{x^2(t)+y^2(t)}{2}$ where $(x(t), y(t))$ is a solution. Now, taking a derivative in t and plugging in the equation, we obtain

$$V'(t) = xx' + yy' = [1 - (x^2 + y^2)](x^2 + y^2). \tag{3.10}$$

If we select C_{in} to be a circle centered at $(0,0)$ with radius $\frac{1}{2}$ and C_{out} a circle centered at $(0,0)$ with radius 2, then

$$V'(t) = [1 - (x^2 + y^2)](x^2 + y^2) > 0 \quad \text{on} \quad C_{in}, \tag{3.11}$$

and

$$V'(t) = [1 - (x^2 + y^2)](x^2 + y^2) < 0 \ \text{ on } \ C_{out}. \tag{3.12}$$

Thus any solution that enters the interior of R formed by C_{in} and C_{out} from the boundary will stay in the interior of R as t increases, hence, from Theorem 8.3.9, there exists a limit cycle.   ♠

Next, let's apply the Poincaré-Bendixson annular region theorem to the famous **Lienard-type equation**

$$x'' + g(x)x' + x = 0, \tag{3.13}$$

which has important applications in physics concerning the sustained oscillations. (See Chapter 1 for a brief introduction of the Lienard-type equations.) Eq. (3.13) is equivalent to

$$\begin{cases} x' &= y - G(x), \quad G(x) = \int_0^x g(s)ds, \\ y' &= -x, \end{cases} \tag{3.14}$$

and includes the well-known **van der Pol equation**

$$x'' + (x^2 - 1)x' + x = 0, \tag{3.15}$$

which can be used to model the voltage in a triode circuit and also the human heartbeat.

For Eq. (3.13), we make the following assumptions.

Hypothesis H. The function g in Eq. (3.13) is continuous, and

1. $G(x) = \int_0^x g(s)ds$ is odd in x,

2. $G(x) \to \infty$ as $x \to \infty$ and there is an $a > 0$ such that $G(x) > 0$ and is increasing monotonically for $x > a$,

3. there is a $b > 0$ such that $G(x) < 0$ for $0 < x < b$.

Theorem 8.3.11 *Under the Hypothesis H, the Lienard equation (3.14) has a limit cycle.*

Proof. Note first that the only critical point of Eq. (3.14) is at $(0,0)$. Define $V(t) = \frac{x^2(t) + y^2(t)}{2}$ where $(x(t), y(t))$ is a solution, then

$$\frac{dV}{dt} = xx' + yy' = x(y - G(x)) + y(-x) = -xG(x).$$

Consider a circle L_r centered at $(0,0)$ with radius r. From the Hypothesis H(1) and H(3), if $0 < r < b$, then $\frac{dV}{dt} > 0$ on L_r except when L_r intersects the y-axis. Thus L_r can be used as C_{in}.

Next, we construct C_{out}. Note that $(-x(t), -y(t))$ is also a solution of Eq. (3.14) (see an exercise), thus if we can construct a trajectory $(x(t), y(t))$ for $x \geq 0$, then $(-x(t), -y(t))$ gives a trajectory for $x \leq 0$, using the symmetry. From Eq. (3.14), we have

$$\frac{dy}{dx} = \frac{-x}{y - G(x)}, \tag{3.16}$$

which is zero if and only if $x = 0$, or the tangent of a trajectory is zero only when the trajectory intersects the y-axis. Also, $x = x(t)$ increases for $y > G(x)$ (since $x' = y - G(x)$) and $y = y(t)$ decreases for $x > 0$ (since $y' = -x$). Thus a trajectory Γ starting from $A = (0, y_0)$, $y_0 > 0$, will go to right and go down, and then intersects $y = G(x)$ since $G(x) \to \infty$, $x \to \infty$. The tangent at the interception is vertical since $x'(t) = y(t) - G(x(t)) = 0$ and hence $\frac{dy}{dx}$ is undefined there. Thus Γ gets below $y = G(x)$, and then $x = x(t)$ decreases and $y = y(t)$ decreases, or Γ goes to left and goes down. If Γ doesn't cross the negative y-axis, then $y \to -\infty$ and hence $x \to -\infty$ from Eq. (3.14), a contradiction because now x is bounded. Consequently, Γ intersects the y-axis at $E = (0, -y_1)$, $y_1 > 0$, and Γ has the shape shown in **Figure 8.16** when y_0 is sufficiently large.

Next, we verify that $y_1 \leq y_0$. When (x, y) is a solution of Eq. (3.14), we have

$$\frac{dV}{dx} = x + y\frac{dy}{dx} = \frac{xG(x)}{G(x) - y}, \tag{3.17}$$

$$\frac{dV}{dy} = x\frac{dx}{dy} + y = G(x). \tag{3.18}$$

Accordingly, the line integral along Γ is such that

$$V(E) - V(A) = \int_{ABCDE} dV = \left(\int_{AB} + \int_{DE}\right)\frac{xG(x)}{G(x) - y}\,dx + \int_{BCD} G(x)\,dy.$$

Now,

$$\left(\int_{AB} + \int_{DE}\right)\frac{xG(x)}{G(x) - y}\,dx \to 0, \quad \text{as } y \to \infty,$$

and $\int_{BCD} G(x)\,dy = -\int_{DCB} G(x)\,dy = q(y_0)$ is a function of y_0. Let m be any point between $(a, 0)$ and C in Figure 8.16, then, using the Hypothesis

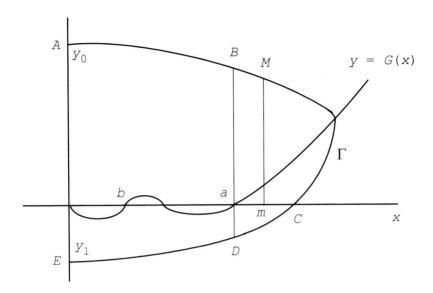

Figure 8.16: The graph of $y = G(x)$ and the shape of trajectory Γ

H(2),

$$\int_{DCB} G(x)dy \geq \int_{CM} G(x)dy \geq G(m)|m - M|.$$

For the fixed m, $\int_{DCB} G(x)dy \geq G(m)|m - M| \to \infty$ as $y_0 \to \infty$. Therefore, there is a $y_0 > 0$ such that $V(E) \leq V(A)$, or $y_1 \leq y_0$. (Note that if $y_1 = y_0$ then Γ is periodic.)

Now, let C_{out} be the union of Γ for $x \geq 0$, its reflection through the origin, and the segments on the y-axis connecting these curves, as shown in **Figure 8.17**. Then C_{in} and C_{out} satisfy the conditions of Theorem 8.3.9, which implies that the Lienard equation (3.14) has a limit cycle. This completes the proof. ♠

For the van der Pol equation given in Eq. (3.15), we can regard it as a special case of the Lienard-type equation (3.13) with $g(x) = x^2 - 1$, and then apply Theorem 8.3.11 to obtain limit cycles. However, in this special case, we would like to construct concrete C_{in} and C_{out} here. To do so, let's introduce **Lienard's graphing method** under the Hypothesis H, given in **Figure 8.18**.

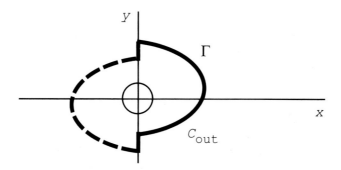

Figure 8.17: Construction of C_{out} using Γ and its reflection through the origin for the Lienard equation

---

In Figure 8.18, the point (x, y) is P, the straight line $\overline{PR}$ is parallel to the y-axis and R is on the curve of $y = G(x)$, and the straight line $\overline{QR}$ is parallel to the x-axis and Q is on the y-axis. Thus $R = (x, G(x))$ and $Q = (0, G(x))$. The straight line $\overline{PS}$ is perpendicular to the straight line $\overline{PQ}$, hence the angle formed by $\overline{PS}$ and $\overline{PR}$ is the same as the angle formed by $\overline{PQ}$ and $\overline{QR}$. Accordingly, the slope of the straight line $\overline{PS}$ in the (x, y) plane is given by

$$- \tan \angle OSP = - \tan(\frac{\pi}{2} - \phi) = - \frac{x}{y - G(x)} = \frac{-x}{y - G(x)} = \frac{dy}{dx}.$$

That is, the derivative $\frac{dy}{dx}$ at the point (x, y), or the slope of the trajectory at (x, y), is determined by the slope of the straight line $\overline{PS}$, or the trajectory moves in the direction of the straight line $\overline{PS}$.

Corollary 8.3.12 *The van der Pol equation*

$$x'' + (x^2 - 1)x' + x = 0, \tag{3.19}$$

has a limit cycle.

Proof. We will construct a concrete C_{out} in this case (C_{in} is the same as in the proof of Theorem 8.3.11). Now, as shown in **Figure 8.19**, $G(x) = \int_0^x (s^2 - 1)ds = \frac{x^3}{3} - x$ has a minimum value $y = -\frac{2}{3}$ for $x > 0$.

Let's construct C_{out} and the corresponding region R only for $x \geq 0$, that is, $\overset{\frown}{AB} \cup \overline{BC} \cup \overset{\frown}{CD}$ in Figure 8.19, then the symmetry through the origin

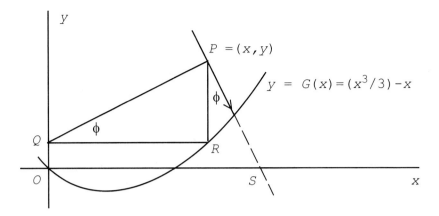

Figure 8.18: Lienard's graphing method under the Hypothesis H

will give the other half. The arc $\overset{\frown}{CD}$ is on the circle centered at $(0, -\frac{2}{3})$ with radius x_1, the vertical straight line $\overline{BC}$ is $x = x_1$, and the arc $\overset{\frown}{AB}$ is on the circle centered at $(0, -\frac{2}{3})$ with radius $x_1 + \frac{4}{3}$. For x_1 sufficiently large, $x_1 + \frac{2}{3} < \frac{x_1^3}{3} - x_1$, thus the y component of the point A is below the graph of $y = G(x) = \frac{x^3}{3} - x$ at x_1. Hence the point B is below $y = \frac{x^3}{3} - x$ at x_1. A solution intersecting $\overline{BC}$ will go to left and go down since $x'(t) = y - G(x) < 0$ and $y'(t) = -x < 0$. For any point E on $\overset{\frown}{AB}$, draw a right triangle according to Lienard's graphing method. Now, the tangent of arc $\overset{\frown}{AB}$ at E is perpendicular to the straight line from E to the center $(0, -\frac{2}{3})$, and the point F is above $(0, -\frac{2}{3})$, hence the direction of a trajectory at E, which is perpendicular to the straight line $\overline{EF}$, must move inside the region R (or below the arc $\overset{\frown}{AB}$). The same is true for the arc $\overset{\frown}{CD}$. This completes the construction of C_{out} and hence the proof.   ♠

From the above, we see that if an annular region can be constructed so as to apply the Poincaré-Bendixson annular region theorem 8.3.9, then the existence of limit cycles is guaranteed. However, constructing annular regions is difficult and requires a very good understanding of both the direction field of the equation under study and the planar geometry in order to find certain relationships between trajectories and the functions involved.

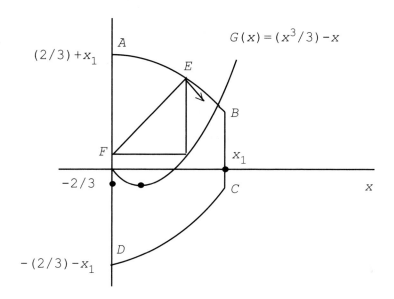

Figure 8.19: Construction of C_{out} for the van der Pol equation

At the beginning of this chapter, we claimed that critical points and limit cycles and their positions can help us determine the global properties of trajectories in $\Re^2$. Now, we will use some examples to demonstrate how they are done.

Example 8.3.13 Consider

$$\begin{cases} x' & = & 2x + 2x^3 - 4xy^2, \\ y' & = & -y + 4x^2y - 3y^3. \end{cases} \tag{3.20}$$

This system has five critical points $C_1 = (0,0), C_2 = (1,1), C_3 = (1,-1), C_4 = (-1,1), C_5 = (-1,-1)$. For $C_1 = (0,0)$, the origin is a saddle point for the linearization, thus C_1 is also a saddle point for Eq. (3.20). For $C_2 = (1,1)$, we change the variables $x_1 = x - 1$, $x_2 = y - 1$ and obtain

$$\begin{cases} x_1' & = & 4x_1 - 8x_2 + 6x_1^2 - 8x_1x_2 - 4x_2^2 + 2x_1^3 - 4x_1x_2^2, \\ x_2' & = & 8x_1 - 6x_2 + 4x_1^2 + 8x_1x_2 - 9x_2^2 + 4x_1^2x_2 - 3x_2^3. \end{cases} \tag{3.21}$$

Now, denote A the coefficient matrix of the linearization of Eq. (3.21), then $p = trA = -2$, $q = \det A = 40 > \frac{p^2}{4}$. Thus, the origin $(0,0)$ for the linearization of Eq. (3.21) is a stable spiral point. Accordingly, $C_1 = (1,1)$ is a stable spiral point for Eq. (3.20). Similarly, one finds that C_i, $i = 3,4,5$, are all stable spiral points for Eq. (3.20), see an exercise.

Since Eq. (3.20) has solutions on the x-axis or on the y-axis, no periodic orbits can intersect the x-axis or the y-axis. Next, if (x,y) is a solution, then so are $(-x,y)$, $(x,-y)$, and $(-x,-y)$ (see an exercise), thus we only need to check the existence of periodic orbits for Eq. (3.20) in the first quadrant. Let $B(x,y) = x^{-3/2}y^{-2}$, then

$$\frac{\partial(PB)}{\partial x} + \frac{\partial(QB)}{\partial y} = -x^{-3/2}y^{-2}[x^2 + y^2],$$

which will not change sign in the first quadrant. Therefore, using Bendixson-Dulac's criterion 8.3.6, Eq. (3.20) has no periodic orbit in the first quadrant, thus Eq. (3.20) has no periodic orbit in $\Re^2$. A phase portrait in $\Re^2$ is shown in **Figure 8.20**.   ♠

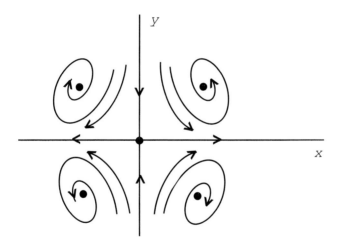

Figure 8.20: A phase portrait in $\Re^2$ for Eq. (3.20)

Example 8.3.14 Consider

$$\begin{cases} x' = 3x - x^2 - 5xy, \\ y' = -y + xy + y^2. \end{cases} \qquad (3.22)$$

This system has four critical points $C_1 = (0,0), C_2 = (0,1), C_3 = (3,0)$, and $C_4 = (\frac{1}{2}, \frac{1}{2})$. Since Eq. (3.22) has solutions on the x-axis or on the y-axis, no periodic orbits can intersect the x-axis or the y-axis. Next, since any periodic orbit contains a critical point in its interior, the only region that may have periodic orbits is the first quadrant.

$C_1 = (0,0)$ is a saddle point for Eq. (3.22) since it is a saddle point for the linearization. One can check that C_2 and C_3 are all saddle points for Eq. (3.22), see an exercise. For $C_4 = (\frac{1}{2}, \frac{1}{2})$, we change the variables $x_1 = x - \frac{1}{2}$, $x_2 = y - \frac{1}{2}$ and obtain

$$\begin{cases} x_1' = -\frac{1}{2}x_1 - \frac{5}{2}x_2 - x_1^2 - 5x_1x_2, \\ x_2' = \frac{1}{2}x_1 + \frac{1}{2}x_2 + x_1x_2 + x_2^2. \end{cases} \tag{3.23}$$

Now, denote A the coefficient matrix of the linearization of Eq. (3.23), then $p = trA = 0$, $q = \det A = 1$. Thus, the origin $(0,0)$ for the linearization of Eq. (3.23) is a center, whose property is, in general, not clear after perturbations. However, Eq. (3.23) is of the second-order polynomials and we can check it with Theorem 4.3.8 in Chapter 4. In order to do this, transform Eq. (3.23) into the standard form in Theorem 4.3.8. Using the analysis in Section 2 of Chapter 4, we find that $-i$ is an eigenvalue and $[-1, 1]^T + i[-2, 0]^T$ is an eigenvector for the matrix A. Then,

$$P = \begin{bmatrix} -1 & -2 \\ 1 & 0 \end{bmatrix}, \quad P^{-1}AP = \begin{bmatrix} 0 & -1 \\ 1 & 0 \end{bmatrix}, \tag{3.24}$$

and $[y_1, y_2]^T = P^{-1}[x_1, x_2]^T$ transforms Eq. (3.23) into

$$\begin{cases} y_1' = -y_2 - 2y_1y_2, \\ y_2' = y_1 - 2y_1^2 - 2y_1y_2 + 2y_2^2. \end{cases} \tag{3.25}$$

Now, applying Theorem 4.3.8, one has $a_{20} = 0, a_{11} = -2, a_{02} = 0, b_{20} = -2, b_{11} = -2, b_{02} = 2$. Then $A = a_{20} + a_{02} = 0, B = b_{20} + b_{02} = 0, \gamma = 0$. Hence $W_1 = W_2 = W_3 = 0$, thus $(0,0)$ for Eq. (3.25) is a center, therefore, $C_4 = (\frac{1}{2}, \frac{1}{2})$ for Eq. (3.22) is also a center. A phase portrait in $\Re^2$ is shown in **Figure 8.21**. ♠

In the above, we briefly discussed the existence and nonexistence of limit cycles, and used examples to demonstrate that they are very important components in determining the global behavior of dynamical systems. Some related studies include the behavior of trajectories at infinity, and uniqueness

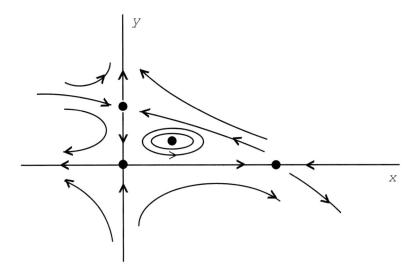

Figure 8.21: A phase portrait in $\Re^2$ for Eq. (3.22)

---

and stability of limit cycles. See Hartman [1964], Hale [1969], and Perko [1991] for additional references.

Exercises 8.3

1. Show that $H_0 = 0$ and $H_1 = 0$ for Hilbert's sixteenth problem.

2. Prove Theorem 8.3.6.

3. Prove the following result: Let $P(x, y)$, $Q(x, y)$, $B(x, y)$, and $F(x, y)$ have continuous first partial derivatives in a simply connected domain $D \subset \Re^2$ and assume that $\frac{\partial(PB)}{\partial x} + \frac{\partial(QB)}{\partial y} + B[P\frac{\partial F}{\partial x} + Q\frac{\partial F}{\partial y}]$ is not identically zero and does not change sign in any open set of D. Then Eq. (3.3) has no periodic orbit in D.

4. Prove that the system

$$x' = y^7, \quad y' = ax + by + cx^2 + dy^2$$

has no limit cycles.

5. Discuss the existence of periodic orbits for

$$x' = y, \quad y' = 2(1 - xy).$$

6. Prove Theorem 8.3.9.

7. Use the Poincaré-Bendixson theorem 8.2.12 to prove that the system

$$\begin{aligned} x' &= x + y - x[x^2 + y^2]\cos^2(x^2 + y^2), \\ y' &= -x + y - y[x^2 + y^2]\cos^2(x^2 + y^2). \end{aligned}$$

has a limit cycle.

8. Verify for Eq. (3.14) that if $(x(t), y(t))$ is a solution, then so is $(-x(t), -y(t))$.

9. Answer the following question: Why must we verify $y_1 \leq y_0$ in the proof of Theorem 8.3.11 ?

10. Verify that the function $g(x) = x^2 - 1$ satisfies the Hypothesis H.

11. Find all the critical points for Example 8.3.13. Then verify that C_i, $i = 3, 4, 5$, are all stable spiral points for Eq. (3.20).

12. In Example 8.3.13, verify that if (x, y) is a solution, then so are $(-x, y)$, $(x, -y)$, and $(-x, -y)$.

13. Find all the critical points for Example 8.3.14. Then verify that C_2 and C_3 are all saddle points for Eq. (3.22).

14. Verify that $y = -\frac{1}{3}x + 1$ is a solution curve for Eq. (3.22).

8.4   An Application: The Lotka-Volterra Equation

Consider two populations, x and y. If they compete for a shared limited resource (space or a nutrient, for example), and each interferes with the other's utilization of it, then a classic model describing the situation is the Lotka-Volterra competition equation

$$\begin{cases} x'(t) = \beta_1 x(K_1 - x - \mu_1 y), \\ y'(t) = \beta_2 y(K_2 - y - \mu_2 x), \\ \quad x(0) \geq 0, \ y(0) \geq 0, \end{cases} \tag{4.1}$$

where β_i, K_i, μ_i, $i = 1, 2$, are positive constants. First, note that a trajectory started from the interior of the first quadrant of the (x, y) plane will remain in the first quadrant (see an exercise). Now, $(0, 0)$, $(0, k_2)$, and $(k_1, 0)$ are critical points of Eq. (4.1). Other possible critical points of Eq. (4.1) are the solutions of

$$\begin{cases} x + \mu_1 y = K_1, \\ y + \mu_2 x = K_2. \end{cases} \qquad (4.2)$$

Note, the solutions of Eq. (4.2) that we are interested in should be inside the interior of the first quadrant of the (x, y) plane because x and y denote populations. Accordingly, we have the four cases shown in **Figure 8.22** based on the straight lines $x + \mu_1 y = K_1$ and $y + \mu_2 x = K_2$.

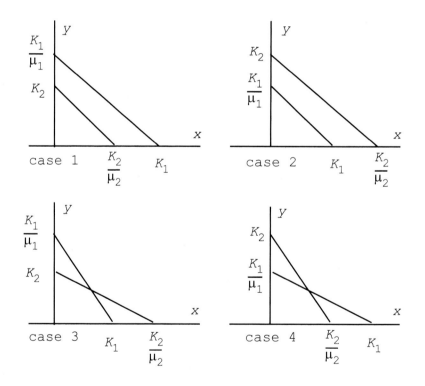

Figure 8.22: Four cases based on the straight lines $x + \mu_1 y = K_1$ and $y + \mu_2 x = K_2$

We also find that when $K_1 < \frac{K_2}{\mu_2}$ and $K_2 < \frac{K_1}{\mu_1}$, it follows that $K_1 > \mu_1 K_2 > \mu_1\mu_2 K_1$, or $1 > \mu_1\mu_2$. Similarly, when $\frac{K_2}{\mu_2} < K_1$ and $\frac{K_1}{\mu_1} < K_2$, one has $K_2 < K_1\mu_2 < \mu_1\mu_2 K_2$, or $1 < \mu_1\mu_2$. In these cases, the solution of Eq. (4.2) given by

$$\left(\frac{K_1 - \mu_1 K_2}{1 - \mu_1\mu_2}, \ \frac{K_2 - \mu_2 K_1}{1 - \mu_1\mu_2}\right) \tag{4.3}$$

is the fourth critical point and is inside the interior of the first quadrant of the (x, y) plane. Next, we shift these critical points to the origin $(0,0)$ in order to apply Theorem 4.3.1 in Chapter 4. Thus we denote by (x_c, y_c) one of those four critical points, and change the variables $\overline{x} = x - x_c$, $\overline{y} = y - y_c$ to derive the equation

$$\begin{cases} \overline{x}'(t) = \beta_1(\overline{x} + x_c)[K_1 - (\overline{x} + x_c) - \mu_1(\overline{y} + y_c)], \\ \overline{y}'(t) = \beta_2(\overline{y} + y_c)[K_2 - (\overline{y} + y_c) - \mu_2(\overline{x} + x_c)], \end{cases} \tag{4.4}$$

such that $(0,0)$ is a critical point for Eq. (4.4), which corresponds to (x_c, y_c) for Eq. (4.1).

The matrix of the linearization of Eq. (4.4) is

$$\begin{bmatrix} \beta_1(K_1 - 2x_c - \mu_1 y_c) & -\beta_1\mu_1 x_c \\ -\beta_2\mu_2 y_c & \beta_2(K_2 - 2y_c - \mu_2 x_c) \end{bmatrix}, \tag{4.5}$$

therefore the local qualitative properties are determined by its eigenvalues λ_i. Next, we look at all different cases. At $(x_c, y_c) = (0,0)$, $\lambda_i = \beta_i K_i > 0$, hence $(0,0)$ is an unstable node. At $(x_c, y_c) = (0, K_2)$, $\lambda_1 = \beta_1(K_1 - \mu_1 K_2)$ and $\lambda_2 = -\beta_2 K_2 < 0$. Hence $(0,0)$ for the linearization of Eq. (4.4), or $(0, K_2)$ for the original equation (4.1), is a saddle if $\beta_1(K_1 - \mu_1 K_2) > 0$ or a stable node if $\beta_1(K_1 - \mu_1 K_2) < 0$. At $(x_c, y_c) = (K_1, 0)$, $\lambda_1 = -\beta_1 K_1 < 0$ and $\lambda_2 = \beta_2(K_2 - \mu_2 K_1)$. Thus $(0,0)$ for the linearization of Eq. (4.4), or $(K_1, 0)$ for the original equation (4.1), is a saddle if $\beta_2(K_2 - \mu_2 K_1) > 0$ or a stable node if $\beta_2(K_2 - \mu_2 K_1) < 0$.

Next, let's look at the four cases shown in Figure 8.22 separately.

Case 1. $\frac{K_2}{\mu_2} < K_1$ and $K_2 < \frac{K_1}{\mu_1}$. Now, $(0, K_2)$ is a saddle and $(K_1, 0)$ is a stable node. In this case, the fourth critical point in (4.3) does not occur; or, there is no critical point of Eq. (4.1) inside the interior of the first quadrant of the (x, y) plane. Then there is no periodic orbit in the interior of the first quadrant due to Theorem 8.2.13. Now, if a point (x, y) is in the first quadrant and above the straight line $x + \mu_1 y = K_1$ (and hence also above

$y + \mu_2 x = K_2$), then one has $x + \mu_1 y > K_1$ and $y + \mu_2 x > K_2$, or $x' < 0$ and $y' < 0$. Therefore, the solutions in the first quadrant must be bounded. Accordingly, every trajectory in the first quadrant, except those started with $x = 0$, will tend to the stable node $(K_1, 0)$ by using the Poincaré-Bendixson theorem 8.2.12 (otherwise, periodic orbits will occur from bounded solutions away from critical points). Therefore, in this case the population x wins and y loses: $\lim_{t \to \infty} x(t) = K_1$ and $\lim_{t \to \infty} y(t) = 0$.

Case 2. $K_1 < \frac{K_2}{\mu_2}$ and $\frac{K_1}{\mu_1} < K_2$. The analysis is similar to Case 1 and is left as an exercise.

Case 3. $K_1 < \frac{K_2}{\mu_2}$ and $K_2 < \frac{K_1}{\mu_1}$, and hence $1 > \mu_1 \mu_2$. Both $(0, K_2)$ and $(K_1, 0)$ are saddles, and the fourth critical point in (4.3) now occurs inside the interior of the first quadrant. The matrix of the linearization (4.5) when (x_c, y_c) is given by (4.3) becomes

$$\begin{bmatrix} -\beta_1 x_c & -\beta_1 \mu_1 x_c \\ -\beta_2 \mu_2 y_c & -\beta_2 y_c \end{bmatrix}, \tag{4.6}$$

with its characteristic equation

$$\lambda^2 + [\beta_1 x_c + \beta_2 y_c]\lambda + [1 - \mu_1 \mu_2]\beta_1 \beta_2 x_c y_c = 0. \tag{4.7}$$

The eigenvalues are given by

$$\lambda = \frac{-[\beta_1 x_c + \beta_2 y_c] \pm \sqrt{[\beta_1 x_c + \beta_2 y_c]^2 - 4[1 - \mu_1 \mu_2]\beta_1 \beta_2 x_c y_c}}{2}, \tag{4.8}$$

and are all negative because

$$\begin{aligned}
& [\beta_1 x_c + \beta_2 y_c]^2 - 4[1 - \mu_1 \mu_2]\beta_1 \beta_2 x_c y_c \\
&= (\beta_1 x_c)^2 + 2\beta_1 x_c \beta_2 y_c + (\beta_2 y_c)^2 - 4\beta_1 \beta_2 x_c y_c + 4\mu_1 \mu_2 \beta_1 \beta_2 x_c y_c \\
&= (\beta_1 x_c)^2 - 2\beta_1 x_c \beta_2 y_c + (\beta_2 y_c)^2 + 4\mu_1 \mu_2 \beta_1 \beta_2 x_c y_c \\
&= [\beta_1 x_c - \beta_2 y_c]^2 + 4\mu_1 \mu_2 \beta_1 \beta_2 x_c y_c > 0,
\end{aligned} \tag{4.9}$$

and, as $1 - \mu_1 \mu_2 > 0$,

$$[\beta_1 x_c + \beta_2 y_c]^2 - 4[1 - \mu_1 \mu_2]\beta_1 \beta_2 x_c y_c < [\beta_1 x_c + \beta_2 y_c]^2. \tag{4.10}$$

Thus the critical point given by (4.3) is a stable node. In this case, since the interior of the first quadrant has a critical point, the existence of limit cycles is possible. We apply Bendixson-Dulac's criterion 8.3.6 and choose the domain D to be the interior of the first quadrant and let $B(x,y) = \frac{1}{xy}$, then, for Eq. (4.1),

$$
\begin{aligned}
\frac{\partial(PB)}{\partial x} &= \frac{\partial}{\partial x}\left[\frac{1}{xy}\beta_1 x(K_1 - x - \mu_1 y)\right] \\
&= \frac{\beta_1}{y}\frac{\partial}{\partial x}\left[K_1 - x - \mu_1 y\right] = -\frac{\beta_1}{y},
\end{aligned}
\tag{4.11}
$$

and

$$
\begin{aligned}
\frac{\partial(QB)}{\partial y} &= \frac{\partial}{\partial y}\left[\frac{1}{xy}\beta_2 y(K_2 - y - \mu_2 x)\right] \\
&= \frac{\beta_2}{x}\frac{\partial}{\partial y}\left[K_2 - y - \mu_2 x\right] = -\frac{\beta_2}{x},
\end{aligned}
\tag{4.12}
$$

hence

$$
\frac{\partial(PB)}{\partial x} + \frac{\partial(QB)}{\partial y} = -\frac{\beta_1}{y} - \frac{\beta_2}{x} < 0, \quad (x,y) \in D.
$$

Therefore, we conclude that there is no periodic orbit in the interior of the first quadrant. Similar to Case 1, the solutions in the first quadrant are bounded and every trajectory, except those started with $x = 0$ or $y = 0$, will tend to the stable node given by the critical point (4.3), by using the Poincaré-Bendixson theorem 8.2.12. Therefore, populations x and y may coexist and tend to the "shared" status.

Case 4. $\frac{K_2}{\mu_2} < K_1$ and $\frac{K_1}{\mu_1} < K_2$, and hence $1 < \mu_1\mu_2$. Both $(0, K_2)$ and $(K_1, 0)$ are stable nodes, and the fourth critical point in (4.3) now occurs inside the interior of the first quadrant. Similar to the calculations in Case 3, we find that now the two eigenvalues have the opposite signs. Thus the critical point given by (4.3) is a saddle. Again, there is no periodic orbit in the interior of the first quadrant. Now, because both $(0, K_2)$ and $(K_1, 0)$ are stable nodes and the critical point given by (4.3) is a saddle, then initial conditions determine who wins the competition.

The phase portraits are given in **Figure 8.23**.

Exercises 8.4

1. Prove that a trajectory started from the interior of the first quadrant of (x,y) plane will remain in the first quadrant.

2. Complete Case 2 and determine its phase portrait.

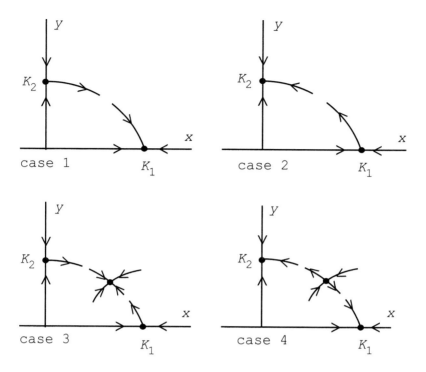

Figure 8.23: Phase portraits of the Lotka-Volterra competition equation

3. Complete Case 4 and determine its phase portrait.

4. Analyze the Volterra equation

$$\begin{cases} x'(t) = & ax(t) - bx^2(t) - cx(t)y(t), \\ y'(t) = & -dy(t) + ex(t)y(t), \end{cases}$$

where a, b, c, d, e are positive constants.

8.5 Manifolds and the Hartman-Grobman Theorem

In Chapter 4, we studied planar autonomous differential equations and found that most properties of those equations can be determined by the eigenvalues of their linearization matrices, and that in a small neighborhood of the origin, solutions of both the original nonlinear equations and their linearizations have essentially the same qualitative properties. In this section, we will extend these ideas to general $\Re^n$ and investigate the relationship between the linear differential equation

$$x' = Ax \qquad (5.1)$$

with a constant $n \times n$ matrix A and its nonlinear perturbation

$$x' = Ax + f(x) \qquad (5.2)$$

where $|f(x)|$ is small when $|x|$ is small. We will show that near the origin, most properties of solutions of Eq. (5.2) can still be determined by the eigenvalues of the matrix A, which is called the stable, unstable, and center manifolds theorem. We will also show (with some additional conditions on the matrix A) that near the origin, the trajectories of Eq. (5.1) and of Eq. (5.2) are very similar. This is called the Hartman-Grobman theorem.

To begin, we notice from linear algebra that there is a real nonsingular $n \times n$ matrix P which transforms Eq. (5.1) into

$$\begin{bmatrix} u \\ v \\ w \end{bmatrix}' = \begin{bmatrix} A_s & 0 & 0 \\ 0 & A_u & 0 \\ 0 & 0 & A_c \end{bmatrix} \begin{bmatrix} u \\ v \\ w \end{bmatrix}, \qquad (5.3)$$

where $(u, v, w) \in \Re^s \times \Re^u \times \Re^c$ with $s + u + c = n$, and the real parts of eigenvalues of the square matrices A_s, A_u, and A_c are negative, positive, and zero respectively. See Coddington and Levinson [1955] or Hirsch and Smale [1974] for a proof. This result is also called the Jordan canonical form theorem, but it is different from the one introduced in Chapter 3 in that for Eq. (5.3), we do not require the eigenvalues to lie on the diagonal, so that the transformation matrix P can be chosen to be a real matrix, while the transformation matrix in Chapter 3 may be complex valued.

Therefore, in the rest of this section, we assume that such a transformation has been made so that we can use $x = (u, v, w)$ and Eq. (5.3) to replace Eq. (5.1). Now, for Eq. (5.3), we find that

$$E^s = \{x = (u, v, w) \in \Re^n : v = 0, w = 0\} \qquad (5.4)$$

is a subspace in $\Re^n$ such that for any initial point $x_0^s = (u_0, 0, 0)$ in E^s, the corresponding solution $x(t, 0, x_0^s)$ of Eq. (5.3) is also in E^s for $t \in \Re$; that is,

E^s is invariant for Eq. (5.3). (This is defined in the same way as in Section 8.2 with $\Re^2$ replaced by $\Re^n$.) Moreover, from Chapter 5, we also know that

$$|x(t, 0, x_0^s)| \longrightarrow 0, \quad t \to \infty. \tag{5.5}$$

Next, for any initial point x_0^u in the subspace

$$E^u = \{x = (u, v, w) \in \Re^n : u = 0, w = 0\}, \tag{5.6}$$

the corresponding solution $x(t, 0, x_0^u)$ of Eq. (5.3) is also in E^u for $t \in \Re$, and

$$|x(t, 0, x_0^u)| \longrightarrow 0, \quad t \to -\infty. \tag{5.7}$$

Finally, for any initial point x_0^c in the subspace

$$E^c = \{x = (u, v, w) \in \Re^n : u = 0, v = 0\}, \tag{5.8}$$

the corresponding solution $x(t, 0, x_0^c)$ of Eq. (5.3) is also in E^c for $t \in \Re$, but (5.5) and (5.7) do not hold (for arbitrary x_0^c in E^c).

Accordingly, the invariant subspaces E^s, E^u, and E^c are called the **stable subspace**, the **unstable subspace**, and the **center subspace** of Eq. (5.3).

Next, let's use an example to illustrate the counterparts of these invariant subspaces of linear equations when small nonlinear perturbations are applied. The example will also be used to introduce the corresponding terminology.

Example 8.5.1 Consider

$$\begin{cases} x_1' &= -x_1, \\ x_2' &= x_2 + x_1^2, \\ x_3' &= 0, \end{cases} \tag{5.9}$$

whose solutions can be found by solving the first equation and then plugging in the second equation, and are given by

$$\begin{cases} x_1 &= c_1 e^{-t}, \\ x_2 &= (c_2 + \frac{c_1^2}{3})e^t - \frac{c_1^2}{3}e^{-2t}, \\ x_3 &= c_3, \end{cases} \tag{5.10}$$

where $x(0) = (c_1, c_2, c_3)$. Now, to find the counterpart of the invariant stable subspace E^s of the linearization of Eq. (5.9), which is the x_1-axis now, we need to find $c = (c_1, c_2, c_3)$ such that $|x(t, 0, c)| \to 0$, $t \to \infty$. From (5.10), it is given by

$$M^s = \{c = (c_1, c_2, c_3) \in \Re^3 : c_2 = -\frac{c_1^2}{3}, \ c_3 = 0\}, \tag{5.11}$$

which defines a curve that is tangent to E^s (the x_1-axis) at the origin. See **Figure 8.24** (where we mix x_i with c_i, $i = 1, 2, 3$).

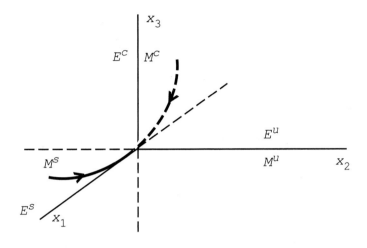

Figure 8.24: E^s, E^u, E^c, and M^s, M^u, M^c, for Eq. (5.9)

Moreover, if an initial point $c = (c_1, c_2, c_3)$ is in M^s, then from (5.10), the corresponding solution $x(t, 0, c) = (x_1, x_2, x_3)$ satisfies $x_3 = c_3 = 0$ and

$$x_2 = (c_2 + \frac{c_1^2}{3})e^t - \frac{c_1^2}{3}e^{-2t} = -\frac{1}{3}(c_1 e^{-t})^2 = -\frac{x_1^2}{3},$$

that is, M^s is invariant for Eq. (5.9).

For the counterpart of the invariant unstable subspace E^u of the linearization of Eq. (5.9), which is the x_2-axis now, we need to find $c = (c_1, c_2, c_3)$ such that $|x(t, 0, c)| \to 0$, $t \to -\infty$. It is given by

$$M^u = \{c = (c_1, c_2, c_3) \in \Re^3 : c_1 = 0, \ c_3 = 0\}, \tag{5.12}$$

or the x_2-axis itself, and it is also invariant for Eq. (5.9). Since M^u is now the same as E^s, it is of course tangent to E^s at the origin. See Figure 8.24.

Similarly, we find that E^c for the linearization of Eq. (5.9) is the x_3-axis, and that

$$M^c = \{c = (c_1, c_2, c_3) \in \Re^3 : c_1 = 0, \ c_2 = 0\} \qquad (5.13)$$

is tangent to E^c at the origin (since they are the same now, see Figure 8.24), and M^c is also invariant for Eq. (5.9). ♠

For the subset M^s given in (5.11), it is not a one-dimensional subspace of $\Re^3$ (in fact, it is not even a *subspace* of $\Re^3$), but M^s is composed with many small pieces such that every small piece is an image of an open interval on the x_1-axis. Accordingly, M^s in (5.11) is called a **one-dimensional manifold**. Moreover, M^s is called a **one-dimensional invariant stable manifold** of Eq. (5.9) due to the properties that M^s is invariant and when initial points are in M^s, the corresponding solutions approach the origin. Similarly, M^u and M^c are called **one-dimensional invariant unstable manifold** and **one-dimensional invariant center manifold** of Eq. (5.9) respectively. (They are identical with E^u and E^c in Example 8.5.1.)

The following is based on Coddington and Levinson [1955], Hartman [1964], Kelley [1967], Guckenheimer and Holmes [1986], Wiggins [1990], and Perko [1991]. We only consider the situation near the origin, and sometimes we only outline the major steps without details, since some of the details are not within the scope of this book. First, we list some definitions.

Definition 8.5.2 *Two subsets A and B of a metric space X are said to be* **homeomorphic** *if there is a continuous map $h : A \to B$ that is one-to-one and onto, and h^{-1} is also continuous. In this case, it is denoted by $A \xrightarrow{h} B$ and h is called a* **homeomorphism**.

Definition 8.5.3 *A* **k-dimensional differentiable manifold** *M is a connected metric space with an open covering $M = \bigcup_\alpha V_\alpha$ such that*

1. *for all α, V_α is homeomorphic to the open unit ball in $\Re^k$, and*

2. *if $V_\alpha \cap V_\beta \neq \varnothing$ (empty set), and $V_\alpha \xrightarrow{h_\alpha} B \subset \Re^k$, $V_\beta \xrightarrow{h_\beta} B$, then the map*

$$h = h_\alpha(h_\beta^{-1}) : \quad h_\beta(V_\alpha \cap V_\beta) \longrightarrow h_\alpha(V_\alpha \cap V_\beta)$$

is a differentiable function of $x \in h_\beta(V_\alpha \cap V_\beta) \subset \Re^k$, with $\det \frac{\partial h(x)}{\partial x} \neq 0$.

That is, a manifold is a set which *locally* has the structure of Euclidean space, and for our purpose here, we can think of a manifold as a curve or a surface embedded in $\Re^n$. See, for example, M^s, M^u, and M^c in Figure 8.24 as one-dimensional differentiable manifolds. The following subset of $\Re^3$,

$$\{x = (x_1, x_2, x_3) \in \Re^3 : x_3 = x_1^2\} \tag{5.14}$$

defines a two-dimensional differentiable manifold that is tangent to the (x_1, x_2) plane at the origin, see **Figure 8.25**.

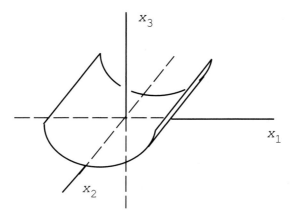

Figure 8.25: A two-dimensional differentiable manifold in $\Re^3$

Definition 8.5.4 *Consider Eq. (5.2) in a small neighborhood D of the origin of $\Re^n$ where the matrix A is given in Eq. (5.3). A manifold $M^s \subset D$ is said to be a* **stable manifold of Eq. (5.2) at the origin** *if M^s is positively invariant for Eq. (5.2), tangent to E^s of the linearization at the origin, and for any $x_0^s \in M^s$, the corresponding solution $x(t, 0, x_0^s)$ of Eq. (5.2) satisfies $|x(t, 0, x_0^s)| \to 0, t \to \infty$. A manifold $M^u \subset D$ is said to be an* **unstable manifold of Eq. (5.2) at the origin** *if M^u is negatively invariant for Eq. (5.2), tangent to E^u of the linearization at the origin, and for any $x_0^u \in M^u$, the corresponding solution $x(t, 0, x_0^u)$ of Eq. (5.2) satisfies $|x(t, 0, x_0^u)| \to 0, t \to -\infty$. A manifold $M^c \subset D$ is said to be a* **center manifold of Eq. (5.2) at the origin** *if M^c is invariant for Eq. (5.2) and tangent to E^c of the linearization at the origin.*

We find from Example 8.5.1 that stable and unstable manifolds are uniquely determined, but center manifolds may not be, see the following example.

Example 8.5.5 Consider

$$\begin{cases} x' = x^2, \\ y' = -y, \end{cases} \tag{5.15}$$

whose solutions are given by $x(t) = \frac{x_0}{1-tx_0}$ and $y(t) = y_0 e^{-t}$. Then we get $t = \frac{1}{x_0} - \frac{1}{x}$ from $x(t) = \frac{x_0}{1-tx_0}$ and hence obtain solution curves which are graphs of the functions $y(x) = (y_0 e^{-1/x_0})e^{1/x}$. Now, for each (x_0, y_0) with $x_0 < 0$, the graph of $y(x) = (y_0 e^{-1/x_0})e^{1/x}$ will start with (x_0, y_0) and approach the origin $(x, y) = (0, 0)$ as $x \to 0^-$. Therefore, we can patch such a graph with the positive x-axis at the origin and still call it $y(x)$, see **Figure 8.26**.

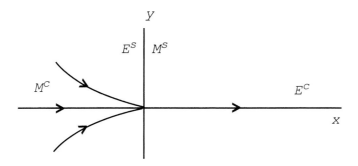

Figure 8.26: The nonuniqueness of center manifolds for Eq. (5.15)

Now, it can be verified (see an exercise) that $\frac{dy}{dx}(0) = 0$, that is, such a graph forms a one-dimensional differentiable center manifold of Eq. (5.15) at the origin. Since such a point (x_0, y_0) with $x_0 < 0$ is arbitrary, center manifolds of Eq. (5.15) at the origin are not uniquely determined, see Figure 8.26. ♠

Next, we state and prove the following result confirming the existence of stable and unstable manifolds of Eq. (5.2) with a small nonlinear perturbation $f(x)$, where we assume

(H). f is continuously differentiable, $f(0) = 0$, $\dfrac{\partial f}{\partial x}(0) = 0$. $\tag{5.16}$

Under the assumption (H), for any $\varepsilon > 0$ there is a $\delta > 0$ such that

$$|f(x) - f(y)| \le \varepsilon |x - y|, \quad \text{for } |x|, |y| \le \delta. \tag{5.17}$$

Theorem 8.5.6 (Stable and Unstable Manifolds Theorem) *Consider Eq. (5.2) where the matrix A is given in Eq. (5.3) and the function f satisfies the hypothesis (H) in (5.16). Let $s = dim(\Re^s) > 0$ (the matrix A has s eigenvalues with negative real parts), $u = dim(\Re^u) > 0$ (u eigenvalues with positive real parts), and $c = dim(\Re^c) = 0$ (no eigenvalues with zero real parts). Then there exists an s-dimensional differentiable stable manifold M^s and a u-dimensional differentiable unstable manifold M^u of Eq. (5.2).*

Proof. Let

$$S(t) = \begin{bmatrix} e^{tA_s} & 0 \\ 0 & 0 \end{bmatrix}, \quad U(t) = \begin{bmatrix} 0 & 0 \\ 0 & e^{tA_u} \end{bmatrix}, \tag{5.18}$$

then $e^{tA} = S(t) + U(t)$ and $S' = AS$, $U' = AU$. Let $\alpha > 0$ be chosen so that the real parts of the eigenvalues of A_s are less than $-\alpha$ (since they are all negative). Then we can find positive constants K and σ such that

$$|S(t)| \le Ke^{-(\alpha+\sigma)t}, \ t \ge 0; \quad |U(t)| \le Ke^{\sigma t}, \ t \le 0. \tag{5.19}$$

For $a \in \Re^n$, consider the integral equation

$$\begin{aligned} v(t,a) &= S(t)a + \int_0^t S(t-h)f(v(h,a))dh \\ &\quad - \int_t^\infty U(t-h)f(v(h,a))dh, \quad t \ge 0, \end{aligned} \tag{5.20}$$

whose construction guarantees that the integrals are convergent using (5.19). It is true (see an exercise) that a solution of Eq. (5.20) is also a solution of Eq. (5.2). We will prove the existence of solutions of Eq. (5.20) and then use them to define a stable manifold. Start with the successive approximations

$$v_{(0)}(t,a) = 0,$$

$$\begin{aligned} v_{(j+1)}(t,a) &= S(t)a + \int_0^t S(t-h)f(v_{(j)}(h,a))dh \\ &\quad - \int_t^\infty U(t-h)f(v_{(j)}(h,a))dh, \quad t \ge 0, \ j \ge 0, \end{aligned} \tag{5.21}$$

we find that

$$|v_{(1)}(t,a)| = |S(t)a| \le K|a|e^{-(\alpha+\sigma)t} \le K|a|, \quad t \ge 0. \tag{5.22}$$

In order to use condition (5.17), we choose $\varepsilon < \frac{\sigma}{4K}$ and then choose a such that $|a| \leq \frac{\delta}{2K}$. Then let's use an induction to prove

$$\begin{cases} |v_{(j)}(t,a)| \leq (\frac{1}{2} + \frac{1}{2^2} + \cdots + \frac{1}{2^j})\delta, \\ |v_{(j)}(t,a) - v_{(j-1)}(t,a)| \leq \frac{K|a|e^{-\alpha t}}{2^{j-1}}, \end{cases} \quad t \geq 0, \ j \geq 1, \quad (5.23)$$

which is true for $j = 1$ already using (5.22). If (5.23) is true for $j = 1, 2, \cdots, m$, then $|v_{(j)}(t,a)| \leq (\frac{1}{2} + \frac{1}{2^2} + \cdots + \frac{1}{2^j})\delta < \delta$, and hence condition (5.17) can be applied to obtain, for $t \geq 0$,

$$|v_{(m+1)}(t,a) - v_{(m)}(t,a)| \leq \int_0^t |S(t-h)|\varepsilon|v_{(m)}(h,a) - v_{(m-1)}(h,a)|dh$$

$$+ \int_t^\infty |U(t-h)|\varepsilon|v_{(m)}(h,a) - v_{(m-1)}(h,a)|dh$$

$$= \varepsilon \int_0^t K e^{-(\alpha+\sigma)(t-h)} \frac{K|a|e^{-\alpha h}}{2^{m-1}}dh + \varepsilon \int_t^\infty K e^{\sigma(t-h)} \frac{K|a|e^{-\alpha h}}{2^{m-1}}dh$$

$$\leq \frac{\varepsilon K^2|a|e^{-\alpha t}}{2^{m-1}} \int_0^t e^{\sigma(h-t)}dh + \frac{\varepsilon K^2|a|e^{-\alpha t}}{2^{m-1}} \int_t^\infty e^{\sigma(t-h)}dh$$

$$\leq \frac{\varepsilon K^2|a|e^{-\alpha t}}{\sigma 2^{m-1}} + \frac{\varepsilon K^2|a|e^{-\alpha t}}{\sigma 2^{m-1}} = \left(\frac{2\varepsilon K}{\sigma}\right)\frac{K|a|e^{-\alpha t}}{2^{m-1}} \leq \frac{K|a|e^{-\alpha t}}{2^m},$$

which in turn implies, for $t \geq 0$,

$$|v_{(m+1)}(t,a)| \leq |v_{(m)}(t,a)| + \frac{K|a|e^{-\alpha t}}{2^m} \leq (\frac{1}{2} + \frac{1}{2^2} + \cdots + \frac{1}{2^m})\delta + \frac{\delta}{2^{m+1}},$$

and hence completes the induction of (5.23). Now, for $j > m > N$ and $t \geq 0$,

$$|v_{(j)}(t,a) - v_{(m)}(t,a)| \leq \sum_{i=N}^\infty |v_{(i+1)}(t,a) - v_{(i)}(t,a)| \leq K|a| \sum_{i=N}^\infty \frac{1}{2^i} = \frac{K|a|}{2^{N-1}},$$

which implies that $\{v_{(i)}(t,a)\}_{i\geq 1}$ is a Cauchy sequence, and that

$$\lim_{i\to\infty} v_{(i)}(t,a) = v(t,a)$$

exists uniformly for $|a| \leq \frac{\delta}{2K}$ and $t \geq 0$, which gives rise to a solution of Eq. (5.20) and of Eq. (5.2). Moreover, since each $v_{(i)}(t,a)$ is a differentiable function of $|a| \leq \frac{\delta}{2K}$ for $t \geq 0$, $v(t,a)$ (as a uniform limit) is also a differentiable function of $|a| \leq \frac{\delta}{2K}$ for $t \geq 0$. The estimate (5.23) also implies that

$$|v(t,a)| \leq 2K|a|e^{-\alpha t}, \quad |a| \leq \frac{\delta}{2K}, \quad t \geq 0. \quad (5.24)$$

Next, let's look at how to define a stable manifold from solutions of Eq. (5.20). First, note from the successive approximations in (5.21) and the structure of the matrix $S(t)$ that the last u components of the vector a do not enter into the solution of Eq. (5.20) and may be taken as zero, which we assume has been done. From Eq. (5.20), the components $v_i(t, a)$ of the solution $v(t, a)$ satisfy the initial conditions

$$
\begin{cases}
v_i(0, a) &= a_i, \quad i = 1, 2, \cdots, s \\
v_i(0, a) &= -\Big(\int_0^\infty U(-h) f(v(h, a)) dh \Big)_i, \quad i = s+1, s+2, \cdots, n,
\end{cases}
\tag{5.25}
$$

where $(\cdot)_i$ denotes the ith component. If the functions ψ_i are defined by

$$
\psi_i(a_1, a_2, \cdots, a_s) = -\Big(\int_0^\infty U(-h) f(v(h, a)) dh \Big)_i, \quad i = s+1, s+2, \cdots, n,
$$

where the first s components of a are $a_1, a_2, \cdots, a_s$ and other components of a are zero, and $v(\cdot, a)$ is the corresponding solution (which is uniquely determined due to the Lipschitz condition), then the initial values $x_i = v_i(0, a)$, $i = 1, 2, \cdots, n$, satisfy the equations

$$
x_i = \psi_i(x_1, x_2, \cdots, x_s), \quad i = s+1, s+2, \cdots, n,
\tag{5.26}
$$

in x space $\Re^n$. This defines an s-dimensional differentiable manifold M^s for $|x^s| \leq \frac{\delta}{2K}$ where $x^s = (x_1, x_2, \cdots, x_s)$. If $x(t)$ is a solution of Eq. (5.2) near the origin with $x(0) \in M^s$, then $x(0) = v(0, a)$ where the first s components of a are the same as those of $x(0)$ and other components of a are zero, and v is the corresponding solution. From uniqueness, we have $x(t) = v(t, a)$, then $x(t) \in M^s$ for $t \geq 0$ and $|x(t)| \to 0$ as $t \to \infty$ using (5.24). Moreover, from Coddington and Levinson [1955] p. 333, one has

$$
\frac{\partial \psi_i}{\partial x_l} = 0, \quad i = s+1, s+2, \cdots, n, \quad l = 1, 2, \cdots, s,
$$

at $x_1 = x_2 = \cdots = x_s = 0$, hence M^s is tangent to $E^s = \{x \in \Re^n : x_{s+1} = \cdots = x_n = 0\}$ of the linearization. Therefore, M^s is an s-dimensional differentiable stable manifold of Eq. (5.2).

To derive an unstable manifold of Eq. (5.2), we can make a change of $t \to -t$ in Eq. (5.2) to get

$$
x'(t) = -Ax(t) - f(x(t)),
\tag{5.27}
$$

and replace the vector $x = (x_1, \cdots, x_s, x_{s+1}, \cdots, x_n)$ by $(x_{s+1}, \cdots, x_n, x_1, \cdots, x_s)$ and then obtain a stable manifold of the resulting equation, which in turn gives an unstable manifold of Eq. (5.2). This completes the proof. ♠

Next, we use an example to illustrate the successive approximations of $v_{(j)}(t, a)$ in the above proof and the approximations of stable and unstable manifolds.

Example 8.5.7 Consider

$$\begin{cases} x_1' & = & -x_1 - x_2^2, \\ x_2' & = & x_2 + x_1^2, \end{cases} \tag{5.28}$$

for which we have

$$A = \begin{bmatrix} -1 & 0 \\ 0 & 1 \end{bmatrix}, \ f(x) = \begin{bmatrix} -x_2^2 \\ x_1^2 \end{bmatrix}, \ S(t) = \begin{bmatrix} e^{-t} & 0 \\ 0 & 0 \end{bmatrix}, \ U(t) = \begin{bmatrix} 0 & 0 \\ 0 & e^t \end{bmatrix}.$$

Then the successive approximations of $v_{(j)}(t, a)$ in (5.21) with $a = [a_1, 0]^T$ give

$$v_{(0)}(t, a) = \begin{bmatrix} 0 \\ 0 \end{bmatrix},$$

$$v_{(1)}(t, a) = S(t)a = \begin{bmatrix} e^{-t}a_1 \\ 0 \end{bmatrix},$$

$$v_{(2)}(t, a) = S(t)a + \int_0^t S(t-h)f(v_{(1)}(h, a))dh - \int_t^\infty U(t-h)f(v_{(1)}(h, a))dh$$

$$= \begin{bmatrix} e^{-t}a_1 \\ 0 \end{bmatrix} - \int_t^\infty \begin{bmatrix} 0 \\ e^{t-h}e^{-2h}a_1^2 \end{bmatrix} dh = \begin{bmatrix} e^{-t}a_1 \\ -\frac{1}{3}e^{-2t}a_1^2 \end{bmatrix},$$

$$v_{(3)}(t, a) = \begin{bmatrix} e^{-t}a_1 \\ 0 \end{bmatrix} - \int_0^t \begin{bmatrix} \frac{e^{-(t-h)}e^{-4h}a_1^4}{9} \\ 0 \end{bmatrix} dh - \int_t^\infty \begin{bmatrix} 0 \\ e^{t-h}e^{-2h}a_1^2 \end{bmatrix} dh$$

$$= \begin{bmatrix} e^{-t}a_1 + \frac{1}{27}(e^{-4t} - e^{-t})a_1^4 \\ -\frac{1}{3}e^{-2t}a_1^2 \end{bmatrix}.$$

Accordingly, we have $v(t, a) \approx \begin{bmatrix} e^{-t}a_1 \\ -\frac{1}{3}e^{-2t}a_1^2 \end{bmatrix}$ when $|a| = |[a_1, 0]^T|$ is small, or when a_1 is small. Thus the stable manifold is approximated by

$$\psi_2(a_1) = -\left(\int_0^\infty U(-h)f(v(h, a))dh \right)_2$$

$$\approx -\left(\int_0^\infty \begin{bmatrix} 0 & 0 \\ 0 & e^{-h} \end{bmatrix} \begin{bmatrix} -\frac{1}{9}e^{-4h}a_1^4 \\ e^{-2h}a_1^2 \end{bmatrix} dh \right)_2$$

$$= -\int_0^\infty e^{-h}e^{-2h}a_1^2 dh = -\frac{1}{3}a_1^2.$$

That is, near the origin, the stable manifold M^s is approximated by

$$\{(x_1, x_2) \in \Re^2 : x_2 = -\frac{1}{3}x_1^2\}.$$

The unstable manifold can be approximated in the same way as above with $t \to -t$ and x_1 and x_2 interchanged, whose stable manifold will be the unstable manifold M^u of Eq. (5.28), and is approximated by

$$\{(x_1, x_2) \in \Re^2 : x_1 = -\frac{1}{3}x_2^2\},$$

see **Figure 8.27**. ♠

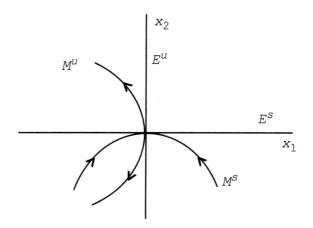

Figure 8.27: Local approximations of M^s and M^u for Eq. (5.28)

Next, we state a result that also concerns center manifolds. Refer to the previous references for a proof and additional remarks.

Theorem 8.5.8 (Stable, Unstable, and Center Manifolds Theorem)
Consider Eq. (5.2) where the matrix A is given in Eq. (5.3) and the function

f satisfies the hypothesis (H) in (5.16). Let $s = dim(\Re^s) > 0$ (the matrix A has s eigenvalues with negative real parts), $u = dim(\Re^u) > 0$ (u eigenvalues with positive real parts), and $c = dim(\Re^c) > 0$ (c eigenvalues with zero real parts). Then there exists an s-dimensional differentiable stable manifold M^s, a u-dimensional differentiable unstable manifold M^u, and a c-dimensional differentiable center manifold M^c of Eq. (5.2). ♠

That is, for Eq. (5.2) near the origin, we can find those invariant manifolds M^s, M^u, and M^c which are the counterparts of the invariant subspaces E^s, E^u, and E^c of the linearization, and M^s, M^u, M^c have the asymptotic properties of E^s, E^u, E^c respectively. That is, solutions of Eq. (5.2) with initial points in M^s (M^u) approach the origin at an exponential rate asymptotically as $t \to \infty$ ($-\infty$), and there exist initial points in M^c such that the corresponding solutions do not approach the origin as $|t| \to \infty$.

Next, we look at the relationship between the trajectories of Eq. (5.2) and its linearization near the origin. Recall from Chapter 4 that a center can be changed to a spiral point after a small perturbation. For example, look at

$$x_1' = -x_2 + x_1(x_1^2 + x_2^2), \quad x_2' = x_1 + x_2(x_1^2 + x_2^2), \tag{5.29}$$

which is Example 4.3.4 in Chapter 4. We know that the origin is a center for the linearization of Eq. (5.29) but a spiral point for Eq. (5.29). Since the trajectories for centers and for spiral points are **qualitatively** different, we find that we must restrict our study to the cases where the real parts of eigenvalues of linearization matrices are nonzero.

Let's also recall that we have seen some cases where trajectories of one equation are transformed to become trajectories of another equation. For example, the linear equation with the matrix

$$B = \begin{bmatrix} -1 & -2 \\ -2 & -1 \end{bmatrix} \tag{5.30}$$

is Example 4.2.5 in Chapter 4, and we know that if we let

$$P = \begin{bmatrix} 1 & 1 \\ -1 & 1 \end{bmatrix}, \tag{5.31}$$

then

$$PBP^{-1} = \begin{bmatrix} -3 & 0 \\ 0 & 1 \end{bmatrix} = A, \tag{5.32}$$

and $y = Px$ transforms $x' = Bx$ into $y' = Ay$. Now, if we define a mapping $H(x) = Px$ on $\Re^2$, then H is a rotation of $-45°$ and hence a homeomorphism on $\Re^2$. Moreover, for $x = e^{tB}x_0$ and $y = e^{tA}y_0$ with $y_0 = Px_0$, we have $e^{tA}Px_0 = e^{tA}y_0 = y = Px = Pe^{tB}x_0$, or

$$e^{tA}H = He^{tB}. \qquad (5.33)$$

That is, H maps trajectories of $x' = Bx$ onto trajectories of $y' = Ay$ and, due to (5.33), we say that H **preserves the parameterization**. See **Figure 8.28**.

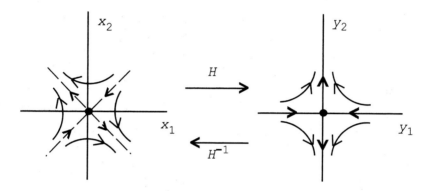

Figure 8.28: A homeomorphism preserving parameterization

Now, we introduce the Hartman-Grobman theorem, proved independently by Hartman in 1960 and Grobman in 1959, showing that near the origin, the trajectories of Eq. (5.2) and its linearization are related by a homeomorphism preserving parameterization.

Theorem 8.5.9 (Hartman-Grobman theorem) *Consider Eq. (5.2) where the matrix A is given in Eq. (5.3) and the function f satisfies the hypothesis (H) in (5.16). Let $s = dim(\Re^s) > 0$ (the matrix A has s eigenvalues with negative real parts), $u = dim(\Re^u) > 0$ (u eigenvalues with positive real parts), and $c = dim(\Re^c) = 0$ (no eigenvalues with zero real parts). Let $T^t : x_0 \to x(t, 0, x_0)$ denote the solution (flow) of Eq. (5.2). Then there exists a homeomorphism H of a neighborhood of the origin onto a neighborhood of the origin such that for x_0 near the origin and t in some interval containing zero,*

$$e^{tA}H = HT^t \quad (\text{or } e^{tA}H(x_0) = H(x(t, 0, x_0))). \qquad (5.34)$$

That is, near the origin, H maps trajectories of Eq. (5.2) onto trajectories of the linearization and preserves the parameterization.

Proof. The following steps are based on Hartman [1964] p. 228–251. See also Perko [1991].

Step 1. Based on Chapter 2, we can assume without loss of generality that solutions (flows) $x(t, 0, x_0)$ of Eq. (5.2) exist on $t \in [-1, 1]$ for x_0 near the origin. The solution $x(t, 0, x_0)$ of Eq. (5.2) can be written as a linear term in x_0 and a higher order term in x_0, given by

$$x(t, 0, x_0) = e^{tA}x_0 + K(t, x_0), \tag{5.35}$$

which can be done if we define K to be the difference of $x(t, 0, x_0)$ and $e^{tA}x_0$.

From Corollary 2.3.11 in Chapter 2, the $n \times n$ matrix $J(t, x_0) = \frac{\partial x(t,0,x_0)}{\partial x_0} = e^{tA} + \frac{\partial K(t,x_0)}{\partial x_0}$ is the solution of

$$\frac{d}{dt}J(t, x_0) = [A + \frac{\partial f}{\partial x}(x(t, 0, x_0))]J(t, x_0), \quad J(0, x_0) = E,$$

where E is the $n \times n$ unit or identity matrix. Since $x(t, 0, 0) = 0$ and $\frac{\partial f}{\partial x}(0) = 0$, we have

$$\frac{d}{dt}J(t, 0) = AJ(t, 0), \quad J(0, 0) = E,$$

thus $J(t, 0) = e^{tA}$, or $e^{tA} + \frac{\partial K}{\partial x_0}(t, 0) = e^{tA}$. Therefore

$$K(t, 0) = 0, \quad \frac{\partial K}{\partial x_0}(t, 0) = 0. \tag{5.36}$$

Now, write $x(1, 0, x_0)$ (at $t = 1$) as

$$x(1, 0, x_0) = e^A x_0 + K(1, x_0) = \begin{bmatrix} e^{A_s} & 0 \\ 0 & e^{A_u} \end{bmatrix} \begin{bmatrix} y_0 \\ z_0 \end{bmatrix} + K(1, x_0), \tag{5.37}$$

where $x_0 = (y_0, z_0)$ and the matrices A_s and A_u are given in Eq. (5.3). From (5.36), we have

$$K(1, 0) = 0, \quad \frac{\partial K}{\partial x_0}(1, 0) = 0. \tag{5.38}$$

Step 2. Based on (5.38), we claim that for any $\theta > 0$ there exists a number $s = s(\theta) > 0$ (which tends to zero with θ) and a continuously differentiable function $G(x)$ defined on $\Re^n$ satisfying $G(x) = K(1, x)$ for $|x| \leq \frac{s}{2}$, $G(x) = 0$ for $|x| \geq s$, and $|\frac{\partial G(x)}{\partial x}| \leq \theta$ for all x.

To prove the claim, we let $\theta > 0$. Then we can let $s > 0$ be so small that $|\frac{\partial K(1,x)}{\partial x}| \leq \frac{\theta}{8}$ when $|x| \leq s$, hence from the mean value theorem, $|K(1, x)| \leq \frac{|x|\theta}{8}$. Then we let $\phi(t)$ be a differentiable real-valued function of $t \geq 0$ such that $\phi(t) = 1$ for $t \leq (\frac{s}{2})^2$, $0 < \phi(t) < 1$ for $(\frac{s}{2})^2 < t < s^2$, $\phi(t) = 0$ for $t \geq s^2$, and $0 \leq -\phi'(t) \leq \frac{2}{s^2}$ for all $t \geq 0$. Define $G(x) = K(1, x)\phi(|x|^2)$ for $|x| \leq s$ and $G(x) = 0$ for $|x| \geq s$ (where $|\cdot|$ is defined using the square root). For $|x| \leq s$, $\frac{\partial G(x)}{\partial x} = \frac{\partial K(1,x)}{\partial x}\phi(|x|^2) + 2\phi'(|x|^2)K(1, x)x^T$ and hence $|\frac{\partial G(x)}{\partial x}| \leq \frac{\theta}{8} + 2\frac{2}{s^2}\frac{|x|^2\theta}{8} \leq \theta$, which proves the claim.

Accordingly, since we are only concerned with the properties near the origin, we may replace $K(1, x)$ by $G(x)$ if necessary. That is, we can assume without loss of generality that

$$
\begin{cases}
K(1, 0) = 0, \quad K(1, x) = 0 \text{ for } |x| \geq s; \\
\frac{\partial K(1,0)}{\partial x} = 0, \quad |\frac{\partial K(1,x)}{\partial x}| \leq \theta \text{ for all } x,
\end{cases}
\tag{5.39}
$$

where $s = s(\theta)$ and $s = s(\theta) \to 0$ as $\theta \to 0$.

Now write

$$
K(1, x) = (Y(x), Z(x))
\tag{5.40}
$$

as components such that

$$
\begin{aligned}
T^1 x_0 &= x(1, 0, x_0) = \begin{bmatrix} e^{A_s} & 0 \\ 0 & e^{A_u} \end{bmatrix} \begin{bmatrix} y_0 \\ z_0 \end{bmatrix} + \begin{bmatrix} Y(x_0) \\ Z(x_0) \end{bmatrix} \\
&= \begin{bmatrix} e^{A_s} y_0 + Y(x_0) \\ e^{A_u} z_0 + Z(x_0) \end{bmatrix}, \quad x_0 = \begin{bmatrix} y_0 \\ z_0 \end{bmatrix}.
\end{aligned}
$$

From (5.39), we have

$$
\begin{cases}
Y(0) = 0, \; Z(0) = 0, \; Y(x) = 0, \text{ and } Z(x) = 0 \text{ for } |x| \geq s; \\
\frac{\partial Y}{\partial x}(0) = 0, \; \frac{\partial Z}{\partial x}(0) = 0, \; |\frac{\partial Y(x)}{\partial x}| \leq \theta \text{ and } |\frac{\partial Z(x)}{\partial x}| \leq \theta \text{ for all } x.
\end{cases}
\tag{5.41}
$$

And we follow a normalization in Hartman [1964] p. 233 and assume without loss of generality that

$$
|e^{A_s}| < 1, \quad |e^{-A_u}| < 1.
\tag{5.42}
$$

Step 3. We show that there exists a homeomorphism

$$H_0(x) = \begin{bmatrix} \Phi(x) \\ \Psi(x) \end{bmatrix} \tag{5.43}$$

on $x \in \Re^n$ such that

$$e^A H_0 = H_0 T^1, \tag{5.44}$$

which is, in components, the same as solving $\Phi(x)$ and $\Psi(x)$ from (for $x = (y, z)$)

$$e^{A_s}\Phi(x) = \Phi(e^{A_s}y + Y(x), e^{A_u}z + Z(x)), \tag{5.45}$$
$$e^{A_u}\Psi(x) = \Psi(e^{A_s}y + Y(x), e^{A_u}z + Z(x)). \tag{5.46}$$

First, let's derive a continuous solution $\Psi(x)$ for Eq. (5.46) using successive approximations. Define

$$\begin{cases} \Psi_{(0)}(x) = z, \quad x = (y, z) \in \Re^n, \\ \Psi_{(j)}(x) = e^{-A_u}\Psi_{(j-1)}(e^{A_s}y + Y(x), e^{A_u}z + Z(x)), \quad j \geq 1, \end{cases} \tag{5.47}$$

so that $\Psi_{(j)}(x)$ is continuous in $x \in \Re^n$. From $Z(x) = 0$, $|x| \geq s$, and $|z| \leq |e^{-A_u}||e^{A_u}z| < |e^{A_u}z|$, we can use an induction to show (see an exercise) that

$$\Psi_{(j)}(x) = z \quad \text{if} \quad |z| \geq s, \quad j \geq 0. \tag{5.48}$$

To prove the convergence of $\Psi_{(j)}(x)$, define

$$D_j(x) = \Psi_{(j)}(x) - \Psi_{(j-1)}(x),$$

so that

$$D_1(x) = \Psi_{(1)}(x) - \Psi_{(0)}(x) = e^{-A_u}Z(x), \tag{5.49}$$
$$D_j(x) = e^{-A_u}D_{j-1}(e^{A_s}y + Y(x), e^{A_u}z + Z(x)), \quad j \geq 2. \tag{5.50}$$

Since $|e^{-A_u}| < 1$, we can find $\delta \in (0, 1)$ such that

$$r = |e^{-A_u}|\left(|e^{A_s}| + |e^{A_u}| + \theta\right)^\delta < 1, \tag{5.51}$$

and then we can define

$$M = \frac{1}{r}|e^{-A_u}|\theta s^{1-\delta}, \tag{5.52}$$

where θ and s are from (5.41). From (5.41), (5.49), and the mean value theorem, we get, for all $x \in \Re^n$,

$$
\begin{aligned}
|D_1(x)| &\leq |e^{-A_u}Z(x)| \leq |e^{-A_u}|\theta|x| = |e^{-A_u}|\theta|x|^{1-\delta}|x|^\delta \\
&\leq |e^{-A_u}|\theta s^{1-\delta}|x|^\delta = Mr|x|^\delta
\end{aligned}
\tag{5.53}
$$

(where $|x| \leq s$ is used because $Z(x) = 0$ if $|x| \geq s$ and then (5.53) is true automatically). Next, we use an induction to prove

$$
|D_j(x)| \leq Mr^j|x|^\delta, \quad x \in \Re^n,
\tag{5.54}
$$

which is true for $j = 1$ from (5.53). If (5.54) is true for $j - 1$, then from (5.39), (5.40), (5.50), and (5.51), we get

$$
\begin{aligned}
|D_j(x)| &\leq |e^{-A_u}||D_{j-1}(e^{A_s}y + Y(x), \; e^{A_u}z + Z(x))| \\
&\leq |e^{-A_u}|Mr^{j-1}|(e^{A_s}y + Y(x), \; e^{A_u}z + Z(x))|^\delta \\
&= |e^{-A_u}|Mr^{j-1}|\begin{bmatrix} e^{A_s}y \\ e^{A_u}z \end{bmatrix} + \begin{bmatrix} Y(x) \\ Z(x) \end{bmatrix}|^\delta \\
&= |e^{-A_u}|Mr^{j-1}|\begin{bmatrix} e^{A_s} & 0 \\ 0 & e^{A_u} \end{bmatrix} x + K(1, x)|^\delta \\
&\leq |e^{-A_u}|Mr^{j-1}\Big[(|e^{A_s}| + |e^{A_u}| + \theta)|x|\Big]^\delta \\
&= |e^{-A_u}|Mr^{j-1}\Big(|e^{A_s}| + |e^{A_u}| + \theta\Big)^\delta|x|^\delta \\
&= Mr^j|x|^\delta,
\end{aligned}
$$

which completes the induction for (5.54). Thus, similar to the proof of the stable and unstable manifolds theorem, $\Psi_{(j)}(x)$ is a Cauchy sequence of continuous functions and hence converges uniformly as $j \to \infty$ to a continuous function $\Psi(x)$, which gives rise to a solution of Eq. (5.46), and $\Psi(x) = z$ if $|z| \geq s$.

To get a continuous solution Φ of Eq. (5.45), we follow Hartman [1964] p. 246 and write the inverse of

$$
T^1 : \bar{y} = e^{A_s}y + Y(x), \quad \bar{z} = e^{A_u}z + Z(x),
\tag{5.55}
$$

as

$$
T^{-1} : y = e^{-A_s}\bar{y} + \overline{Y}(\bar{x}), \quad z = e^{-A_u}\bar{z} + \overline{Z}(\bar{x}),
$$

which exists near the origin. Now, $x = (y, z) = (e^{-A_s}\overline{y} + \overline{Y}(\overline{x}),\ e^{-A_u}\overline{z} + \overline{Z}(\overline{x}))$, hence from (5.55), Eq. (5.45) becomes

$$e^{A_s}\Phi(e^{-A_s}\overline{y} + \overline{Y}(\overline{x}),\ e^{-A_u}\overline{z} + \overline{Z}(\overline{x})) = \Phi(\overline{y},\ \overline{z}), \qquad (5.56)$$

and moreover, from the structures of T^1 and T^{-1}, we have

$$Y(x) = -e^{A_s}\overline{Y}(\overline{x}), \quad Z(x) = -e^{A_u}\overline{Z}(\overline{x}), \qquad (5.57)$$

for x and $\overline{x}$ near the origin. Therefore, $\overline{Y}(\overline{x})$ and $\overline{Z}(\overline{x})$ satisfy the analogous conditions for $Y(x)$ and $Z(x)$ in (5.41) and Eq. (5.56) can be treated in exactly the same way Eq. (5.46) was treated since we assume $|e^{A_s}| < 1$. Thus we obtain a continuous solution Φ of Eq. (5.45) with $\Phi(x) = y$ if $|y| \geq s$, and then obtain the continuous map H_0 in (5.43) satisfying (5.44).

Finally, H_0 is one-to-one and onto; that is, H_0 is a homeomorphism on $\Re^n$. For example, when $|y| \geq s$ and $|z| \geq s$, we have $\Phi(x) = y$ and $\Psi(x) = z$ for $x = (y, z)$, thus $H_0(x) = (\Phi(x), \Psi(x)) = x$, which is one-to-one and onto. If $x_0 \neq \overline{x}_0$ exist near the origin such that $H_0(x_0) = H_0(\overline{x}_0)$, then an induction shows (see an exercise) that

$$H_0\Big[T^j(x_0)\Big] = H_0\Big[T^j(\overline{x}_0)\Big], \quad j \geq 1. \qquad (5.58)$$

In one case, x_0 is in the stable manifold and $\overline{x}_0$ is in the unstable manifold, then $|T^j(x_0)| \to 0$ and $|T^j(\overline{x}_0)| \to \infty$ as $j \to \infty$. Thus, as $j \to \infty$, $|H_0\Big[T^j(x_0)\Big]| \to 0$ since $H_0(0) = 0$, and $|H_0\Big[T^j(\overline{x}_0)\Big]| \to \infty$ since $H_0 = (\Phi, \Psi)$ and $\Phi(x) = y$ when y of $x = (y, z)$ is big, and $\Psi(x) = z$ when z of $x = (y, z)$ is big. This contradicts (5.58). See Hartman [1964] p. 248–249 for other cases.

Step 4. Use the homeomorphism H_0 derived above to define

$$H = \int_0^1 e^{-hA} H_0 T^h dh. \qquad (5.59)$$

Since e^{tA} and T^t satisfy the properties of dynamical systems, and $H_0 = e^{-A}H_0T^1$, we obtain

$$\begin{aligned}
e^{tA}H &= \Big(\int_0^1 e^{(t-h)A} H_0 T^{h-t} dh\Big) T^t = \Big(\int_{-t}^{1-t} e^{-\tau A} H_0 T^\tau d\tau\Big) T^t \\
&= \Big(\int_{-t}^0 e^{-\tau A} H_0 T^\tau d\tau + \int_0^{1-t} e^{-\tau A} H_0 T^\tau d\tau\Big) T^t
\end{aligned}$$

$$
\begin{aligned}
&= \left(\int_{-t}^{0} e^{-\tau A} [e^{-A} H_0 T^1] T^\tau d\tau + \int_{0}^{1-t} e^{-\tau A} H_0 T^\tau d\tau \right) T^t \\
&= \left(\int_{-t}^{0} e^{-(\tau+1)A} H_0 T^{\tau+1} d\tau + \int_{0}^{1-t} e^{-\tau A} H_0 T^\tau d\tau \right) T^t \\
&= \left(\int_{1-t}^{1} e^{-hA} H_0 T^h dh + \int_{0}^{1-t} e^{-hA} H_0 T^h dh \right) T^t \\
&= \left(\int_{0}^{1} e^{-hA} H_0 T^h dh \right) T^t = H T^t,
\end{aligned}
$$

which is (5.34). Since e^{-hA}, H_0, and T^h are all homeomorphisms, H in (5.59) is also a homeomorphism. See Hartman [1964] p. 250–251 for details.

This completes the proof of the Hartman-Grobman theorem. ♠

Next, we use an example to illustrate how the successive approximations in the above proof are used to approximate the homeomorphism in the Hartman-Grobman theorem.

Example 8.5.10 Consider

$$
\begin{cases}
y' &= -y, \\
z' &= z + y^2,
\end{cases}
\tag{5.60}
$$

which comes from the first two equations of Eq. (5.9) in Example 8.5.1, hence solutions are given by

$$
\begin{cases}
y &= y_0 e^{-t}, \\
z &= z_0 e^t + \frac{y_0^2}{3}(e^t - e^{-2t}),
\end{cases}
\tag{5.61}
$$

where $x = (y, z)$, $x(0) = (y_0, z_0)$. Thus, at $t = 1$, we have

$$
e^{A_s} = e^{-1}, \quad e^{A_u} = e, \quad Y(x) = 0, \quad Z(x) = \frac{e - e^{-2}}{3} y^2, \quad x = (y, z). \tag{5.62}
$$

Therefore,

$$
e^{A_u} \Psi(x) = \Psi(e^{A_s} y + Y(x), \ e^{A_u} z + Z(x))
$$

becomes

$$
e \Psi(x) = \Psi(e^{-1} y, \ ez + \frac{e - e^{-2}}{3} y^2), \tag{5.63}
$$

and the successive approximations are given by

$$
\begin{cases}
\Psi_{(0)}(x) &= z, \quad x = (y, z) \in \Re^2, \\[2mm]
\Psi_{(1)}(x) &= e^{-1}\Psi_{(0)}(e^{-1}y,\ ez + \frac{e-e^{-2}}{3}y^2) = z + \frac{1-e^{-3}}{3}y^2, \\[2mm]
\Psi_{(2)}(x) &= e^{-1}\Psi_{(1)}(e^{-1}y,\ ez + \frac{e-e^{-2}}{3}y^2) \\[2mm]
&= e^{-1}\left(ez + \frac{e-e^{-2}}{3}y^2 + \frac{1-e^{-3}}{3}e^{-2}y^2\right) = z + \frac{1-e^{-3}}{3}\left(1 + e^{-3}\right)y^2, \\[2mm]
\Psi_{(3)}(x) &= e^{-1}\Psi_{(2)}(e^{-1}y,\ ez + \frac{e-e^{-2}}{3}y^2) \\[2mm]
&= e^{-1}\left(ez + \frac{e-e^{-2}}{3}y^2 + \frac{1-e^{-3}}{3}\left(1 + e^{-3}\right)e^{-2}y^2\right) \\[2mm]
&= z + \frac{1-e^{-3}}{3}\left(1 + e^{-3} + e^{-6}\right)y^2, \\[2mm]
&\quad \cdots \\[2mm]
\Psi_{(j)}(x) &= z + \frac{1-e^{-3}}{3}\left(1 + e^{-3} + [e^{-3}]^2 + \cdots + [e^{-3}]^{j-1}\right)y^2.
\end{cases}
$$

Accordingly, we obtain

$$
\lim_{j\to\infty} \Psi_{(j)}(x) = z + \frac{1 - e^{-3}}{3}\frac{1}{1 - e^{-3}}y^2 = z + \frac{y^2}{3} = \Psi(x),
$$

uniformly for $x = (y, z) \in \Re^2$, and $\Psi(x)$ satisfies (5.63).

Similarly, Φ can be solved from

$$
e^{A_s}\Phi(e^{-A_s}\overline{y} + \overline{Y}(\overline{x}),\ e^{-A_u}\overline{z} + \overline{Z}(\overline{x})) = \Phi(\overline{y},\ \overline{z}), \tag{5.64}
$$

where from (5.55), (5.57), and (5.62),

$$
\begin{aligned}
\overline{Y}(\overline{x}) &= 0, \\[2mm]
\overline{Z}(\overline{x}) &= -e^{-1}Z(x) = -e^{-1}\frac{e - e^{-2}}{3}y^2 = -\frac{1 - e^{-3}}{3}\left[e^{-A_s}\overline{y} + \overline{Y}(\overline{x})\right]^2 \\[2mm]
&= -\frac{1 - e^{-3}}{3}e^2\overline{y}^2 = -\frac{e^2 - e^{-1}}{3}\overline{y}^2.
\end{aligned}
$$

Thus Eq. (5.64) can be replaced by

$$
e^{-1}\Phi(e\overline{y},\ e^{-1}\overline{z} - \frac{e^2 - e^{-1}}{3}\overline{y}^2) = \Phi(\overline{y},\ \overline{z}), \tag{5.65}
$$

or

$$
e^{-1}\Phi(ey,\ e^{-1}z - \frac{e^2 - e^{-1}}{3}y^2) = \Phi(y,\ z). \tag{5.66}
$$

With the successive approximations

$$\begin{cases} \Phi_{(0)}(x) &= y, \quad x = (y, z) \in \Re^2, \\ \Phi_{(1)}(x) &= e^{-1}\Phi_{(0)}(ey, \ e^{-1}z - \frac{e^2 - e^{-1}}{3}y^2) = y, \\ \quad \cdots\cdots \end{cases}$$

we obtain

$$\Phi(x) = y, \quad x = (y, z) \in \Re^2.$$

Therefore, the homeomorphism H_0 is given by

$$H_0(x) = \begin{bmatrix} \Phi(x) \\ \Psi(x) \end{bmatrix} = \begin{bmatrix} y \\ z + \frac{y^2}{3} \end{bmatrix}, \quad x = (y, z).$$

Since now we have, for $x = (y, z) \in \Re^2$,

$$e^{tA}x = \begin{bmatrix} e^{-t}y \\ e^t z \end{bmatrix}, \quad T^t(x) = \begin{bmatrix} ye^{-t} \\ ze^t + \frac{y^2}{3}(e^t - e^{-2t}) \end{bmatrix},$$

the homeomorphism H is given by

$$\begin{aligned} H(x) &= \int_0^1 e^{-hA} H_0 T^h(x) dh = \int_0^1 e^{-hA} H_0 \begin{bmatrix} ye^{-h} \\ ze^h + \frac{y^2}{3}(e^h - e^{-2h}) \end{bmatrix} dh \\ &= \int_0^1 e^{-hA} \begin{bmatrix} ye^{-h} \\ ze^h + \frac{y^2}{3}(e^h - e^{-2h}) + \frac{1}{3}y^2 e^{-2h} \end{bmatrix} dh \\ &= \int_0^1 e^{-hA} \begin{bmatrix} ye^{-h} \\ ze^h + \frac{y^2}{3}e^h \end{bmatrix} dh \\ &= \int_0^1 \begin{bmatrix} e^h ye^{-h} \\ e^{-h}[ze^h + \frac{y^2}{3}e^h] \end{bmatrix} dh = \begin{bmatrix} y \\ z + \frac{y^2}{3} \end{bmatrix}, \end{aligned}$$

and

$$e^{tA} H(x) = \begin{bmatrix} e^{-t}y \\ e^t z + e^t \frac{y^2}{3} \end{bmatrix} = HT^t(x).$$

To find the subset that gets mapped onto the stable subspace E^s of the linearization, we need to find (y, z) such that

$$H(x) = \begin{bmatrix} y \\ z + \frac{y^2}{3} \end{bmatrix} \in E^s = \{(y, z) \in \Re^2 : z = 0\},$$

thus, the subset is given by

$$\{(y, z) \in \Re^2 : z = -\frac{1}{3}y^2\},$$

which, from Example 8.5.1, is the stable manifold of Eq. (5.60). That is, H maps the stable manifold M^s onto E^s. Similarly, H maps the unstable manifold $M^u = \{(y, z) \in \Re^2 : y = 0\}$ onto $E^u = \{(y, z) \in \Re^2 : y = 0\} = M^u$. Also, to find the curve that gets mapped onto the curve $z = \frac{c}{y}$ of the linearization, we need $H(x) = (y, \frac{c}{y})$, or $(y, z + \frac{1}{3}y^2) = (y, \frac{c}{y})$, thus $z = \frac{c}{y} - \frac{1}{3}y^2$. Finally, note that H preserves the parameterization. See **Figure 8.29**. ♠

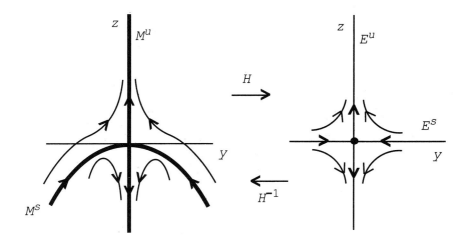

Figure 8.29: The map H and M^s, M^u, E^s, E^u, of Eq. (5.60)

Exercises 8.5

1. Prove that M^s in Figure 8.24 is a one-dimensional differentiable manifold.

2. Prove that the subset defined in (5.14) is a two-dimensional differentiable manifold that is tangent to the (x_1, x_2) plane at the origin.

3. In Example 8.5.5, prove that $y(x) = (y_0 e^{-1/x_0})e^{1/x} \to 0$ as $x \to 0^-$ and then prove that $\frac{dy}{dx}(0) = 0$ for $y(x)$ (together with the positive x-axis at the origin). Then prove that such a graph forms a one-dimensional differentiable center manifold of Eq. (5.15) at the origin.

4. Prove (5.17).

5. Verify in the proof of Theorem 8.5.6 that a solution of Eq. (5.20) is also a solution of Eq. (5.2).

6. Verify that a change of $t \to -t$ will change Eq. (5.2) to Eq. (5.27), and then complete the proof of Theorem 8.5.6 by deriving an unstable manifold.

7. Verify (5.26) in the proof of Theorem 8.5.6.

8. Prove that $x_i = \psi_i(x_1, x_2, \cdots, x_s)$, $i = s+1, s+2, \cdots, n$, in (5.26) of Theorem 8.5.6 defines an s-dimensional differentiable manifold M^s for $|x^s| \le \frac{\delta}{2K}$ where $x^s = (x_1, x_2, \cdots, x_s)$.

9. Find the approximations $v_{(4)}(t, a)$ and $v_{(5)}(t, a)$ for Example 8.5.7. Next, approximate the unstable manifold near the origin.

10. Find the first four successive approximations $v_{(1)}(t, a)$, $v_{(2)}(t, a)$, $v_{(3)}(t, a)$, and $v_{(4)}(t, a)$ for

$$\begin{cases} x_1' &= -x_1 - x_2^3, \\ x_2' &= x_2 + x_1^3, \end{cases} \tag{5.67}$$

and then approximate the stable manifold near the origin. Next, approximate the unstable manifold near the origin.

11. Find the first three successive approximations $v_{(1)}(t, a)$, $v_{(2)}(t, a)$, and $v_{(3)}(t, a)$ for

$$\begin{cases} x_1' &= -x_1, \\ x_2' &= -x_2 + x_1^2, \\ x_3' &= x_3, \end{cases} \tag{5.68}$$

and then approximate the stable manifold near the origin. Next, approximate the unstable manifold near the origin.

12. Use an induction to show (5.48).

13. Verify (5.57).

14. Verify that the continuous map H_0 in (5.43) of Theorem 8.5.9 satisfies $H_0(0) = 0$.

15. Verify (5.58).

16. Prove for each fixed t that e^{tA} and T^t (defined in Theorem 8.5.9) are homeomorphisms on $\Re^n$.

17. Approximate the homeomorphism H and the stable and unstable manifolds near the origin for

$$(a) \begin{cases} y_1' &= -y_1, \\ y_2' &= -y_2 + y_1^3, \\ z' &= z + y_1^3. \end{cases} \qquad (b) \begin{cases} z_1' &= z_1, \\ z_2' &= 2z_2 + z_1^3. \end{cases}$$

$$(c) \begin{cases} y' &= -y, \\ z_1' &= z_1 + y^2, \\ z_2' &= z_2. \end{cases} \qquad (d) \begin{cases} y_1' &= -y_1, \\ y_2' &= -3y_2 + y_1^2. \end{cases}$$

Chapter 9

Stability. Part II

9.1 Introduction

In Chapter 5, we studied stabilities for autonomous differential equations and linear differential equations with constant or periodic coefficients, and those results are applied in chapters 6, 7, and 8. In this chapter, we will extend the study of stabilities to general nonautonomous differential equations, which, of course, include autonomous differential equations as special cases.

Thus, we will look at the general differential equation

$$x'(t) = f(t, x(t)), \tag{1.1}$$

in $D = [0, \infty) \times Q$, where $Q \subset \Re^n$ is a domain containing the zero vector. Similar to Chapter 5, when we study stability properties, we are concerned with "long-term" behavior of solutions for "future time," so we will make the following assumption throughout this chapter.

(H). *For any $(t_0, x_0) \in [0, \infty) \times Q$, Eq.(1.1) has a unique solution $x(t, t_0, x_0)$ existing on $[t_0, \infty)$ with $x(t_0) = x_0$.*

We will still use the definitions regarding stability properties in the sense of Liapunov given in Section 1 of Chapter 5. In particular, if $x(t) = 0$, $t \geq 0$, is a solution of Eq. (1.1), or equivalently when $f(t, 0) = 0$, $t \geq 0$, those definitions give the corresponding definitions concerning stability properties for the zero solution $x = 0$. Also recall from Chapter 5 that to consider stabilities of a solution $\phi(t)$ of Eq. (1.1), we may look at $x(t) = y(t) - \phi(t)$ where y is also a solution of Eq. (1.1), and obtain

$$
\begin{aligned}
x'(t) &= y'(t) - \phi'(t) = f(t, y(t)) - f(t, \phi(t)) \\
&= f(t, x(t) + \phi(t)) - f(t, \phi(t)).
\end{aligned} \tag{1.2}
$$

Now we can define

$$\overline{f}(t, x) = f(t, x + \phi(t)) - f(t, \phi(t)),$$

then $\overline{f}(t, 0) = f(t, \phi(t)) - f(t, \phi(t)) = 0$, and $x(t) = y(t) - \phi(t)$ is a solution of

$$x'(t) = \overline{f}(t, x(t)). \tag{1.3}$$

That is, the zero is now a solution of Eq. (1.3) and it corresponds to the solution $\phi(t)$ of Eq. (1.1). Therefore, in most cases, we can simply assume $f(t, 0) = 0$ in Eq. (1.1) and study stability properties of the zero solution of Eq. (1.1).

In Chapter 5 we stated without proof that "stability" (or "asymptotic stability") for autonomous or periodic differential equations is equivalent to "uniform stability" (or "uniform asymptotic stability"). Now, we provide a proof.

Theorem 9.1.1 *Assume that Eq. (1.1) is autonomous or periodic in t, and satisfies a Lipschitz condition (or weak or local Lipschitz) with respect to x on D. Assume further that $\phi(t) = 0$, $t \geq 0$, is a solution of Eq. (1.1). If $\phi = 0$ is stable, then it is uniformly stable. If $\phi = 0$ is asymptotically stable, then it is uniformly asymptotically stable.*

Proof. We only prove the statement for the case when Eq. (1.1) is autonomous, that is, $f(t, x) = f(x)$. The proof for the case when Eq. (1.1) is periodic can be found in Yoshizawa [1966].

Assume that $\phi = 0$ is stable. Note that $\phi = 0$ is defined on $[t_\phi, \infty) = [0, \infty)$, then for $\overline{t}_0 = 0$ and any $\varepsilon > 0$, there exists a $\delta = \delta(\varepsilon, \overline{t}_0) = \delta(\varepsilon, 0) > 0$, such that $|x_0| \leq \delta(\varepsilon, 0)$ implies $|x(t, 0, x_0)| \leq \varepsilon$ for $t \geq 0$. Now, for any solution $x(t, t_0, \overline{x}_0)$ with $|\overline{x}_0| \leq \delta(\varepsilon, 0)$ and $t_0 \geq 0$, we define $y(t) = x(t + t_0, t_0, \overline{x}_0)$, $t \geq 0$. Then, similar to the "shifting" used in Chapter 4 for autonomous differential equations, we have

$$y'(t) = x'(t + t_0, t_0, \overline{x}_0) = f(x(t + t_0, t_0, \overline{x}_0)) = f(y(t)), \quad y(0) = \overline{x}_0. \tag{1.4}$$

That is, we can shift $x(t, t_0, \overline{x}_0)$ to become a solution starting at 0. Thus, by uniqueness, $y(t) = x(t, 0, \overline{x}_0)$, that is, $y = x(t + t_0, t_0, \overline{x}_0)$ is the unique solution of Eq. (1.1) with the initial data $(0, \overline{x}_0)$. According to the stability of $\phi = 0$, $|\overline{x}_0| \leq \delta(\varepsilon, 0)$ implies $|y(t)| = |x(t, 0, \overline{x}_0)| \leq \varepsilon$ for $t \geq 0$. Thus $t_0 \geq 0$ and $|\overline{x}_0| \leq \delta(\varepsilon, 0)$ imply $|x(t + t_0, t_0, \overline{x}_0)| \leq \varepsilon$ for $t \geq 0$. That is, $t_0 \geq 0$ and $|\overline{x}_0| \leq \delta(\varepsilon, 0)$ imply $|x(t, t_0, \overline{x}_0)| \leq \varepsilon$ for $t \geq t_0$, proving the uniform stability of $\phi = 0$ because $\delta(\varepsilon, 0)$ is independent of t_0.

Next, assume that $\phi = 0$ is asymptotically stable. Then it is stable, and hence uniformly stable according to what we have just proved. Using the asymptotic stability of $\phi = 0$, for $t_0 = 0$, there exists $r(0) > 0$ such that $|x_0| \leq r(0)$ implies $\lim_{t \to \infty} |x(t, 0, x_0)| = 0$. From the above, we see that solutions starting at t_0 can be shifted to become solutions starting at 0. Therefore, to prove the uniform asymptotic stability of $\phi = 0$, we only need to verify that for any $\varepsilon > 0$, there exists a $T = T(\varepsilon) > 0$ such that $\{|x_0| \leq r(0), \ t \geq T\}$ imply $|x(t, 0, x_0)| \leq \varepsilon$. Suppose this is not true, then there exists a small $\varepsilon_0 > 0$, and a sequence x_k and t_k with $|x_k| \leq r(0)$, $t_k \to \infty$ as $k \to \infty$, such that

$$|x(t_k, 0, x_k)| > \varepsilon_0, \quad k \geq 1. \tag{1.5}$$

From the stability of $\phi = 0$, for the above $\varepsilon_0 > 0$, there exists a $\delta = \delta(\varepsilon_0, 0) > 0$ such that $|x_0| \leq \delta(\varepsilon_0, 0)$ implies $|x(t, 0, x_0)| \leq \varepsilon_0$ for $t \geq 0$. Now, the sequence x_k is bounded, thus it has a convergent subsequence. By using a subindex if necessary, we may denote this subsequence by x_k again, so that $x_k \to x^0$, $k \to \infty$, for some $|x^0| \leq r(0)$. Therefore, $\lim_{t \to \infty} |x(t, 0, x^0)| = 0$ according to the asymptotic stability of $\phi = 0$. For the above $\delta(\varepsilon_0, 0) > 0$, there exists a $T_0 > 0$ such that

$$|x(T_0, 0, x^0)| \leq \frac{\delta(\varepsilon_0, 0)}{2}.$$

Now recall from Chapter 2 that a solution is continuous with respect to initial data on a finite interval, therefore, since $x_k \to x^0$ as $k \to \infty$, there is a $k_0 > 0$ such that $|x(T_0, 0, x_k)| \leq \delta(\varepsilon_0, 0)$, $k \geq k_0$.

Next, using the fact that the differential equation is autonomous again, we can start a solution from $x(T_0, 0, x_k)$. That is, we define $y(t) = x(t + T_0, 0, x_k)$, $t \geq 0$, then y is a solution and $y(t) = x(t, 0, x(T_0, 0, x_k))$ by uniqueness. Therefore, as the initial value for y is such that $|y(0)| = |x(T_0, 0, x_k)| \leq \delta(\varepsilon_0, 0)$, $k \geq k_0$, it follows from the definition of $\delta(\varepsilon_0, 0)$ that $|y(t)| \leq \varepsilon_0$ for $t \geq 0$. This means that $|x(t + T_0, 0, x_k)| \leq \varepsilon_0$ for

$k \geq k_0$, $t \geq 0$. Thus we can find $k \geq k_0$ and choose $t > 0$ with $t + T_0 = t_k$ to derive

$$|x(t_k, 0, x_k)| \leq \varepsilon_0,$$

which contradicts (1.5), and hence completes the proof.   ♠

This chapter is organized as follows: In Section 2, we use the results from Chapter 3 to derive stability properties for general linear differential equations, and prove that they are determined by the fundamental matrix solutions. The results here include those derived in Chapter 5 for linear differential equations with constant or periodic coefficients as special cases. Stability properties of general linear differential equations with linear or nonlinear perturbations are also studied using the variation of parameters formula and Gronwall's inequality. In Section 3, we introduce Liapunov's method for general (nonautonomous) differential equations and derive their stability properties, which extends the study of stabilities in Chapter 5 for autonomous differential equations.

Exercises 9.1

1. Examine the stabilities of the zero solution for the scalar differential equation $x'(t) = a(t)x(t)$.

2. Verify in the proof of Theorem 9.1.1 that to prove the uniform asymptotic stability of $\phi = 0$, we only need to verify that for any $\varepsilon > 0$, there exists a $T = T(\varepsilon) > 0$ such that $\{|x_0| \leq r(0), \ t \geq T\}$ imply $|x(t, 0, x_0)| \leq \varepsilon$.

9.2 General Linear Differential Equations

Here, we study the linear differential equation

$$x'(t) = A(t)x(t) + f(t), \quad x(t_0) = x_0, \quad t \geq t_0 \geq 0, \quad x \in \Re^n, \qquad (2.1)$$

where $A(t)$ and $f(t)$ are continuous on $\Re^+ = [0, \infty)$.

Using the results of Chapter 3, we know that the unique solution of Eq. (2.1) is given by the variation of parameters formula

$$
\begin{aligned}
x(t) &= U(t, t_0)\Big[x_0 + \int_{t_0}^{t} U^{-1}(s, t_0)f(s)ds\Big] \\
&= U(t, t_0)x_0 + \int_{t_0}^{t} U(t, s)f(s)ds, \quad t \geq t_0, \qquad (2.2)
\end{aligned}
$$

where the matrix $U(t, t_0)$ is the fundamental matrix solution of Eq. (2.1) when $f = 0$. The following "evolution system property" is also from Chapter 3: for $t, t_0, t_1 \in \Re^+ = [0, \infty)$,

$$\begin{cases} U(t, t_1) U^{-1}(t_0, t_1) = U(t, t_0), \\ U^{-1}(t_0, t_1) = U(t_1, t_0), \\ U(t, t_1) U(t_1, t_0) = U(t, t_0). \end{cases} \tag{2.3}$$

We first study the linear homogeneous differential equation

$$x'(t) = A(t)x(t), \tag{2.4}$$

where the unique solution is given by

$$x(t) = U(t, t_0)x_0$$

for the initial data (t_0, x_0). Note that now $\phi = 0$ is a solution of Eq. (2.4).

The following theorem generalizes the corresponding results for autonomous linear differential equations given in Chapter 5.

Theorem 9.2.1 *Assume that $A(t)$ is continuous on $\Re^+$ and let U be the fundamental matrix solution of Eq. (2.4). The zero solution $\phi = 0$ of Eq. (2.4) is*

(A). *stable if and only if there is an (independent or generic) constant $C > 1$ such that*

$$|U(t, 0)| \leq C, \quad 0 \leq t < \infty. \tag{2.5}$$

(B). *uniformly stable if and only if there is an (independent or generic) constant $C > 1$ such that*

$$|U(t, s)| \leq C, \quad 0 \leq s \leq t < \infty. \tag{2.6}$$

(C). *asymptotically stable if and only if*

$$|U(t, 0)| \to 0, \quad t \to \infty. \tag{2.7}$$

(D). *uniformly asymptotically stable if and only if there are (independent or generic) constants $C > 1$ and $\alpha > 0$ such that*

$$|U(t, s)| \leq C e^{-\alpha(t-s)}, \quad 0 \leq s \leq t < \infty. \tag{2.8}$$

(This case is also called exponentially stable.)

Proof. (A): Suppose that $\phi = 0$ is stable. Then for $t_0 = 0$ and $\varepsilon = 1$, there exists a $\delta = \delta(\varepsilon, t_0) = \delta(1, 0) \in (0, 1)$ such that $|x_0| \leq \delta(1, 0)$ implies $|x(t, 0, x_0)| = |U(t, 0)x_0| \leq \varepsilon = 1$ for $t \geq 0$. Now take $x_0 = \delta(1, 0)e_i$, where $e_i, 1, 2, \cdots, n$, form the standard unit basis for $\Re^n$. Then $|x_0| = |\delta e_i| = \delta$ and hence $|U(t, 0)\delta e_i| \leq 1$ or $|U(t, 0)e_i| \leq \frac{1}{\delta}$. But $U(t, 0)e_i$ is the ith column of $U(t, 0)$, therefore

$$|U(t, 0)| \leq n\frac{1}{\delta} \stackrel{\text{def}}{=} C, \quad t \geq 0.$$

This proves (2.5) because $C > 1$ is an independent constant.

Conversely, if (2.5) is true, then for any $t_0 \geq 0$, we have, from (2.3), $|x(t, t_0, x_0)| = |U(t, t_0)x_0| = |U(t, 0)U(0, t_0)x_0| \leq C|U(0, t_0)||x_0|$. Thus for any $t_0 \geq 0$ and any $\varepsilon > 0$, we can choose $\delta(\varepsilon, t_0) = \frac{\varepsilon}{C|U(0, t_0)|}$, so that $|x_0| \leq \delta(\varepsilon, t_0)$ implies $|x(t, t_0, x_0)| \leq C|U(0, t_0)|\delta(\varepsilon, t_0) = \varepsilon$, proving the stability of $\phi = 0$.

(B): Suppose that $\phi = 0$ is uniformly stable. Then for $\varepsilon = 1$, there exists a $\delta = \delta(1) > 0$, such that $s \geq 0$ and $|x_0| \leq \delta(1)$ imply $|x(t, s, x_0)| = |U(t, s)x_0| \leq 1$ for $t \geq s \geq 0$. Now, similar to the proof of (A) given above, there is a constant $C > 1$ such that $|U(t, s)| \leq C$, $0 \leq s \leq t < \infty$.

Next, if (2.6) is true, then $|x(t, t_0, x_0)| = |U(t, t_0)x_0| \leq C|x_0|$, $0 \leq t_0 \leq t < \infty$. Thus for any $\varepsilon > 0$, we can choose $\delta = \frac{\varepsilon}{C}$, such that $t_0 \geq 0$ and $|x_0| \leq \delta$ imply $|x(t, t_0, x_0)| = |U(t, t_0)x_0| \leq C|x_0| \leq C\delta = \varepsilon$. This proves that $\phi = 0$ is uniformly stable.

(C): Suppose that $\phi = 0$ is asymptotically stable. Then for $t_0 = 0$, there exists an $r(0) > 0$ such that $|x_0| \leq r(0)$ implies $\lim_{t \to \infty} |x(t, 0, x_0)| = \lim_{t \to \infty} |U(t, 0)x_0| = 0$. Then similar to the proof of (A), we see that every entry in $U(t, 0)$ goes to zero, thus $\lim_{t \to \infty} |U(t, 0)| = 0$.

Now, assume that (2.7) is true. Then $|U(t, 0)|$ is bounded and hence from (A), $\phi = 0$ is stable. Next, $|x(t, t_0, x_0)| = |U(t, t_0)x_0| = |U(t, 0)U(0, t_0)x_0| \leq |U(t, 0)||U(0, t_0)||x_0|$. Thus for any $t_0 \geq 0$, we can choose $r(t_0) = 1$ such that $|x_0| \leq r(t_0)$ implies $|x(t, t_0, x_0)| \leq |U(t, 0)||U(0, t_0)| \to 0$ as $t \to \infty$ (since $|U(0, t_0)|$ is fixed). This proves the asymptotic stability of $\phi = 0$.

(D): Suppose that $\phi = 0$ is uniformly asymptotically stable. Then it is uniformly stable and in addition, there exists an independent constant $r > 0$ such that for $\varepsilon = \frac{r}{2n}$, where n is the dimension of $\Re^n$, there exists a $T = T(\varepsilon) = T(r) > 0$ such that $\{t_0 \geq 0, |x_0| \leq r, t \geq t_0 + T\}$ imply $|x(t, t_0, x_0)| = |U(t, t_0)x_0| \leq \varepsilon$. By letting $x_0 = re_i$, $i = 1, 2, \cdots, n$, we see that

$$|U(t, t_0)| \leq n\frac{\varepsilon}{r} = \frac{1}{2}, \quad t_0 \geq 0, \quad t \geq t_0 + T. \tag{2.9}$$

In particular, letting $h = t_0$, $t = t_0 + T = h + T$, we obtain

$$|U(h + T, h)| \le \frac{1}{2}, \quad h \ge 0. \tag{2.10}$$

Using the uniform stability of $\phi = 0$ and the result in (B) we have just proved, there is an independent constant $M > 1$ such that $|U(t, s)| \le M$, $0 \le s \le t < \infty$. For any $t \ge s \ge 0$, there is an integer $k \ge 0$ such that $t = s + kT + d$, $0 \le d < T$. Thus, from (2.3) and (2.10),

$$
\begin{aligned}
|U(t, s)| &= |U(t, s + T)U(s + T, s)| \\
&= |U(s + kT + d, s + kT)U(s + kT, s + (k - 1)T) \cdots U(s + T, s)| \\
&\le |U(s + kT + d, s + kT)| \cdots |U(s + T, s)| \\
&\le M|U(s + kT, s + (k - 1)T)| \cdots |U(s + 2T, s + T)||U(s + T, s)| \\
&\le M|U(s + kT, s + (k - 1)T)| \cdots |U(s + 2T, s + T)|\frac{1}{2} \\
&\le M(\frac{1}{2})^k, \quad 0 \le s \le t < \infty. \tag{2.11}
\end{aligned}
$$

Now, there is an $\alpha = \alpha(T) = \alpha(r) > 0$ such that $\frac{1}{2} = e^{-\alpha T}$. Thus (2.11) becomes, for $0 \le s \le t < \infty$,

$$
\begin{aligned}
|U(t, s)| &\le M(\frac{1}{2})^k = Me^{-\alpha kT} = Me^{-\alpha(t - s - d)} \\
&= Me^{\alpha d}e^{-\alpha(t - s)} \le Me^{\alpha T}e^{-\alpha(t - s)} \overset{\text{def}}{=} Ce^{-\alpha(t - s)}, \tag{2.12}
\end{aligned}
$$

where $\alpha > 0$ and $C = Me^{\alpha T} > 1$ are independent constants because they are determined by the independent constant r.

Conversely, if (2.8) is true, then first, $|U(t, s)|$ is bounded and hence from (B), $\phi = 0$ is uniformly stable. Next, $|x(t, t_0, x_0)| = |U(t, t_0)x_0| \le |U(t, t_0)||x_0| \le Ce^{-\alpha(t - t_0)}|x_0|$, $0 \le t_0 \le t < \infty$. Thus we may let $r = 1$; and for any $0 < \varepsilon < C$, solve $T = T(\varepsilon) > 0$ from $Ce^{-\alpha T} = \varepsilon$. Then $\{t_0 \ge 0, |x_0| \le r = 1, t \ge t_0 + T\}$ imply $|x(t, t_0, x_0)| \le Ce^{-\alpha(t - t_0)}|x_0| \le Ce^{-\alpha T} = \varepsilon$, proving the uniform asymptotic stability of $\phi = 0$.

This completes the proof of the theorem. ♠

Example 9.2.2 Consider the scalar differential equation $x'(t) = a(t)x(t)$ with $a(t)$ real and continuous on $\Re^+$. Then the fundamental (scalar) solution is given by

$$U_1(t, s) = \exp\left(\int_s^t a(h)dh\right).$$

Therefore, the stability properties of the zero solution $\phi = 0$ are determined by $\int_s^t a(h)dh$. If there is a constant $M > 0$ such that $\int_0^t a(h)dh \leq M$, $0 \leq t < \infty$, then $\phi = 0$ is stable. If $\int_s^t a(h)dh \leq M$, $0 \leq s \leq t < \infty$, then $\phi = 0$ is uniformly stable. If $\int_0^t a(h)dh \to -\infty$, then $\phi = 0$ is asymptotically stable. Finally, if $a(t) \leq -\alpha$, $t \geq 0$, for some constant $\alpha > 0$, then $\phi = 0$ is uniformly asymptotically stable. (In an exercise, you are asked to construct an example for which $\phi = 0$ is asymptotically stable but not uniformly stable.)   ♠

For the linear nonhomogeneous differential equation (2.1) with $f \neq 0$, the zero is not a solution of Eq. (2.1). So now we must look at stabilities of nonzero solutions of Eq. (2.1). These stabilities can be reduced to those of the zero solution of Eq. (2.4), due to the affine structure of Eq. (2.1). Similar to Theorem 5.2.6 in Chapter 5, we have

Theorem 9.2.3 *Assume that $A(t)$ and $f(t)$ are continuous on $\Re^+$. The zero solution of Eq. (2.4) is stable if and only if every solution of Eq. (2.1) is stable. The same statement is true for uniform stability, asymptotic stability, and uniform asymptotic stability.*   ♠

Next, we use the variation of parameters formula and Gronwall's inequality to derive stability properties for some perturbed differential equations. These results are similar to those in Chapter 5.

Theorem 9.2.4 *Assume that $A(t)$ is continuous on $\Re^+$. If the zero solution of Eq. (2.4) is uniformly stable, and if the $n \times n$ continuous matrix function $B(t)$ satisfies $\int_0^\infty |B(t)|dt < \infty$, then the zero solution of*

$$x'(t) = A(t)x(t) + B(t)x(t) = [A(t) + B(t)]x(t) \qquad (2.13)$$

is also uniformly stable.

Proof. First, since $A(t)$ and $B(t)$ are continuous, the existence and uniqueness for Eq. (2.13) is guaranteed. Let $x(t)$ be a solution of Eq. (2.13) and treat $B(t)x(t)$ as $f(t)$ in Eq. (2.1), then $x(t)$ is given by the variation of parameters formula

$$x(t) = U(t, t_0)x(t_0) + \int_{t_0}^t U(t, s)B(s)x(s)ds, \quad t \geq t_0 \geq 0, \qquad (2.14)$$

where U is the fundamental matrix solution of Eq. (2.4). From Theorem 9.2.1, the uniform stability of the zero solution of Eq. (2.4) implies that there is an (independent) constant $C > 1$ such that $|U(t, t_0)| \le C$, $0 \le t_0 \le t < \infty$. Therefore, we have

$$|x(t)| \le C|x(t_0)| + \int_{t_0}^{t} C|B(s)||x(s)|ds, \quad t \ge t_0 \ge 0. \tag{2.15}$$

Now, Gronwall's inequality implies that

$$|x(t)| \le C|x(t_0)| \exp\left(\int_{t_0}^{t} C|B(s)|ds\right) \le \left\{C \exp\left(\int_{0}^{\infty} C|B(s)|ds\right)\right\}|x(t_0)|$$

$$\stackrel{\text{def}}{=} C_1|x(t_0)|, \quad t \ge t_0 \ge 0, \tag{2.16}$$

where C_1 is an independent constant, from which we can derive the uniform stability of the zero solution for Eq. (2.13). This completes the proof. ♠

Theorem 9.2.4.a *Assume that $A(t)$ is continuous on $\Re^+$. If the zero solution of Eq. (2.4) is uniformly stable, and if the continuous function $f(t, x)$ satisfies a weak Lipschitz condition in x and $f(t, 0) = 0$, which implies $|f(t, x)| \le k(t)|x|$, where $k(t)$ is from the weak Lipschitz condition. If $\int_{0}^{\infty} k(t)dt < \infty$, then the zero solution of*

$$x'(t) = A(t)x(t) + f(t, x(t)) \tag{2.17}$$

is also uniformly stable. ♠

When the zero solution of Eq. (2.4) is uniformly asymptotically stable, the condition on $B(t)$ in Theorem 9.2.4 can be relaxed.

Theorem 9.2.5 *Assume that $A(t)$ is continuous on $\Re^+$. If the zero solution of Eq. (2.4) is uniformly asymptotically stable, and if the $n \times n$ continuous matrix function $B(t)$ satisfies*

$$\int_{t_0}^{t} |B(s)|ds \le m(t - t_0) + r, \quad t \ge t_0 \ge 0, \tag{2.18}$$

for some positive constants m and r, then there is an $m_0 > 0$ such that if $m \le m_0$, then the zero solution of Eq. (2.13) is also uniformly asymptotically stable.

Proof. Similar to the beginning part in the proof of Theorem 9.2.4, any solution of Eq. (2.13) is given by

$$x(t) = U(t, t_0)x(t_0) + \int_{t_0}^{t} U(t, s)B(s)x(s)ds, \quad t \ge t_0 \ge 0. \tag{2.19}$$

From Theorem 9.2.1, the uniform asymptotic stability of the zero solution of Eq. (2.4) implies that there are (independent) constants $C > 1$ and $\alpha > 0$ such that $|U(t, t_0)| \leq Ce^{-\alpha(t-t_0)}$. Therefore, we have

$$|x(t)| \leq Ce^{-\alpha(t-t_0)}|x(t_0)| + \int_{t_0}^{t} Ce^{-\alpha(t-s)}|B(s)||x(s)|ds, \quad t \geq t_0 \geq 0, \quad (2.20)$$

or

$$|x(t)|e^{\alpha t} \leq Ce^{\alpha t_0}|x(t_0)| + \int_{t_0}^{t} Ce^{\alpha s}|B(s)||x(s)|ds, \quad t \geq t_0 \geq 0. \quad (2.21)$$

Define $u(t) = |x(t)|e^{\alpha t}$, then $u(t)$ satisfies

$$u(t) \leq Cu(t_0) + \int_{t_0}^{t} C|B(s)|u(s)ds, \quad t \geq t_0 \geq 0, \quad (2.22)$$

hence Gronwall's inequality implies that

$$
\begin{aligned}
u(t) &\leq Cu(t_0)\exp\left(\int_{t_0}^{t} C|B(s)|ds\right) \leq Cu(t_0)\exp\left(C[m(t-t_0)+r]\right) \\
&= Cu(t_0)e^{Cr}\exp\left[Cm(t-t_0)\right].
\end{aligned}
\quad (2.23)
$$

This means

$$|x(t)| \leq Ce^{Cr}|x(t_0)|\exp\left[-(\alpha - Cm)(t - t_0)\right]. \quad (2.24)$$

If we let $m_0 = \frac{\alpha}{2C}$, then $m \leq m_0$ implies that

$$|x(t)| \leq Ce^{Cr}|x(t_0)|\exp\left[-\frac{1}{2}\alpha(t - t_0)\right]. \quad (2.25)$$

This guarantees the uniform asymptotic stability of the zero solution for Eq. (2.13), and completes the proof. ♠

Theorem 9.2.5.a *Assume that $A(t)$ is continuous on $\Re^+$. If the zero solution of Eq. (2.4) is uniformly asymptotically stable, and if the continuous function $f(t,x)$ satisfies a weak Lipschitz condition in x and $f(t,0) = 0$, which implies $|f(t,x)| \leq k(t)|x|$, where $k(t)$ is from the weak Lipschitz condition. If*

$$\int_{t_0}^{t} |k(s)|ds \leq m(t - t_0) + r, \quad t \geq t_0 \geq 0, \quad (2.26)$$

for some positive constants m and r, then there is an $m_0 > 0$ such that if $m \le m_0$, then the zero solution of Eq. (2.17) is also uniformly asymptotically stable. ♠

The following is another result concerning uniform asymptotic stability with nonlinear perturbations.

Theorem 9.2.6 *Assume that $A(t)$ is continuous on $\Re^+$. If the zero solution of Eq. (2.4) is uniformly asymptotically stable, and if the continuous function $f(t, x)$ satisfies a Lipschitz condition (or weak or local Lipschitz) with respect to x on D, and*

$$\lim_{x \to 0} \frac{|f(t, x)|}{|x|} = 0, \quad \text{uniformly for } t \in [0, \infty), \tag{2.27}$$

then the zero solution of

$$x'(t) = A(t)x(t) + f(t, x(t)) \tag{2.28}$$

is also uniformly asymptotically stable.

Proof. Now, the conditions guarantee the existence and uniqueness of solutions, and they also imply that the zero is a solution of Eq. (2.28). Let U be the fundamental matrix solution of Eq. (2.4). By the variation of parameters formula, any solution of Eq. (2.28) is given by

$$x(t) = U(t, t_0)x(t_0) + \int_{t_0}^{t} U(t, s)f(s, x(s))ds, \quad t \ge t_0 \ge 0. \tag{2.29}$$

From Theorem 9.2.1, the uniform asymptotic stability of the zero solution of Eq. (2.4) implies that there are (independent) constants $C > 1$ and $\alpha > 0$ such that $|U(t, t_0)| \le Ce^{-\alpha(t-t_0)}$. From (2.27), for any η with $0 < \eta < \frac{\alpha}{C}$, there is a $\Delta = \Delta(\eta) > 0$ such that if $|x| \le \Delta$, then $|f(t, x)| \le \eta|x|$ uniformly for $t \ge 0$.

We first verify that if $|x(t_0)| \le \frac{\Delta}{C}$, then $|x(t)| < \Delta$ for $t \ge t_0$. If this is not true, then as $C > 1$ and $|x(t_0)| \le \frac{\Delta}{C} < \Delta$, there is a $t_1 > t_0$ such that $|x(t)| < \Delta$ for $t \in [t_0, t_1)$ and $|x(t_1)| = \Delta$. Then $|f(t, x(t))| \le \eta|x(t)|$ for $t \in [t_0, t_1]$. Thus, for $t \in [t_0, t_1]$, we have, from (2.29),

$$|x(t)| \le Ce^{-\alpha(t-t_0)}|x(t_0)| + \int_{t_0}^{t} Ce^{-\alpha(t-s)}|f(s, x(s))|ds$$

$$\le Ce^{-\alpha(t-t_0)}|x(t_0)| + \int_{t_0}^{t} Ce^{-\alpha(t-s)}\eta|x(s)|ds. \tag{2.30}$$

Similar to the final part of the proof of Theorem 9.2.5, we obtain

$$|x(t)| \leq C|x(t_0)| \exp\left[-(\alpha - C\eta)(t - t_0)\right], \quad t \in [t_0, t_1]. \qquad (2.31)$$

But then,

$$
\begin{aligned}
|x(t_1)| &\leq C|x(t_0)| \exp\left[-(\alpha - C\eta)(t_1 - t_0)\right] \\
&\leq \Delta \exp\left[-(\alpha - C\eta)(t_1 - t_0)\right] < \Delta,
\end{aligned}
$$

contradicting $|x(t_1)| = \Delta$. Thus $|x(t_0)| \leq \frac{\Delta}{C}$ implies $|x(t)| < \Delta$ for $t \geq t_0$. Consequently, if $|x(t_0)| \leq \frac{\Delta}{C}$, then for $t \geq t_0$, one has $|f(t, x(t))| \leq \eta|x(t)|$. Now, similar to (2.30) and (2.31), we have, for $t \geq t_0 \geq 0$,

$$|x(t)| \leq C|x(t_0)| \exp\left[-(\alpha - C\eta)(t - t_0)\right], \quad \text{if } |x(t_0)| \leq \frac{\Delta}{C}. \qquad (2.32)$$

Then, similar to the last part of the proof of Theorem 5.3.3 in Chapter 5, the zero solution of Eq. (2.28) is uniformly asymptotically stable. This completes the proof. ♠

Theorem 9.2.6.a *In Theorem 9.2.6, the condition (2.27) can be relaxed to $|f(t, x)| \leq \eta|x|$ with $0 < \eta < \frac{\alpha}{C}$, where α and C are from Theorem 9.2.1(D).* ♠

Exercises 9.2

1. Determine the stability properties of the zero solution for $x'(t) = A(t)x(t)$.

   (a) $A(t) = \begin{bmatrix} 0 & 0 & 3 \\ 0 & 0 & 1 \\ 0 & 0 & -2t \end{bmatrix}$; (b) $A(t) = \begin{bmatrix} 0 & 1 & 3 \\ 0 & 0 & 1 \\ 0 & 0 & -2t \end{bmatrix}$;

   (c) $A(t) = \begin{bmatrix} 1 & 1 & 3 \\ 0 & -7 & 1 \\ 0 & 0 & -2t \end{bmatrix}$.

2. Rewrite $x^{(n)}(t) + a_1(t)x^{(n-1)}(t) + \cdots + a_{n-1}(t)x'(t) + a_n(t)x(t) = 0$ as a system and then state and prove a theorem similar to Theorem 9.2.1.

3. Determine the stability properties of the zero solution for

(a) $x'' + 3x' + 4tx = 0$.

(b) $x'' - 3x' - 4tx = 0$.

(c) $x^{(3)} + 3x'' + 4tx = 0$.

(d) $x^{(4)} + 5x'' - 2tx = 0$.

4. In Example 9.2.2, construct a function $a(t)$ such that $\phi = 0$ is asymptotically stable but not uniformly stable.

5. Theorem 9.2.3 can be improved in the following way: If Eq. (2.1) has a stable solution on $[0, \infty)$, then the zero solution of Eq. (2.4) is stable. Prove this claim and extend it to uniform stability, asymptotic stability and uniform asymptotic stability.

6. Assume that the fundamental matrix solution of $x'(t) = A(t)x(t)$ satisfies $|U(t, t_0)| \leq Ce^{-\alpha(t-t_0)}$ for some constants $C > 1$ and $\alpha > 0$. If $B(t)$ is continuous and $|B(t)| < \frac{\alpha}{C}$, then prove that the zero solution of $x'(t) = A(t)x(t) + B(t)x(t)$ is uniformly asymptotically stable.

7. Prove Theorem 9.2.4.a.

8. Prove Theorem 9.2.5.a.

9. Prove Theorem 9.2.6.a.

9.3 Liapunov's Method for General Equations

In this section, we extend Liapunov's method for autonomous differential equations studied in Chapter 5 to the general differential equation

$$x'(t) = f(t, x(t)), \tag{3.1}$$

and derive its stability properties. We assume that Eq. (3.1) is defined on $D = [0, \infty) \times Q$ where the domain $Q \subset \Re^n$ contains the zero vector, and that for any $(t_0, x_0) \in D$, Eq. (3.1) has a unique solution $x(t, t_0, x_0)$ existing on $[t_0, \infty)$ with $x(t_0) = x_0$.

The treatment here is similar to that given in Chapter 5 for autonomous differential equations. However, we will see that since $f(t, x)$ now also depends on the time variable t, some new definitions and conditions must be imposed. Additional related results, remarks, and examples can be found in some reference books, including Yoshizawa [1966] and Burton [1985].

Definition 9.3.1 *Consider Eq. (3.1). For a continuous function $V : D = [0, \infty) \times Q \to [0, \infty)$ that is Lipschitz (or weak or local Lipschitz) with respect to x on D, define*

$$V'_{(3.1)}(t, x) \stackrel{\text{def}}{=} \limsup_{h \to 0^+} \frac{V(t+h,\, x + hf(t,x)) - V(t,x)}{h}, \quad (t,x) \in D. \quad (3.2)$$

Next, for a solution $x(t)$ of Eq. (3.1), define

$$V'(t, x(t)) \stackrel{\text{def}}{=} \limsup_{h \to 0^+} \frac{V(t+h,\, x(t+h)) - V(t, x(t))}{h}. \quad (3.3)$$

A version of the following result was stated without proof for autonomous differential equations in Chapter 5. Now we state and prove it for general cases.

Lemma 9.3.2 *Let $x = x(t)$ be a solution of Eq. (3.1), then*

$$V'_{(3.1)}(t, x) = V'(t, x(t)). \quad (3.4)$$

Moreover, if $V'(t, x(t)) \leq 0$, then

$$V(b, x(b)) - V(a, x(a)) \leq \int_a^b V'(t, x(t)) dt, \quad 0 \leq a \leq b. \quad (3.5)$$

Proof. First, as $x = x(t)$ is a solution of Eq. (3.1), we have

$$x(t + h) = x(t) + hf(t, x(t)) + o(h), \quad (3.6)$$

where

$$\frac{|o(h)|}{h} \to 0 \quad \text{as} \quad h \to 0.$$

Next, assume a Lipschitz condition for V with a Lipschitz constant $k > 0$, then

$$-k|y_1 - y_2| \leq V(s, y_1) - V(s, y_2) \leq k|y_1 - y_2|, \quad s \geq 0, \quad y_1, y_2 \in Q. \quad (3.7)$$

(If we only assume a weak or local Lipschitz condition, then a small domain can be found for the same proof to go through.) Then

$$\begin{aligned}
&V(t+h,\, x(t+h)) - V(t,\, x(t)) \\
&= V(t+h,\, x(t) + hf(t,x(t)) + o(h)) - V(t,\, x(t)) \\
&\leq V(t+h,\, x(t) + hf(t,x(t))) + k|o(h)| - V(t,\, x(t)) \\
&= V(t+h,\, x + hf(t,x)) + k|o(h)| - V(t,\, x),
\end{aligned} \quad (3.8)$$

therefore,

$$
\begin{aligned}
V'(t, x(t)) &= \limsup_{h \to 0^+} \frac{V(t+h, x(t+h)) - V(t, x(t))}{h} \\
&\leq \limsup_{h \to 0^+} \frac{V(t+h, x+hf(t,x)) + k|o(h)| - V(t, x)}{h} \\
&= \limsup_{h \to 0^+} \left[\frac{V(t+h, x+hf(t,x)) - V(t, x)}{h} + \frac{k|o(h)|}{h} \right] \\
&= \limsup_{h \to 0^+} \frac{V(t+h, x+hf(t,x)) - V(t, x)}{h} \\
&= V'_{(3.1)}(t, x).
\end{aligned}
\tag{3.9}
$$

On the other hand, we have

$$
\begin{aligned}
&V(t+h, x(t+h)) - V(t, x(t)) \\
&= V(t+h, x(t) + hf(t, x(t)) + o(h)) - V(t, x(t)) \\
&\geq V(t+h, x(t) + hf(t, x(t))) - k|o(h)| - V(t, x(t)) \\
&= V(t+h, x+hf(t,x)) - k|o(h)| - V(t, x),
\end{aligned}
\tag{3.10}
$$

therefore,

$$
\begin{aligned}
V'(t, x(t)) &= \limsup_{h \to 0^+} \frac{V(t+h, x(t+h)) - V(t, x(t))}{h} \\
&\geq \limsup_{h \to 0^+} \frac{V(t+h, x+hf(t,x)) - k|o(h)| - V(t, x)}{h} \\
&= \limsup_{h \to 0^+} \left[\frac{V(t+h, x+hf(t,x)) - V(t, x)}{h} - \frac{k|o(h)|}{h} \right] \\
&= \limsup_{h \to 0^+} \frac{V(t+h, x+hf(t,x)) - V(t, x)}{h} \\
&= V'_{(3.1)}(t, x).
\end{aligned}
\tag{3.11}
$$

The inequality (3.5) is a standard result from advanced calculus. This completes the proof. ♠

Lemma 9.3.2 says that the derivative of V with respect to Eq. (3.1) defined by (3.2) is actually the derivative of V along a solution of Eq. (3.1) defined by (3.3). That is, to find $V'_{(3.1)}(t, x)$, we can take a derivative of $V(t, x(t))$ in t by plugging in the differential equation (3.1).

Next we introduce some terminology that will be used in Liapunov's method.

Definition 9.3.3 (Wedge) *A strictly increasing continuous function* $W :$ $[0, \infty) \to [0, \infty)$ *satisfying* $W(0) = 0$ *is called a wedge.*

If we compare the treatment for autonomous differential equations, we find that the notion of "wedge" is not needed for autonomous differential equations. Recall that for $x = [x_1, x_2, \cdots, x_n]^T$ in $\Re^n$, $|x| = \sum_{i=1}^n |x_i|$ is equivalent to $r(x) = \sqrt{x_1^2 + x_2^2 + \cdots + x_n^2}$.

Definition 9.3.4 *Let Q be a domain in $\Re^n$ that contains the zero vector. A continuous function* $V : D = [0, \infty) \times Q \to [0, \infty)$ *is said to be*

(a). **Positive definite** *(or bounded below by a wedge) if there is a wedge W_1 such that $W_1(|x|) \le V(t, x)$, $(t, x) \in D$; or equivalently, $W_1^\star(r(x)) \le V(t, x)$, $(t, x) \in D$, for a wedge $W_1^\star$.*

(b). **Decrescent** *(or bounded above by a wedge) if there is a wedge W_2 such that $V(t, x) \le W_2(|x|)$, $(t, x) \in D$; or equivalently, $V(t, x) \le W_2^\star(r(x))$, $(t, x) \in D$, for a wedge $W_2^\star$.*

This definition is also different from the one for autonomous differential equations, because now the function $V(t, x)$ is dependent on the variable t, thus, "positive definite" and "decrescent" in Definition 9.3.4 are used to reduce the dependence of $V(t, x)$ on the variable t.

Definition 9.3.5 (Liapunov function) *Let Q be a domain in $\Re^n$ that contains the zero vector. A function* $V : D = [0, \infty) \times Q \to [0, \infty)$ *is called a Liapunov function if $V(t, 0) = 0$, $t \ge 0$, V is positive definite and has continuous first partial derivatives.*

Example 9.3.6 Based on $V(t) = \frac{1}{2}[x_1^2(t) + x_2^2(t)]$ in the proof of Theorem 4.3.1(a) in Chapter 4, if we define

$$V(t, x) = \frac{1}{2}[x_1^2 + x_2^2], \quad t \ge 0, \ x = [x_1, x_2]^T \in \Re^2,$$

then

$$V(t, x) = \frac{1}{2}[x_1^2 + x_2^2] = \frac{1}{2}\left(\sqrt{x_1^2 + x_2^2}\right)^2 = \frac{1}{2}[r(x)]^2 \stackrel{\text{def}}{=} W(r(x)), \quad (3.12)$$

where $W(r) = \frac{1}{2}r^2$ is a wedge. Therefore, $V(t,x)$ is positive definite and decrescent. Moreover, in the proof of Theorem 4.3.1(a) in Chapter 4,

$$V'(t,x(t)) \leq -\left[\frac{-\alpha}{2}\left(x_1^2(t) + x_2^2(t)\right)\right] = -\left[\frac{-\alpha}{2}\left(\sqrt{x_1^2(t) + x_2^2(t)}\right)^2\right]$$
$$\stackrel{\text{def}}{=} -W_3(r(x(t))),$$

where $W_3(r) = \frac{-\alpha}{2}r^2$ is a wedge, since $\alpha < 0$. Thus $-V'(t,x)$ is also positive definite. $\spadesuit$

The following theorems utilize Liapunov functions and wedges to derive stability properties for general differential equations. Without loss of generality, we assume in the following proofs that $|x| \leq 1$ implies $x \in Q$, also, the parameters used in the proofs, such as ε and δ, are assumed to be less than or equal to 1.

Theorem 9.3.7 *Let Q be a domain in $\Re^n$ that contains the zero vector. Consider Eq. (3.1) on $[0,\infty) \times Q$ with $f(t,0) = 0$ so that $\phi = 0$ is a solution of Eq. (3.1). Assume that V is a Liapunov function. If $V'_{(3.1)}(t,x) \leq 0$, then $\phi = 0$ is stable.*

Proof. Let $t_0 \geq 0$ and let $x(t) = x(t, t_0, x_0)$ be a solution of Eq. (3.1). Now we have $V'(t,x(t)) \leq 0$, $t \geq t_0$; and there is a wedge W_1 such that $W_1(|x(t)|) \leq V(t,x(t))$. For any $\varepsilon > 0$, we must find a $\delta = \delta(\varepsilon, t_0) > 0$ such that $\{|x_0| \leq \delta(\varepsilon, t_0), t \geq t_0\}$ imply $|x(t, t_0, x_0)| \leq \varepsilon$. As $V(t,x(t))$ is decreasing, one has $W_1(|x(t)|) \leq V(t,x(t)) \leq V(t_0, x_0)$, $t \geq t_0$. Now, if we can control $V(t_0, x_0)$ by W_1 at some value, then $|x(t)|$ can be controlled since W_1 is strictly increasing. To this end, we observe that $V(t_0, x)$ is continuous in x and $V(t_0, 0) = 0$, thus there is a $\delta(\varepsilon, t_0) > 0$ such that $|x| \leq \delta(\varepsilon, t_0)$ implies $V(t_0, x) \leq W_1(\varepsilon)$. Therefore, if $|x_0| \leq \delta(\varepsilon, t_0)$ and $t \geq t_0$, then

$$W_1(|x(t)|) \leq V(t,x(t)) \leq V(t_0, x_0) \leq W_1(\varepsilon), \qquad (3.13)$$

which implies $|x(t)| \leq \varepsilon$ since W_1 is strictly increasing. This verifies the stability of $\phi = 0$ and completes the proof. $\spadesuit$

In the proof of Theorem 9.3.7, δ is determined from the continuity of $V(t_0, x)$, thus δ is dependent on t_0 and we only get the stability of the zero solution. To obtain the uniform stability of the zero solution, we need an additional condition which can remove the dependence of δ on t_0.

Theorem 9.3.8 *Let Q be a domain in $\Re^n$ that contains the zero vector. Consider Eq. (3.1) on $[0, \infty) \times Q$ with $f(t, 0) = 0$ so that $\phi = 0$ is a solution of Eq. (3.1). Assume that V is a Liapunov function and is decrescent. If $V'_{(3.1)}(t, x) \leq 0$, then $\phi = 0$ is uniformly stable.*

Proof. In this case, for $x(t) = x(t, t_0, x_0)$, there are wedges W_1 and W_2 such that $W_1(|x(t)|) \leq V(t, x(t)) \leq V(t_0, x_0) \leq W_2(|x_0|)$, $t \geq t_0$. Now, let's control $W_2(|x_0|)$ by W_1 at some value so as to control $|x(t)|$. For any $\varepsilon > 0$, we choose $\delta = \delta(\varepsilon) > 0$ such that $W_2(\delta) \leq W_1(\varepsilon)$. Then for $|x_0| \leq \delta(\varepsilon)$ and $t \geq t_0$,

$$W_1(|x(t)|) \leq V(t, x(t)) \leq V(t_0, x_0) \leq W_2(|x_0|) \leq W_2(\delta) \leq W_1(\varepsilon), \quad (3.14)$$

therefore, $|x(t, t_0, x_0)| \leq \varepsilon$ since W_1 is strictly increasing, proving the uniform stability of $\phi = 0$.   ♠

In the above, the condition of $V'_{(3.1)}(t, x) \leq 0$ is used to guarantee that $V(t, x(t))$ is not increasing, which means roughly that the solution is not increasing. Therefore, stability and uniform stability are obtained. To derive asymptotic and uniform asymptotic stabilities, we need to "drag" solutions to the zero. Thus we require solutions to be decreasing, or require $V'_{(3.1)}(t, x) < 0$, see the following.

Theorem 9.3.9 *Let Q be a domain in $\Re^n$ that contains the zero vector. Consider Eq. (3.1) on $[0, \infty) \times Q$ with $f(t, 0) = 0$ so that $\phi = 0$ is a solution of Eq. (3.1). Assume that V is a Liapunov function. If $-V'_{(3.1)}(t, x)$ is positive definite, and if $f(t, x)$ is bounded for x bounded, then $\phi = 0$ is asymptotically stable.*

Proof. From Theorem 9.3.7, $\phi = 0$ is stable. Suppose that $\phi = 0$ is not asymptotically stable. Then there is a $t'_0 \geq 0$ such that the $r(t'_0) > 0$ in the definition of asymptotic stability cannot be found. Now, from the stability of the zero solution, for the given t'_0 and $\varepsilon = 1$, there exists a $\delta = \delta(\varepsilon, t'_0) = \delta(t'_0) > 0$ such that $|x_0| \leq \delta(t'_0)$ implies $|x(t)| = |x(t, t'_0, x_0)| \leq \varepsilon = 1$ for $t \geq t'_0$. Since $\delta(t'_0)$ cannot be used as the $r(t'_0)$ in the definition of asymptotic stability, there is an x^* such that $|x^*| \leq \delta(t'_0)$ but $|x(t, t'_0, x^*)| \nrightarrow 0$, $t \to \infty$. Therefore there exists an $\varepsilon_0 > 0$ and $t_m \to \infty$ such that $|x(t_m, t'_0, x^*)| \geq \varepsilon_0$. Now, $|x^*| \leq \delta(t'_0)$ implies $|x(t)| = |x(t, t'_0, x^*)| \leq 1$ for $t \geq t'_0$, or $x(t) = x(t, t'_0, x^*)$ is bounded. From the assumption, $f(t, x)$ is bounded for

x bounded, thus, for $x(t) = x(t, t_0', x^*)$,

$$|\frac{d}{dt}x(t)| = |f(t, x(t))| \leq \frac{K}{n},$$
(3.15)

for some constant $K > 0$ (n is the dimension of $\Re^n$). By taking a subsequence of t_m if necessary, we may assume that the intervals

$$\left[t_m - \frac{\varepsilon_0}{2K}, \ t_m + \frac{\varepsilon_0}{2K}\right], \quad m \geq 1,$$
(3.16)

are disjoint and $t_1 - \frac{\varepsilon_0}{2K} > t_0'$. Now, for t in (3.16), we can examine every component of $x(t)$ and apply the mean value theorem and then use (3.15) to obtain, for $t \in \left[t_m - \frac{\varepsilon_0}{2K}, \ t_m + \frac{\varepsilon_0}{2K}\right]$,

$$|x(t) - x(t_m)| \leq n\frac{K}{n}|t - t_m| \leq K\frac{\varepsilon_0}{2K} = \frac{\varepsilon_0}{2}.$$

Hence, as

$$\varepsilon_0 \leq |x(t_m)| \leq |x(t_m) - x(t)| + |x(t)| \leq \frac{\varepsilon_0}{2} + |x(t)|,$$

we have

$$|x(t)| = |x(t, t_0', x^*)| \geq \frac{\varepsilon_0}{2}$$

for t in (3.16).

From the assumption, there is a wedge W_2 such that $V'(t, x(t)) \leq -W_2(|x(t)|)$. Then $V'(t, x(t)) \leq -W_2(\frac{\varepsilon_0}{2})$ on the intervals given in (3.16) and $V'(t, x(t)) \leq 0$ elsewhere. Therefore

$$V(t_m + \frac{\varepsilon_0}{2K}, \ x(t_m + \frac{\varepsilon_0}{2K})) \leq V(t_0', x^*) + \int_{t_0'}^{t_m + \frac{\varepsilon_0}{2K}} V'(t, x(t))dt$$

$$\leq V(t_0', x^*) - W_2(\frac{\varepsilon_0}{2})\frac{\varepsilon_0}{K}m \to -\infty, \quad m \to \infty,$$

which contradicts $V(t, x(t)) \geq 0$. Thus $\phi = 0$ is asymptotically stable. This completes the proof. ♠

In the proof of Theorem 9.3.9, the stability of the zero solution is used, or $\delta(t_0')$ is used, thus the uniform asymptotic stability is not expected. Similar to Theorem 9.3.8, an additional condition is needed to remove the dependence of δ on t_0.

Theorem 9.3.10 *Let Q be a domain in $\Re^n$ that contains the zero vector. Consider Eq. (3.1) on $[0, \infty) \times Q$ with $f(t, 0) = 0$ so that $\phi = 0$ is a solution of Eq. (3.1). Assume that V is a Liapunov function and is decrescent. If $-V'_{(3.1)}(t, x)$ is positive definite, then $\phi = 0$ is uniformly asymptotically stable.*

Proof. From Theorem 9.3.8, $\phi = 0$ is uniformly stable. Next, take $r = 1$. Then for any $\varepsilon > 0$, we need to find $T = T(\varepsilon) > 0$ such that $\{|x_0| \le 1, t_0 \ge 0, t \ge t_0 + T\}$ imply $|x(t, t_0, x_0)| \le \varepsilon$. Now, there are wedges W_1 and W_2 such that for $t \ge s \ge 0$, one has $W_1(|x(t)|) \le V(t, x(t)) \le V(s, x(s)) \le W_2(|x(s)|)$. Thus to control $W_2(|x(s)|)$ by W_1 at some value so as to control $|x(t)|$, we choose $\gamma = \gamma(\varepsilon) > 0$ such that $W_2(\gamma) \le W_1(\varepsilon)$. For any solution $x(t) = x(t, t_0, x_0)$ with $|x_0| \le 1$ and $t_0 \ge 0$, as long as $|x(t)| > \gamma$ for $t \ge t_0$, then, as $V'(t, x(t)) \le -W_3(|x(t)|)$ for some wedge W_3, we have

$$
\begin{aligned}
0 \le V(t, x(t)) &\le V(t_0, x_0) - \int_{t_0}^{t} W_3(|x(s)|)ds \\
&< W_2(|x_0|) - W_3(\gamma)(t - t_0) \\
&\le W_2(1) - W_3(\gamma)(t - t_0), \quad\quad\quad (3.17)
\end{aligned}
$$

which fails when $W_2(1) - W_3(\gamma)(t - t_0) = 0$, or when

$$
t = t_0 + \frac{W_2(1)}{W_3(\gamma)} \stackrel{\text{def}}{=} t_0 + T, \quad \left(T = \frac{W_2(1)}{W_3(\gamma)}\right) \quad\quad (3.18)
$$

where $T = T(\gamma) = T(\varepsilon)$. Therefore, there is a $\bar{t}_0 \in [t_0, t_0 + T]$ such that $|x(\bar{t}_0, t_0, x_0)| \le \gamma$. Then for $t \ge \bar{t}_0$,

$$
W_1(|x(t)|) \le V(t, x(t)) \le V(\bar{t}_0, x(\bar{t}_0)) \le W_2(|x(\bar{t}_0)|) \le W_2(\gamma) \le W_1(\varepsilon),
$$

hence, $|x(t)| \le \varepsilon$ for $t \ge \bar{t}_0$. Therefore, $\{|x_0| \le 1, t_0 \ge 0, t \ge t_0 + T\}$ imply $|x(t, t_0, x_0)| \le \varepsilon$ since $t_0 + T \ge \bar{t}_0$, thus $\phi = 0$ is uniformly asymptotically stable. This completes the proof. ♠

So far, we have seen that the notion of wedge as defined in Definition 9.3.3 has made the above proofs concerning stabilities possible. However, we will see from some examples that in practice, it is not easy to construct these kind of wedges.

Example 9.3.11 Consider the scalar differential equation

$$
u'' + (1 - e^{-2t})h(u')g(u) = 0,
$$

where the functions h and g are continuous, $ug(u) > 0$ for $u \neq 0$, and h is positive on $\Re$. We note that $u = 0$ is a solution and write the equation as a system

$$\begin{cases} x_1' &= x_2, \\ x_2' &= -(1 - e^{-2t})h(x_2)g(x_1). \end{cases} \tag{3.19}$$

Now try, for $x = (x_1, x_2) \in \Re^2$,

$$V(t, x) = (1 - e^{-t})^{-1} \int_0^{x_2} \frac{s}{h(s)} ds + (1 + e^{-t}) \int_0^{x_1} g(s) ds. \tag{3.20}$$

For $t \geq 1$, we have

$$\int_0^{x_2} \frac{s}{h(s)} ds + \int_0^{x_1} g(s) ds \leq V(t, x) \leq (1 - e^{-1})^{-1} \int_0^{x_2} \frac{s}{h(s)} ds + 2 \int_0^{x_1} g(s) ds.$$

Therefore, if we define

$$W_1(x) = \int_0^{x_2} \frac{s}{h(s)} ds + \int_0^{x_1} g(s) ds,$$

and

$$W_2(x) = (1 - e^{-1})^{-1} \int_0^{x_2} \frac{s}{h(s)} ds + 2 \int_0^{x_1} g(s) ds,$$

then, for $t \geq 1$, we have

$$W_1(x) \leq V(t, x) \leq W_2(x). \tag{3.21}$$

Moreover, from the assumptions, $W_i(0) = 0$ and $W_i(x) > 0$ for $x \neq 0$, $i = 1, 2$.

However, if you look at W_i carefully, you will find that $W_i, i = 1, 2$, are defined on $\Re^2$, and without further assumptions on the functions h and g it is not easy to construct **scalar wedges** as defined in Definition 9.3.3 from W_i. (Try it to see why.) ♠

Thus, we conclude that scalar wedges as defined in Definition 9.3.3 have made theoretical proofs easy but practical examples hard. Therefore, based on Example 9.3.11, we ask if scalar wedges as defined in Definition 9.3.3 can be constructed when we have two functions $W_i, i = 1, 2$, defined on $\Re^n$, such that $W_i(0) = 0$, $W_i(x) > 0$ for $x \neq 0$, and (3.21) is satisfied for some function $V(t, x)$ defined on $\Re \times \Re^n$.

The following construction from Burton [1985] shows a way to construct a scalar wedge on the left-hand side of (3.21), which is related to "positive definiteness." Assume that $W : \{x \in \Re^n : |x| \le 1\} \to [0, \infty)$, $W(0) = 0$, $W(x) > 0$ for $x \ne 0$, and $W(x) \le V(t, x)$ for some function $V(t, x)$. Define

$$\alpha(r) = \min_{r \le |x| \le 1} W(x),$$

so that $\alpha : [0, 1] \to [0, \infty)$ is nondecreasing. Then define

$$\omega_1(r) = \int_0^r \alpha(s) ds, \quad r \in [0, 1],$$

so that $\omega_1(0) = 0$ and $\omega_1'(r) = \alpha(r) > 0$, $r > 0$. Therefore, ω_1 is strictly increasing, hence ω_1 is a scalar wedge as defined in Definition 9.3.3. Using the mean value theorem for integration, $\omega_1(r) \le r\alpha(r) \le \alpha(r)$ for $r \le 1$. Thus, if $|x| \le 1$, then

$$V(t, x) \ge W(x) \ge \min_{|x| \le |p| \le 1} W(p) = \alpha(|x|) \ge \omega_1(|x|),$$

or $V(t, x)$ is positive definite if we only consider $|x| \le 1$, which is good enough for stabilities of the zero solution.

Based on this construction, we next extend the idea and come up with the following construction of a scalar wedge on the right-hand side of (3.21), which is related to "decrescentness." Assume that $W : \Re^n \to [0, \infty)$, $W(0) = 0$, $W(x) > 0$ for $x \ne 0$, and $V(t, x) \le W(x)$ for some function $V(t, x)$. Define

$$\beta(r) = \max_{0 \le |x| \le r} W(x), \quad r \ge 0,$$

then $\beta : [0, \infty) \to [0, \infty)$ is nondecreasing. Next define

$$\omega_2(r) = \beta(r) + r, \quad r \in \Re^+,$$

so that $\omega_2(0) = 0$ and ω_2 is strictly increasing on $\Re^+$. Therefore, ω_2 is a scalar wedge as defined in Definition 9.3.3. Moreover,

$$V(t, x) \le W(x) \le \max_{0 \le |p| \le |x|} W(p) = \beta(|x|) \le \omega_2(|x|),$$

or $V(t, x)$ is decrescent on $[0, \infty) \times \Re^n$.

According to the above constructions, we find that if (3.21) is satisfied, then the corresponding scalar wedges as defined in Definition 9.3.3 can be constructed. Therefore, we can weaken the conditions required in Definition 9.3.3. That is, instead of requiring wedges to be scalar wedges defined on $[0, \infty)$, we can simply ask that wedges are defined on a domain $Q \subset \Re^n$ containing the zero vector. This will make things easier to check for the examples where wedges need to be constructed.

Consequently, we can make the following definition.

Definition 9.3.12 ($\Re^n$-Wedge) *Let Q be a domain in $\Re^n$ containing the zero vector. A continuous function $W : Q \to [0, \infty)$ satisfying $W(0) = 0$ and $W(x) > 0$ for $x \neq 0$ is called a $\Re^n$-wedge.*

The corresponding definitions concerning "positive definiteness" and "decrescentness" can also be revised.

Definition 9.3.13 *Let Q be a domain in $\Re^n$ containing the zero vector. A continuous function $V : D = [0, \infty) \times Q \to [0, \infty)$ is said to be*

*(a). $\Re^n$-**positive definite** (or bounded below by a $\Re^n$-wedge) if there is a $\Re^n$-wedge W_1 such that $W_1(x) \leq V(t, x)$, $(t, x) \in D$.*

*(b). $\Re^n$-**decrescent** (or bounded above by a $\Re^n$-wedge) if there is a $\Re^n$-wedge W_2 such that $V(t, x) \leq W_2(x)$, $(t, x) \in D$.*

After these discussions, we restate Theorems 9.3.7–9.3.10 using $\Re^n$-wedges, which will be much easier to use in applications.

Theorem 9.3.14 *Let Q be a domain in $\Re^n$ containing the zero vector. Consider Eq. (3.1) on $[0, \infty) \times Q$ with $f(t, 0) = 0$ so that $\phi = 0$ is a solution of Eq. (3.1). Assume that V is a Liapunov function where the wedge is replaced by a $\Re^n$-wedge, that is, $W_1(x) \leq V(t, x)$ for some $\Re^n$-wedge W_1. If $V'_{(3.1)}(t, x) \leq 0$, then $\phi = 0$ is stable.* ♠

Theorem 9.3.15 *Let Q be a domain in $\Re^n$ containing the zero vector. Consider Eq. (3.1) on $[0, \infty) \times Q$ with $f(t, 0) = 0$ so that $\phi = 0$ is a solution of Eq. (3.1). Assume that V is a Liapunov function and is decrescent where the wedges are replaced by $\Re^n$-wedges, that is, $W_1(x) \leq V(t, x) \leq W_2(x)$ for some $\Re^n$-wedges W_1 and W_2. If $V'_{(3.1)}(t, x) \leq 0$, then $\phi = 0$ is uniformly stable.* ♠

Theorem 9.3.16 *Let Q be a domain in $\Re^n$ containing the zero vector. Consider Eq. (3.1) on $[0, \infty) \times Q$ with $f(t, 0) = 0$ so that $\phi = 0$ is a solution* *of Eq. (3.1). Assume that V is a Liapunov function where the wedge is replaced by a $\Re^n$-wedge. If $-V'_{(3.1)}(t,x)$ is $\Re^n$-positive definite, and if $f(t,x)$ is bounded for x bounded, then $\phi = 0$ is asymptotically stable.* ♠

Theorem 9.3.17 *Let Q be a domain in $\Re^n$ containing the zero vector. Consider Eq. (3.1) on $[0,\infty) \times Q$ with $f(t,0) = 0$ so that $\phi = 0$ is a solution of Eq. (3.1). Assume that V is a Liapunov function and is decrescent where the wedges are replaced by $\Re^n$-wedges. If $-V'_{(3.1)}(t,x)$ is $\Re^n$-positive definite, then $\phi = 0$ is uniformly asymptotically stable.* ♠

Example 9.3.18 Let's revisit Example 9.3.11,

$$u'' + (1 - e^{-2t})h(u')g(u) = 0, \tag{3.22}$$

where the functions h and g are continuous, $ug(u) > 0$ for $u \neq 0$, and h is positive on $\Re$. From the function V given in (3.20), we know from (3.21) that $V(t,x)$ is $\Re^n$-positive definite and $\Re^n$-decrescent for $t \geq 1$. Next, from (3.19),

$$\frac{d}{dt}V(t,x) = -(1 - e^{-t})^{-2}e^{-t}\int_0^{x_2}\frac{s}{h(s)}ds - e^{-t}\int_0^{x_1}g(s)ds$$

$$+ (1 + e^{-t})g(x_1)x_2 - (1 - e^{-t})^{-1}\frac{x_2}{h(x_2)}(1 - e^{-2t})h(x_2)g(x_1)$$

$$= -(1 - e^{-t})^{-2}e^{-t}\int_0^{x_2}\frac{s}{h(s)}ds - e^{-t}\int_0^{x_1}g(s)ds \leq 0. \tag{3.23}$$

Therefore, using Theorem 9.3.15, the solution $\phi = 0$ of Eq. (3.22) is uniformly stable. ♠

Compare the corresponding results in Chapter 5. We now see that the Liapunov theory concerning stabilities of autonomous differential equations are special cases of theorems 9.3.15 and 9.3.17.

The above is a brief coverage of the Liapunov theory concerning stability properties for general differential equations. Roughly speaking, it reduces the study of stability properties to the problem of constructing appropriate Liapunov functions. However, constructing appropriate Liapunov functions is difficult. See Chapter 5 for some constructions for autonomous differential equations, and refer to Yoshizawa [1966] and Burton [1985] for treatment of certain nonautonomous differential equations.

Finally, we close this section by introducing some results concerning the **converse of stabilities**. That is, we now know that the existence of Liapunov functions imply certain stabilities; the converse of stabilities asks if stabilities imply the existence of certain Liapunov functions. Here, we provide a result for the linear differential equation

$$x'(t) = Ax(t), \tag{3.24}$$

where A is an $n \times n$ constant matrix. See Yoshizawa [1966] for related results for general differential equations.

Theorem 9.3.19 *Assume that the zero solution of Eq. (3.24) is uniformly asymptotically stable, then there exists a Liapunov function $V : \Re^n \to [0, \infty)$ such that V is decrescent and $-V'_{(3.24)}(x)$ is positive definite.*

Proof. From Theorem 5.2.1(B) in Chapter 5, there are (independent or generic) constants $C > 1$ and $\alpha > 0$ such that

$$|e^{tA}| \leq Ce^{-\alpha t}, \quad 0 \leq t < \infty. \tag{3.25}$$

Define an $n \times n$ matrix

$$B = \int_0^\infty e^{A^T t} e^{At} dt,$$

(where T means the transpose). Because of (3.25), the matrix B is well defined. Moreover,

$$
\begin{aligned}
B^T &= \int_0^\infty \left[e^{A^T t} e^{At} \right]^T dt = \int_0^\infty \left[e^{At} \right]^T \left[e^{A^T t} \right]^T dt \\
&= \int_0^\infty e^{A^T t} e^{At} dt = B,
\end{aligned} \tag{3.26}
$$

thus B is symmetric. Define

$$V(x) = x^T B x, \quad x \in \Re^n,$$

then $V : \Re^n \to \Re$, and

$$
\begin{aligned}
V(x) &= x^T B x = \int_0^\infty x^T e^{A^T t} e^{At} x \, dt \\
&= \int_0^\infty \left[e^{At} x \right]^T \left[e^{At} x \right] dt \geq 0,
\end{aligned} \tag{3.27}
$$

and $V(x) > 0$ for $x \neq 0$. Thus, V itself is a $\Re^n$-wedge, hence V is positive definite and decrescent, therefore V is a Liapunov function. Next, for a solution $x = x(t)$,

$$\frac{d}{dt}V(x(t)) = (x^T)'Bx + x^T Bx' = x^T A^T Bx + x^T BAx$$

$$= x^T[A^T B + BA]x = x^T\left[\int_0^\infty A^T e^{A^T t}e^{At}dt + \int_0^\infty e^{A^T t}e^{At}A\,dt\right]x$$

$$= x^T\left[\int_0^\infty \frac{d}{dt}\left(e^{A^T t}e^{At}\right)dt\right]x$$

$$= x^T\left[\left(e^{A^T t}e^{At}\right)\big|_0^\infty\right]x = -x^T x,$$

therefore, $-V'_{(3.24)}(x)$ is positive definite since $x^T x$ is a $\Re^n$-wedge. This completes the proof.   ♠

Note that Theorem 9.3.19 is the converse of Theorem 9.3.17 for linear cases.

Exercises 9.3

1. Determine if $V : [0,\infty) \times Q \to [0,\infty)$ is positive definite and/or decrescent on some domain Q containing the zero vector.

   (a) $V(t,x) = x^2 + \cos(tx)$.
   (b) $V(t,x) = x^2 + \cos^2(tx)$.
   (c) $V(t,x_1,x_2) = x_1^2 + t^2 x_2^4$.
   (d) $V(t,x_1,x_2) = x_1^2 + (1 + t^4)x_2^2$.
   (e) $V(t,x_1,x_2) = x_1^2 + \frac{2}{1+t^4}x_2^2$.
   (f) $V(t,x_1,x_2) = t[x_1^4 + x_2^2]$.
   (g) $V(t,x_1,x_2) = x_1^2 + (2 + \sin^4 t)x_2^2$.
   (h) $V(t,x_1,x_2) = x_1^2 + (\sin^4 t)x_2^2$.
   (i) $V(t,x_1,x_2,x_3) = x_1^2 + tx_2^2 + t^2 x_3^4$.
   (j) $V(t,x_1,x_2,x_3) = x_1^2 + x_2^2 - tx_3^4$.
   (k) $V(t,x_1,x_2,x_3) = x_1^2 + x_2^2 + x_3^2 + x_1 x_2 \sin tx_3$.

2. Prove Theorem 9.3.14.

3. Prove Theorem 9.3.15.

4. Prove Theorem 9.3.16.

5. Prove Theorem 9.3.17.

6. Verify in Example 9.3.11 that $u = 0$ is a solution.

7. Let b be a constant and let $a(t)$, $c(t)$ be continuous functions on $\Re^+$ such that $a(t) \leq d$, $c(t) \leq d$ for some negative constant d. Use $V(t, x_1, x_2) = x_1^2 + x_2^2$ to prove that the zero solution of

$$\begin{cases} x_1'(t) = a(t)x_1(t) - bx_2(t), \\ x_2'(t) = bx_1(t) + c(t)x_2(t), \end{cases}$$

is uniformly asymptotically stable.

8. Let $a(t)$, b, $c(t)$ be the same as above and let $f_i(t, x_1, x_2)$ be such that $\frac{f_i(t, x_1, x_2)}{\sqrt{x_1^2 + x_2^2}} \to 0$ as $x_1^2 + x_2^2 \to 0$. Use $V(t, x_1, x_2) = x_1^2 + x_2^2$ to prove that the zero solution of

$$\begin{cases} x_1'(t) = a(t)x_1(t) - bx_2(t) + f_1(t, x_1, x_2), \\ x_2'(t) = bx_1(t) + c(t)x_2(t) + f_2(t, x_1, x_2), \end{cases}$$

is uniformly asymptotically stable.

9. Consider the second-order equation $x''(t) = f(t, x)$. Assume that on $(t, x) \in [0, \infty) \times \Re$, $x f_t(t, x) \geq 0$; $f(t, x) \leq k(x) < 0$ if $x > 0$ and $f(t, x) \geq k(x) > 0$ if $x < 0$ for some continuous function $k(x)$. Write the equation as a system and use $V(t, x_1, x_2) = \frac{x_2^2}{2} - \int_0^{x_1} f(t, s)ds$ to prove that the zero solution is stable.

10. Discuss the stability properties of the zero solution for $x''(t) = a(t)x^4$ where $a(t)$ is a continuous function.

11. Discuss the stability properties of the zero solution for $x''(t) = a(t)b(x)$ where $a(t)$ and $b(x)$ are continuous functions.

12. Discuss the stability properties of the zero solution for $x'' + a(t)x' + x = 0$ where $a(t)$ is a continuous function and $a(t) \geq d$ on $\Re^+$ for some positive constant d.

13. Discuss the stability properties of the zero solution for $x'' + a(t)x' + f(x) = 0$ where $a(t)$ is the same as above and f is continuously differentiable with $x f(x) > 0$ for $x \neq 0$.

14. In the proof of Theorem 9.3.19, verify that the matrix B is well defined, and $V(x) > 0$ for $x \neq 0$.

Chapter 10

Bounded Solutions

10.1 Introduction

We have studied stability properties in chapters 5 and 9, where we can determine if a small change in the initial data for a system will cause a small change of the behavior for future time. Another notion that is very closely related to the notion of stability is the notion of "boundedness," which in applications indicates whether the future behavior of a system can be controlled when initial values are being controlled. See Chapter 1 for a brief discussion of boundedness for scalar differential equations.

Now, we study boundedness for general differential equations. We assume that $f(t, x)$ in

$$x'(t) = f(t, x(t)) \tag{1.1}$$

is defined on $D = [0, \infty) \times \Re^n$ in order to allow solutions to become large. This is different from the study of the stabilities of the zero solution, where we look at the solutions that are close to the zero, in which case the set Q in $D = [0, \infty) \times Q$ could be a small domain containing the zero vector in $\Re^n$. Next, since boundedness is a "long-term" behavior of solutions for "future time," we will assume that for any $(t_0, x_0) \in [0, \infty) \times \Re^n$, Eq. (1.1) has a unique solution $x(t, t_0, x_0)$ existing on $[t_0, \infty)$ with $x(t_0) = x_0$.

Now, we make the following definitions.

470

Definition 10.1.1 *Consider Eq. (1.1) on $D = [0, \infty) \times \Re^n$.*

(a). *A solution $x(t, t_0, x_0)$ of Eq. (1.1) is said to be **bounded** if there exists a $B(t_0, x_0) > 0$ such that $|x(t, t_0, x_0)| \leq B(t_0, x_0)$ for $t \geq t_0$.*

(b). *The solutions of Eq. (1.1) are said to be **equi-bounded** if for any $t_0 \geq 0$ and any $B_1 > 0$, there exists a $B_2 = B_2(t_0, B_1) > 0$ such that $\{|x_0| \leq B_1, t \geq t_0\}$ imply $|x(t, t_0, x_0)| \leq B_2$.*

(c). *The solutions of Eq. (1.1) are said to be **uniformly bounded** if the B_2 in the definition of equi-boundedness can be chosen to be independent of $t_0 \geq 0$. That is, for any $B_1 > 0$, there exists a $B_2 = B_2(B_1) > 0$ such that $\{|x_0| \leq B_1, t_0 \geq 0, t \geq t_0\}$ imply $|x(t, t_0, x_0)| \leq B_2$.*

(d). *The solutions of Eq. (1.1) are said to be **equi-ultimately bounded** if there is an (independent or generic) constant $B > 0$ such that for any $t_0 \geq 0$ and any $B_1 > 0$, there exists a $T = T(t_0, B_1) > 0$ such that $\{|x_0| \leq B_1, t \geq t_0 + T\}$ imply $|x(t, t_0, x_0)| \leq B$.*

(e). *The solutions of Eq. (1.1) are said to be **uniformly ultimately bounded** if the T in the definition of equi-ultimate boundedness can be chosen to be independent of $t_0 \geq 0$. That is, there is an (independent or generic) constant $B > 0$ such that for any $B_1 > 0$, there exists a $T = T(B_1) > 0$ such that $\{|x_0| \leq B_1, t_0 \geq 0, t \geq t_0 + T\}$ imply $|x(t, t_0, x_0)| \leq B$.*

Definition 10.1.1(a) is concerned with a single solution, while all other cases in Definition 10.1.1 are for a set of solutions. Notice the difference between boundedness and ultimate boundedness. For boundedness, initial values of the solutions are confined in a bounded set, then a bound is selected to control the solutions. For ultimate boundedness, however, a bound is specified first and initial values of the solutions are free, thus in general the specified bound cannot be used to control the solutions. Therefore, ultimate boundedness asks for a sufficiently large t value after which the solutions are controlled by the specified bound. The graphs showing the ideas and differences are given in **Figure 10.1**.

Recall from Chapter 2 that on a finite interval, a solution is continuous with respect to initial data. Therefore, we have the following result.

Theorem 10.1.2 *Assume that $f(t, x)$ is continuous and satisfies a weak Lipschitz condition with respect to x on $D = [0, \infty) \times \Re^n$. If the solutions of Eq. (1.1) are equi-ultimately bounded, then they are equi-bounded.*

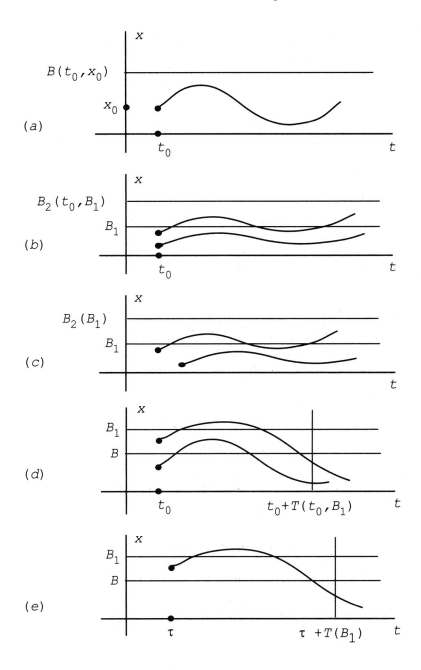

Figure 10.1: Different types of boundedness properties according to Definition 10.1.1

Proof. Let $B > 0$ be the bound in the definition of equi-ultimate bounded-ness, then for any $t_0 \geq 0$ and any $B_1 > 0$, there is a $T = T(t_0, B_1) > 0$ such that $\{|x_0| \leq B_1, t \geq t_0 + T\}$ imply $|x(t, t_0, x_0)| \leq B$. Next, on the interval $[t_0, t_0 + T]$, an argument similar to the proof of Theorem 2.3.3 in Chapter 2 shows that, for $t \in [t_0, t_0 + T]$,

$$|x(t, t_0, x_0) - x(t, t_0, y_0)| \leq |y_0 - x_0| e^{\int_{t_0}^{t} k(s)ds} \leq |y_0 - x_0| e^{\int_{t_0}^{t_0+T} k(s)ds}, \quad (1.2)$$

where y_0 is any fixed vector in $\Re^n$ and $k(\cdot)$ is from the weak Lipschitz condition. This implies, for $t \in [t_0, t_0 + T]$ and $|x_0| \leq B_1$,

$$
\begin{aligned}
|x(t, t_0, x_0)| &\leq |x(t, t_0, x_0) - x(t, t_0, y_0)| + |x(t, t_0, y_0)| \\
&\leq |y_0 - x_0| e^{\int_{t_0}^{t_0+T} k(s)ds} + |x(t, t_0, y_0)| \\
&\leq |x(t, t_0, y_0)| + (|y_0| + B_1) e^{\int_{t_0}^{t_0+T} k(s)ds}. \quad (1.3)
\end{aligned}
$$

As $x(t, t_0, y_0)$ is a fixed solution on $[t_0, t_0 + T]$, the inequality (1.3) implies that for $|x_0| \leq B_1$, the solutions $x(t, t_0, x_0)$ are bounded on $[t_0, t_0 + T]$. That is, there is a $B_2 = B_2(t_0, B_1, T) = B_2(t_0, B_1) > 0$ such that $|x(t, t_0, x_0)| \leq B_2(t_0, B_1)$ for $t \in [t_0, t_0 + T]$ and $|x_0| \leq B_1$. Now, we see that $\{|x_0| \leq B_1, t \geq t_0\}$ imply $|x(t, t_0, x_0)| \leq \max\{B_2, B\}$, which verifies the equi-boundedness and completes the proof. ♠

In the above proof, the equi-ultimate boundedness is used to control solutions for large t, and the weak Lipschitz condition together with that fixed solution $x(t, t_0, y_0)$ are used to control solutions on $[t_0, t_0 + T]$. To extend this to the uniform boundedness, the argument in the above proof should be independent of t_0. If the solutions of Eq. (1.1) are uniformly ultimately bounded, then the time T in the above proof can be chosen to be independent of t_0, and then $\int_{t_0}^{t_0+T} k(s)ds$ is bounded independently of t_0 if a (global) Lipschitz condition is assumed. But in general, $|x(t, t_0, y_0)|$ in the above proof may not be bounded independently of t_0. Therefore, from the proof of Theorem 10.1.2 and the previous remarks, we have the following result.

Theorem 10.1.3 *Assume that $f(t, x)$ is continuous and satisfies a Lipschitz condition with respect to x on $D = [0, \infty) \times \Re^n$, and that the solutions of Eq. (1.1) are uniformly ultimately bounded. If there is a $y_0 \in \Re^n$ such that for any $T > 0$, $x(t, t_0, y_0)$ is bounded on $[t_0, t_0 + T]$ independently of $t_0 \geq 0$ (for example, when the zero or some constants are solutions), then the solutions are uniformly bounded.* ♠

Theorem 10.1.3 says that uniform ultimate boundedness implies uniform boundedness under certain conditions, including a (global) Lipschitz condition. Later, we will see that when those conditions are not satisfied, we do not have the same conclusion.

This chapter is organized as follows: In Section 2, we derive boundedness results for general linear differential equations by using the results from Chapter 9. It will be seen that stability and boundedness are almost equivalent for linear homogeneous differential equations, and they are determined by the fundamental matrix solutions. For nonlinear differential equations, examples will be given to show that the concepts of stability and boundedness are not equivalent. In Section 3, we look at the case when the coefficient matrix is a constant matrix, and verify that the eigenvalues of the coefficient matrix determine boundedness properties. In Section 4, the case of a periodic coefficient matrix is treated. The Floquet theory from Chapter 3 is used to transform the equation with a periodic coefficient matrix into an equation with a constant coefficient matrix. Therefore, the results from Section 3 can be applied. In Section 5, we use Liapunov's method to study boundedness properties for general nonlinear differential equations.

Exercises 10.1

1. Prove Theorem 10.1.3.

2. Determine the boundedness properties for the scalar differential equation $x'(t) = a(t)x(t)$.

10.2 General Linear Differential Equations

Here, we study the linear differential equation

$$x'(t) = A(t)x(t) + f(t), \quad x(t_0) = x_0, \quad t \geq t_0 \geq 0, \quad x \in \Re^n, \qquad (2.1)$$

where $A(t)$ and $f(t)$ are continuous on $\Re^+ = [0, \infty)$.

Using the results from Chapter 3, we know that the unique solution of Eq. (2.1) for the initial data (t_0, x_0) is given by the variation of parameters formula

$$
\begin{aligned}
x(t) &= U(t, t_0)\left[x_0 + \int_{t_0}^t U^{-1}(s, t_0)f(s)ds\right] \\
&= U(t, t_0)x_0 + \int_{t_0}^t U(t, s)f(s)ds, \quad t \geq t_0. \qquad (2.2)
\end{aligned}
$$

We first study the linear homogeneous differential equation

$$x'(t) = A(t)x(t), \qquad (2.3)$$

where the unique solution is given by $x(t) = U(t, t_0)x_0$ for the initial data (t_0, x_0). Based on Theorem 9.2.1 in Chapter 9, the following result says that for linear homogeneous differential equations, stability properties and boundedness properties are almost equivalent. For nonlinear differential equations, examples will be given to show that the concepts are not equivalent.

Theorem 10.2.1 *Assume that $A(t)$ is continuous on $\Re^+$. Then for Eq. (2.3),*

(A). the following statements are equivalent:

   (1). The solutions are equi-bounded;

   (2). $\phi = 0$ is stable;

   (3). There is an (independent or generic) constant $C > 1$ such that

$$|U(t, 0)| \le C, \quad 0 \le t < \infty. \qquad (2.4)$$

(B). the following statements are equivalent:

   (1). The solutions are uniformly bounded;

   (2). $\phi = 0$ is uniformly stable;

   (3). There is an (independent or generic) constant $C > 1$ such that

$$|U(t, s)| \le C, \quad 0 \le s \le t < \infty. \qquad (2.5)$$

(C). the following statements are equivalent:

   (1). The solutions are equi-ultimately bounded;

   (2). $\phi = 0$ is asymptotically stable;

   (3). $|U(t, 0)| \to 0, \quad t \to \infty.$

(D). If $\phi = 0$ is uniformly asymptotically stable, then the solutions are uniformly ultimately bounded. If, on the other hand, the solutions are uniformly ultimately bounded, then under some additional conditions, such as when $A(t)$ is bounded on $\Re^+$, $\phi = 0$ is uniformly asymptotically stable.

Proof. (A): Suppose that $\phi = 0$ is stable. Then from Theorem 9.2.1 in Chapter 9, there is an independent constant $C > 1$ such that $|U(t, 0)| \leq C$, $t \geq 0$. Now, the solutions satisfy

$$|x(t, t_0, x_0)| = |U(t, t_0)x_0| = |U(t, 0)U(0, t_0)x_0| \leq C|U(0, t_0)||x_0|.$$

For any $t_0 \geq 0$ and $B_1 > 0$, we can choose $B_2 = B_2(t_0, B_1) = C|U(0, t_0)|B_1$, such that $\{t \geq t_0, |x_0| \leq B_1\}$ imply $|x(t, t_0, x_0)| \leq B_2$. This proves the equi-boundedness.

Conversely, if the solutions are equi-bounded, then for $t_0 = 0$ and $B_1 = 1$, there exists a $B_2 = B_2(t_0, B_1) = B_2(0, 1) > 0$ such that $\{|x_0| \leq 1, t \geq 0\}$ imply $|U(t, 0)x_0| = |x(t, 0, x_0)| \leq B_2$. Then similar to the proof of Theorem 9.2.1, $|U(t, 0)| \leq C$ for some independent constant $C > 1$.

(B): Suppose that $\phi = 0$ is uniformly stable. Then from Theorem 9.2.1, $|U(t, t_0)| \leq C$, $0 \leq t_0 \leq t < \infty$, for some independent constant $C > 1$. Now, the solutions satisfy $|x(t, t_0, x_0)| = |U(t, t_0)x_0| \leq C|x_0|$, which implies the uniform boundedness.

Next, if the solutions are uniformly bounded, then for $t_0 \geq 0$ and $B_1 = 1$, there exists a $B_2 = B_2(B_1) = B_2(1) > 0$ such that $\{|x_0| \leq 1, t \geq t_0\}$ imply $|U(t, t_0)x_0| = |x(t, t_0, x_0)| \leq B_2$. Similar to the proof of Theorem 9.2.1, $|U(t, t_0)| \leq C$ for some independent constant $C > 1$.

(C): Suppose that $\phi = 0$ is asymptotically stable. Then from Theorem 9.2.1, $|U(t, 0)| \to 0$, $t \to \infty$. Hence for any $t_0 \geq 0$ and $B_1 > 0$, we can find $T = T(t_0, B_1) > 0$ such that $t \geq t_0 + T$ implies $|U(t, 0)| \leq \frac{1}{|U(0, t_0)|B_1}$. Let $B = 1$. Then $\{|x_0| \leq B_1, t \geq t_0 + T\}$ imply

$$
\begin{aligned}
|x(t, t_0, x_0)| &= |U(t, t_0)x_0| = |U(t, 0)U(0, t_0)x_0| \\
&\leq |U(t, 0)||U(0, t_0)|B_1 \leq 1 = B,
\end{aligned}
$$

proving the equi-ultimate boundedness.

Conversely, if the solutions are equi-ultimately bounded, then there is a constant $B > 0$, such that for $t_0 = 0$ and any $B_1 > 0$, there exists a $T = T(0, B_1) = T(B_1) > 0$ such that $\{|x_0| \leq B_1, t \geq t_0 + T = T\}$ imply $|U(t, 0)x_0| = |x(t, 0, x_0)| \leq B$. Similar to the proof of Theorem 9.2.1, this implies that for any $B_1 > 0$, there is a $T = T(0, B_1) = T(B_1) > 0$ such that $|U(t, 0)| \leq \frac{nB}{B_1}$ for $t \geq T$, (n is the dimension of $\Re^n$). Now, for any $\varepsilon > 0$, choose $B_1 = \frac{nB}{\varepsilon}$, then there is a $T = T(B_1) = T(\varepsilon) > 0$ such that $t \geq T$ implies $|U(t, 0)| \leq \frac{nB}{B_1} = \varepsilon$. Thus $|U(t, 0)| \to 0$, $t \to \infty$.

(D): Suppose that $\phi = 0$ is uniformly asymptotically stable. Then from Theorem 9.2.1, $|x(t, t_0, x_0)| = |U(t, t_0)x_0| \leq |U(t, t_0)||x_0| \leq Ce^{-\alpha(t - t_0)}|x_0|$

for some independent constants $C > 1$, $\alpha > 0$. Now let $B = 1$ and, for any $B_1 > 1$, solve $T = T(B_1) > 0$ from $CB_1 e^{-\alpha T} = 1$. Then $\{t_0 \geq 0, |x_0| \leq B_1, t \geq t_0 + T\}$ imply

$$|x(t, t_0, x_0)| = |U(t, t_0)x_0| \leq Ce^{-\alpha(t-t_0)}|x_0| \leq CB_1 e^{-\alpha T} = 1 = B,$$

proving the uniform ultimate boundedness.

On the other hand, assume that the solutions are uniformly ultimately bounded and that $A(t)$ is bounded on $\Re^+$. Then Eq. (2.3) satisfies a Lipschitz condition and $\phi = 0$ is a solution, hence from Theorem 10.1.3, the solutions of Eq. (2.3) are uniformly bounded. Therefore, from part (B) we just proved, $\phi = 0$ is uniformly stable. Next, from the uniform ultimate boundedness, there is a constant $B > 0$, such that for any $t_0 \geq 0$ and any $B_1 > 0$, there exists a $T = T(B_1) > 0$ such that $\{|x_0| \leq B_1, t \geq t_0 + T\}$ imply $|U(t, t_0)x_0| = |x(t, t_0, x_0)| \leq B$. Similar to the proof of Theorem 9.2.1, this implies that for any $t_0 \geq 0$ and $B_1 = 2nB$ (n is the dimension of $\Re^n$), there is a $T = T(B_1) = T(B) > 0$ such that

$$|U(t, t_0)| \leq \frac{nB}{B_1} = \frac{1}{2}, \quad t_0 \geq 0, \quad t \geq t_0 + T. \tag{2.6}$$

As $\phi = 0$ is uniformly stable, the same proof of part (D) in Theorem 9.2.1 can be used to show that

$$|U(t, s)| \leq Ce^{-\alpha(t-s)}, \quad 0 \leq s \leq t < \infty, \tag{2.7}$$

for some independent constants $\alpha > 0$, $C > 1$. Thus $\phi = 0$ is uniformly asymptotically stable. This completes the proof. ♠

From Theorem 10.2.1, we find that for Eq. (2.3), most stability and boundedness properties are equivalent. The exception is uniform asymptotic stability and uniform ultimate boundedness, where an additional condition that $A(t)$ being bounded on $\Re^+$ is used.

Next, we provide an example where $A(t)$ is unbounded on $\Re^+$ and the result in Theorem 10.2.1(D) is not true; that is, the solutions are uniformly ultimately bounded, but $\phi = 0$ is not uniformly stable (hence not uniformly asymptotically stable).

Example 10.2.2 Consider the scalar linear differential equation

$$x'(t) = a(t)x(t),$$

where $a(t)$ is given in **Figure 10.2** and is continuous, and for $i = 0, 1, 2, \cdots,$

(a). $a(i) = 0$; $a(t) > 0$ for $t \in (2i, 2i+1)$; $a(t) < 0$ for $t \in (2i+1, 2i+2)$.

(b). $\int_{2i}^{2i+1} a(s)ds = i+1$; $\int_{2i+1}^{2i+2} a(s)ds = -(2i+3)$.

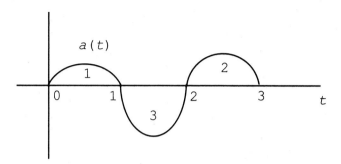

Figure 10.2: The function $a(t)$

This example is formulated from an exercise in Burton [1985] where no solution is given. Since several cases should be considered, we provide a detailed analysis here. Now, the solutions are given by

$$x(t, t_0, x_0) = x_0 \exp[\int_{t_0}^{t} a(s)ds], \quad t \geq t_0.$$

To get the uniform ultimate boundedness, we let $B = 1$ and need to show that for any $B_1 > 0$, there exists a $T = T(B_1) > 0$ such that $\{|x_0| \leq B_1, t_0 \geq 0, t \geq t_0 + T\}$ imply

$$|x(t, t_0, x_0)| = |x_0 \exp[\int_{t_0}^{t} a(s)ds]| \leq B_1 \exp[\int_{t_0}^{t} a(s)ds] \leq 1 = B,$$

or equivalently,

$$\int_{t_0}^{t} a(s)ds \leq \ln \frac{1}{B_1}, \quad t_0 \geq 0, \ t \geq t_0 + T. \tag{2.8}$$

To this end, observe from the condition (b) that,

$$\int_{2i}^{2i+3} a(s)ds = \int_{2i}^{2i+1} a(s)ds + \int_{2i+1}^{2i+2} a(s)ds + \int_{2i+2}^{2i+3} a(s)ds$$
$$= (i+1) - (2i+3) + (i+1+1) = 0, \tag{2.9}$$

and

$$\int_{2k+1}^{2k+3} a(s)ds \; = \; \int_{2k+1}^{2k+2} a(s)ds + \int_{2k+2}^{2k+3} a(s)ds$$

$$= \; -(2k+3) + (k+1+1) = -(k+1). \qquad (2.10)$$

Then, from (2.9) and (2.10),

$$\int_{2i}^{2i+2m+1} a(s)ds \; = \; \int_{2i}^{2i+3} a(s)ds + \int_{2i+3}^{2i+5} a(s)ds + \cdots + \int_{2i+2m-1}^{2i+2m+1} a(s)ds$$

$$= \; 0 + \int_{2(i+1)+1}^{2(i+1)+3} a(s)ds + \cdots + \int_{2(i+m-1)+1}^{2(i+m-1)+3} a(s)ds$$

$$= \; -(i+1+1) - (i+2+1) - \cdots - (i+m-1+1)$$

$$= \; -(i+2) - (i+3) - \cdots - (i+m). \qquad (2.11)$$

Thus, for any $B_1 > 0$, we choose $N = N(B_1) > 3$ such that

$$-2 - 3 - \cdots - N \le \ln\frac{1}{B_1},$$

and define $T = T(N) = T(B_1) = 2N + 1$. Then for $t_0 \ge 0$ and $t \ge t_0 + T$ we have $t_0 \in [2i, 2i+1)$ or $t_0 \in [2i+1, 2i+2)$ for some $i = 0, 1, 2, \cdots$, and $t \in [2s, 2s+1)$ or $t \in [2s+1, 2s+2)$ for some $s = 0, 1, 2, \cdots$.

Case 1: $t_0 \in [2i+1, 2i+2)$, $t \in [2s+1, 2s+2)$ for some $i \ge 0$ and $s \ge 0$. Now, from the condition (a),

$$\int_{t_0}^{t} a(s)ds \le \int_{2i+2}^{2s+1} a(s)ds. \qquad (2.12)$$

Since $t - t_0 \ge T = 2N+1$, one has $(2s+2) - (2i+1) \ge t - t_0 \ge T = 2N+1$, or $s - i \ge N$. Then from (2.11) and (2.12), we obtain

$$\int_{t_0}^{t} a(s)ds \; \le \; \int_{2i+2}^{2s+1} a(s)ds = \int_{2(i+1)}^{2(i+1)+2(s-i-1)+1} a(s)ds$$

$$= \; -(i+1+2) - (i+1+3) - \cdots - (i+1+s-i-1)$$

$$\le \; -(i+1+2) - (i+1+3) - \cdots - (i+1+N-1)$$

$$\le \; -2 - 3 - \cdots - N \le \ln\frac{1}{B_1}, \qquad (2.13)$$

therefore, (2.8) is true.

Case 2: $t_0 \in [2i, 2i+1)$, $t \in [2s, 2s+1)$ for some $i \geq 0$ and $s \geq 0$. Now, from the condition (a),

$$\int_{t_0}^{t} a(s)ds \leq \int_{2i}^{2s+1} a(s)ds. \tag{2.14}$$

Since $t - t_0 \geq T = 2N+1$, one has $(2s+1) - (2i) \geq t - t_0 \geq T = 2N+1$, or $s - i \geq N$. Then, as $\int_{2k+1}^{t} a(s)ds \leq 0$ for $t \geq 2k+1$, we have, from (2.11) and (2.14),

$$
\begin{aligned}
\int_{t_0}^{t} a(s)ds &\leq \int_{2i}^{2s+1} a(s)ds = \int_{2i}^{2i+2N+1} a(s)ds + \int_{2i+2N+1}^{2s+1} a(s)ds \\
&\leq \int_{2i}^{2i+2N+1} a(s)ds = -(i+2) - (i+3) - \cdots - (i+N) \\
&\leq -2 - 3 - \cdots - N \leq \ln \frac{1}{B_1}. \tag{2.15}
\end{aligned}
$$

Case 3: $t_0 \in [2i+1, 2i+2)$, $t \in [2s, 2s+1)$ for some $i \geq 0$ and $s \geq 0$. Now, from the condition (a),

$$\int_{t_0}^{t} a(s)ds \leq \int_{2i+2}^{2s+1} a(s)ds. \tag{2.16}$$

Since $t - t_0 \geq T = 2N+1$, one has $(2s+1) - (2i+1) \geq t - t_0 \geq T = 2N+1$, or $s - i \geq N + \frac{1}{2}$. As $s - i$ is an integer, we must have $s - i \geq N + 1$. From (2.11) and (2.16),

$$
\begin{aligned}
\int_{t_0}^{t} a(s)ds &\leq \int_{2i+2}^{2s+1} a(s)ds = \int_{2(i+1)}^{2(i+1)+2(s-i-1)+1} a(s)ds \\
&= -(i+1+2) - (i+1+3) - \cdots - (i+1+s-i-1) \\
&\leq -(i+1+2) - (i+1+3) - \cdots - (i+1+N) \\
&\leq -2 - 3 - \cdots - N \leq \ln \frac{1}{B_1}. \tag{2.17}
\end{aligned}
$$

Case 4: $t_0 \in [2i, 2i+1)$, $t \in [2s+1, 2s+2)$ for some $i \geq 0$ and $s \geq 0$. This case is similar to Case 3 and is left as an exercise.

From the above, we see that (2.8) is true for all cases, thus we have the uniform ultimate boundedness. To see that $\phi = 0$ is not uniformly stable, we only need to note that

$$\lim_{i \to \infty} \exp[\int_{2i}^{2i+1} a(s)ds] = \infty. \qquad \spadesuit \tag{2.18}$$

Remark 10.2.3 According to Theorem 10.2.1(B), the solutions in Example 10.2.2 are not uniformly bounded since the zero solution is not uniformly stable. Therefore, Example 10.2.2 is also an example where the solutions are uniformly ultimately bounded, but not uniformly bounded. Compare this case with Theorem 10.1.3 (which says that if a Lipschitz condition is satisfied and if the zero or some constants are solutions, then uniform ultimate boundedness implies uniform boundedness). We find for Example 10.2.2 that the zero is indeed a solution, but a weak Lipschitz condition, rather than a Lipschitz condition, is satisfied, which is the reason why the solutions are not uniformly bounded. Example 10.2.2 also indicates that in general, the Lipschitz condition in Theorem 10.1.3 cannot be reduced to a weak Lipschitz condition.

Remark 10.2.4 The second conclusion in Theorem 10.2.1(D) reveals the relationship between uniform ultimate boundedness and uniform asymptotic stability for linear homogeneous differential equations. Moreover, from Example 10.2.2, uniform ultimate boundedness does not imply uniform asymptotic stability if $A(t)$ is unbounded. Therefore, the condition that $A(t)$ being bounded in Theorem 10.2.1(D) is probably the best condition.

Next, we illustrate that for nonlinear differential equations, boundedness and stability are not equivalent concepts.

Example 10.2.5 Consider the scalar nonlinear differential equation

$$x' = -x + x^2,$$

which is given as Example 5.3.5 in Chapter 5, from where we know that $\phi = 0$ is uniformly asymptotically stable, but for $x_0 > 1$, $x(t, 0, x_0) \to \infty$ as $t \nearrow \ln[x_0/(x_0 - 1)]$. Therefore the solutions of $x' = -x + x^2$ are not equi-bounded. See **Figure 10.3**.

Next, consider the scalar nonlinear differential equation

$$x' = x(1 - x)(1 + x).$$

First, $x = -1$, $x = 0$, $x = 1$ are three constant solutions. For $x > 1$, $x' = x(1-x)(1+x) < 0$, thus the solutions are decreasing when $x(0) > 1$. Similar observation can be made for $x \in (0, 1)$, $x \in (-1, 0)$, and $x < -1$ to come up with the following solutions in **Figure 10.4**.

Now we find that the solutions are equi-bounded, but $\phi = 0$ is not stable.
♠

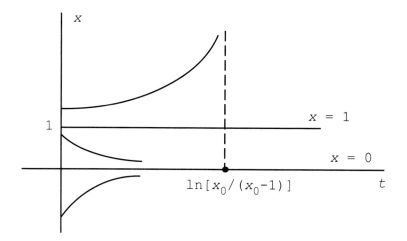

Figure 10.3: The solutions of Example 10.2.5 are not equi-bounded

---

Similar to the study of stabilities, the results concerning the boundedness of equations with linear or nonlinear perturbations can also be derived.

Theorem 10.2.6 *Assume that $A(t)$ is continuous on $\Re^+$. If the solutions of Eq. (2.3) are uniformly bounded, and if the continuous function $f(t)$ satisfies $\int_0^\infty |f(t)|dt < \infty$, then the solutions of*

$$x'(t) = A(t)x(t) + f(t) \tag{2.19}$$

are also uniformly bounded.

Proof. First, since $A(t)$ and $f(t)$ are continuous, existence and uniqueness of the solutions for Eq. (2.19) is guaranteed. Then, for U being the fundamental matrix solution of Eq. (2.3), the solutions of Eq. (2.19) are given by the variation of parameters formula

$$x(t) = U(t,t_0)x(t_0) + \int_{t_0}^t U(t,s)f(s)ds, \ \ t \geq t_0 \geq 0. \tag{2.20}$$

From Theorem 10.2.1, uniform boundedness of the solutions of Eq. (2.3) implies that there is an (independent) constant $C > 1$ such that $|U(t,t_0)| \leq C$, $0 \leq t_0 \leq t < \infty$. Therefore, we have

$$|x(t)| \ \leq \ C|x(t_0)| + \int_{t_0}^t C|f(s)|ds$$

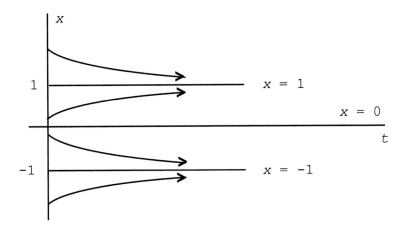

Figure 10.4: The solutions of $x' = x(1-x)(1+x)$

$$\leq\ C|x(t_0)| + C\int_0^\infty |f(s)|ds, \quad t \geq t_0 \geq 0. \qquad (2.21)$$

Now, for any $B_1 > 0$, we can choose $B_2 = CB_1 + C\int_0^\infty |f(s)|ds = B_2(B_1)$, such that $\{|x(t_0)| \leq B_1, t \geq t_0 \geq 0\}$ imply $|x(t)| \leq B_2$. Thus the solutions of Eq. (2.19) are uniformly bounded. This completes the proof. ♠

The proofs of the following results are left as exercises.

Theorem 10.2.7 *Assume that $A(t)$ is continuous on $\Re^+$. If the solutions of Eq. (2.3) are uniformly bounded, and if the $n \times n$ continuous matrix function $B(t)$ satisfies $\int_0^\infty |B(t)|dt < \infty$, then the solutions of*

$$x'(t) = A(t)x(t) + B(t)x(t) = [A(t) + B(t)]x(t) \qquad (2.22)$$

are also uniformly bounded. ♠

Theorem 10.2.7.a *Assume that $A(t)$ is continuous on $\Re^+$. If the solutions of Eq. (2.3) are uniformly bounded, and if the continuous function $f(t,x)$ satisfies a weak Lipschitz condition in x and $|f(t,x)| \leq b(t)|x|$, where $b(t) \geq 0$ satisfies $\int_0^\infty b(t)dt < \infty$, then the solutions of*

$$x'(t) = A(t)x(t) + f(t, x(t)) \qquad (2.23)$$

are also uniformly bounded. ♠

The previous statements concern uniform boundedness. For uniform ultimate boundedness, we have

Theorem 10.2.8 *Assume that $A(t)$ is continuous on $\Re^+$. If the zero solution of Eq. (2.3) is uniformly asymptotically stable, (or if the solutions of Eq. (2.3) are uniformly ultimately bounded and $A(t)$ is bounded on $\Re^+$; see Theorem 10.2.1(D)), and if the $n \times n$ continuous matrix function $B(t)$ satisfies*

$$\int_{t_0}^t |B(s)|ds \le m(t - t_0) + r, \quad t \ge t_0 \ge 0, \tag{2.24}$$

for some positive constants m and r, then there is an $m_0 > 0$ such that if $m \le m_0$, then the solutions of Eq. (2.22) are uniformly ultimately bounded.
♠

Theorem 10.2.8.a *Assume that $A(t)$ is continuous on $\Re^+$. If the zero solution of Eq. (2.3) is uniformly asymptotically stable, (or if the solutions of Eq. (2.3) are uniformly ultimately bounded and $A(t)$ is bounded on $\Re^+$; see Theorem 10.2.1(D)), and if the continuous function $f(t, x)$ satisfies a weak Lipschitz condition in x and $|f(t, x)| \le b(t)|x|$, where $b(t) \ge 0$ satisfies (2.24), that is*

$$\int_{t_0}^t b(s)ds \le m(t - t_0) + r, \quad t \ge t_0 \ge 0,$$

for some positive constants m and r, then there is an $m_0 > 0$ such that if $m \le m_0$, then the solutions of Eq. (2.23) are uniformly ultimately bounded.
♠

The following is a result concerning uniform boundedness and uniform ultimate boundedness for nonlinear perturbations.

Theorem 10.2.9 *Assume that $A(t)$ is continuous on $\Re^+$. If the zero solution of Eq. (2.3) is uniformly asymptotically stable, (or if the solutions of Eq. (2.3) are uniformly ultimately bounded and $A(t)$ is bounded on $\Re^+$; see Theorem 10.2.1(D)), and if the continuous function $f(t, x)$ satisfies a weak Lipschitz condition in x and $|f(t, x)| \le M$ for some constant $M > 0$, then the solutions of Eq. (2.23) are uniformly bounded and uniformly ultimately bounded.*

Proof. The solutions of Eq. (2.23) are given by

$$x(t) = U(t, t_0)x(t_0) + \int_{t_0}^t U(t, s)f(s, x(s))ds, \quad t \ge t_0 \ge 0. \qquad (2.25)$$

From Theorem 9.2.1 in Chapter 9, uniform asymptotic stability of the zero solution of Eq. (2.3) implies that there are (independent) constants $C > 1$ and $\alpha > 0$ such that $|U(t, t_0)| \le Ce^{-\alpha(t-t_0)}$. Then,

$$
\begin{aligned}
|x(t)| &\le Ce^{-\alpha(t-t_0)}|x(t_0)| + \int_{t_0}^t Ce^{-\alpha(t-s)}|f(s, x(s))|ds \\
&\le Ce^{-\alpha(t-t_0)}|x(t_0)| + \int_0^t Ce^{-\alpha(t-s)}Mds \\
&\le Ce^{-\alpha(t-t_0)}|x(t_0)| + \frac{CM}{\alpha}, \quad t \ge t_0 \ge 0. \qquad (2.26)
\end{aligned}
$$

Now, we can let $B = 1 + \frac{CM}{\alpha}$, such that for any $B_1 > 1$, we choose $T = T(B_1) > 0$ from $Ce^{-\alpha T}B_1 = 1$. Then $\{|x(t_0)| \le B_1, t_0 \ge 0, t \ge t_0 + T\}$ imply

$$|x(t)| \le Ce^{-\alpha T}B_1 + \frac{CM}{\alpha} = 1 + \frac{CM}{\alpha} = B.$$

This verifies uniform ultimate boundedness. The inequality (2.26) also implies uniform boundedness. This completes the proof. ♠

Exercises 10.2

1. Give the details for Case 4 in Example 10.2.2.

2. In Example 10.2.2, provide some details why $\phi = 0$ is not uniformly stable.

3. Derive Figure 10.4.

4. Prove Theorem 10.2.7.

5. Prove Theorem 10.2.7.a.

6. Prove Theorem 10.2.8.

7. Prove Theorem 10.2.8.a.

10.3   Linear Equations with Constant Coefficients

In this section, we study the linear differential equation with constant coefficients,

$$x'(t) = Ax(t) + f(t), \quad x(t_0) = x_0, \quad t \geq t_0 \geq 0, \quad x \in \Re^n, \tag{3.1}$$

where $f(t)$ is continuous on $\Re^+ = [0, \infty)$.

Using the results from Chapter 3, we know that the unique solution of Eq. (3.1) is given by the variation of parameters formula

$$
\begin{aligned}
x(t) &= e^{(t-t_0)A}\left[x_0 + \int_{t_0}^t \{e^{(s-t_0)A}\}^{-1} f(s)ds\right] \\
&= e^{(t-t_0)A}x_0 + \int_{t_0}^t e^{(t-s)A} f(s)ds, \quad t \geq t_0,
\end{aligned}
\tag{3.2}
$$

with the fundamental matrix solution of Eq. (3.1) (when $f = 0$) given by $U(t, t_0) = e^{(t-t_0)A}$.

First we study the linear homogeneous differential equation

$$x'(t) = Ax(t), \tag{3.3}$$

where the unique solution is given by $x(t) = e^{(t-t_0)A}x_0$ for the initial data (t_0, x_0).

Now, A is a constant matrix and hence is bounded, therefore, based on Theorem 10.2.1(D), we have the following result.

Theorem 10.3.1 *For Eq. (3.3), the solutions are uniformly ultimately bounded if and only if $\phi = 0$ is uniformly asymptotically stable.*

With Theorem 10.3.1, we conclude that boundedness properties and stability properties are equivalent for linear homogeneous and autonomous differential equations. Therefore, we can use the eigenvalues of matrix A to determine the boundedness properties.

Theorem 10.3.2 *Let λ be a complex number and denote by $R(\lambda)$ the real part of λ. Then for Eq. (3.3),*

(A). the following statements are equivalent:

    (1). The solutions are equi-bounded or uniformly bounded;

(2). For each eigenvalue λ of the matrix A, either $R(\lambda) < 0$, or $R(\lambda) = 0$ but in this case λ appears only in matrices J_i (in the Jordan canonical form for A) such that J_i is a 1×1 matrix;

(3). There is an (independent or generic) constant $M > 1$ such that

$$|e^{tA}| \leq M, \quad 0 \leq t < \infty. \tag{3.4}$$

(B). the following statements are equivalent:

(1). The solutions are equi-ultimately bounded or uniformly ultimately bounded;

(2). Each eigenvalue of the matrix A has a negative real part;

(3). There are (independent or generic) constants $M > 1$ and $\alpha > 0$ such that

$$|e^{tA}| \leq M e^{-\alpha t}, \quad 0 \leq t < \infty. \tag{3.5}$$

Example 10.3.3 Let's revisit Example 5.2.2 in Chapter 5 and look at the boundedness properties for the equations with the following matrices

$$A_1 = \begin{bmatrix} 0 & 0 \\ 0 & 0 \end{bmatrix}, \quad A_2 = \begin{bmatrix} 0 & 1 \\ 0 & 0 \end{bmatrix}, \quad A_3 = \begin{bmatrix} -1 & 1 \\ 0 & -1 \end{bmatrix}, \quad A_4 = \begin{bmatrix} -1 & 0 \\ 0 & -1 \end{bmatrix}.$$

The matrix A_1 has a repeated eigenvalue 0, and the corresponding J_0 in the Jordan canonical form are two 1×1 matrices, thus the solutions of the linear equation with the coefficient matrix A_1 are uniformly bounded. The matrix A_2 has a repeated eigenvalue 0, and the corresponding J_0 in the Jordan canonical form is A_2 itself, a 2×2 matrix, thus the solutions are not equi-bounded. For the matrix A_3 or A_4, the solutions are uniformly bounded and uniformly ultimately bounded because the eigenvalues are all negative. ♠

Exercises 10.3

1. If every solution of $x'(t) = Ax(t)$ is bounded and if $f(t)$ is continuous and $\int_0^\infty |f(t)|dt < \infty$, then show that every solution of $x'(t) = Ax(t) + f(t)$ is also bounded.

2. If every solution of $x'(t) = Ax(t)$ is bounded and if $B(t)$ is continuous and $\int_0^\infty |B(t)|dt < \infty$, then show that every solution of $x'(t) = Ax(t) + B(t)x(t)$ is also bounded.

3. In $x'(t) = Ax(t) + f(t)$, if $|f(t)| \le Me^{at}$, $t \ge t_0$, for some $M \ge 0$ and $a \in \Re$, then show that for some $M_1 \ge 0$ and $a_1 \in \Re$,

$$|x(t)|, \ |x'(t)| \le M_1 e^{a_1 t}, \quad t \ge t_0.$$

4. In $x^{(n)}(t) + b_1 x^{(n-1)}(t) + \cdots + a_n x(t) = f(t) \in \Re$, if $|f(t)| \le Me^{at}$, $t \ge t_0$, for some $M \ge 0$ and $a \in \Re$, show that for some $M_1 \ge 0$ and $a_1 \in \Re$,

$$|x^{(i)}(t)| \le M_1 e^{a_1 t}, \quad i = 0, 1, \cdots, n, \ \ t \ge t_0.$$

5. For the linear differential equations with constant coefficients, derive the boundedness results with linear or nonlinear perturbations.

10.4   Linear Equations with Periodic Coefficients

In this section, we study the linear differential equation

$$x'(t) = A(t)x(t) + f(t), \quad x(t_0) = x_0, \quad t \ge t_0 \ge 0, \quad x \in \Re^n, \qquad (4.1)$$

where $A(t)$, $f(t)$ are continuous on $\Re^+ = [0, \infty)$, and $A(t + T) = A(t)$, $t \in \Re^+$, for some constant $T > 0$.

We first study the linear homogeneous equation

$$x'(t) = A(t)x(t). \qquad (4.2)$$

From the Floquet theory in Chapter 3, we know that a constant matrix C and a nonsingular continuous T-periodic matrix $P(t)$ exist, such that $x(t) = P(t)y(t)$ transforms Eq. (4.2) into

$$y'(t) = Cy(t). \qquad (4.3)$$

Since $P(t)$ is periodic and continuous, $P(t)$ is bounded. Thus the transformation $x(t) = P(t)y(t)$ will reduce the boundedness properties of Eq. (4.2) to those of Eq. (4.3), for which the results from the previous section can be applied. Therefore we have the following results based on Theorem 10.3.2.

Theorem 10.4.1 *Let λ be a complex number and denote by $R(\lambda)$ the real part of λ. Then for Eq. (4.2),*

(A). the following statements are equivalent:

(1). The solutions are equi-bounded or uniformly bounded;

(2). For each eigenvalue λ of the matrix C, either $R(\lambda) < 0$, or $R(\lambda) = 0$ but in this case λ appears only in matrices J_i (in the Jordan canonical form for C) such that J_i is a 1×1 matrix;

(3). There is an (independent or generic) constant $M > 1$ such that

$$|e^{tC}| \leq M, \quad 0 \leq t < \infty.$$

(B). the following statements are equivalent:

(1). The solutions are equi-ultimately bounded or uniformly ultimately bounded;

(2). Each eigenvalue of the matrix C has a negative real part;

(3). There are (independent or generic) constants $M > 1$ and $\alpha > 0$ such that

$$|e^{tC}| \leq Me^{-\alpha t}, \quad 0 \leq t < \infty. \qquad \spadesuit \qquad (4.4)$$

These results can also be stated using characteristic exponents and characteristic multipliers, which are the eigenvalues of the matrices C and e^{TC} respectively. See the related results in Chapter 3.

Exercises 10.4

1. For the linear differential equations with periodic coefficients, derive the boundedness results with linear or nonlinear perturbations.

10.5 Liapunov's Method for General Equations

In this section, we study boundedness of solutions for the nonlinear differential equation

$$x'(t) = f(t, x(t)) \qquad (5.1)$$

in $D = [0, \infty) \times \Re^n$. We assume that for any $(t_0, x_0) \in [0, \infty) \times \Re^n$, Eq. (5.1) has a unique solution $x(t, t_0, x_0)$ existing on $[t_0, \infty)$ with $x(t_0) = x_0$.

Note that to derive boundedness, we only need to control the solutions when they become large, so we will see that in most cases, the conditions on the wedges and on the Liapunov functions are only imposed for $|x| \geq M$, where $M > 0$ may be a big number. Of course, if those conditions are satisfied for all $x \in \Re^n$, then they are automatically satisfied for $|x| \geq M$ with any $M > 0$. Let

$$H = \{x \in \Re^n : |x| \geq M\},$$

where $M > 0$ is a constant, and let W_i be the scalar wedges as defined in Chapter 9.

Theorem 10.5.1 *Assume there exists a function $V(t, x)$ that is continuous and satisfies a Lipschitz condition in x, and, on $[0, \infty) \times H$,*

(a). $W_1(|x|) \leq V(t, x)$ with $W_1(r) \to \infty$ as $r \to \infty$, and

(b). $V'_{(5.1)}(t, x) \leq 0$.

If there is a constant $P > 0$ such that $V(t, x) \leq P$ for $|x| = M$ and $t \geq 0$ (M is from the definition of the set H), then every solution of Eq. (5.1) is bounded.

Proof. Let $x(t) = x(t, t_0, x_0)$ be any given solution, then we need to prove that $x(t)$ is bounded. Now, $V(t_0, x_0)$ is fixed and hence, as $W_1(r) \to \infty$, $r \to \infty$, there is an $r > 0$ such that $V(t_0, x_0) \leq W_1(r)$. If $|x(t)| \leq M$ for $t \geq t_0$, then $x(t)$ is bounded. If $|x(t)| \geq M$ for $t \geq t_0$, then the conditions (a) and (b) are satisfied, so we obtain

$$W_1(|x(t)|) \leq V(t, x(t)) \leq V(t_0, x_0) \leq W_1(r), \tag{5.2}$$

where W_1 is monotone, hence $x(t)$ is bounded because $|x(t)| \leq r$ for $t \geq t_0$. Finally, if there is an interval $[t_1, T]$ such that $|x(t_1)| = M$ and $|x(t)| \geq M$ on $[t_1, T]$, then on this interval, the conditions (a) and (b) imply

$$W_1(|x(t)|) \leq V(t, x(t)) \leq V(t_1, x(t_1)) \leq P, \tag{5.3}$$

therefore $|x(t)| \leq W_1^{-1}(P)$ on $[t_1, T]$. This argument can be repeated on any such interval $[t_1, T]$, therefore, $x(t)$ is also bounded in this case. This completes the proof.   ♠

Theorem 10.5.1 is a result for a single solution. When the set H in Theorem 10.5.1 is replaced by $\Re^n$, we have the following result for a set of solutions.

Theorem 10.5.2 *Assume there exists a function $V(t, x)$ that is continuous and satisfies a Lipschitz condition in x, and, on $[0, \infty) \times \Re^n$,*

(a). $W_1(|x|) \leq V(t, x)$ with $W_1(r) \to \infty$ as $r \to \infty$, and

(b). $V'_{(5.1)}(t, x) \leq 0$.

Then the solutions of Eq. (5.1) are equi-bounded.

Proof. For any $t_0 \geq 0$ and any $B_1 > 0$, we need to find a $B_2 = B_2(t_0, B_1) > 0$ such that $\{|x_0| \leq B_1, t \geq t_0\}$ imply $|x(t, t_0, x_0)| \leq B_2$. As $V(t_0, x)$ is continuous in x, there is a $K(t_0, B_1) > 0$ such that $|x| \leq B_1$ implies $V(t_0, x) \leq K(t_0, B_1)$. Now, we can find $B_2 = B_2(t_0, B_1) \geq B_1$ such that $K(t_0, B_1) \leq W_1(B_2)$. Then for $t \geq t_0$ and $|x_0| \leq B_1$,

$$W_1(|x(t)|) \leq V(t, x(t)) \leq V(t_0, x_0) \leq K(t_0, B_1) \leq W_1(B_2), \qquad (5.4)$$

hence $|x(t, t_0, x_0)| \leq B_2$ for $t \geq t_0$ and $|x_0| \leq B_1$, this verifies equi-boundedness and completes the proof. ♠

Similar to the results in Chapter 9, to get uniform boundedness, an additional condition is needed to reduce the dependence on t_0.

Theorem 10.5.3 *Assume there is a function $V(t, x)$ that is continuous and satisfies a Lipschitz condition in x, and, on $[0, \infty) \times H$,*

(a). $W_1(|x|) \leq V(t, x) \leq W_2(|x|)$ with $W_1(r) \to \infty$ as $r \to \infty$, and

(b). $V'_{(5.1)}(t, x) \leq 0$.

Then the solutions of Eq. (5.1) are uniformly bounded.

Proof. For any $B_1 \geq M$ (M is from the definition of the set H), we need to find a $B_2 = B_2(B_1) > 0$ such that $\{|x_0| \leq B_1, t \geq t_0 \geq 0\}$ imply $|x(t, t_0, x_0)| \leq B_2$. Now, we can find a $B_2 = B_2(B_1) \geq B_1$ such that $W_2(B_1) \leq W_1(B_2)$. Let $t_0 \geq 0$ and $|x_0| \leq B_1$. If $|x(t, t_0, x_0)| \leq B_1$ for $t \geq t_0$, then $|x(t, t_0, x_0)| \leq B_2$ for $t \geq t_0$. Next, if there is an interval $[t_1, T]$ such that $|x(t_1)| = B_1$ and $|x(t)| \geq B_1$ on $[t_1, T]$, then on this interval, $|x(t)| \geq B_1 \geq M$, thus the conditions (a) and (b) imply

$$W_1(|x(t)|) \leq V(t, x(t)) \leq V(t_1, x(t_1)) \leq W_2(|x(t_1)|) = W_2(B_1) \leq W_1(B_2),$$

hence $|x(t, t_0, x_0)| \leq B_2$ for $t \in [t_1, T]$. This argument can be repeated on any such interval $[t_1, T]$, therefore, $|x(t)| \leq B_2$ for $t \geq t_0$ and $|x_0| \leq B_1$, this verifies uniform boundedness and completes the proof. ♠

To obtain ultimate boundedness, the condition $V'_{(5.1)}(t, x) \leq 0$ will be replaced by some stronger conditions.

Theorem 10.5.4 *Assume there exists a function $V(t, x)$ that is continuous and satisfies a Lipschitz condition in x, and,*

(a). $W_1(|x|) \leq V(t, x)$ is satisfied on $[0, \infty) \times H$, with $W_1(r) \to \infty$ as $r \to \infty$, and

(b). $V'_{(5.1)}(t, x) \leq -cV(t, x)$ is satisfied on $[0, \infty) \times \Re^n$, where $c > 0$ a constant.

Then the solutions of Eq. (5.1) are equi-ultimately bounded.

Proof. Choose $B = M$ (M is from the definition of the set H). We need to show that for any $t_0 \geq 0$ and any $B_1 > 0$, there exists a $T = T(t_0, B_1) > 0$ such that $\{|x_0| \leq B_1, t \geq t_0+T\}$ imply $|x(t, t_0, x_0)| \leq B$. Now let $t_0 \geq 0$ and $|x_0| \leq B_1$. As $V(t_0, x)$ is continuous in x, there is a $K(t_0, B_1) > W_1(B) > 0$ such that $|x| \leq B_1$ implies $V(t_0, x) \leq K(t_0, B_1)$. From the condition (b), it follows that

$$V(t, x(t)) \leq V(t_0, x_0)e^{-c(t-t_0)} \leq K(t_0, B_1)e^{-c(t-t_0)}.$$

Solve $T = T(t_0, B_1) > 0$ from $K(t_0, B_1)e^{-cT} = W_1(B)$, that is, let

$$T = \frac{1}{c} \ln \frac{K(t_0, B_1)}{W_1(B)}.$$

Now, if $|x(t^*, t_0, x_0)| > B$ for some $t^* \geq t_0 + T$, then there is an interval $[t_1, T_1]$ with $t_1 > t_0+T$ such that $|x(t)| = |x(t, t_0, x_0)| > B = M$ on $[t_1, T_1]$. Then on this interval, the conditions (a) and (b) imply

$$W_1(B) < W_1(|x(t)|) \leq V(t, x(t)) \leq V(t_0, x_0)e^{-c(t-t_0)} \leq K(t_0, B_1)e^{-c(t-t_0)}$$
$$< K(t_0, B_1)e^{-cT} = W_1(B), \tag{5.5}$$

which is a contradiction. Therefore, $|x(t, t_0, x_0)| \leq B$ for $t \geq t_0 + T$ and $|x_0| \leq B_1$, this verifies equi-ultimate boundedness and completes the proof. ♠

Finally, let's use an additional condition to reduce the dependence on t_0.

Theorem 10.5.5 *Assume there exists a function $V(t, x)$ that is continuous and satisfies a Lipschitz condition in x, and, on $[0, \infty) \times H$,*

(a). $W_1(|x|) \leq V(t, x) \leq W_2(|x|)$ with $W_1(r) \to \infty$ as $r \to \infty$, and

(b). $V'_{(5.1)}(t, x) \leq -W_3(|x|)$.

Then the solutions of Eq. (5.1) are uniformly bounded and uniformly ultimately bounded.

Proof. First, we have uniform boundedness from Theorem 10.5.3. Next, we need to find a constant $B > 0$ such that for any $B_1 \geq M$ (M is from the definition of the set H), there exists a $T = T(B_1) > 0$ such that $\{|x_0| \leq B_1, t_0 \geq 0, t \geq t_0 + T\}$ imply $|x(t, t_0, x_0)| \leq B$. The following is similar to the proof of a corresponding result in Chapter 9. For any solution $x(t) = x(t, t_0, x_0)$ with $|x_0| \leq B_1$ and $t_0 \geq 0$, as long as $|x(t)| > M$ for $t \geq t_0$, then, as $V'(t, x(t)) \leq -W_3(|x(t)|)$ for some wedge W_3, we have

$$
\begin{aligned}
0 \leq V(t, x(t)) \ &\leq \ V(t_0, x_0) - \int_{t_0}^{t} W_3(|x(s)|)ds \\
&< \ W_2(|x_0|) - W_3(M)(t - t_0) \\
&\leq \ W_2(B_1) - W_3(M)(t - t_0), \quad (5.6)
\end{aligned}
$$

which leads to a contradiction when $W_2(B_1) - W_3(M)(t - t_0) = 0$, or when

$$
t = t_0 + \frac{W_2(B_1)}{W_3(M)} \overset{\text{def}}{=} t_0 + T, \quad \left(T = \frac{W_2(B_1)}{W_3(M)}\right) \quad (5.7)
$$

where $T = T(B_1)$. Therefore, there is a $\bar{t} \in [t_0, t_0 + T]$ such that $|x(\bar{t})| = |x(\bar{t}, t_0, x_0)| \leq M$. Now, we choose a constant $B > M$ such that $W_2(M) \leq W_1(B)$, and claim that $\{|x_0| \leq B_1, t \geq t_0 + T\}$ imply $|x(t, t_0, x_0)| \leq B$. If this is not true, then there is a $t^* \geq t_0 + T$ with $|x(t^*)| > B > M$. As $|x(\bar{t})| \leq M, \bar{t} \leq t_0 + T \leq t^*$, and $\bar{t} \neq t^*$, there is a $t_1, \bar{t} \leq t_1 < t^*$, such that $|x(t_1)| = M$ and $|x(t)| \geq M$ on $[t_1, t^*]$. Now, on this interval, the conditions (a) and (b) imply, for $t \in [t_1, t^*]$,

$$
W_1(|x(t)|) \leq V(t, x(t)) \leq V(t_1, x(t_1)) \leq W_2(|x(t_1)|) = W_2(M) \leq W_1(B).
$$

In particular, $W_1(B) < W_1(|x(t^*)|) \leq W_1(B)$, a contradiction. Therefore, $\{|x_0| \leq B_1, t_0 \geq 0, t \geq t_0 + T\}$ imply $|x(t, t_0, x_0)| \leq B$, which verifies uniform ultimate boundedness and completes the proof. ♠

Remark 10.5.6 Note that $W_1(|x|) \le V(t,x)$ and $V'_{(5.1)}(t,x) \le -cV(t,x)$ imply $V'_{(5.1)}(t,x) \le -cW_1(|x|)$, and cW_1 is a scalar wedge when $c > 0$. Thus the conditions in Theorem 10.5.4 are stronger than the conditions that $V(t,x)$ and $-V'_{(5.1)}(t,x)$ are positive definite.

Finally, we look at how to use Liapunov's method to derive boundedness. Note the difference that for the stabilities of the zero solution, we construct Liapunov functions for small $|x|$; while for boundedness, we construct Liapunov functions for large $|x|$.

Example 10.5.7 We modify Example 5.6.7 in Chapter 5 and consider the scalar differential equation

$$u'' + u' + g(u) = f(t), \tag{5.8}$$

for a continuous and bounded function $f(t)$. As we did in Example 5.6.7, we still try the Liapunov function

$$V(t, x_1, x_2) = \frac{1}{2}x_2^2 + \frac{1}{2}[x_1 + x_2]^2 + 2\int_0^{x_1} g(s)ds, \tag{5.9}$$

where $x_1 = u$ and $x_2 = u'$. This example is similar to Example 9.3.11 in Chapter 9 since it is not easy to construct the scalar wedges as defined there without further conditions on the function g.   ♠

Accordingly, we ask whether the constructions of scalar wedges from functions defined on $\Re^n$ (as given in Chapter 9 for stabilities) can also be extended here for boundedness. Note that for the stabilities of the zero solution, the wedges for $|x| \le 1$ are good enough. However, for boundedness, we need to construct wedges for $|x| \ge M$, where $M > 0$ may be a big number. Thus, the method of constructing a wedge in Chapter 9 related to positive definiteness is no longer valid for boundedness.

Let's make the following modification: Assume that $W : \{x \in \Re^n : |x| \ge M\} \to [0, \infty)$, where $M > 0$ is a constant, $W(x) > 0$ for $x \ne 0$, $\lim_{|x| \to \infty} W(x) = \infty$, and $W(x) \le V(t,x)$ for some function $V(t,x)$. Define

$$\alpha(r) = \min_{r \le |x|} W(x), \tag{5.10}$$

such that $\alpha : [M, \infty) \to (0, \infty)$ is nondecreasing, and $\lim_{r \to \infty} \alpha(r) = \infty$. Then, (see an exercise), a function $\omega_1(r)$ can be constructed on $[M, \infty) \to$

$(0, \infty)$ in such a way that $\omega_1(r) \leq \alpha(r)$ and ω_1 is continuous and strictly increasing. Thus, if $|x| \geq M$, then

$$V(t, x) \geq W(x) \geq \min_{|x| \leq |p|} W(p) = \alpha(|x|) \geq \omega_1(|x|).$$

Moreover, if necessary, $\omega_1(r)$ can be defined on $[0, \infty)$ with $\omega_1(0) = 0$, and hence bocomes a scalar wedge. Note that the method of constructing a wedge in Chapter 9 related to decrescentness is still valid for boundedness because the function is defined on $\Re^+$.

Consequently, we can weaken the conditions in Theorems 10.5.1–10.5.5 by replacing the scalar wedges with the $\Re^n$-wedges as defined in Chapter 9. That is, in the following theorems, we can let W_i be the $\Re^n$-wedges. We still denote $H = \{x \in \Re^n : |x| \geq M\}$ for a constant $M > 0$.

Theorem 10.5.8 *Assume there exists a function $V(t, x)$ that is continuous and satisfies a Lipschitz condition in x, and, on $[0, \infty) \times H$,*

(a). *$W_1(x) \leq V(t, x)$ with $W_1(x) \rightarrow \infty$ as $|x| \rightarrow \infty$, and*

(b). *$V'_{(5.1)}(t, x) \leq 0$.*

If there is a constant $P > 0$ such that $V(t, x) \leq P$ for $|x| = M$ and $t \geq 0$, then every solution of Eq. (5.1) is bounded. ♠

Theorem 10.5.9 *Assume there exists a function $V(t, x)$ that is continuous and satisfies a Lipschitz condition in x, and, on $[0, \infty) \times \Re^n$,*

(a). *$W_1(x) \leq V(t, x)$ with $W_1(x) \rightarrow \infty$ as $|x| \rightarrow \infty$, and*

(b). *$V'_{(5.1)}(t, x) \leq 0$.*

Then the solutions of Eq. (5.1) are equi-bounded. ♠

Theorem 10.5.10 *Assume there exists a function $V(t, x)$ that is continuous and satisfies a Lipschitz condition in x, and, on $[0, \infty) \times H$,*

(a). *$W_1(x) \leq V(t, x) \leq W_2(x)$ with $W_1(x) \rightarrow \infty$ as $|x| \rightarrow \infty$, and*

(b). *$V'_{(5.1)}(t, x) \leq 0$.*

Then the solutions of Eq. (5.1) are uniformly bounded. ♠

Theorem 10.5.11 *Assume there exists a function $V(t,x)$ that is continuous and satisfies a Lipschitz condition in x, and,*

(a). *$W_1(x) \leq V(t,x)$ is satisfied on $[0,\infty) \times H$, with $W_1(x) \to \infty$ as $|x| \to \infty$, and*

(b). *$V'_{(5.1)}(t,x) \leq -cV(t,x)$ is satisfied on $[0,\infty) \times \Re^n$, where $c > 0$ a constant.*

Then the solutions of Eq. (5.1) are equi-ultimately bounded.   ♠

Theorem 10.5.12 *Assume there is a function $V(t,x)$ that is continuous and satisfies a Lipschitz condition in x, and, on $[0,\infty) \times H$,*

(a). *$W_1(x) \leq V(t,x) \leq W_2(x)$ with $W_1(x) \to \infty$ as $|x| \to \infty$, and*

(b). *$V'_{(5.1)}(t,x) \leq -W_3(x)$.*

Then the solutions of Eq. (5.1) are uniformly bounded and uniformly ultimately bounded.   ♠

Example 10.5.13 Let's revisit Example 10.5.7, and consider

$$V(t,x) = V(t,x_1,x_2) = \frac{1}{2}x_2^2 + \frac{1}{2}[x_1 + x_2]^2 + 2\int_0^{x_1} g(s)ds, \qquad (5.11)$$

where $x = (x_1, x_2) \in \Re^2$. Now, if we define

$$W(x) = W(x_1,x_2) = \frac{1}{2}x_2^2 + \frac{1}{2}[x_1 + x_2]^2 + 2\int_0^{x_1} g(s)ds, \qquad (5.12)$$

then $W(0) = 0$, $W(x) > 0$ for $x \neq 0$, and $W(x) \to \infty$ as $|x| \to \infty$. Thus $V(t,x)$ is $\Re^2$-positive definite and $\Re^2$-decrescent. Next, we have

$$
\begin{aligned}
V'_{(5.8)}(t,x) &= x_2 x_2' + [x_1 + x_2][x_1' + x_2'] + 2g(x_1)x_1' \\
&= x_2[-x_2 - g(x_1) + f(t)] + [x_1 + x_2]x_2 \\
&\quad + [x_1 + x_2][-x_2 - g(x_1) + f(t)] + 2g(x_1)x_2 \\
&= -x_2^2 - x_1 g(x_1) + (x_1 + 2x_2)f(t). \qquad (5.13)
\end{aligned}
$$

Now, the function f is bounded, thus to get an $\Re^2$-wedge for $V'_{(5.8)}(t,x)$, we assume further that

$$\lim_{x_1 \to \infty} g(x_1) > 2\max_{t \in \Re} |f(t)|.$$

Then,

$$\lim_{x_1^2 + x_2^2 \to \infty} \frac{|(x_1 + 2x_2)f(t)|}{x_1 g(x_1) + x_2^2}$$

$$\leq \lim_{x_1^2 + x_2^2 \to \infty} \frac{|x_1 f(t)|}{x_1 g(x_1) + x_2^2} + \lim_{x_1^2 + x_2^2 \to \infty} \frac{|2x_2 f(t)|}{x_1 g(x_1) + x_2^2} \leq \frac{1}{2}. \quad (5.14)$$

Therefore, there is an $M > 0$ such that $|(x_1 + 2x_2)f(t)| \leq \frac{1}{2}[x_1 g(x_1) + x_2^2]$ when $|x| \geq M$. Hence, if $|x| \geq M$, then

$$V'_{(5.8)}(t, x) \quad \leq \quad -x_2^2 - x_1 g(x_1) + (x_1 + 2x_2)f(t)$$

$$\leq \quad -\frac{1}{2}[x_1 g(x_1) + x_2^2], \quad (5.15)$$

where $W_1(x) = \frac{1}{2}[x_1 g(x_1) + x_2^2]$ is an $\Re^2$-wedge. Thus, Theorem 10.5.12 can be applied to derive uniform boundedness and uniform ultimate boundedness. ♠

The uniform boundedness and uniform ultimate boundedness in Example 10.5.13 is derived under the condition that $\lim_{x_1 \to \infty} g(x_1) > 2 \max_{t \in \Re} |f(t)|$. However, the same conclusions can also be derived if we assume

$$\lim_{x_1 \to \infty} g(x_1) > \max_{t \in \Re} |f(t)|. \quad (5.16)$$

See an exercise. Next, if we compare Eq. (5.8) in Example 10.5.7 (or Example 10.5.13) with the general second-order differential equation (2.8) in Chapter 1 derived using Newton's second law of motion, we find for Eq. (5.8) that the function $f(t)$ is the externally applied force. Thus, the interpretation in physics for the condition (5.16) and the result in Example 10.5.13 is that if the restoring force, in terms of the function g, is greater than the externally applied force, then the solutions will be uniformly bounded and uniformly ultimately bounded.

Exercises 10.5

1. Verify that $\alpha(r)$ defined in (5.10) satisfies $\lim_{r \to \infty} \alpha(r) = \infty$.

2. Verify that for $\alpha(r)$ defined in (5.10), a function $\omega_1(r)$ can be constructed on $[M, \infty) \to (0, \infty)$ in such a way that $\omega_1(r) \leq \alpha(r)$ and ω_1 is continuous and strictly increasing.

3. Prove Theorem 10.5.8.

4. Prove Theorem 10.5.9.

5. Prove Theorem 10.5.10.

6. Prove Theorem 10.5.11.

7. Prove Theorem 10.5.12.

8. For $V(t,x)$ given in (5.11), one has

$$V(t,x_1,x_2) \geq \frac{1}{2}x_2^2 + \frac{1}{2}[x_1^2 + 2x_1x_2 + x_2^2] = \frac{1}{2}x_1^2 + x_1x_2 + x_2^2. \quad (5.17)$$

From $2|AB| \leq A^2 + B^2$ and let $A = \sqrt{r}x_1$, $B = \frac{x_2}{\sqrt{r}}$ for $r > 0$, one obtains $2|x_1x_2| \leq rx_1^2 + \frac{x_2^2}{r}$, or $x_1x_2 \geq -\frac{r}{2}x_1^2 - \frac{1}{2r}x_2^2$. Then,

$$V(t,x_1,x_2) \geq \frac{1}{2}x_1^2 + x_1x_2 + x_2^2 \geq (\frac{1}{2} - \frac{r}{2})x_1^2 + (1 - \frac{1}{2r})x_2^2. \quad (5.18)$$

Select r such that V is positive definite with a scalar wedge.

9. Verify that for the $W(x)$ defined in (5.12), one has $W(0) = 0$, $W(x) > 0$ for $x \neq 0$, and $W(x) \to \infty$ as $|x| \to \infty$.

10. Prove in Example 10.5.13 that

$$\lim_{x_1^2+x_2^2 \to \infty} \frac{|(x_1 + 2x_2)f(t)|}{x_1g(x_1) + x_2^2} \leq \frac{1}{2}.$$

11. In Example 10.5.13, derive uniform boundedness and uniform ultimate boundedness under the condition that $\lim_{x_1 \to \infty} g(x_1) > \max_{t \in \Re} |f(t)|$.

12. When the function V is autonomous, state and prove the corresponding results concerning boundedness properties without using the notion of wedges. (Look at the corresponding results in Chapter 5 as a reference.)

Chapter 11

Periodic Solutions

11.1 Introduction

So far, we have already seen some discussions about periodic solutions. For example, for planar autonomous differential equations, when the origin is a center, there are periodic orbits around the origin. These periodic orbits are periodic solutions for planar autonomous differential equations. Examples in applications include an undamped simple pendulum, where the pendulum keeps "oscillating" indefinitely (that is, periodically) around the equilibrium after being perturbed. Other models include radio circuits, temperature distribution, or chemical and biological oscillations. Therefore, the study of periodic solutions and their applications is a very important subject in differential equations.

The importance of periodic solutions can also be seen from the Poincaré-Bendixson theorem and the Lorenz system. That is, if there are some stable periodic solutions, then typically other solutions will pile up at these stable periodic solutions, such that the system will behave in an orderly fashion, as stated in the Poincaré-Bendixson theorem. Otherwise, if a system has bounded solutions but has no stable periodic solutions, then it is an indication that these bounded solutions may behave in a strange and chaotic fashion. This can be seen from the Lorenz system.

Now, we study periodic solutions for the general differential equation

$$x'(t) = f(t, x(t)), \quad t \geq 0, \tag{1.1}$$

in $D = [0, \infty) \times \Re^n$, where $f(t, x)$ is continuous and satisfies at least a local
Lipschitz condition with respect to x on D. We assume, for simplicity, that
solutions start from $t_0 = 0$. We also assume that for any $x_0 \in \Re^n$, Eq. (1.1)
has a unique solution $x(t, 0, x_0)$ existing on $[0, \infty)$ with $x(0, 0, x_0) = x_0$. In
order to obtain periodic solutions, it is also necessary to assume that the
function f is periodic in t, that is, there is a constant $T > 0$ such that

$$f(t + T, x) = f(t, x),$$

for $(t, x) \in D$. Note that this includes the case when f is autonomous,
for which we may obtain periodic solutions with periods that are not pre-
determined. For example, for the planar autonomous differential equations
in Chapter 4, the solutions for Case $(IIIb)$ are given by $a_1 \cos \beta t + a_2 \sin \beta t$,
hence they are periodic of periods $\frac{2\pi}{\beta}$, which are determined by β. Another
example of autonomous differential equations is an undamped simple pen-
dulum, where the period is determined by how far the pendulum is moved
away initially from the equilibrium.

   To begin, we first give some basic results concerning the search of periodic
solutions.

Lemma 11.1.1 *Assume that for a constant $T > 0$, $f(t+T, x) = f(t, x)$ for
$(t, x) \in D$.*

   (a). If $x(t)$ is a solution of Eq. (1.1), then so is $x(t + T)$, $t \geq 0$.

   *(b). Let $x(t, 0, x_0)$ be a solution of Eq. (1.1) with $x(0, 0, x_0) = x_0$. Then
   $x(t, 0, x_0)$ is T-periodic if and only if $x(T, 0, x_0) = x_0$.*

Proof. (a): Let $y(t) = x(t + T)$, $t \geq 0$. Then for $t \geq 0$,

$$y'(t) = x'(t + T) = f(t + T, x(t + T)) = f(t, x(t + T)) = f(t, y(t)),$$

thus $x(t + T)$ is also a solution of Eq. (1.1).

   (b): If $x(t, 0, x_0)$ is a T-periodic solution of Eq. (1.1), then $x(T, 0, x_0) = x(0, 0, x_0) = x_0$. On the other hand, if $x(T, 0, x_0) = x_0$, then, from (a), $y(t) = x(t + T, 0, x_0)$ is also a solution of Eq. (1.1) with $y(0) = x(T, 0, x_0) = x_0$.
By uniqueness, $y(t) = x(t, 0, x_0)$, or $x(t + T, 0, x_0) = x(t, 0, x_0)$, $t \geq 0$, thus
$x(t, 0, x_0)$ is T-periodic. This completes the proof.   ♠

   Recall from Chapter 2 that a continuous function $x(t, 0, x_0)$ is a solution
of Eq. (1.1) if and only if

$$x(t) = x_0 + \int_0^t f(s, x(s))ds, \quad t \geq 0. \tag{1.2}$$

Accordingly, we can define a mapping $P : \Re^n \to \Re^n$ such that for $x_0 \in \Re^n$ and for the unique solution $x(t) = x(t, 0, x_0)$ with $x(0, 0, x_0) = x_0$, define

$$P(x_0) = x(T) = x_0 + \int_0^T f(s, x(s))ds, \qquad (1.3)$$

see **Figure 11.1**.

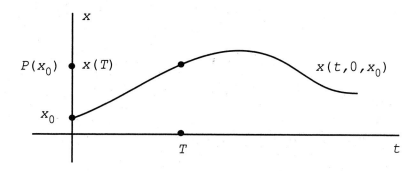

Figure 11.1: The mapping P

Now, Lemma 11.1.1(b) indicates that Eq. (1.1) has a T-periodic solution if and only if there exists an $x_0 \in \Re^n$ such that $P(x_0) = x_0$; that is, the mapping P has a **fixed point**. We formulate this as follows.

Lemma 11.1.2 *Assume that for a constant $T > 0$, $f(t + T, x) = f(t, x)$ for $(t, x) \in D$. Then Eq. (1.1) has a T-periodic solution if and only if the mapping $P : \Re^n \to \Re^n$ defined in (1.3) has a fixed point.* ♠

Lemma 11.1.2 provides a very useful approach for deriving periodic solutions since it reduces the search of periodic solutions to that of fixed points of the mapping P, for which some well-known fixed point theorems from functional analysis can be applied.

This chapter is organized as follows: In Section 2, we derive the existence of periodic solutions for general linear differential equations. First, we derive periodic solutions using the eigenvalues of $U(T, 0)$, where $U(t, s)$ is the fundamental matrix solution of linear homogeneous differential equations. Then we derive periodic solutions from the bounded solutions. Periodic solutions of linear differential equations with linear and nonlinear perturbations are also given. In Section 3, we look at general nonlinear differential equations.

Since using eigenvalues is not applicable now, we extend the idea of deriving periodic solutions using the boundedness. First, we present some Massera-type results for one-dimensional and two-dimensional differential equations, whose proofs are generally not extendible to higher dimensional cases. Then, for general n-dimensional differential equations, we apply Horn's fixed point theorem to the mapping P defined by (1.3) and obtain fixed points, and hence periodic solutions, under the assumption that the solutions are equi-ultimate bounded.

Exercises 11.1

1. Prove Lemma 11.1.2.

2. Let P be a mapping. Prove that if for some positive integer m, the mapping P^m has a unique fixed point, then P has a fixed point.

11.2 Linear Differential Equations

Here, we study the linear differential equation

$$x'(t) = A(t)x(t) + f(t), \quad t \geq 0, \quad x \in \Re^n, \tag{2.1}$$

where $A(t)$, $f(t)$ are continuous on $\Re^+ = [0, \infty)$, and $A(t+T) = A(t)$, $f(t+T) = f(t)$, $t \geq 0$, for some constant $T > 0$.

Using the results from Chapter 3, we know that the unique solution of Eq. (2.1) for the initial data $(0, x_0)$ is given by the variation of parameters formula

$$
\begin{aligned}
x(t) &= U(t,0)\left[x_0 + \int_0^t U^{-1}(s,0)f(s)ds\right] \\
&= U(t,0)x_0 + \int_0^t U(t,s)f(s)ds, \quad t \geq 0.
\end{aligned}
\tag{2.2}
$$

First we study the linear homogeneous differential equation

$$x'(t) = A(t)x(t), \tag{2.3}$$

where the unique solution is given by $x(t) = U(t,0)x_0$ for the initial data $(0, x_0)$. Now, for the mapping P defined by (1.3) to have a fixed point x_0, we need

$$x_0 = U(T,0)x_0, \quad \text{or} \quad [E - U(T,0)]x_0 = 0, \tag{2.4}$$

where E is the $n \times n$ identity or unit matrix. Thus we have the following result, whose proof is left as an exercise.

Theorem 11.2.1 *Let $A(t)$ be continuous and T-periodic. Then Eq. (2.3) has a nonzero T-periodic solution if and only if 1 is an eigenvalue of $U(T,0)$.*

♠

For the nonhomogeneous differential equation (2.1), note that the zero is not a solution when $f \neq 0$. Therefore, any periodic solution of Eq. (2.1) will be nonzero when $f \neq 0$. Now, the mapping P has a fixed point x_0 if and only if

$$x_0 = U(T,0)x_0 + \int_0^T U(T,s)f(s)ds, \tag{2.5}$$

or

$$[E - U(T,0)]x_0 = \int_0^T U(T,s)f(s)ds. \tag{2.6}$$

Accordingly, we have the following result.

Theorem 11.2.2 *Let $A(t)$ be continuous and T-periodic. Then Eq. (2.1) has a T-periodic solution for any continuous and T-periodic function f if and only if 1 is not an eigenvalue of $U(T,0)$.*

Proof. If 1 is not an eigenvalue of $U(T,0)$, then the matrix $[E - U(T,0)]$ is nonsingular. Hence for any continuous and T-periodic function f, (2.6) has a unique solution x_0, which is a fixed point of the mapping P. Therefore, Lemma 11.1.2 can be applied to derive a periodic solution of Eq. (2.1).

On the other hand, assume that Eq. (2.1) has a T-periodic solution for any continuous and T-periodic function f. For every $y \in \Re^n$, let $f(s) = U(s,0)yg(s)$ for $s \in [0,T]$, where $g : [0,T] \to \Re$ is continuous such that $g(0) = g(T) = 0$, and

$$\int_0^T g(s)ds = 1.$$

Then f can be extended to become a T-periodic continuous function on $\Re$ which we still denote by f. Now, Eq. (2.1) has a T-periodic solution for this f, thus (2.6) has a solution x_0 for this f. Therefore,

$$[E - U(T,0)]x_0 = \int_0^T U(T,s)f(s)ds$$

$$= \int_0^T U(T,s)U(s,0)yg(s)ds$$

$$= \int_0^T U(T,0)yg(s)ds$$

$$= U(T,0)y\int_0^T g(s)ds$$

$$= U(T,0)y. \tag{2.7}$$

Since $U(T,0)$ is nonsingular and $y \in \Re^n$ is arbitrary, the right-hand side of (2.7) covers the whole $\Re^n$ when y takes all vectors in $\Re^n$ as values. Thus the range of $[E-U(T,0)]$ is $\Re^n$; therefore, $[E-U(T,0)]$ is nonsingular, hence 1 is not an eigenvalue of $U(T,0)$. This completes the proof.   ♠

From the above proof we see that the periodic solution in Theorem 11.2.2 is uniquely determined, and we state it as follows.

Theorem 11.2.2.a *Let $A(t)$ be continuous and T-periodic. Then Eq. (2.1) has a unique T-periodic solution for any continuous and T-periodic function f if and only if 1 is not an eigenvalue of $U(T,0)$.*   ♠

Consequently, we have

Theorem 11.2.2.aa *Let $A(t)$ be continuous and T-periodic. Then the following statements are equivalent.*

1. *1 is not an eigenvalue of $U(T,0)$.*

2. *Eq. (2.1) has a unique T-periodic solution for any continuous and T-periodic function f.*

3. *Eq. (2.1) has a T-periodic solution for any continuous and T-periodic function f.*   ♠

Next, we look at Eq. (2.3) with linear or nonlinear perturbations,

$$x'(t) = A(t)x(t) + f(t,x(t)), \quad t \geq 0, \quad x \in \Re^n, \tag{2.8}$$

where $f(t,x)$ is also T-periodic in t. To derive periodic solutions for Eq. (2.8), or for general nonlinear differential equations, we should use some fixed point theorems from functional analysis that are listed in the Appendix.

Theorem 11.2.3 *Assume that Eq. (2.8) is T-periodic in t and that $f(t, x)$ is continuous on $[0, \infty) \times \Re^n$ and satisfies a weak Lipschitz condition with respect to x. If 1 is not an eigenvalue of $U(T, 0)$ and if $|f(t, x)| \leq C$, $(t, x) \in [0, \infty) \times \Re^n$, for some constant $C > 0$, then Eq. (2.8) has a T-periodic solution.*

Proof. First, the conditions guarantee that solutions of Eq. (2.8) exist on $[0, \infty)$ and are uniquely determined. Next, since 1 is not an eigenvalue of $U(T, 0)$, we can define a mapping $F : \Re^n \to \Re^n$ by

$$F(x_0) = [E - U(T, 0)]^{-1}\{x(T, 0, x_0) - U(T, 0)x_0\}, \quad x_0 \in \Re^n, \qquad (2.9)$$

where x is the unique solution through $(0, x_0)$ given by

$$x(t) = x(t, 0, x_0) = U(t, 0)x_0 + \int_0^t U(t, s)f(s, x(s))ds, \quad t \geq 0. \qquad (2.10)$$

Consequently,

$$x(T, 0, x_0) - U(T, 0)x_0 = \int_0^T U(T, s)f(s, x(s))ds,$$

and then from (2.9) we obtain

$$
\begin{aligned}
|F(x_0)| &\leq \ |[E - U(T, 0)]^{-1}||x(T, 0, x_0) - U(T, 0)x_0| \\
&\leq \ |[E - U(T, 0)]^{-1}| \int_0^T |U(T, s)f(s, x(s))|ds \\
&\leq \ |[E - U(T, 0)]^{-1}|CT \max_{s \in [0, T]} |U(T, s)| \\
&\stackrel{\text{def}}{=} \ M, \qquad\qquad\qquad\qquad\qquad\qquad\qquad\qquad (2.11)
\end{aligned}
$$

where $M > 0$ is a constant. Now, $B = \{x \in \Re^n : |x| \leq M\}$ is nonempty, convex, and compact. Also, we have $F : B \to B$ and F is continuous. Thus Brouwer's fixed point theorem (see the Appendix) can be applied to the mapping F on B to get a fixed point $x_0 = F(x_0)$. Hence,

$$x_0 = [E - U(T, 0)]^{-1}\{x(T, 0, x_0) - U(T, 0)x_0\}, \qquad (2.12)$$

which implies

$$[E - U(T, 0)]x_0 = x(T, 0, x_0) - U(T, 0)x_0, \qquad (2.13)$$

or,

$$x_0 = x(T, 0, x_0).$$

Therefore, from Lemma 11.1.2, Eq. (2.8) has a T-periodic solution. This completes the proof.   ♠

The following is a generalization of Theorem 11.2.3, see Amann [1990].

Theorem 11.2.4 *Assume that Eq. (2.8) is T-periodic in t and that $f(t, x)$ is continuous on $[0, \infty) \times \Re^n$ and satisfies a weak Lipschitz condition with respect to x. If 1 is not an eigenvalue of $U(T, 0)$ and if*

$$\lim_{|x| \to \infty} \frac{|f(t, x)|}{|x|} = 0 \quad \text{uniformly for } t \in [0, T], \tag{2.14}$$

then Eq. (2.8) has a T-periodic solution.

Proof. Similar to the proof of Theorem 11.2.3, we will find a set to apply Brouwer's fixed point theorem to the mapping F defined in (2.9). Define

$$\alpha = \max\{|U(t, s)| : 0 \le s \le t \le T\}$$

and let $\varepsilon \in (0, 1)$ be arbitrary. From (2.14), there exists a $\beta = \beta(\varepsilon) > 0$ such that

$$|f(t, x)| \le \beta + \varepsilon|x|, \quad (t, x) \in [0, T] \times \Re^n. \tag{2.15}$$

Then, from (2.10), we obtain

$$|x(t, 0, x_0)| \le \alpha|x_0| + \alpha\beta T + \varepsilon\alpha \int_0^t |x(s)|ds, \ \ 0 \le t \le T, \tag{2.16}$$

hence by Gronwall's inequality we get

$$|x(t, 0, x_0)| \le [\alpha|x_0| + \alpha\beta T] \exp(\varepsilon\alpha T), \ \ 0 \le t \le T. \tag{2.17}$$

Then, using (2.10) and (2.15) again, we obtain

$$
\begin{aligned}
|x(T, 0, x_0) - U(T, 0)x_0| &\le \left| \int_0^T U(T, s)f(s, x(s))ds \right| \\
&\le \alpha T\Big[\beta + \varepsilon\big([\alpha|x_0| + \alpha\beta T]\exp(\varepsilon\alpha T)\big)\Big],
\end{aligned}
$$

hence

$$\limsup_{|x_0| \to \infty} \frac{|x(T,0,x_0) - U(T,0)x_0|}{|x_0|} \le \alpha T \varepsilon \alpha \exp(\varepsilon \alpha T). \qquad (2.18)$$

Now, $\varepsilon > 0$ is arbitrary, so we have

$$\lim_{|x_0| \to \infty} \frac{|x(T,0,x_0) - U(T,0)x_0|}{|x_0|} = 0, \qquad (2.19)$$

which implies

$$\lim_{|x_0| \to \infty} \frac{|F(x_0)|}{|x_0|} = 0, \qquad (2.20)$$

where F is the mapping defined in (2.9). Consequently, there exists a $\rho > 0$ such that

$$|F(x)| \le \rho + \frac{|x|}{2}, \quad x \in \Re^n.$$

Therefore, solving k from $\rho + \frac{k}{2} = k$, we find that F maps the set $\{x \in \Re^n : |x| \le 2\rho\}$ into itself, hence Brouwer's fixed point theorem can be applied to the mapping F to obtain a fixed point. This completes the proof. ♠

When $A(t) = A$ is a constant matrix, $U(T,0) = e^{TA}$. Thus the corresponding results can be stated by using the matrix e^{TA}. Moreover, for the set of all eigenvalues of e^{TA}, denoted by $\sigma(e^{TA})$, we have, from the spectral mapping theorem 3.3.14 in Chapter 3,

$$\sigma(e^{TA}) = \{e^{T\lambda} : \lambda \in \sigma(A)\}. \qquad (2.21)$$

Now, note that $1 = e^{T\lambda}$ implies $T\lambda = 2\pi k i$ or $\lambda = \frac{2\pi k i}{T}$, $k = 0, \pm 1, \pm 2, \cdots$, $(i = \sqrt{-1})$. Thus from (2.21), we obtain the following result concerning the eigenvalues of the matrix e^{TA}.

Lemma 11.2.5 1 *is not an eigenvalue of e^{TA} if and only if*

$$\sigma(A) \cap \{\frac{2\pi k i}{T} : k = 0, \pm 1, \pm 2, \cdots\} = \varnothing \quad \text{(empty set)}. \quad ♠ \qquad (2.22)$$

Therefore, periodic solutions for

$$x'(t) = Ax(t) + f(t) \qquad (2.23)$$

and

$$x'(t) = Ax(t) + f(t, x(t)) \tag{2.24}$$

can be obtained by checking the condition (2.22). We list them below.

Theorem 11.2.6 *The following statements are equivalent.*

(a). $\sigma(A) \cap \{\frac{2\pi ki}{T} : k = 0, \pm 1, \pm 2, \cdots\} = \emptyset$.

(b). *Eq. (2.23) has a unique T-periodic solution for any continuous and T-periodic function f.*

(c). *Eq. (2.23) has a T-periodic solution for any continuous and T-periodic function f.*   ♠

Theorem 11.2.7 *Assume that $f(t, x)$ is T-periodic in t, continuous on $[0, \infty) \times \Re^n$ and satisfies a weak Lipschitz condition with respect to x, and satisfies (2.14), or $|f(t, x)| \leq C$, $(t, x) \in [0, \infty) \times \Re^n$, for some constant $C > 0$. If $\sigma(A) \cap \{\frac{2\pi ki}{T} : k = 0, \pm 1, \pm 2, \cdots\} = \emptyset$, then Eq. (2.24) has a T-periodic solution.*   ♠

The above represents some results of deriving periodic solutions using the eigenvalues of $U(T, 0)$ (or of e^{TA}).

Another approach to derive periodic solutions is using the boundedness of solutions.

Theorem 11.2.8 *Let $A(t)$, $f(t)$ be continuous and T-periodic. Then Eq. (2.1) has a T-periodic solution if and only if Eq. (2.1) has a solution that is bounded on $[0, \infty)$.*

Proof. If Eq. (2.1) has a T-periodic solution, then it is bounded on $[0, \infty)$.

On the other hand, if Eq. (2.1) has a solution that is bounded on $[0, \infty)$, then let's prove that Eq. (2.1) has a T-periodic solution. Suppose that Eq. (2.1) has no T-periodic solution, then there is no $x_0 \in \Re^n$ such that

$$x_0 = x(T, 0, x_0) = U(T, 0)x_0 + \int_0^T U(T, s)f(s)ds, \tag{2.25}$$

or, for $y_0 = \int_0^T U(T, s)f(s)ds$,

$$[E - U(T, 0)]x_0 = y_0 \tag{2.26}$$

has no solution x_0 in $\Re^n$. Moreover, if Eq. (2.1) has no T-periodic solution, then from Theorem 11.2.2, we find that 1 is an eigenvalue of $U(T,0)$, and hence of $U(T,0)^{tr}$ (here, because we used T for the period, so we use tr for the transpose). Therefore, from the Fredholm theorem in linear algebra, there is an eigenvector w of $U(T,0)^{tr}$ corresponding to the eigenvalue 1 of $U(T,0)^{tr}$ such that $y_0^{tr} w \neq 0$. That is, there is a $w \in \Re^n$ such that

$$U(T,0)^{tr} w = w, \quad y_0^{tr} w \neq 0. \tag{2.27}$$

Let $x(t) = x(t,0,x_0)$ be any solution of Eq. (2.1). If we can prove that $x(t)$ is unbounded on $[0,\infty)$, then it will contradict the assumption that Eq. (2.1) has a solution that is bounded on $[0,\infty)$, and hence completes the proof. To this end, let's define $x_k(t) = x(t + kT)$, $k = 1, 2, \cdots$, $t \geq 0$. Then from Lemma 11.1.1(a), $x_k(t)$ is a solution of Eq. (2.1) with $x_k(0) = x(kT)$. Therefore,

$$x((k+1)T) = x_k(T) \;=\; U(T,0)x(kT) + \int_0^T U(T,s)f(s)ds$$
$$=\; U(T,0)x(kT) + y_0, \tag{2.28}$$

then an induction using (2.28) shows that

$$x_k(T) = x((k+1)T) = [U(T,0)]^{k+1}x_0 + \sum_{i=0}^{k}[U(T,0)]^i y_0. \tag{2.29}$$

Now, from (2.27) and (2.29), we have

$$w^{tr}x((k+1)T) \;=\; w^{tr}[U(T,0)]^{k+1}x_0 + \sum_{i=0}^{k} w^{tr}[U(T,0)]^i y_0$$
$$=\; x_0^{tr}\{[U(T,0)]^{k+1}\}^{tr}w + \sum_{i=0}^{k} y_0^{tr}\{[U(T,0)]^i\}^{tr}w$$
$$=\; x_0^{tr}[U(T,0)^{tr}]^{k+1}w + \sum_{i=0}^{k} y_0^{tr}[U(T,0)^{tr}]^i w$$
$$=\; x_0^{tr}w + \sum_{i=0}^{k} y_0^{tr}w$$
$$=\; x_0^{tr}w + (k+1)y_0^{tr}w. \tag{2.30}$$

Since $y_0^{tr} w \neq 0$, the right-hand side of (2.30) goes to ∞ as $k \to \infty$. Therefore, $|x((k+1)T)|$ goes to ∞ as $k \to \infty$; that is, $x(t) = x(t,0,x_0)$ is unbounded on $[0,\infty)$. This completes the proof. ♠

Note that if $f(t)$ is continuous and T-periodic, then $|f(t)| \leq M$ for some constant $M > 0$. Now, from Theorem 10.2.9 in Chapter 10 concerning the boundedness, we have

Theorem 11.2.9 *Let $A(t)$, $f(t)$ be continuous and T-periodic. If the zero solution of Eq. (2.3) is uniformly asymptotically stable, (or if the solutions of Eq. (2.3) are uniformly ultimately bounded and $A(t)$ is bounded on $\Re^+$; see Theorem 10.2.1(D)), then Eq. (2.1) has a T-periodic solution.*   ♠

When $A(t) = A$ is a constant matrix, using results in Chapter 10 we derive the following.

Theorem 11.2.9.a *Let $f(t)$ be continuous and T-periodic. If the zero solution of $x'(t) = Ax(t)$ is asymptotically stable, or if the solutions of $x'(t) = Ax(t)$ are equi-ultimately bounded, or if each eigenvalue of the matrix A has a negative real part, then the equation $x'(t) = Ax(t) + f(t)$ has a T-periodic solution.*   ♠

Exercises 11.2

1. Let $A(t) = A(t + T)$ for some constant $T > 0$ and let $A(t)$ be odd, that is, $A(-t) = -A(t)$. Use an induction to prove that the functions $\{x_m(t, 0, x_0)\}_{m \geq 1}$ in Picard's approximations of $x' = A(t)x$ are even and T-periodic. Then show that all solutions $x(t, 0, x_0)$ are even and T-periodic.

2. Prove Theorem 11.2.1.

3. Prove Theorem 11.2.2.a.

4. Prove Theorem 11.2.2.aa.

5. In the proof of Theorem 11.2.3, verify that $B = \{x \in \Re^n : |x| \leq M\}$ is nonempty, compact, and convex, $F : B \to B$, and F is continuous.

6. Prove Lemma 11.2.5.

7. Prove Theorem 11.2.6.

8. Prove Theorem 11.2.7.

9. Use an induction to verify (2.29).

10. Prove Theorem 11.2.9.

11. Prove Theorem 11.2.9.a.

11.3   Nonlinear Differential Equations

In the above, we derived periodic solutions for linear differential equations using two approaches: the eigenvalues and the boundedness. For general nonlinear differential equations, the notion of "eigenvalues" is not applicable. Thus, for nonlinear equations, we will try to extend the idea of deriving periodic solutions from the boundedness. First we present a result of Massera [1950] for scalar equations.

Theorem 11.3.1 (Massera) *Assume that Eq. (1.1) is T-periodic in t and that $f(t,x)$ is continuous on $[0,\infty) \times \Re^n$ and satisfies a weak Lipschitz condition with respect to x. If $n = 1$ (that is, Eq. (1.1) is a scalar equation) and if Eq. (1.1) has a solution that is bounded on $[0,\infty)$, then Eq. (1.1) has a T-periodic solution.*

Proof. Let $x(t)$ be the bounded solution on $[0,\infty)$ with $|x(t)| \leq C$ for some constant $C > 0$. Then from Lemma 11.1.1(a), for $m = 1, 2, \cdots$, $x_m(t) \overset{\text{def}}{=} x(t + mT)$ is also a solution on $[0,\infty)$, and $|x_m(t)| \leq C$. If $x_1(0) = x(0)$, then from Lemma 11.1.1(b), $x(t)$ is T-periodic. If $x_1(0) \neq x(0)$, then, since $x \in \Re$, we may assume $x_1(0) > x(0)$, (the case for $x_1(0) < x(0)$ is similar). Now, by uniqueness, $x_1(t) > x(t)$, $t \geq 0$. Plugging in $t = mT$, we obtain $x_{m+1}(0) = x((m+1)T) = x_1(mT) > x(mT) = x_m(0)$. By uniqueness again, $x_{m+1}(t) > x_m(t)$, $t \geq 0$. Thus, for each $t \geq 0$ fixed, $\{x_m(t)\}_{m \geq 1}$ is an increasing bounded sequence of numbers, hence $\{x_m(t)\}_{m \geq 1}$ converges to some $y(t)$ as $m \to \infty$. Next, f is T-periodic in t, then $|f(t,x)| \leq C_1$ (a constant) for $t \geq 0$ and $|x| \leq C$. Thus

$$|x'_m(t)| = |f(t, x_m(t))| \leq C_1,$$

and then the mean value theorem implies

$$|x_m(t) - x_m(s)| \leq C_1 |t - s|, \quad t, s \geq 0, \ m \geq 1.$$

Accordingly, on any compact t-interval in $[0,\infty)$, the sequence of functions $\{x_m(t)\}_{m \geq 1}$ is uniformly bounded and equi-continuous. Then from Arzela-Ascoli's theorem (see the Appendix), $\{x_m(t)\}_{m \geq 1}$ has a uniformly convergent subsequence. But since $\{x_m(t)\}_{m\geq 1}$ is monotone, it is itself uniformly convergent to $y(t)$ on any compact interval. Consequently, $y(t)$ is continuous on $[0, \infty)$. Next, for any fixed $t \geq 0$, we can use the weak Lipschitz condition to obtain

$$
\begin{aligned}
&\left| \int_0^t f(s, x_m(s))ds - \int_0^t f(s, y(s))ds \right| \\
&\leq \int_0^t |f(s, x_m(s)) - f(s, y(s))|ds \\
&\leq \int_0^t k(s)|x_m(s) - y(s)|ds \\
&\leq t \max_{s \in [0,t]} k(s)|x_m(s) - y(s)| \to 0, \quad m \to \infty,
\end{aligned}
\tag{3.1}
$$

(where $k(s)$ is from the weak Lipschitz condition). Now, from Chapter 2,

$$
x_m(t) = x_m(0) + \int_0^t f(s, x_m(s))ds, \quad t \geq 0.
\tag{3.2}
$$

Thus, if we let $m \to \infty$ in (3.2), then (3.1) implies, for each $t \geq 0$,

$$
y(t) = y(0) + \int_0^t f(s, y(s))ds,
\tag{3.3}
$$

which implies that $y(t)$ is a solution of Eq. (1.1). Finally, observe that

$$
y(T) = \lim_{m \to \infty} x_m(T) = \lim_{m \to \infty} x(T + mT) = \lim_{m \to \infty} x_{m+1}(0) = y(0),
\tag{3.4}
$$

therefore $y(t)$ is a T-periodic solution of Eq. (1.1). This completes the proof.
♠

Note that the above proof depends on the condition that $n = 1$, or $x \in \Re$, because the argument that "if $x_1(0) \neq x(0)$, then $x_1(0) > x(0)$ or $x_1(0) < x(0)$" is used. Massera has a similar result for $n = 2$.

Theorem 11.3.2 (Massera) *Assume that Eq. (1.1) is T-periodic in t and that $f(t, x)$ is continuous on $[0, \infty) \times \Re^n$ and satisfies a weak Lipschitz condition with respect to x. If $n = 2$ and if Eq. (1.1) has a solution that is bounded on $[0, \infty)$, then Eq. (1.1) has a T-periodic solution.* ♠

For Theorem 11.3.2, the proof depends on the geometry in $\Re^2$ and in general, cannot be extended to higher dimensional cases.

These indicate that in order to obtain periodic solutions for differential equations in general $\Re^n, n \geq 1$, other ideas that are dimension-independent must be exploited. To this end, note that if x_0 is in some set, then $P(x_0) = x(T, 0, x_0)$ may not be in the same set. Therefore, many fixed point theorems in functional analysis are not applicable. However, Horn's fixed point theorem (see the Appendix) can be used very well here.

Theorem 11.3.3 (Horn's fixed point theorem) *Let $E_0 \subset E_1 \subset E_2$ be convex subsets of a Banach space Z, with E_0 and E_2 compact subsets and E_1 open relative to E_2. Let $P : E_2 \to Z$ be a continuous mapping such that for some integer m, one has*

$$P^j(E_1) \quad \subset \quad E_2, \quad 1 \leq j \leq m - 1, \tag{3.5}$$

$$P^j(E_1) \quad \subset \quad E_0, \quad m \leq j \leq 2m - 1, \tag{3.6}$$

then P has a fixed point in E_2. ♠

Horn's fixed point theorem is also called "Horn's asymptotic fixed point theorem" since it uses the idea that if P^k has a unique fixed point for some positive integer k, then P itself has a fixed point. If we analyze conditions (3.5) and (3.6), we find that (3.5) is related to the boundedness, and (3.6) is related to the ultimate boundedness studied in Chapter 10. Therefore, we will see that Horn's fixed point theorem allows us to derive periodic solutions from the boundedness properties for general (linear or nonlinear) n-dimensional differential equations, $n \geq 1$.

Theorem 11.3.4 *Assume that Eq. (1.1) is T-periodic in t and that $f(t, x)$ is continuous on $[0, \infty) \times \Re^n$ and satisfies a weak Lipschitz condition with respect to x. If the solutions of Eq. (1.1) are equi-ultimately bounded, then Eq. (1.1) has a T-periodic solution.*

Proof. Using the weak Lipschitz condition, we know from Theorem 10.1.2 in Chapter 10 that the solutions of Eq. (1.1) are also equi-bounded.

Let B be the bound in the definition of the equi-ultimate boundedness, then by the equi-boundedness, there is a $B_1 > B$ such that $|x_0| \leq B$ implies $|x(t, 0, x_0)| \leq B_1$, $t \geq 0$. Furthermore, there is a $B_2 > B_1$ such that $|x_0| \leq B_1$ implies $|x(t, 0, x_0)| \leq B_2$, $t \geq 0$. Now, using the equi-ultimate boundedness, there is a positive integer $m = m(B_1)$ such that $|x_0| \leq B_1$ implies $|x(t, 0, x_0)| \leq B$ for $t \geq mT$.

Now, let

$$E_0 = \{x \in \Re^n : |x| \le B\}, \tag{3.7}$$

$$E_1 = \{x \in \Re^n : |x| < B_1\}, \tag{3.8}$$

$$E_2 = \{x \in \Re^n : |x| \le B_2\}, \tag{3.9}$$

and define a mapping $P : E_2 \to \Re^n$ by

$$P(x_0) = x_0 + \int_0^T f(s, x(s))ds = x(T, 0, x_0), \tag{3.10}$$

where $x(t) = x(t, 0, x_0)$ is the unique solution of Eq. (1.1) with $x(0) = x_0$. We first note that P is continuous in x_0, because, from Chapter 2, a solution is continuous with respect to initial data on any finite t-interval.

Next, let's compare $w_1(t) = x(t, 0, x(T, 0, x_0))$ and $w_2(t) = x(t+T, 0, x_0)$. We have

$$w_1(0) = x(0, 0, x(T, 0, x_0)) = x(T, 0, x_0),$$
$$w_2(0) = x(T, 0, x_0),$$

and, since f is T-periodic,

$$w_1'(t) = \frac{d}{dt}x(t, 0, x(T, 0, x_0)) = f(t, x(t, 0, x(T, 0, x_0))) = f(t, w_1(t)),$$
$$w_2'(t) = \frac{d}{dt}x(t + T, 0, x_0) = f(t + T, x(t + T, 0, x_0)) = f(t, x(t + T, 0, x_0))$$
$$= f(t, w_2(t)).$$

Therefore, both w_1 and w_2 are solutions of Eq. (1.1) with the same initial data $(0, x(T, 0, x_0))$, thus, using uniqueness, $w_1(t) = w_2(t)$, $t \ge 0$. In particular, $w_1(T) = w_2(T)$, or

$$P^2 x_0 = x(2T, 0, x_0).$$

Similarly, we have (see an exercise)

$$P^m x_0 = x(mT, 0, x_0), \quad m = 1, 2, \cdots. \tag{3.11}$$

Then we obtain

$$|P^j x_0| = |x(jT, 0, x_0)| \le B_2, \quad j \ge 1, \ |x_0| \le B_1, \tag{3.12}$$

$$|P^j x_0| = |x(jT, 0, x_0)| \le B, \quad j \ge m, \ |x_0| \le B_1, \tag{3.13}$$

or,

$$P^j(E_1) \quad \subset \quad E_2, \quad j \geq 1, \qquad (3.14)$$

$$P^j(E_1) \quad \subset \quad E_0, \quad j \geq m. \qquad (3.15)$$

Next, observe that E_0, E_1, and E_2 are convex subsets of $\Re^n$ with E_0 and E_2 compact subsets and $E_1 = E_1 \cap E_2$ open relative to E_2, and $E_0 \subset E_1 \subset E_2$. Consequently, Horn's fixed point theorem can be applied to obtain a fixed point x^* for the mapping P, that is, $P(x^*) = x^*$. Now, Lemma 11.1.2 implies that Eq. (1.1) has a periodic solution. This completes the proof. ♠

Remark 11.3.5 Theorem 11.3.4 is similar to a result in Yoshizawa [1966] where Browder's fixed point theorem (see the Appendix) is used. They both improve a result in Burton [1985], where the existence of mT-periodic solutions ($m \geq 1$ is an integer) is proved under the assumption that the solutions are uniformly ultimately bounded. The existence of T-periodic solutions is proved in Burton [1985] under the assumption that the solutions are uniformly bounded and uniformly ultimately bounded. ♠

Next, note from Chapter 10 that Liapunov's method can be used to obtain the boundedness, thus, from Theorem 11.3.4, Liapunov's method can also be used to obtain periodic solutions.

Theorem 11.3.6 *Assume that Eq. (1.1) is T-periodic in t and that $f(t,x)$ is continuous on $[0,\infty) \times \Re^n$ and satisfies a weak Lipschitz condition with respect to x. Let $H = \{x \in \Re^n : |x| \geq M\}$ where $M > 0$ is a constant, and let W_i be the $\Re^n$-wedges.*

(A). *Assume there exists a function $V(t,x)$ that is continuous and satisfies a Lipschitz condition in x, and*

> *(a). $W_1(x) \leq V(t,x)$ is satisfied on $[0,\infty) \times H$, with $W_1(x) \to \infty$ as $|x| \to \infty$, and*
>
> *(b). $V'_{(1.1)}(t,x) \leq -cV(t,x)$ is satisfied on $[0,\infty) \times \Re^n$, where $c > 0$ a constant.*

Then Eq. (1.1) has a T-periodic solution.

(B). *Assume there exists a function $V(t,x)$ that is continuous and satisfies a Lipschitz condition in x, and, on $[0,\infty) \times H$,*

> *(a). $W_1(x) \leq V(t,x) \leq W_2(x)$ with $W_1(x) \to \infty$ as $|x| \to \infty$, and*

(b). $V'_{(1.1)}(t, x) \leq -W_3(x)$.

Then Eq. (1.1) has a T-periodic solution.   ♠

In the above, we used ultimate boundedness and Horn's fixed point theorem to derive periodic solutions where the ultimate boundedness is used to eventually bring solutions back such that Horn's fixed point theorem can be applied. Accordingly, we have the following definition.

Definition 11.3.7 *Assume that Eq. (1.1) is T-periodic in t. The solutions of Eq. (1.1) are said to be T-**strictly** **bounded** if there is a $B > 0$ such that $|x_0| \leq B$ implies $|x(T, 0, x_0)| \leq B$. See* **Figure 11.2**.

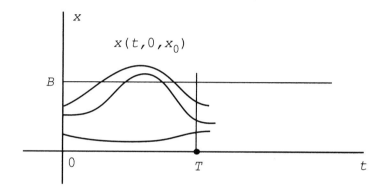

Figure 11.2: The solutions are T-strictly bounded

When the solutions of Eq. (1.1) are T-strictly bounded, Brouwer's fixed point theorem can be applied to the mapping $x_0 \to x(T, 0, x_0)$ on $\{x \in \Re^n : |x| \leq B\}$ to get a fixed point $x_0 = x(T, 0, x_0)$, which gives rise to a T-periodic solution. We formulate this as follows.

Theorem 11.3.8 *Assume that Eq. (1.1) is T-periodic in t and that $f(t, x)$ is continuous on $[0, \infty) \times \Re^n$ and satisfies a weak Lipschitz condition with respect to x. If the solutions of Eq. (1.1) are T-strictly bounded, then Eq. (1.1) has a T-periodic solution.*   ♠

The condition of strictly boundedness is satisfied for a system when, for example, the solutions are nonincreasing, that is, when $|x(t)|$ are nonincreasing in t.

Next, we look at some examples.

Example 11.3.9 Modify Example 10.5.7 in Chapter 10 and consider the scalar equation

$$u'' + u' + g(u) = f(t), \tag{3.16}$$

where $f(t)$ is continuous and T-periodic. Now, f is bounded, thus it is shown in Example 10.5.7 that the solutions of Eq. (3.16) are uniformly bounded and uniformly ultimately bounded. Therefore, according to Theorem 11.3.4, Eq. (3.16) has a T-periodic solution. The periodic solution is nonzero when $f \neq 0$. ♠

Example 11.3.10 Consider

$$x'(t) = A(t)x(t) + f(t, x(t)), \quad t \geq 0, \quad x \in \Re^n, \tag{3.17}$$

where $A(t)$ and $f(t, x)$ are continuous and are T-periodic in t, f is weak Lipschitz in x, and $|f(t, x)| \leq M$ for some constant M. Assume further that all solutions of $y' = A(t)y$ go to zero as $t \to \infty$. Then the solutions of Eq. (3.17) are equi-ultimately bounded, and hence from Theorem 11.3.4, Eq. (3.17) has a T-periodic solution.

We will provide several ways to verify this claim. First, we verify directly that the solutions of Eq. (3.17) are equi-ultimately bounded. We only consider the case when $t_0 = 0$, because the same argument can be used for an arbitrary t_0. From the Floquet theory in Chapter 3, there is a constant matrix C and a nonsingular T-periodic matrix $P(t)$ such that $P(t)e^{tC}$ is the fundamental matrix solution of $y' = A(t)y$. If all solutions of $y' = A(t)y$ go to zero as $t \to \infty$, then so do the solutions of $z' = Cz$. Thus all eigenvalues of C have negative real parts. Then, from Chapter 3, $|e^{tC}| \leq M_1 e^{-\alpha t}$ for some positive constants M_1 and α. Now, the solution $x(t) = x(t, 0, x_0)$ of Eq. (3.17) is given by the variation of parameters formula

$$x(t) = P(t)e^{tC}x_0 + \int_0^t P(t)e^{(t-s)C}P^{-1}(s)f(s, x(s))ds. \tag{3.18}$$

Hence,

$$\begin{aligned}
|x(t)| &\leq |P(t)|M_1 e^{-\alpha t}|x_0| \\
&\quad + \int_0^t \left(\max_{t \in [0,T]} |P(t)| \right) M_1 e^{-\alpha(t-s)} \left(\max_{t \in [0,T]} |P^{-1}(t)| \right) M ds \\
&\leq |P(t)|M_1 e^{-\alpha t}|x_0| + K \int_0^t e^{-\alpha(t-s)} ds \\
&\leq |P(t)|M_1 e^{-\alpha t}|x_0| + \frac{K}{\alpha}, \quad t \geq 0, \tag{3.19}
\end{aligned}$$

where $K = MM_1 \max|P(t)| \max|P^{-1}(t)|$ is a constant. Now, note that for $|x_0|$ bounded, $|P(t)|M_1 e^{-\alpha t}|x_0| \leq 1$ for large t (here t depends on x_0). Thus the solutions of Eq. (3.17) are equi-ultimately bounded with the bound $B = 1 + \frac{K}{\alpha}$, hence Eq. (3.17) has a T-periodic solution.

Another way to see why Eq. (3.17) has a T-periodic solution in this case is as follows: If all solutions of $x' = A(t)x$ go to zero as $t \to \infty$, then from Theorem 11.2.1, we find that 1 is not an eigenvalue of $U(T, 0)$. Thus from Theorem 11.2.3, Eq. (3.17) has a T-periodic solution.

One more way to see this is given below: If all solutions of $x' = A(t)x$ go to zero as $t \to \infty$, then from Chapter 9, the zero solution of $x' = A(t)x$ is asymptotically stable. Hence, from Chapter 9, the zero solution is also uniformly asymptotically stable because the equation is periodic. Now, Theorem 10.2.9 in Chapter 10 implies that the solutions of Eq. (3.17) are uniformly bounded and uniformly ultimately bounded, hence Eq. (3.17) has a T-periodic solution.   ♠

Example 11.3.11 In Example 11.3.10, if we replace $|f(t, x)| \leq M$ by $|f(t, x)| \leq \eta|x|$ with

$$\eta M_1 \max|P(t)| \max|P^{-1}(t)| < \alpha,$$

where $P(t)$, M_1, and α are from Example 11.3.10, then the same conclusion holds. Because now we have

$$
\begin{aligned}
|x(t)| \ &\leq\ |P(t)|M_1 e^{-\alpha t}|x_0| \\
&\quad + \int_0^t \Big(\max_{t \in [0,T]} |P(t)| \Big) M_1 e^{-\alpha(t-s)} \Big(\max_{t \in [0,T]} |P^{-1}(t)| \Big) \eta|x(s)|ds \\
&\leq\ |P(t)|M_1 e^{-\alpha t}|x_0| + K_1 \int_0^t e^{-\alpha(t-s)}|x(s)|ds \\
&\leq\ K_2 M_1 e^{-\alpha t}|x_0| + K_1 \int_0^t e^{-\alpha(t-s)}|x(s)|ds, \quad t \geq 0, \qquad (3.20)
\end{aligned}
$$

where $K_1 = \eta M_1 \max|P(t)| \max|P^{-1}(t)|$ and $K_2 = \max|P(t)|$ are constants. Let $u(t) = x(t)e^{\alpha t}$, then

$$|u(t)| \leq K_2 M_1 |x_0| + K_1 \int_0^t |u(s)|ds, \quad t \geq 0. \qquad (3.21)$$

Now, Gronwall's inequality implies

$$|u(t)| \ \leq\ \Big(K_2 M_1 |x_0| \Big) e^{K_1 t}, \quad t \geq 0, \qquad (3.22)$$

or,

$$|x(t)| \leq \left(K_2 M_1 |x_0|\right) e^{-(\alpha - K_1)t}, \quad t \geq 0, \tag{3.23}$$

with $\alpha - K_1 = \alpha - \eta M_1 \max |P(t)| \max |P^{-1}(t)| > 0$. Now, we can take the bound $B = 1$ to see that the solutions of Eq. (3.17) are equi-ultimately bounded, hence Eq. (3.17) has a T-periodic solution. ♠

This completes a brief account of deriving periodic solutions from the boundedness properties for general differential equations.

Exercises 11.3

1. In Theorem 11.3.1, prove for the case when $x_1(0) < x(0)$.

2. Assume that $f(t, x)$ in

$$x'(t) = f(t, x(t)), \quad x(0) = x_0, \quad t \geq 0, \quad x \in \Re^n,$$

is T-periodic in t, where the existence and uniqueness are also assumed. Define a mapping P such that $Px_0 = x(T, 0, x_0)$, where $x(t, 0, x_0)$ is the unique solution of the equation with $x(0, 0, x_0) = x_0$. Use an induction to prove that $P^m x_0 = x(mT, 0, x_0)$, $m = 1, 2, 3, \cdots$.

3. Prove Theorem 11.3.6.

4. Prove Theorem 11.3.8.

5. Generalize the result in Example 11.3.10 when f satisfies (2.14).

Chapter 12

Some New Types
of Equations

12.1 Introduction

In the previous chapters, we have studied some properties of the differential equation

$$x'(t) = f(t, x(t)), \quad x(t_0) = x_0, \quad t \geq t_0, \quad x(t) \in \Re^n, \tag{1.1}$$

where the fundamental assumption used when modeling a system using a differential equation is that the time rate at time t, given as $x'(t)$, depends only on the current status at time t, given as $f(t, x(t))$. Moreover, the initial condition is given in the form of $x(t_0) = x_0$. In applications, this assumption and the initial condition should be improved so we can model the situations more accurately and therefore derive better results.

One improvement of Eq. (1.1) is to assume that the time rate depends not only on the current status, but also on the status in the past; that is, the past history will contribute to the future development, or, there is a **time-delay** effect. For example, for a university, its current population will affect its population growth, however, its population in the past may also affect its population growth.

In fact, in his study of predator-prey models, Volterra [1928] had inves-

520

tigated the equation

$$\begin{cases} x'(t) = x(t)[a - by(t) - \int_{-r}^{0} F_1(s)y(t+s)ds], \\ y'(t) = y(t)[-a + cx(t) + \int_{-r}^{0} F_2(s)x(t+s)ds], \end{cases} \tag{1.2}$$

where x and y are the number of preys and predators, respectively, and all constants and functions are nonnegative and r is a positive constant. In $\int_{-r}^{0} F_1(s)y(t+s)ds$, the variable s varies in the interval $[-r, 0]$, thus $y(t+s)$ is a function defined on the interval $[t-r, t]$. This says that for Eq. (1.2), the time rate at t, $[x'(t), y'(t)]^T$, depends not only on the status of $x(t)$ and $y(t)$ at t, but also on the past status of $x(t+s)$ and $y(t+s)$ defined on the interval $[t-r, t]$. That is, the history on the interval $[t-r, t]$ will affect the growth rates of the preys and predators at time t.

For similar models, Wangersky and Cunningham [1957] also used the equation

$$\begin{cases} x'(t) = ax(t)\left[\frac{m-x(t)}{m}\right] - bx(t)y(t), \\ y'(t) = -cy(t) + kx(t-r)y(t-r), \end{cases}$$

for the predator-prey models, where r is a positive constant. Here, the assumption is that the current status at t and the past status at $t-r$ will affect the population growth.

In the study of the van der Pol equation, Rubanik [1969] encountered

$$x''(t) + ax'(t) - f(x(t-r))x'(t-r) + x(t) = 0,$$

when taking into account the transmission time in the triode oscillator.

Other physical procedures that possess such time-delay properties include blood moving through arteries, relaxation of materials with memory from bending (e.g., metal), and signals traveling through mediums (e.g., nerves). Differential equations incorporating delay effect, or using information from the past, are called "**delay differential equations.**" They include finite delay differential equations, infinite delay differential equations, and integrodifferential equations. Refer to Hale and Verduyn Lunel [1993] for additional results.

Another improvement of Eq. (1.1) is to allow the system to undergo some abrupt perturbations (such as due to harvesting, diseases, wars, etc.) whose duration can be negligible in comparison with the duration of the process. Therefore, we assume in this case that a solution x of the system may have jump discontinuities, or **impulses**, at times $t_1 < t_2 < \cdots$, given in the form of

$$x(t_i^+) - x(t_i^-) = I_i(x(t_i)), \quad i = 1, 2, \cdots, \tag{1.3}$$

where "+" and "$-$" denote the right and the left limit respectively, and I_i, $i = 1, 2, \cdots$, are some given functions. (Of course, I_i may be identically zero, in which case there are no impulses.)

For example, Freedman, Liu, and Wu [1991] studied models of single species growth with impulsive effects; Zavalishchin [1994] studied impulsive dynamic systems for mathematical economics. Differential equations with impulsive effects are called "**impulsive differential equations.**" Refer to Lakshmikantham, Bainov, and Simeonov [1989] for additional details.

One more improvement of Eq. (1.1) we want to introduce here is the so-called "**equations with nonlocal conditions.**" That is, we extend the initial condition (also called the "local condition")

$$x(t_0) = x_0$$

to the following nonlocal condition

$$x(t_0) + g(x(\cdot)) = x_0, \tag{1.4}$$

where $x(\cdot)$ denotes a solution (that is, $x(\cdot)$ is a function) and g is a mapping defined on some space consisting of certain functions. (Of course, g may be identically zero, in which case it reduces to the local condition or initial condition $x(t_0) = x_0$.) The advantage of using nonlocal conditions is that measurements at more places can be incorporated to get better models.

For example, $g(x(\cdot))$ may be given by

$$g(x(\cdot)) = \sum_{i=1}^{q} c_i x(s_i), \tag{1.5}$$

where c_i, $i = 1, \cdots, q$, are given constants and $t_0 < s_1 < s_2 < \cdots < s_q$. In this case, (1.5) allows the additional measurements at s_i, $i = 1, 2, \cdots, q$. A formula similar to (1.5) is also used in Deng [1993] to describe the diffusion phenomenon of a small amount of gas in a transparent tube. Refer to Byszewski and Lakshmikantham [1990] for additional studies of nonlocal conditions.

Having discussed the impulsive differential equations and differential equations with nonlocal conditions, it is natural to combine them together to obtain "**impulsive differential equations with nonlocal conditions.**"

For example, if a sound wave travels through a nonuniform rod (where nonlocal conditions can be applied), and if the sound wave's amplitude or frequency changes in an impulsive fashion, then the vibration in the rod will also experience impulsive effects. So the merging of nonlocal and impulsive conditions would be helpful in modeling this system.

Finally, we point out that when we model some situations arising from physics and other applied sciences, we typically end up with **partial differential equations**. We can formally formulate these partial differential equations as differential equations we have seen so far. For example, for a one-dimensional heat equation

$$\begin{cases} u_t(t,x) = u_{xx}(t,x), & t \geq 0, \ x \in [0,1], \\ u(t,0) = u(t,1) = 0, & u(0,x) = \phi(x), \end{cases} \tag{1.6}$$

in $L^2([0,1],\Re)$ (space of square integrable functions from $[0,1]$ to $\Re$), we can define $y_0 = \phi$ and $y(t) = u(t,\cdot)$, that is, for each fixed t, $y(t)$ is a function in x, given as $u(t,\cdot)$ or $u(t,x)$. Next, define an operator $A = \frac{\partial^2}{\partial x^2}$ with the domain $D(A) = W_0^{1,2}[0,1] \cap W^{2,2}[0,1]$ (a Sobolev space; consult a text on applied functional analysis, such as Pazy [1983]) which is dense in $L^2([0,1],\Re)$, then we can rewrite Eq. (1.6) as the differential equation

$$y'(t) = Ay(t), \quad y(0) = y_0. \tag{1.7}$$

Formally, Eq. (1.7) looks exactly the same as a linear differential equation with a constant matrix A studied in Chapter 3. However, now $y(t)$ in Eq. (1.7) is no longer a vector in $\Re^n$ and A in Eq. (1.7) is no longer a matrix of numbers; instead, $y(t)$ is an element in the infinite-dimensional function space $L^2([0,1],\Re)$, and A is an operator acting on the infinite-dimensional space $L^2([0,1],\Re)$ and is unbounded. Thus, we call the differential equations in function spaces, or in general, abstract spaces, such as Eq. (1.7), as "**abstract differential equations**." For abstract differential equations, if the operators, such as A in Eq. (1.7), are nondensely defined, then we get "**abstract differential equations with nondensely defined operators**." For example, in Eq. (1.6), if we change $L^2([0,1],\Re)$ to $C([0,1],\Re)$ with the sup-norm, then the domain $D(A) = \{\phi \in C^2([0,1],\Re) : \phi(0) = \phi(1) = 0\}$ is not dense in $C([0,1],\Re)$ with the sup-norm.

Next, for abstract differential equations, delay effect, impulsive, and nonlocal conditions can also be imposed in the same way as we mentioned above; thus we obtain "**abstract differential equations with delay, impulsive, and/or nonlocal conditions**."

Most properties of $x'(t) = f(t, x(t))$ in $\Re^n$ have been extended to the new types of differential equations introduced above. Refer to Pazy [1983], Burton [1985], Lakshmikantham, Bainov, and Simeonov [1989], Gripenberg, Londen, and Staffans [1990], Hino, Murakami, and Naito [1991], and Hale and Verduyn Lunel [1993] for additional details.

There are still so many important and interesting properties of $x'(t) = f(t, x(t))$ in $\Re^n$ that are yet to be extended to the more advanced settings. This makes the research in differential equations an active, exciting, and fruitful area where it is full of opportunities for making contributions. New ideas, new methods, and even new research areas can also be discovered in such a process of advancement.

In the following sections, we briefly describe some important features for each new type of differential equations introduced above. You may use these remarks and references to access some frontline research, and perhaps even start your own research and make contributions in this area.

12.2   Finite Delay Differential Equations

Here, we assume that the time rate of a system at t depends on the status at t and also on the status in the past on a **finite interval** $[t - r, t]$, where $r > 0$ is a constant. For example, the population growth of a university may depend on the current population and also on the population in the past five years.

We first introduce some notations in order to set up the equations. For a fixed $t \in \Re$ and a function x on the interval $[t - r, t]$, we define $x_t(\cdot)$ to be a function on the interval $[-r, 0]$ such that

$$x_t(s) = x(t + s), \quad s \in [-r, 0],$$

and we denote by $C([-r, 0], \Re^n)$ the Banach space of all continuous functions on the interval $[-r, 0]$ with the sup-norm

$$\|\phi\|_C = \sup_{-r \le s \le 0} |\phi(s)|, \quad \phi \in C([-r, 0], \Re^n).$$

And we let f be a continuous function from $\Re \times C([-r, 0], \Re^n)$ to $\Re^n$.

Now, we are ready to set up the **finite delay differential equations**

$$x'(t) = f(t, x_t), \quad x_{t_0} = \phi \in C([-r, 0], \Re^n), \quad t \ge t_0, \quad x(t) \in \Re^n. \qquad (2.1)$$

Definition 12.2.1 *A function* $x(t) : [t_0 - r, t_0 + T] \to \Re^n$, *where* $T > 0$ *is a constant, is said to be a* **solution** *of Eq.* *(2.1) on* $[t_0 - r, t_0 + T]$ *if* $x_{t_0} = \phi$, $x(t)$ *is differentiable on* $[t_0, t_0 + T]$, *and satisfies Eq.* *(2.1) for* $t \in [t_0, t_0 + T]$. *See* **Figure 12.1**.

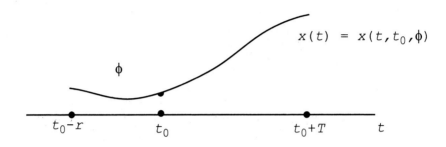

Figure 12.1: A solution of Eq. (2.1)

That is, the initial "value" for a finite delay differential equation is a function ϕ in the Banach space $C([-r, 0], \Re^n)$, and the time rate at t depends on the status at time t and also on the status on the interval $[t - r, t]$. Therefore, to study the finite delay differential equation (2.1), some knowledge of functional analysis is needed, because now we have to deal with the function space $C([-r, 0], \Re^n)$, instead of $\Re^n$ used for differential equations without delay.

Similar to a result in Chapter 2, we note that a continuous function x is a solution of Eq. (2.1) if and only if

$$x(t) = \phi(0) + \int_{t_0}^{t} f(s, x_s) ds, \quad t \geq t_0, \quad x_{t_0} = \phi. \tag{2.2}$$

Now, (2.2) leads us naturally to the following mapping P, such that for $x \in C([t_0 - r, t_0 + T], \Re^n)$ with $x_{t_0} = \phi$, define $(Px)(s) = \phi(s - t_0)$, $s \in [t_0 - r, t_0]$, and define

$$(Px)(t) = \phi(0) + \int_{t_0}^{t} f(s, x_s) ds, \quad t \in [t_0, t_0 + T]. \tag{2.3}$$

Therefore, given some conditions, such as Lipschitz conditions, a fixed point approach similar to those in Chapter 2 can be carried out here to derive existence, uniqueness, and other properties. See Hale and Verduyn Lunel [1993] for additional details.

12.3   Infinite Delay Differential Equations

Here, the assumption is that the time rate of a system at t depends on the status at t and also on the status in the past on the **infinite interval** $(-\infty, t]$. For example, the population growth of a country may depend on the current population and also on the population in the whole history.

Now, for a fixed $t \in \Re$ and a function x on the interval $(-\infty, t]$, we define $x_t(\cdot)$ to be a function on the interval $(-\infty, 0]$ such that

$$x_t(s) = x(t + s), \quad s \in (-\infty, 0]. \tag{3.1}$$

Accordingly, we can set up the **infinite delay differential equations**

$$x'(t) = f(t, x_t), \quad x_{t_0} = \phi, \quad t \geq t_0, \quad x(t) \in \Re^n. \tag{3.2}$$

Definition 12.3.1 *A function* $x(t) : (-\infty, t_0 + T] \to \Re^n$, *where* $T > 0$ *is a constant, is said to be a* **solution** *of Eq. (3.2) on* $(-\infty, t_0 + T]$ *if* $x_{t_0} = \phi$, $x(t)$ *is differentiable on* $[t_0, t_0 + T]$, *and satisfies Eq. (3.2) for* $t \in [t_0, t_0 + T]$. *See* **Figure 12.2**.

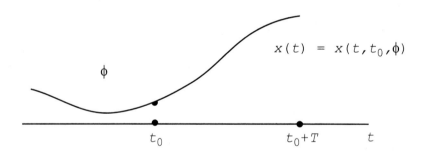

$$x(t) = x(t, t_0, \phi)$$

ϕ

t_0 $t_0 + T$ t

Figure 12.2: A solution of Eq. (3.2)

Note that for the finite delay differential equation (2.1), the initial function ϕ is in $C([-r, 0], \Re^n)$ with the sup-norm $\|\phi\|_C = \sup_{-r \leq s \leq 0} |\phi(s)|$. Now, (3.1) indicates that for the infinite delay differential equation (3.2), the initial function ϕ must be defined on the interval $(-\infty, 0]$. However, assigning norms to functions on $(-\infty, 0]$ is not completely clear, which makes the choice of a space for initial functions, also called a **phase space** in most references, a complex matter. A commonly used phase space is

$$C_g = \{\phi \in C((-\infty, 0], \Re^n) : \|\phi\|_g < \infty\},$$

where

$$\|\phi\|_g \overset{\text{def}}{=} \sup\{\frac{|\phi(s)|}{g(s)} : -\infty < s \le 0\},$$

and g is a continuous positive function on $(-\infty, 0]$. To derive additional properties, you can also use the following phase spaces,

$$UC_g = \{\phi \in C_g : \frac{\phi}{g} \text{ is uniformly continuous on } (-\infty, 0]\},$$

$$LC_g = \{\phi \in C_g : \lim_{s \to -\infty} \frac{\phi(s)}{g(s)} \text{ exists in } \Re^n\},$$

$$LC_g^0 = \{\phi \in C_g : \lim_{s \to -\infty} \frac{\phi(s)}{g(s)} = 0\}.$$

The function g in the above may be a positive constant, thus the phase spaces mentioned above include, respectively, the usual spaces of bounded continuous functions, of bounded uniformly continuous functions, and of bounded continuous functions such that the limits, as $s \to -\infty$, exist.

An important thing to note is that for the finite delay differential equation (2.1), the initial function ϕ on $[-r, 0]$ may not be differentiable, but the corresponding solution $x(t)$ has derivatives for $t \ge t_0$. Therefore, for $h \ge t_0 + r$, since $x_h(\cdot)$ is a function defined on $[h - r, h] \subset [t_0, \infty)$, $x_h(\cdot)$ is differentiable. That is, for Eq. (2.1), one may start with a "bad" initial function ϕ in $C([-r, 0], \Re^n)$, but after a while ($h \ge t_0 + r$), one gets a "good" function x_h in $C([-r, 0], \Re^n)$. Therefore, we say that "Eq. (2.1) will smooth things out." See **Figure 12.3**.

However, for the infinite delay differential equation (3.2), the initial function ϕ on $(-\infty, 0]$ will be carried to $x_h(s) = x(h + s)$ for any $h \ge t_0$. That is, for $s \in (-\infty, 0]$, one has $h + s \in (-\infty, h]$, hence

$$x_h(s) = x(h + s) = \phi(h + s - t_0)$$

when $h + s - t_0 \le 0$, or when $s \le -(h - t_0)$. Thus, if ϕ has certain property, then $x_h(\cdot)$ must have the same property because ϕ is a segment of $x_h(\cdot)$. In particular, if ϕ is not differentiable, then x_h cannot be differentiable. Therefore, we say that "Eq. (3.2) will not smooth things out." See **Figure 12.4**.

To overcome this difficulty and other difficulties associated with the infinite delay, and also to include more general cases, an abstract and axiomatic approach was considered. For example, to keep certain continuity properties of solutions, the following two axioms are imposed for a general phase space B of functions from $(-\infty, 0]$ to $\Re^n$ with a seminorm $\|\cdot\|_B$.

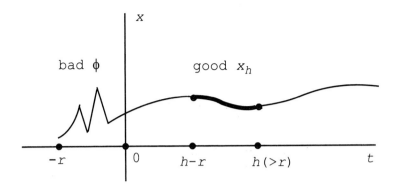

Figure 12.3: Eq. (2.1) will smooth things out

(A1). There is a positive constant H and functions K, $M : \Re^+ \to \Re^+$, with K continuous and M locally bounded, such that for any $t_0 \in \Re$ and $T > 0$, if $x : (-\infty, t_0 + T] \to \Re^n$, $x_{t_0} \in B$, and x is continuous on $[t_0, t_0+T]$, then for every $t \in [t_0, t_0+T]$, the following conditions hold:

    a. $x_t \in B$,

    b. $|x(t)| \leq H\|x_t\|_B$,

    c. $\|x_t\|_B \leq K(t - t_0) \sup_{t_0 \leq s \leq t} |x(s)| + M(t - t_0)\|x_{t_0}\|_B$.

(A2). For the function x in (A1), x_t is a B-valued continuous function for $t \in [t_0, t_0 + T]$.

Under these assumptions, some fundamental results concerning existence, uniqueness, and other properties have been derived. Refer to Hale and Kato [1978], Hino, Murakami, and Naito [1991], and Hale and Verduyn Lunel [1993] for additional details. Compared to finite delay differential equations, many questions for infinite delay differential equations are still open and under investigation.

12.4   Integrodifferential Equations

Here, to simplify the notation, we let $t_0 = 0$ and consider, for $t \geq 0$,

$$x'(t) = Ax(t) + \int_0^t B(t - s)x(s)ds + f(t), \quad x(0) = x_0 \in \Re^n, \qquad (4.1)$$

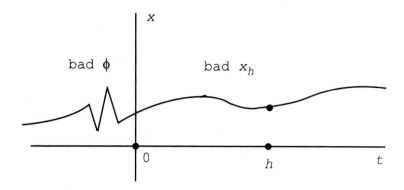

Figure 12.4: Eq. (3.2) will not smooth things out

and

$$x'(t) = Ax(t) + \int_{-\infty}^{t} B(t-s)x(s)ds + f(t), \quad x(s) = \phi(s), \ s \le 0, \quad (4.2)$$

where A, $B(\cdot)$ are $n \times n$ matrices. Since eqs. (4.1) and (4.2) contain both integral and derivative, they are called "**integrodifferential equations**."

Eq. (4.1) describes a system where solutions start at $t = 0$ and the time rate at t is determined by the status at t and also by the status on the interval $[0, t]$. Since $t \ge 0$ is a variable, the size of $[0, t]$ is not fixed and is not infinite, thus this case is different from the finite delay or infinite delay differential equations mentioned in the previous sections. However, for Eq. (4.2), the time rate at t is determined by the status on the interval $(-\infty, t]$, thus Eq. (4.2) can be regarded as a special case of the infinite delay differential equations described in the previous section.

To study Eq. (4.1), one method is to use an idea of Miller [1975] (see also Desch, Grimmer, and Schappacher [1988]) and reformulate integrodifferential equations into differential equations.

Let F be a space of functions from $[0, \infty)$ to $\Re^n$ and consider the product space $\Re^n \times F$. We have

$$\frac{d}{dt}x(t+h) = Ax(t+h) + \int_0^{t+h} B(t+h-s)x(s)ds + f(t+h)$$

$$= Ax(t+h) + \int_0^{t} B(t+h-s)x(s)ds + f(t+h)$$

$$+ \int_t^{t+h} B(t + h - s)x(s)ds + f(t + h)$$

$$= Ax(t + h) + \int_0^h B(h - s)x(s + t)ds + g(t)(h), \quad (4.3)$$

where $g(t)(h) = \int_0^t B(t + h - s)x(s)ds + f(t + h)$. For $h = 0$, we get

$$\frac{d}{dt}x(t) = Ax(t) + g(t)(0), \quad (4.4)$$

and

$$\begin{aligned}
\frac{d}{dt}g(t)(h) &= B(h)x(t) + \int_0^t \left[\frac{d}{dt}B(t + h - s)x(s)\right]ds + \frac{d}{dt}f(t + h) \\
&= B(h)x(t) + \int_0^t \left[\frac{d}{dh}B(t + h - s)x(s)\right]ds + \frac{d}{dh}f(t + h) \\
&= B(h)x(t) + \frac{d}{dh}g(t)(h). \quad (4.5)
\end{aligned}$$

Then, as shown in Miller [1975], a solution x of Eq. (4.1) is the first component of a solution of the differential equation

$$\frac{d}{dt}\begin{bmatrix} x(t) \\ g(t)(\cdot) \end{bmatrix} = \begin{bmatrix} Ax(t) + g(t)(0) \\ \left(\overline{B}x(t)\right)(\cdot) + \frac{d}{dh}g(t)(\cdot) \end{bmatrix} \quad (4.6)$$

in the product space $\Re^n \times F$, where $\overline{B}$ is an operator $\Re^n \to F$ such that for $x \in \Re^n$ and $h \geq 0$, $(\overline{B}x)(h) = B(h)x$.

Therefore, we can formulate the following differential equation in the product space $\Re^n \times F$,

$$\frac{d}{dt}\begin{bmatrix} x(t) \\ g(t)(\cdot) \end{bmatrix} = \begin{bmatrix} A & \delta_0 \\ B & \frac{d}{dh} \end{bmatrix}\begin{bmatrix} x(t) \\ g(t)(\cdot) \end{bmatrix}, \quad (4.7)$$

where $\delta_0 \phi = \phi(0)$ and $\frac{d}{dh}\phi = \phi'$ for $\phi \in F$ that is differentiable. Now, if we let

$$y(t) = \begin{bmatrix} x(t) \\ g(t)(\cdot) \end{bmatrix} \in \Re^n \times F, \quad y_0 = \begin{bmatrix} x_0 \\ f(\cdot) \end{bmatrix} \in \Re^n \times F, \quad \overline{A} = \begin{bmatrix} A & \delta_0 \\ B & \frac{d}{dh} \end{bmatrix},$$

then Eq. (4.7) becomes

$$y'(t) = \overline{A}y(t), \quad y(0) = y_0. \quad (4.8)$$

Thus, the study of abstract differential equations (see Section 12.8) can be applied to treat eqs. (4.7) and (4.8), which in turn will give the corresponding results for the integrodifferential equation (4.1).

For Eq. (4.2), one approach is to assume that x is known on the interval $(-\infty, 0]$, thus Eq. (4.2) can be reduced to Eq. (4.1). But other cases can also be considered. See Desch, Grimmer, and Schappacher [1988], and Grimmer and Liu [1994](1)(2) for additional related results.

12.5 Impulsive Differential Equations

Here, we consider the differential equation

$$
\begin{cases}
x'(t) = f(t, x(t)), \ 0 \le t \le T, \ t \ne t_i, \ x \in \Re^n, \\
x(0) = x_0, \\
\Delta x(t_i) = I_i(x(t_i)), \ i = 1, 2, \cdots, p, \ 0 < t_1 < t_2 < \cdots < t_p < T,
\end{cases}
\tag{5.1}
$$

where $\Delta x(t_i) = x(t_i^+) - x(t_i^-)$, and I_i's are some given functions. The equation is called an **impulsive differential equation** because solutions are allowed to have jump discontinuities, or impulses, at points t_i, $i = 1, 2, \cdots, p$. It can be used to model more physical phenomena than just using the classical initial value problem with $x(0) = x_0$, where solutions are assumed to be continuous.

To make things more precise, we let

$$
PC([0,T], \Re^n) = \Big\{ x : x \text{ is a mapping from } [0, T] \text{ into } \Re^n \text{ such that } x(t) \text{ is}
$$

continuous at $t \ne t_i$ and left continuous at $t = t_i$, and the

right limit $x(t_i^+)$ exists (finite) for $i = 1, 2, \cdots, p \Big\}$.

Then, one can verify that $PC([0,T], \Re^n)$ is a Banach space with the sup-norm

$$
\|x\|_{PC} = \sup_{t \in [0,T]} |x(t)|.
\tag{5.2}
$$

Definition 12.5.1 *A* **solution** *of Eq. (5.1) is a function*

$$
x(\cdot) \in PC([0,T], \Re^n) \cap C^1([0,T] \setminus \{t_1, t_2, \cdots, t_p\}, \Re^n),
$$

which satisfies Eq. (5.1) on $[0,T]$.

Now, observe that if x is a solution of Eq. (5.1), then for $t \in (t_j, t_{j+1}]$,

$$\int_0^t f(s, x(s))ds = \int_0^t x'(s)ds$$

$$= \int_0^{t_1} x'(s)ds + \int_{t_1}^{t_2} x'(s)ds + \cdots + \int_{t_j}^t x'(s)ds$$

$$= [x(t_1^-) - x(0^+)] + [x(t_2^-) - x(t_1^+)] + \cdots + [x(t^-) - x(t_j^+)]$$

$$= [x(t_1^-) - x(0)] + [x(t_2^-) - x(t_1^+)] + \cdots + [x(t) - x(t_j^+)]$$

$$= -x(0) - [x(t_1^+) - x(t_1^-)] - [x(t_2^+) - x(t_2^-)] - \cdots$$

$$\quad -[x(t_j^+) - x(t_j^-)] + x(t),$$

hence

$$x(t) = x(0) + \int_0^t f(s, x(s))ds$$

$$\quad +[x(t_1^+) - x(t_1^-)] + [x(t_2^+) - x(t_2^-)] + \cdots + [x(t_j^+) - x(t_j^-)]$$

$$= x(0) + \int_0^t f(s, x(s))ds + \sum_{0 < t_i < t} \Delta x(t_i)$$

$$= x_0 + \int_0^t f(s, x(s))ds + \sum_{0 < t_i < t} I_i(x(t_i)). \tag{5.3}$$

On the other hand, let $x(\cdot) \in PC([0, T], \Re^n)$ be a function satisfying Eq. (5.3). First, note that for $t \in (t_j, t_{j+1})$, $\sum_{0 < t_i < t} I_i(x(t_i)) = \sum_{i=1}^j I_i(x(t_i))$ is independent of t, thus $\frac{d}{dt} \sum_{0 < t_i < t} I_i(x(t_i)) = 0$ for $t \neq t_i$, $i = 1, 2, \cdots, p$. Hence, we deduce from (5.3) that $x'(t) = f(t, x(t))$, $t \neq t_i$, $x(0) = x_0$, and

$$\Delta x(t_i) = x(t_i^+) - x(t_i^-)$$

$$= \left[x(0) + \int_0^{t_i} f(s, x(s))ds + \sum_{j=1}^i I_j(x(t_j))\right]$$

$$\quad -\left[x(0) + \int_0^{t_i} f(s, x(s))ds + \sum_{j=1}^{i-1} I_j(x(t_j))\right]$$

$$= I_i(x(t_i)).$$

Therefore, we conclude that a function x in $PC([0, T], \Re^n)$ is a solution of Eq. (5.1) if and only if

$$x(t) = x_0 + \int_0^t f(s, x(s))ds + \sum_{0 < t_i < t} I_i(x(t_i)). \tag{5.4}$$

This leads to the definition of a mapping P on $PC([0,T], \Re^n)$ given by

$$(Px)(t) = x_0 + \int_0^t f(s, x(s))ds + \sum_{0 < t_i < t} I_i(x(t_i)). \tag{5.5}$$

Accordingly, conditions can be given to establish existence, uniqueness, and other properties. Refer to Guo and Liu [1993], Rogovchenko [1997], and Liu [1999] for additional related results.

12.6 Equations with Nonlocal Conditions

Here, we consider

$$\begin{cases} x'(t) = f(t, x(t)), \ 0 \le t \le T, \ x \in \Re^n, \\ x(0) + g(x(\cdot)) = x_0, \end{cases} \tag{6.1}$$

where $x(\cdot)$ denotes a solution and g is a mapping acting on some space of functions defined on $[0, T]$.

Since $g(x(\cdot))$ in Eq. (6.1) is defined on the interval $[0, T]$ rather than at a single point, Eq. (6.1) is called a "**differential equation with nonlocal conditions**," and can be applied with better effect than just using the classical initial value problem with $x(0) = x_0$, because now measurements at more places are allowed, thus more information is available.

For Eq. (6.1), observe that if x is a solution, then

$$\begin{aligned} x(t) &= x(0) + \int_0^t f(s, x(s))ds \\ &= [x_0 - g(x(\cdot))] + \int_0^t f(s, x(s))ds. \end{aligned} \tag{6.2}$$

On the other hand, assume that a function x on the interval $[0, T]$ satisfies (6.2). For this fixed function x, $g(x(\cdot))$ is a fixed element in $\Re^n$, hence $\frac{d}{dt}g(x(\cdot)) = 0$. Therefore, if we take a derivative in t, then x is a solution of Eq. (6.1). That is, we conclude that a continuous function x is a solution of Eq. (6.1) if and only if

$$x(t) = [x_0 - g(x(\cdot))] + \int_0^t f(s, x(s))ds. \tag{6.3}$$

This leads to a mapping P on $C([0, T], \Re^n)$ such that

$$(Px)(t) = [x_0 - g(x(\cdot))] + \int_0^t f(s, x(s))ds. \tag{6.4}$$

Based on this, one can derive existence, uniqueness, and other properties. Refer to Byszewski and Lakshmikantham [1990] and Lin and Liu [1996] for additional related results.

12.7 Impulsive Equations with Nonlocal Conditions

Here, we combine impulsive differential equations and differential equations with nonlocal conditions and consider **impulsive differential equations with nonlocal conditions**,

$$
\begin{cases}
x'(t) = f(t, x(t)), \ \ 0 \leq t \leq T, \ \ t \neq t_i, \ \ x \in \Re^n, \\
x(0) + g(x(\cdot)) = x_0, \\
\Delta x(t_i) = I_i(x(t_i)), \ \ i = 1, 2, \cdots, p, \ \ 0 < t_1 < t_2 < \cdots < t_p < T.
\end{cases}
\tag{7.1}
$$

The remarks for impulsive differential equations and differential equations with nonlocal conditions indicate that we should define a solution of Eq. (7.1) to be a function

$$
x(\cdot) \in PC([0, T], \Re^n) \cap C^1([0, T] \setminus \{t_1, t_2, \cdots, t_p\}, \Re^n),
$$

that satisfies Eq. (7.1). Similar to the discussions in the previous sections, it can be verified that a function x in $PC([0, T], \Re^n)$ is a solution of Eq. (7.1) if and only if

$$
x(t) = [x_0 - g(x(\cdot))] + \int_0^t f(s, x(s))ds + \sum_{0 < t_i < t} I_i(x(t_i)),
\tag{7.2}
$$

which leads to a mapping P on $PC([0, T], \Re^n)$ such that

$$
(Px)(t) = [x_0 - g(x(\cdot))] + \int_0^t f(s, x(s))ds + \sum_{0 < t_i < t} I_i(x(t_i)).
\tag{7.3}
$$

Based on this, some related results can be obtained.

12.8 Abstract Differential Equations

We have seen in Section 12.1 that a one-dimensional heat equation

$$
\begin{cases}
u_t(t, x) = u_{xx}(t, x), \ \ t \geq 0, \ \ x \in [0, 1], \\
u(t, 0) = u(t, 1) = 0, \ \ u(0, x) = \phi(x),
\end{cases}
\tag{8.1}
$$

can be written as an abstract autonomous homogeneous differential equation

$$y'(t) = Ay(t), \quad y(0) = y_0, \quad t \geq 0, \tag{8.2}$$

in $L^2([0,1], \Re)$, where $y(t) = u(t, \cdot)$, $y_0 = \phi$, and $A = \frac{\partial^2}{\partial x^2}$ with the domain $D(A) = W_0^{1,2}[0,1] \cap W^{2,2}[0,1]$.

Next, look at the following one-dimensional heat equation for material with memory, see, for example, Gripenberg, Londen, and Staffans [1990],

$$\begin{cases} q(t,x) = -Eu_x(t,x) - \int_0^t b(t-s)u_x(s,x)ds, & t \geq 0, \ x \in [0,1], \\ u_t(t,x) = -\frac{\partial q(t,x)}{\partial x} + f(t,x), \\ u(t,0) = u(t,1) = 0, \ u(0,x) = \phi(x). \end{cases} \tag{8.3}$$

The first equation gives the heat flux and the second is the balance equation. Eq. (8.3) can be written as (assuming $E = 1$)

$$u_t(t,x) = \frac{\partial^2}{\partial x^2} \left[u(t,x) + \int_0^t b(t-s)u(s,x)ds \right] + f(t,x), \quad u(0,x) = \phi(x).$$

Similar to Eq. (8.2), we define $y(t) = u(t, \cdot)$, $y_0 = \phi$, $f(t) = f(t, \cdot)$, and $A = \frac{\partial^2}{\partial x^2}$ with the domain $D(A) = W_0^{1,2}[0,1] \cap W^{2,2}[0,1]$ in $L^2([0,1])$, then we derive the following integrodifferential equation

$$y'(t) = A\left[y(t) + \int_0^t b(t-s)y(s)ds\right] + f(t), \quad y(0) = y_0, \ t \geq 0, \tag{8.4}$$

or

$$y'(t) = Ay(t) + \int_0^t b(t-s)Ay(s)ds + f(t), \quad y(0) = y_0, \ t \geq 0, \tag{8.5}$$

in $L^2([0,1], \Re)$. These indicate that differential equations from physics or other applied sciences can be reformulated as abstract differential or integrodifferential equations, such that solutions take values in abstract infinite-dimensional spaces, and the matrices consisting of numbers used before will now be replaced, in general, by unbounded operators acting on some abstract infinite-dimensional spaces. Therefore, to study these abstract differential equations, the knowledge of functional analysis, such as linear or nonlinear semigroup theory, should be used. Sometimes, infinite-dimensional spaces impose some special difficulties. For example, in infinite-dimensional spaces, a bounded set may not be precompact, thus the fixed point theorems requiring compactness are not applicable sometimes and new ideas must be exploited.

Next, we briefly outline some basic theory concerning abstract differential equation (8.2) and abstract integrodifferential equation (8.4) (or Eq. (8.5)) in a general Banach space X. (In the case of the one-dimensional heat equation (8.1), the Banach space is $X = L^2([0,1], \Re)$.)

Definition 12.8.1 *Let X be a Banach space. A one-parameter family $T(t)$, $0 \leq t < \infty$, of bounded linear operators from X into X is a* **semigroup of bounded linear operators on** X *if*

1. $T(0) = I$, (I is the identity operator on X).

2. $T(t + s) = T(t)T(s)$ for all t, $s \geq 0$.

A semigroup $T(\cdot)$ is called a "C_0 semigroup" if it is strongly continuous, that is,

$$\lim_{t \searrow 0} T(t)x = x \quad \text{for every } x \in X.$$

The linear operator A defined on the domain

$$D(A) = \{x \in X : \lim_{t \searrow 0} \frac{T(t)x - x}{t} \text{ exists}\}$$

such that

$$Ax = \lim_{t \searrow 0} \frac{T(t)x - x}{t} = \frac{d^+ T(t)x}{dt}\Big|_{t=0} \quad \text{for } x \in D(A)$$

is called the (infinitesimal) **generator** *of the semigroup $T(\cdot)$.*

Definition 12.8.2 *Let X be a Banach space. A continuous function $y : [0, \infty) \to X$ is called a* **solution** *of (8.2) if $y : [0, \infty) \to X$ is differentiable, $y(t) \in D(A)$, $t \geq 0$, and $y(\cdot)$ satisfies Eq. (8.2).*

We have the following fundamental results (see, for example, Pazy [1983]).

Theorem 12.8.3 *Let A be a densely defined linear and closed operator with a nonempty resolvent set. The following three statements are equivalent:*

1. *The linear operator A is the generator of a C_0 semigroup $T(\cdot)$ satisfying $|T(t)| \leq Me^{\omega t}$ for some constants $M \geq 1$ and $\omega \in \Re$.*

2. *The resolvent set of A contains (ω, ∞) for some $\omega \in \Re$ and the resolvent satisfies*

$$|R(\lambda, A)^n| \leq \frac{M}{(\lambda - \omega)^n}, \quad \text{for } \lambda > \omega, \quad n = 1, 2, \cdots.$$

3. *For any initial value $y_0 \in D(A)$, Eq. (8.2) has a unique solution on $[0, \infty)$; and the unique solution is given by $y(t) = T(t)y_0$, where $T(\cdot)$ is the C_0 semigroup generated by the operator A.* ♠

Moreover, for the abstract nonhomogeneous differential equation

$$y'(t) = Ay(t) + f(t), \quad y(0) = y_0, \quad t \geq 0, \tag{8.6}$$

the solution is given by

$$y(t) = T(t)y_0 + \int_0^t T(t - s)f(s)ds, \tag{8.7}$$

where $T(\cdot)$ is the C_0 semigroup generated by the operator A. The formula (8.7), which resembles formula (1.11) or (1.12) in Chapter 1 and also formula (3.3) in Chapter 3, is also called a "variation of parameters formula." Therefore, the bounded operator $T(t)$ plays the same role as the fundamental matrix solutions do for the corresponding equations in $\Re^n$. Accordingly, e^{kt} in (1.11) of Chapter 1 and e^{tA} in (3.3) of Chapter 3 are the semigroups generated by the number k and the matrix A respectively.

In the same way, for the abstract nonautonomous nonhomogeneous differential equation

$$y'(t) = A(t)y(t) + f(t), \quad y(0) = y_0, \quad t \geq 0, \tag{8.8}$$

where for each fixed t, $A(t)$ is a generator of a C_0 semigroup and satisfies some other conditions, one can show that there is an **evolution system** $U(t, s)$, $0 \leq s \leq t < \infty$, satisfying the **evolution system property** listed in Chapter 1 and Chapter 3. The solution of Eq. (8.8) is also given by another variation of parameters formula

$$y(t) = U(t, 0)y(0) + \int_0^t U(t, s)f(s)ds, \tag{8.9}$$

which resembles formula (1.14) in Chapter 1 where $U(t, s) = e^{\int_s^t k(h)dh}$, and formula (2.19) in Chapter 3.

Formulas (8.7) and (8.9) indicate that for abstract infinite-dimensional differential equations, we sometimes can still derive results that are similar to those for differential equations in $\Re^n$. They also lead to the definitions of mappings such that for Eq. (8.6),

$$(Py)(t) = T(t)y(0) + \int_0^t T(t - s)f(s)ds, \tag{8.10}$$

and for Eq. (8.8),

$$(Py)(t) = U(t,0)y(0) + \int_0^t U(t,s)f(s)ds. \tag{8.11}$$

Therefore, fixed point approaches can be applied to derive certain properties.

For the abstract integrodifferential equation (8.4) (or Eq. (8.5)), the counterpart of a semigroup $T(\cdot)$ of Eq. (8.2) is now a **resolvent operator** $R(\cdot)$ of Eq. (8.4), which we briefly define as an operator for each t, such that for $t \geq 0$,

$$\frac{d}{dt}R(t)y = A\Big[R(t)y + \int_0^t b(t-s)R(s)yds\Big]$$

$$= R(t)Ay + \int_0^t R(t-s)Ab(s)yds, \quad y \in D(A). \tag{8.12}$$

See Grimmer and Liu [1994](2) and Liu [1994] for additional details showing that the solution of Eq. (8.4) (or Eq. (8.5)) is given by

$$y(t) = R(t)y(0) + \int_0^t R(t-s)f(s)ds, \tag{8.13}$$

provided a resolvent operator $R(\cdot)$ exists. This is similar to the semigroup approach for the differential equation (8.2), and leads to the definition of a mapping P such that

$$(Py)(t) = R(t)y(0) + \int_0^t R(t-s)f(s)ds. \tag{8.14}$$

Therefore, some results can be obtained along this line.

Many results for differential equations $x'(t) = f(t,x(t))$ in $\Re^n$ have been carried to the new types of differential equations in abstract infinite-dimensional spaces. For example, Miller [1975] and Desch, Grimmer, and Schappacher [1988] examined the wellposedness and other properties for abstract integrodifferential equations; Grimmer and Liu [1992] studied stability and boundedness of abstract integrodifferential equations using Liapunov-Razumikhin's methods; Naito and Minh [1999] obtained almost periodic solutions for abstract evolution equations; and, Henriquez [1974], Burton and Zhang [1991], and Liu [2000] derived periodic solutions for abstract evolution equations with infinite delay.

However, many important and interesting questions for differential equations and integrodifferential equations in abstract infinite-dimensional spaces are still open and under investigation. You are encouraged to join this active research in differential equations and make some contributions.

Appendix

1. Linear Algebra

Definition. *A set of vectors* $\{v_1, v_2, \cdots, v_k\}$ *in* $\Re^n$ *is said to be* **linearly dependent** *if there exist constants* $\{c_1, c_2, \cdots, c_k\}$, *not all zero, such that*

$$c_1 v_1 + c_2 v_2 + \cdots + c_k v_k = 0.$$

A set of vectors $\{v_1, v_2, \cdots, v_k\}$ *in* $\Re^n$ *is said to be* **linearly independent** *if it is not linearly dependent; that is, there do not exist constants* $\{c_1, c_2, \cdots, c_k\}$, *not all zero, such that* $\sum_{i=1}^{k} c_i v_i = 0$.

Theorem. *If the* $n \times n$ *matrices* A *and* B *are similar, then* A *and* B *have the same eigenvalues.*

Proof. Assume $C^{-1}AC = B$ for some nonsingular $n \times n$ matrix C, then

$$
\begin{aligned}
\det(B - \lambda E) &= \det(C^{-1}AC - \lambda E) = \det\left(C^{-1}(A - \lambda E)C\right) \\
&= [\det C^{-1}][\det(A - \lambda E)][\det C] \\
&= [\det(C^{-1}C)][\det(A - \lambda E)] = [\det E][\det(A - \lambda E)] \\
&= \det(A - \lambda E), \tag{1}
\end{aligned}
$$

therefore, A and B have the same eigenvalues. ♠

Theorem. *If the eigenvalues* $\lambda_1, \lambda_2, \cdots, \lambda_n$ *of an* $n \times n$ *matrix* A *are distinct, then the corresponding eigenvectors* $v_1, v_2, \cdots, v_n$ *are linearly independent.*

Proof. If $v_1, v_2, \cdots, v_n$ are linearly dependent, then there is a least index p such that $v_1, v_2, \cdots, v_p$ are linearly independent and v_{p+1} is a linear combination of $v_1, v_2, \cdots, v_p$. Thus there exist numbers $c_1, c_2, \cdots, c_p$ such that

$$c_1 v_1 + c_2 v_2 + \cdots + c_p v_p = v_{p+1}. \tag{2}$$

Applying the matrix A to both sides of (2) and using $A v_i = \lambda_i v_i$, we obtain

$$c_1 \lambda_1 v_1 + c_2 \lambda_2 v_2 + \cdots + c_p \lambda_p v_p = \lambda_{p+1} v_{p+1}. \tag{3}$$

Now, multiplying λ_{p+1} to (2) and subtracting the result from (3), we get

$$c_1(\lambda_1 - \lambda_{p+1})v_1 + c_2(\lambda_2 - \lambda_{p+1})v_2 + \cdots + c_p(\lambda_p - \lambda_{p+1})v_p$$
$$= \lambda_{p+1}v_{p+1} - \lambda_{p+1}v_{p+1} = 0. \tag{4}$$

Since $v_1, v_2, \cdots, v_p$ are linearly independent, we must have $c_1(\lambda_1 - \lambda_{p+1}) = c_2(\lambda_2 - \lambda_{p+1}) = \cdots = c_p(\lambda_p - \lambda_{p+1}) = 0$. Because the eigenvalues are distinct, then $c_1 = c_2 = \cdots = c_p = 0$. Using (2), we obtain $v_{p+1} = 0$, which contradicts the fact that an eigenvector is a nonzero vector. This completes the proof. ♠

Theorem (Cayley-Hamilton). *Let A be an $n \times n$ matrix and let $P(\lambda) = \det(\lambda E - A)$ be the characteristic polynomial of matrix A, then $P(A) = 0$ (the 0 matrix).*

Proof. From linear algebra, the matrix $\lambda E - A$ has an $n \times n$ adjoint matrix $B(\lambda)$ such that

$$B(\lambda)(\lambda E - A) = [\det(\lambda E - A)]E = P(\lambda)E, \tag{5}$$

and, in this case, every entry of $B(\lambda)$ is a polynomial in λ with order $\leq n-1$. Accordingly, using the matrix operations, $B(\lambda)$ can be written as

$$B(\lambda) = \lambda^{n-1}B_0 + \lambda^{n-2}B_1 + \cdots + \lambda B_{n-2} + B_{n-1}, \tag{6}$$

where $B_0, B_1, \cdots, B_{n-1}$ are all $n \times n$ matrices consisting of numbers. Also, we can write $P(\lambda) = \lambda^n + a_1\lambda^{n-1} + \cdots + a_{n-1}\lambda + a_n$, then

$$P(\lambda)E = \lambda^n E + a_1\lambda^{n-1}E + \cdots + a_{n-1}\lambda E + a_n E. \tag{7}$$

On the other hand, from (6),

$$\begin{aligned}
B(\lambda)(\lambda E - A) &= [\lambda^{n-1}B_0 + \lambda^{n-2}B_1 + \cdots + \lambda B_{n-2} + B_{n-1}](\lambda E - A) \\
&= \lambda^n B_0 + \lambda^{n-1}(B_1 - B_0 A) + \lambda^{n-2}(B_2 - B_1 A) \\
&\quad + \cdots + \lambda(B_{n-1} - B_{n-2}A) - B_{n-1}A.
\end{aligned} \tag{8}$$

Now, from (5), (7), and (8),

$$\begin{cases}
B_0 &= E, \\
B_1 - B_0 A &= a_1 E, \\
B_2 - B_1 A &= a_2 E, \\
\quad \cdots \cdots \\
B_{n-1} - B_{n-2}A &= a_{n-1}E, \\
-B_{n-1}A &= a_n E.
\end{cases} \tag{9}$$

Multiplying A^n to the first equality in (9) from the right, multiplying A^{n-1} to the second equality in (9) from the right, $\cdots$, and multiplying A to the nth equality in (9) from the right, we obtain

$$
\begin{cases}
\phantom{B_1 A^{n-1} - {}} B_0 A^n &= A^n, \\
B_1 A^{n-1} - B_0 A^n &= a_1 A^{n-1}, \\
B_2 A^{n-2} - B_1 A^{n-1} &= a_2 A^{n-2}, \\
\phantom{B_2 A^{n-2}} \cdots\cdots \\
B_{n-1} A - B_{n-2} A^2 &= a_{n-1} A, \\
\phantom{B_{n-1} A {}} - B_{n-1} A &= a_n E.
\end{cases}
\tag{10}
$$

Add them and we see that the left-hand side becomes zero, and the right-hand side is $P(A)$. Thus $P(A) = 0$, which completes the proof. ♠

2. Functions

Definition. *A pair (X, ρ) is called a* **metric space** *if X is a set (not necessarily a linear space) and $\rho : X \times X \to [0, \infty)$ such that for $x, y, z \in X$, one has*

(a). $\rho(x, y) \geq 0$; $\rho(x, x) = 0$; $\rho(x, y) = 0$ implies $x = y$,

(b). $\rho(x, y) = \rho(y, x)$,

(c). $\rho(x, y) \leq \rho(x, z) + \rho(z, y)$.

Definition. *A sequence $\{x_n\}$ in a metric space (X, ρ) is called a* **Cauchy sequence** *if for any $\varepsilon > 0$ there exists N such that $n, m > N$ implies $\rho(x_n, x_m) < \varepsilon$. A metric space (X, ρ) is called* **complete** *if every Cauchy sequence in X converges to a limit in X.*

Definition. *A triple $(L, +, \cdot)$ is called a* **linear space** *(or vector space) over the real numbers $\Re$ (or a field F) if L is a set, and for $x, y \in L$ and $a \in \Re$, $x + y$ and $a \cdot x = ax$ are in L, and*

(1). $x + y = y + x$, $x, y \in L$,

(2). $x + (y + z) = (x + y) + z$, $x, y, z \in L$,

(3). there is a unique $0 \in L$ with $0 + x = x$ for all $x \in L$,

(4). for each $x \in L$, there is a unique $-x \in L$ with $x + (-x) = 0$,

(5). $a(bx) = (ab)x$, $a, b \in \Re$, $x \in L$,

(6). $1 \cdot x = x$, $x \in L$,

(7). $(a + b)x = ax + bx$, $a, b \in \Re$, $x \in L$,

(8). $a(x + y) = ax + ay$, $a \in \Re$, $x, y \in L$.

Definition. *A linear space L is called a **normed space** if for each $x \in L$ there is a nonnegative real number $|x|$, called the **norm** of x, such that*

(1). $|x| = 0$ if and only if $x = 0$,

(2). $|ax| = |a||x|$ for each $a \in \Re$, $x \in L$,

(3). $|x + y| \leq |x| + |y|$.

Definition. *A linear normed space $(L, |\cdot|)$ is called a **Banach space** if it is complete in the metric $\rho(x, y) = |x - y|$.*

Definition. *A sequence of functions $\{f_m(t)\}_{m \geq 1}$ on an interval $[a, b]$ is called **uniformly bounded** on $[a, b]$ if for some constant $C > 0$, $|f_m(t)| \leq C$ on $[a, b]$ for all $m \geq 1$; the sequence is called **equi-continuous** on $[a, b]$ if for any $\varepsilon > 0$, there exists a $\delta = \delta(\varepsilon) > 0$ such that $t_1, t_2 \in [a, b]$ and $|t_1 - t_2| \leq \delta$ imply $|f_m(t_1) - f_m(t_2)| \leq \varepsilon$ for all $m \geq 1$.*

Theorem (Cauchy's criterion). *Let $\{f_m(x)\}$ be a sequence of functions from a domain $D \subset \Re$ to $\Re^n$. The series $\sum_{m=1}^{\infty} f_m(x)$ converges uniformly to a function on D if and only if for any $\varepsilon > 0$ there exists an $M(\varepsilon) > 0$ such that if $m \geq k \geq M(\varepsilon)$, then*

$$|f_k(x) + \cdots + f_m(x)| \leq \varepsilon,$$

for all $x \in D$.

Theorem. *Let $\{f_m(x)\}$ be a sequence of continuous functions from a domain $D \subset \Re$ to $\Re^n$ such that the series $\sum_{m=1}^{\infty} f_m(x)$ converges uniformly to a function f on D, then f is continuous on D.*

Theorem (Weierstrass M-test). *Let $\{M_m\}$ be a sequence of positive real numbers and let $\{f_m(x)\}$ be a sequence of functions from a domain $D \subset \Re$ to $\Re^n$ such that $|f_m(x)| \leq M_m$ for $x \in D$ and $m \geq 1$. If the series $\sum_{m=1}^{\infty} M_m$ is convergent, then the series $\sum_{m=1}^{\infty} f_m(x)$ converges uniformly to a function on D. If in addition each $\{f_m(x)\}$ is continuous on D, then the series $\sum_{m=1}^{\infty} f_m(x)$ converges uniformly to a continuous function on D.*

Theorem (Arzela-Ascoli theorem). *Let $\{f_m(t)\}_{m \geq 1}$ be a sequence of continuous functions on a compact interval $[a, b]$. If the sequence is uniformly bounded and equi-continuous on $[a, b]$, then $\{f_m(t)\}_{m \geq 1}$ has a subsequence, say $\{f_{m_k}(t)\}_{k \geq 1}$, that converges uniformly on $[a, b]$ to some continuous function $y(t)$. (That is, for any $\varepsilon > 0$, there exists a $K = K(\varepsilon) > 0$ such that $k \geq K$ implies $|f_{m_k}(t) - y(t)| \leq \varepsilon$ for all $t \in [a, b]$.)*

3. Fixed Point Theorems

The following fixed point theorems can be found in Smart [1980], Burton [1985], and Hale and Verduyn Lunel [1993].

Definition. *A mapping P on a metric space (X, ρ) is called a* **contraction mapping** *if there is an $r \in (0, 1)$ such that*

$$\rho(Px, \, Py) \leq r\rho(x, \, y).$$

Theorem (Contraction mapping principle). *Let P be a contraction mapping on a complete metric space X, then there is a unique $x \in X$ with $Px = x$. Moreover, $x = \lim_{n \to \infty} x_n$, where x_0 is any element of X and $x_{j+1} = Px_j$, $j = 0, 1, \cdots$.*

Proof. Now, for some $0 < r < 1$, we have $\rho(Py, Pz) \leq r\rho(y, z)$ when $y, z \in X$. Let x_0 be any element of X and define $x_{j+1} = Px_j$, $j = 0, 1, \cdots$. Then $x_1 = Px_0$, $x_2 = Px_1 = P^2 x_0, \cdots, x_j = Px_{j-1} = \cdots = P^j x_0$, $j = 1, 2, \cdots$. Thus, for $m > n$,

$$
\begin{aligned}
\rho(x_n, \, x_m) &= \rho(P^n x_0, \, P^m x_0) \\
&\leq r\rho(P^{n-1} x_0, \, P^{m-1} x_0) \\
&\vdots
\end{aligned}
$$

$$
\begin{aligned}
&\leq \ r^n \rho(x_0,\, P^{m-n}x_0) = r^n \rho(x_0,\, x_{m-n}) \\
&\leq \ r^n \Big[\rho(x_0,\, x_1) + \rho(x_1,\, x_2) + \cdots + \rho(x_{m-n-1},\, x_{m-n})\Big] \\
&\leq \ r^n \Big[\rho(x_0,\, x_1) + r\rho(x_0,\, x_1) + \cdots + r^{m-n-1}\rho(x_0,\, x_1)\Big] \\
&= \ r^n \rho(x_0,\, x_1)\Big[1 + r + \cdots + r^{m-n-1}\Big] \\
&\leq \ r^n \rho(x_0,\, x_1)\frac{1}{1-r}. \tag{11}
\end{aligned}
$$

As $0 < r < 1$, the right-hand side goes to zero when $n \to \infty$. Thus $\{x_n\}$ is a Cauchy sequence, and hence has a limit $x \in X$ because X is a complete metric space. Now, it is easily seen that P is continuous, therefore

$$
Px = P\Big(\lim_{n\to\infty} x_n\Big) = \lim_{n\to\infty}\Big(Px_n\Big) = \lim_{n\to\infty} x_{n+1} = x, \tag{12}
$$

and x is a fixed point of P. If y is also a fixed point of P, then

$$
\rho(x, y) = \rho(Px,\, Py) \leq r\rho(x, y), \tag{13}
$$

and, as $0 < r < 1$, we must have $\rho(x, y) = 0$, which implies $x = y$. This completes the proof. ♠

Theorem (Brouwer's fixed point theorem). *Let $B \subset \Re^n$ be nonempty, convex, and compact, and let $F : B \to B$ be a continuous mapping. Then F has a fixed point in B.*

Theorem (Schauder's first fixed point theorem). *Let X be a nonempty, convex, and compact subset of a Banach space Y, and let $P : X \to X$ be a continuous mapping. Then P has a fixed point in X.*

Theorem (Schauder's second fixed point theorem). *Let X be a nonempty, convex subset of a Banach space Y and let $P : X \to X$ be a compact mapping (that is, P is continuous and maps a bounded set into a precompact set). Then P has a fixed point in X.*

The following are called "asymptotic fixed point theorems" since they use the idea that if P^m has a unique fixed point for some positive integer m, then P itself has a fixed point.

Definition. *Let A and B be subsets of a Banach space Z. If $A = B \cap C$ for an open subset C of Z, then A is open relative to B.*

Theorem (Horn's fixed point theorem). *Let $E_0 \subset E_1 \subset E_2$ be convex subsets of a Banach space Z, with E_0 and E_2 compact subsets and E_1 open relative to E_2. Let $P : E_2 \to Z$ be a continuous mapping such that for some integer m, one has*

$$P^j(E_1) \quad \subset \quad E_2, \quad 1 \le j \le m-1, \tag{14}$$

$$P^j(E_1) \quad \subset \quad E_0, \quad m \le j \le 2m-1, \tag{15}$$

then P has a fixed point in E_2.

Theorem (Browder's fixed point theorem). *Let $E_0 \subset E_1 \subset E_2$ be convex subsets of a Banach space Z, with E_0 closed and E_1, E_2 open. Let $P : E_2 \to Z$ be a compact mapping such that for some integer m, one has*

$$P^j(E_0) \quad \subset \quad E_1, \quad 0 \le j \le m, \tag{16}$$

$$P^m(E_1) \quad \subset \quad E_0, \tag{17}$$

then P has a fixed point in E_2.

References

1. R. Abraham and C. Shaw [1988], *Dynamics: The Geometry of Behavior. Part 4: Bifurcation Behavior*, Aerial Press, Santa Cruz, California.

2. V. Afraimovich, V. Bykov, and L. Silnikov [1983], *On structurally unstable attracting limit sets of Lorenz attractor type*, Trans. Moscow Math. Soc., **2**, 153–216.

3. K. Alligood, T. Sauer, and J. Yorke [1997], *Chaos: An Introduction to Dynamical Systems*, Springer-Verlag, New York.

4. H. Amann [1990], *Ordinary Differential Equations*, de Gruyter Studies in Math. 13, Walter de Gruyter, Berlin.

5. A. Andronov [1929], *Application of Poincaré's theorem on "bifurcation points" and "change in stability" to simple autooscillatory systems*, C. R. Acad. Sci., Paris, **189**(15), 559–561.

6. A. Andronov, E. Leontovich, I. Gordon, and A. Maier [1973], *Qualitative Theory of Second-order Dynamical Systems*, John Wiley & Sons, Inc., New York.

7. A. Andronov, E. Leontovich, I. Gordon, and A. Maier [1973](2), *Theory of Bifurcations of Dynamic Systems on a Plane*, John Wiley & Sons, Inc., New York.

8. D. Arrowsmith and C. Place [1990], *An Introduction to Dynamical Systems*, Cambridge University Press, Cambridge.

9. G. Birkhoff [1927], *Dynamical Systems*, Amer. Math. Soc. Colloq. Publ. **9**, Amer. Math. Soc., Providence, Rhode Island.

10. W. Boyce and R. DiPrima [1992], *Elementary Differential Equations*, John Wiley & Sons, Inc., New York.

11. F. Brauer and J. Nohel [1989], *The Qualitative Theory of Ordinary Differential Equations*, Dover Publications, Inc., New York.

12. T. Burton [1985], *Stability and Periodic Solutions of Ordinary Differential Equations and Functional Differential Equations*, Academic Press, New York.

13. T. Burton and B. Zhang [1991], *Periodic solutions of abstract differential equations with infinite delay*, J. Diff. Eq., **90**, 357–396.

14. L. Byszewski and V. Lakshmikantham [1990], *Theorem about the existence and uniqueness of a solution of a nonlocal abstract Cauchy problem in a Banach space*, Applicable Anal., **40**, 11–19.

15. C. Chicone [1999], *Ordinary Differential Equations with Applications*, Springer-Verlag, New York.

16. S. Chow and J. Hale [1982], *Methods of Bifurcation Theory*, Springer-Verlag, New York.

17. E. Coddington and N. Levinson [1955], *Theory of Ordinary Differential Equations*, McGraw-Hill, New York.

18. P. Collet and J. Eckmann [1980], *Iterated Maps on the Interval as Dynamical Systems*, Birkhauser, Boston.

19. J. Cronin [1994], *Differential Equations*, Marcel Dekker, New York.

20. P. Cvitanovic [1984], *Universality in Chaos*, Adam Hilger.

21. W. de Melo and S. van Strien [1993], *One-dimensional Dynamics*, Springer-Verlag, Berlin.

22. K. Deng [1993], *Exponential decay of solutions of semilinear parabolic equations with nonlocal initial conditions*, J. Math. Anal. Appl., **179**, 630–637.

23. W. Desch, R. Grimmer, and W. Schappacher [1988], *Wellposedness and wave propagation for a class of integrodifferential equations in Banach space*, J. Diff. Eq., **74**, 391–411.

24. E. Dowell and M. Ilgamova [1988], *Studies in Nonlinear Aeroelasticity*, Springer-Verlag, New York.

25. P. Drazin and W. Reid [1981], *Hydrodynamic Stability*, Cambridge University Press, Cambridge.

26. T. Erber, S. Guralnik, and H. Latal [1972], *A general phenomenology of hysteresis*, Ann. Phys., **69**, 161–192.

27. L. Euler [1744] (1952), *De curvis elasticis, methodus inveniendi lineas curvas maximi minimive proprietate gaudentes*, Additamentum I. Opera Omnia I, Zurich, **24**, 231–297.

28. M. Feigenbaum [1979], *The universal metric properties of nonlinear transformations*, J. Stat. Phys., **21**, 69.

29. E. Freedman, X. Liu, and J. Wu [1991], *Comparison principles for impulsive parabolic equations with applications to models of single species growth*, J. Australian Math. Soc., Series B, **32**, 382–400.

30. P. Glendinning [1994], *Stability, Instability, and Chaos: An Introduction to the Theory of Nonlinear Differential Equations*, Cambridge University Press, Cambridge.

31. R. Goldberg [1976], *Methods of Real Analysis*, 2d ed., John Wiley & Sons, Inc., New York.

32. M. Golubitsky and D. Schaeffer [1979], *A theory for imperfect bifurcation via singularity theory*, Comm. Pure Appl. Math., **32**, 21–98.

33. R. Grimmer and J. Liu [1992], *Liapunov-Razumikhin methods for integrodifferential equations in Hilbert space*, Delay and Differential Equations, A. Fink, R. Miller, and W. Kliemann (eds.), World Scientific, London, 9–24.

34. R. Grimmer and J. Liu [1994](1), *Limiting equations of integrodifferential equations in Banach space*, J. Math. Anal. Appl., **188**, 78–91.

35. R. Grimmer and J. Liu [1994](2), *Integrated semigroups and integrodifferential equations*, Semigroup Forum, **48**, 79–95.

36. G. Gripenberg, S. Londen, and O. Staffans [1990], *Volterra Integral and Functional Equations*, Cambridge University Press, Cambridge.

37. J. Guckenheimer and P. Holmes [1986], *Nonlinear Oscillations, Dynamical Systems, and Bifurcations of Vector Fields*, 2d ed., Springer-Verlag, New York.

38. D. Guo and X. Liu [1993], *Extremal solutions of nonlinear impulsive integrodifferential equations in Banach spaces*, J. Math. Anal. Appl., **177**, 538–552.

39. H. Haken [1983], *Synergetics*, 3d ed., Springer-Verlag, Berlin.

40. J. Hale [1969], *Ordinary Differential Equations*, John Wiley & Sons, Inc., New York.

41. J. Hale and J. Kato [1978], *Phase space for retarded equations with infinite delay*, Funk. Ekvacioj., **21**, 11–41.

42. J. Hale and H. Kocak [1991], *Dynamics and Bifurcations*, Springer-Verlag, New York.

43. J. Hale and S. Verduyn Lunel [1993], *Introduction to Functional Differential Equations*, Springer-Verlag, New York.

44. P. Halmos [1958], *Finite Dimensional Vector Spaces*, 2d ed., Van Nostrand, Princeton, New Jersey.

45. P. Hartman [1964], *Ordinary Differential Equations*, John Wiley & Sons, Inc., New York.

46. B. Hassard, N. Kazarinoff, and Y. Wan [1980], *Theory and Applications of the Hopf Bifurcation*, Cambridge University Press, Cambridge.

47. H. Henriquez [1974], *Periodic solutions of quasi-linear partial functional differential equations with unbounded delay*, Funk. Ekvacioj., **37**, 329–343.

48. G. Hill [1886], *On the part of the motion of the lunar perigee which is a function of the mean motions of the sun and moon*, Acta Math., **8**, 1–36.

49. Y. Hino, S. Murakami, and T. Naito [1991], *Functional Differential Equations with Infinite Delay*, Lect. Notes Math., **1473**, Springer-Verlag, New York.

50. M. Hirsch and S. Smale [1974], *Differential Equations, Dynamical Systems, and Linear Algebra*, Academic Press, New York.

51. P. Holmes [1979], *A nonlinear oscillator with a strange attractor*, Phil. Trans. of the Royal Soc. London, **A292**, 419–448.

52. E. Hopf [1942], *Abzweigung einer periodischen Losung von einer stationaten Losung eines differential systems*, Ber. Math. Phys. Sachsische Akademie der Wissenschaften Leipzig, **94**, 1–22. (See also the English translation in Marsden and McCracken [1976].)

53. N. Huang and W. Nachbar [1968], *Dynamic snap-through of imperfect viscoelastic shallow arches*, J. Appl. Mech., **35**, 289–296.

54. J. Hubbard and B. West [1995], *Differential Equations: A Dynamical Systems Approach*, Springer-Verlag, New York.

55. M. Jakobsen [1981], *Absolutely continuous invariant measures for one-parameter families of one-dimensional maps*, Comm. Math. Phys., **81**, 39–88.

56. R. Johnson [1987], *m-Functions and Floquet exponents for linear differential systems*, Ann. Mat. Pura Appl., (4) vol. CXLVII, 211–248.

57. A. Kelley [1967], *The stable, center stable, center, center unstable, and unstable manifolds*, J. Diff. Eq., **3**, 546–570.

58. C. Krebs [1972], *Ecology: The Experimental Analysis of Distribution and Abundance*, Harper & Row, New York.

59. V. Lakshmikantham, D. Bainov, and P. Simeonov [1989], *Theory of Impulsive Differential Equations*, World Scientific, Singapore.

60. C. Li [1982], *Two questions concerning the planar second-degree systems*, Sci.Sinica, A, **12**, 1087–1096.

61. T. Li and J. Yorke [1975], *Period three implies chaos*, Amer. Math. Monthly, **82**, 985–992.

62. A. Liapunov [1892], *Probleme General de la Stabilite du Mouvement* (Reproduction of the French translation in 1907 of a Russian memoire, dated 1892), Annals of Mathematics Studies, **17**, Princeton University Press, Princeton, New Jersey, 1947.

63. A. Lienard [1928], *Etude des oscillations entretenues*, Rev. Gen. de l'Electricite **23**, 901–946.

64. Y. Lin and J. Liu [1996], *Semilinear integrodifferential equations with nonlocal Cauchy problem*, Nonlinear Anal., **26**, 1023–1033.

65. J. Liu [1994], *Resolvent operators and weak solutions of integrodifferential equations*, Diff. Integ. Eq., **7**, 523–534.

66. J. Liu [1995], *Uniform asymptotic stability via Liapunov-Razumikhin technique*, Proc. Ameri. Math. Soc., **123**, 2465–2471.

67. J. Liu [1999], *Nonlinear impulsive evolution equations*, Dynam. Conti. Discr. Impul. Sys., **6**, 77–85.

68. J. Liu [2000], *Periodic solutions of infinite delay evolution equations*, J. Math. Anal. Appl., **247**, 627–644.

69. J. Liu, *A remark on the chain rule for exponential matrix functions*, College Math. J., to appear.

70. E. Lorenz [1963], *Deterministic nonperiodic flow*, J. Atmospheric Science, **20**, 130–141. See also: SPIE Milestone Series, 1994, 542–553.

71. L. Marcus and H. Yamabe [1960], *Global stability criteria for differential systems*, Osaka Math. J. **12**, 305–317.

72. M. Marden [1966], *The Geometry of the Zeros of a Polynomial in a Complex Variable*, Mathematical Surveys, No. 3, Amer. Math. Soc., Providence, Rhode Island.

73. J. Marsden and M. McCracken [1976], *The Hopf Bifurcation and its Applications*, Springer-Verlag, New York.

74. J. Massera [1950], *The existence of periodic solutions of differential equations*, Duke Math. J., **17**, 457–475.

75. R. May [1973], *Stability and Complexity in Model Ecosystems*, Princeton University Press, Princeton, New Jersey.

76. R. May [1976], *Simple mathematical models with very complicated dynamics*, Nature, **261**, 459–467.

77. N. Metropolis, M. Stein, and P. Stein [1973], *On finite limit sets for transformation on the unit interval*, J. Combin. Theor., **15**, 25.

78. K. Meyer and G. Hall [1992], *Introduction to Hamiltonian Dynamical Systems and the N-body Problem*, Springer-Verlag, New York.

79. R. Miller [1975], *Volterra integral equations in a Banach space*, Funk. Evacioj., **18**, 163–193.

80. M. Misiurewicz [1981], *The structure of mapping of an interval with zero entropy*, Publ. Math. IHES **53**, 5–16.

81. J. Murray [1989], *Mathematical Biology*, Springer-Verlag, New York.

82. T. Naito and N. Minh [1999], *Evolution semigroups and spectral criteria for almost periodic solutions of periodic evolution equations*, J. Diff. Eq., **152**, 358–376.

83. V. Nemytskii and V. Stepanov [1960], *Qualitative Theory of Differential Equations*, Princeton University Press, Princeton, New Jersey.

84. A. Pazy [1983], *Semigroups of Linear Operators and Applications to Partial Differential Equations*, Springer-Verlag, New York.

85. H. Peitgen, H. Jurgens, and D. Saupe [1992], *Chaos and Fractals: New Frontiers of Sciences*, Springer-Verlag, New York.

86. L. Perko [1991], *Differential Equations and Dynamical Systems*, Springer-Verlag, New York.

87. H. Poincaré [1885], *Sur l'equilibre d'une masse fluide animee d'un mouvement de rotation*, Acta Mathematica, 259–380.

88. H. Poincaré [1892], *Les Methodes Nouvelles de la Mecanique Celeste*, Vol. I, Gauthier-Villars, Paris.

89. E. Putzer [1966], *Avoiding the Jordan canonical form in the discussion of linear systems with constant coefficients*, Amer. Math. Month., **73**, 2–7.

90. Y. Rogovchenko [1997], *Impulsive evolution systems: main results and new trends*, Dynamics Contin. Discr. Impulsive Sys., **3**, 57–88.

91. V. Rubanik [1969], *Oscillations of Quasilinear Systems with Retardation*, (Russian), Nauk, Moscow.

92. D. Sanchez [1968], *Ordinary Differential Equations and Stability*, W. H. Freeman and Company, San Francisco.

93. A. Sarkovskii [1964], *Coexistence of cycles of a continuous map of a line into itself*, Ukrainian Math. J., **16**, 61–71.

94. R. Seydel [1988], *From Equilibrium to Chaos: Practical Bifurcation and Stability Analysis*, Elsevier Science Publishing, New York.

95. S. Shi [1980], *A concrete example of the existence of four limit cycles for plane quadratic systems*, Sci. Sinica, **23**, 153–158.

96. J. Sinai and E. Vul [1981], *Hyperbolicity conditions for the Lorenz model*, Physica **2D**, 3–7.

97. S. Smale [1963], *Diffeomorphisms with many periodic points*, Differential and Combinatorial Topology, S. Cairns (ed.), 63–80, Princeton University Press, Princeton, New Jersey.

98. D. Smart [1980], *Fixed Point Theorems*, Cambridge University Press, Cambridge.

99. J. Smith [1974], *Models in Ecology*, Cambridge University Press, Cambridge.

100. P. Smith [1998], *Explaining Chaos*, Cambridge University Press, Cambridge.

101. C. Sparrow [1982], *The Lorenz Equations: Bifurcations, Chaos, and Strange Attractors*, Springer-Verlag, New York.

102. S. Strogatz [1994], *Nonlinear Dynamics and Chaos*, Addison-Wesley Publishing Company, New York.

103. A. Tsonis [1992], *Chaos: From Theory to Applications*, Plenum Press, New York.

104. B. van der Pol [1927], *Forced oscillations in a circuit with nonlinear resistance*, Phil. Mag., **3**, 65–80; Proc. Inst. Radio Eng., **22**(1934), 1051–1086.

105. V. Volterra [1928], *Sur la theorie mathematique des phenomenes hereditaires*, J. Math. Pures Appl., **7**, 249–298.

106. P. Waltman [1986], *A Second Course in Elementary Differential Equations*, Academic Press, New York.

107. P. Wangersky and W. Cunningham [1957], *Time lag in prey-predation population models*, Ecology, **38**, 136–139.

108. S. Wiggins [1990], *Introduction to Applied Nonlinear Dynamical Systems and Chaos*, Springer-Verlag, New York.

109. T. Yoshizawa [1966], *Stability Theory by Liapunov's Second Method*, Math. Soc. Japan, Tokyo.

110. A. Zavalishchin [1994], *Impulsive dynamic systems and applications to mathematical economics*, Dynam. Sys. Appl., **3**, 443–449.

Index